Hydrogen in Disordered and Amorphous Solids

NATO ASI Series

Advanced Science Institutes Series

A series presenting the results of activities sponsored by the NATO Science Committee, which aims at the dissemination of advanced scientific and technological knowledge, with a view to strengthening links between scientific communities.

The series is published by an international board of publishers in conjunction with the NATO Scientific Affairs Division

A	**Life Sciences**	Plenum Publishing Corporation
B	**Physics**	New York and London
C	**Mathematical and Physical Sciences**	D. Reidel Publishing Company Dordrecht, Boston, and Lancaster
D	**Behavioral and Social Sciences**	Martinus Nijhoff Publishers
E	**Engineering and Materials Sciences**	The Hague, Boston, and Lancaster
F	**Computer and Systems Sciences**	Springer-Verlag
G	**Ecological Sciences**	Berlin, Heidelberg, New York, and Tokyo

Recent Volumes in this Series

Volume 129—Transport in Nonstoichiometric Compounds
edited by George Simkovich and Vladimir S. Stubican

Volume 130—Heavy Ion Collisions: *Cargèse* 1984
edited by Paul Bonche, Maurice Lévy, Philippe Quentin, and Dominique Vautherin

Volume 131—Physics of Plasma-Wall Interactions in Controlled Fusion
edited by D. E. Post and R. Behrisch

Volume 132—Physics of New Laser Sources
edited by Neal B. Abraham, F. T. Arecchi, Aram Mooradian, and Alberto Sona

Volume 133—Scaling Phenomena in Disordered Systems
edited by Roger Pynn and Arne Skjeltorp

Volume 134—Fundamental Processes in Atomic Collision Physics
edited by H. Kleinpoppen, J. S. Briggs, and H. O. Lutz

Volume 135—Frontiers of Nonequilibrium Statistical Physics
edited by Garald T. Moore and Marlan O. Scully

Volume 136—Hydrogen in Disordered and Amorphous Solids
edited by Gust Bambakidis and Robert C. Bowman, Jr.

Series B: Physics

Hydrogen in Disordered and Amorphous Solids

Edited by

Gust Bambakidis

Wright State University
Dayton, Ohio

and

Robert C. Bowman, Jr.

The Aerospace Corporation
Los Angeles, California

Springer Science+Business Media, LLC

Proceedings of a NATO Advanced Study Institute on
Hydrogen in Disordered and Amorphous Solids,
held September 9–19, 1985,
in Rhodes, Greece

ISBN 978-1-4899-2027-0 ISBN 978-1-4899-2025-6 (eBook)
DOI 10.1007/978-1-4899-2025-6

Library of Congress Cataloging in Publication Data

NATO Advanced Study Institute on Hydrogen in Disordered and Amorphous Solids (1985: Rhodes, Greece)
Hydrogen in disordered and amorphous solids.

(NATO ASI series. Series B, Physics; v. 136)
"Proceedings of a NATO Advanced Study Institute on Hydrogen in Disordered and Amorphous Solids, held September 9–19, 1985, in Rhodes Greece"—T.p.
"Published in cooperation with NATO Scientific Affairs Division."
Includes bibliographical references and index.
1. Amorphous semiconductors. 2. Order-disorder models—Congresses. 3. Amorphous substances—Congresses. 4. Hydrogen—Congresses. I. Bambakidis, Gust. II. Bowman, R. C. III. North Atlantic Treaty Organization. Scientific Affairs Division. IV. Title. V. Series.
QC611.8.A5N37 1985 530.4′1 86-8154
ISBN 978-1-4899-2027-0

Originally published by Plenum Press, New York in 1986
Softcover reprint of the hardcover 1st edition 1986

PREFACE

This is the second volume in the NATO ASI series dealing with the topic of hydrogen in solids. The first (V. B76, Metal Hydrides) appeared five years ago and focussed primarily on crystalline phases of hydrided metallic systems. In the intervening period, the amorphous solid state has become an area of intense research activity, encompassing both metallic and non-metallic, e.g. semiconducting, systems. At the same time the problem of storage of hydrogen, which motivated the first ASI, continues to be important.

In the case of metallic systems, there were early indications that metallic glasses and disordered alloys may be more corrosion resistant, less susceptible to embrittlement by hydrogen and have a higher hydrogen mobility than ordered metals or intermetallics. All of these properties are desirable for hydrogen storage. Subsequent research has shown that thermodynamic instability is a severe problem in many amorphous metal hydrides. The present ASI has provided an appropriate forum to focus on these issues.

In the case of semiconducting systems, the discovery that hydrogen, present as a contaminant during the production of amorphous silicon from silane, actually improves the efficiency of solar cells which use amorphous silicon, has spurred a great deal of activity aimed at understanding the mechanism of hydrogen passivation. The sensitivity of the electrical properties of amorphous semiconductors to processing variables underscores the importance of understanding the role played by the local environment around the hydrogen atom and its interaction with other defects. The attractiveness of amorphous semiconductors is their potential for replacing their crystalline counterparts in photovoltaic devices and in electrodes for hydrogen production by photo-assisted dissociation.

As reflected in these Proceedings, by covering both classes of amorphous hydrides the Study Institute provided a unique opportunity to identify common problem areas and for the cross-fertilization of ideas for future research.

This book is divided into two parts. Part A deals with hydrogen in amorphous semiconductors. After a review, by several speakers, of the theoretical understanding of electronic and transport properties in these systems, there follow lectures dealing with the distribution of hydrogen in external and internal surfaces and its effect on defect structure. Finally, attention is given to the formation and trapping of molecular hydrogen in voids in the host matrix.

Part B is concerned with hydrogen in disordered and amorphous metals. It begins with a review of the preparation, structure and properties of these systems. Then a discussion is given of the theory of electronic states in disordered alloy hydrides, followed by presentation of a semi-empirical band structure model for the heat of solution and pressure-composition isotherm. Several lectures deal with the formation and thermal stability of amorphous hydrides. The long-range motion of hydrogen through the amorphous matrix is discussed, as studied by a variety of experimental techniques. Hydrogen site occupancies and dynamics are examined in several subsequent talks, based on elastic and inelastic neutron scattering studies. The next three papers compare the thermodynamic behavior of hydrogen in ordered and disordered alloys. Finally, lectures are given dealing with the use of incoherent X-ray and γ-ray scattering to probe the electronic structure, and with a discussion of disordered alloy tritides.

We wish to express our deep appreciation to the lecturers, seminar speakers and students, all of whom contributed to the success of the Institute. Special thanks go also to Mrs. Elli Bambakidis for help with preparation of the manuscript and to Mrs. Judy Bowman for coordinating activities related to the Institute. The financial support of the NATO Scientific Affairs Division, the Wright State University Research Council and of Energy Conversion Devices, Inc. is gratefully acknowledged.

G. Bambakidis
Dayton, Ohio

R. C. Bowman, Jr.
Los Angeles, California

January 19, 1986

CONTENTS

PART A. HYDROGEN IN AMORPHOUS SEMICONDUCTORS

PART B. HYDROGEN IN DISORDERED AND AMORPHOUS METALS

ELEMENTS OF THE THEORY OF AMORPHOUS SEMICONDUCTORS*

Morrel H. Cohen

Exxon Research and Engineering Company
Annandale, NJ 08801

ABSTRACT

The main elements of the theory of amorphous semiconductors are reviewed. The kinds of disorder occurring in these materials are classified. The consequences of the various kinds of disorder for electronic states and energies at and near the band edges and in the gaps are discussed. The energy spectrum is divided into ranges according to the principal characteristics of the states and the spectrum, localized vs. extended, universal vs. nonuniversal, smooth vs. fractal, etc. The effects of interactions are discussed. The general theory of transport is reviewed. The consequences of the above for the optical and the transport properties are discussed briefly.

I. INTRODUCTION

The investigation of amorphous semiconductors constitutes a large, highly developed field of science. There is a broad range of materials studied. Diverse methods are used to prepare them. Many different physical properties are measured. A wide and flexible set of experimental techniques is used for the measurements. As a consequence vast amounts of data exist.

Nonetheless, much of that information is poorly understood. There are conflicting interpretations found within the literature. Inapplicable concepts are used, sometimes carried over incorrectly from crystalline semiconductor physics and sometimes generated de novo. There is, in my opinion, much confusion in the field.

Why is this so? Again in my opinion, the intrinsic complexity of the materials has inhibited the development of the theory. As a consequence, theory has had little impact on the development of the subject, in contrast to the case of crystalline semiconductors, where there has been a healthy symbiosis between theory and experiment.

*Substantially the same material will appear in Proceedings of the IMA Workshop on Random Media, Springer Verlag (in press).

It is the presence of disorder which leads to the great complexity of amorphous semiconductors; that is, it is the loss of long-range order which makes the theory of amorphous semiconductors so much more difficult than that of crystalline semiconductors. The materials have compositional disorder; that is, they are rarely pure materials, and the constituents are not regularly arranged (e.g., a-SiH_x, a-SiF_x, a-H_xF_y). The materials also contain structural disorder, both geometric and topological in nature. Because the materials are covalently bonded, the geometric and topological disorder have certain characteristic features. The geometric disorder is comprised of random bond-length variation, bond-angle variation, and dihedral-angle variation. The topological disorder derives from the presence of coordination defects, odd rings, and other, subtler features. There may be inhomogeneity in composition, in the range of order (e.g., microcrystallites embedded in an amorphous matrix), in the distribution of defects (e.g., dangling bonds may cluster on internal surfaces, odd rings are threaded by line defects). Another complication is the multiplicity of atomic orbitals required to describe the electronic structure of the material. At the very least, there is needed a set of one s- and three p-orbitals for each host element and possibly an excited s-orbital and 5 d-orbitals for a more accurate treatment of the conduction band. Beyond that, one needs orbitals for any hydrogen and fluorine present. This multiplicity is a major complication; most exact or accurate theoretical results have been obtained only for models containing a single s-orbital. The final complication is the presence of interactions. Electron-phonon interactions are always important, and electron-electron interactions are important in certain specific circumstances.

Despite this complexity some progress has been made recently towards a definitive theory. The purpose of the present paper is to provide a selective review of the present state of the theory, concentrating on the contributions of my colleagues and/or myself. As such, it will serve as an introduction to the papers by Soukoulis, by Economou and by Zdetsis in the present volume.

The paper is organized as follows. Section II contains a discussion of band bounds in disordered materials, distinguishing between band edges and band limits. Section III introduces the mobility-edge concept. Section IV introduces the current version of the simplest band model of an amorphous semiconductor. Section V contains a discussion of the band-edge features in the electronic structure of a disordered material and classifies them according to the degree to which they can be represented as universal. In Section VI the effects of electron-phonon interaction are discussed. In Section VII there is a brief dicussion of fast processes such as the optical absorption and in Section VIII of slow processes such as the dc transport properties. We conclude in Section IX with an overall summary of the present status of the theory.

II. BAND BOUNDS, LIMITS AND EDGES

From the existence of an optical-absorption edge and an activation energy in the d.c. conductivity of amorphous semiconductors, it has been inferred that the materials possess energy gaps in the presence of disorder. At the simplest level, this has been explained in terms of the covalent character of the material. Most atoms have their valence requirements locally satisfied so that it costs energy to break a bond, hence an energy gap. However, it is possible to go beyond such primitive arguments and prove rigorously that for certain simple models of disordered systems, the energy bands can have sharply defined bounds. This was shown first in 1964 by Lifshitz, and Weaire and Thorpe showed in 1971 that gaps can exist.

Consider for example a tightly bound s band for which the states are linear combinations of s-orbitals centered on the sites of a regular lattice. The matrix elements of the Hamiltonian are

$$H_{\ell m} = \varepsilon_\ell \delta_{\ell m} + V_{\ell m}(1-\delta_{\ell m}) \qquad (1)$$

where ℓ, m indicate the s-orbitals on sites ℓ and m of the lattice. Randomness in the diagonal (off-diagonal) elements is termed diagonal (off-diagonal) disorder.

There exist two kinds of bounds. To illustrate the first kind, suppose there is diagonal disorder only, with the ε_ℓ independent random variables each having an average value ε, the same as for the perfect crystal, and with the $V_{\ell m}$ as in the crystal. Let $p(\varepsilon_\ell)$ be the probability distribution of the individual ε_ℓ. If $p(\varepsilon_\ell)$ is bounded, that is

$$p(\varepsilon_\ell) = 0, \ \varepsilon_\ell < \varepsilon_L \text{ or } > \varepsilon_U, \qquad (2)$$

then the energy band has a lower bound $\varepsilon-\varepsilon_L$ below the bottom of the unperturbed energy band and an upper bound $\varepsilon_U-\varepsilon$ above its top. The states just inside the bounds are confined to large regions within which the ε_ℓ are all nearly equal to ε_L or ε_U, respectively. Such regions are highly improbable, and the density of states vanishes strongly as the band bounds are approached,

$$n(E) \propto e^{-A_L/|E-E_L|^{d/2}} \qquad (3)$$

as $E \to E_L^+$, the lower bound, and similarly for E_U, the upper bound. Such bounds we call band limits or Lifshitz limits, after their discoverer.

If $p(\varepsilon_\ell)$ is unbounded, i.e. $\varepsilon_L \to -\infty$ and $\varepsilon_U \to \infty$, as is the case for, e.g., a Gaussian or a Lorentzian, and if $|E-E_{L,U}| >> B$, the unperturbed band width,

$$n(E) \sim p(E) \qquad (4)$$

holds, and n(E) has no bounds or limits.

The second kind of bound is a normal band edge E_N at which

$$n(E) \propto |E-E_N|^{(\frac{d}{2} - 1)} \qquad (5)$$

inside the band and zero outside, where d is the Euclidian dimension of the material. The behavior in (5) is the same as for a crystal and occurs when the disordered material has a hidden symmetry. Consider a tight-binding s-band for which all $\varepsilon_\ell = 0$, all $V_{\ell m} = V < 0$ for ℓm nearest neighbors and zero otherwise, all sites have the fixed coordination number z, and otherwise the structure is arbitrary. Such a structure is an ideal continous random net (CRN), without coordination defects, and the Hamiltonian contains topological disorder only, no quantitative disorder. It is easy to prove that the Hamiltonian (1) has a spectrum bounded by $\pm$ zV. The wave function at the bound zV is of bonding type and is translationally invariant. As a consequence of this hidden symmetry, the bonding bound is a normal band edge with n(E) given by (5). The wave function at the other bound, -zV, is antibonding and of equal amplitude everywhere but of opposite sign for all nearest neighbor pairs. Such a state can be realized only in a net having no odd rings, a bichromatic net. The antibonding bound is thus a normal band edge only in the presence of another hidden symmetry, evenness of all rings in the structure. By rings we mean a closed graph of lines connecting nearest neighbors, an even ring having an even number of lines. In the presence of odd rings, the antibonding bound becomes a Lifshitz limit.

To summarize, if $p(\varepsilon_\ell)$ is unbounded the energy band is unbounded. If the band is bounded and there is sufficient symmetry remaining, the bound is a normal band edge. Otherwise, it is a Lifshitz limit.

Let us turn now to a more complex model, one considered by Singh for the representation of a-Si in a 4-coordinated CRN structure. It is a tight-binding model with a basis set consisting of an s, p_x, p_y, and p_z orbital on each atomic site. If the structure has odd rings, but no bond-length, bond-angle, or dihedral-angle disorder, the top of the valence band is a normal band edge corresponding to a 3-fold degenerate bonding p state. As soon as dihedral-angle disorder is introduced, the top of the valence band turns into a Lifshitz limit. When bond-length and bond-angle disorder is added, the Lifshitz limit is smeared out. Similarly, in the presence of dihedral-angle disorder but no odd rings or bond-length or bond-angle disorder, the bottom of the conduction band is a normal band edge coresponding to a doubly degenerate antibonding σ state (for values of the matrix elements which would put the bottom of the conduction band at the zone boundary for the crystal). Adding odd rings turns the normal band edge into a Lifshitz limit, and including bond-length and bond-angle disorder smears out the Lifshitz limit.

From this simplified but still realistic model, we are led to two conclusions. First, band bounds are smeared out in general, and therefore we must determine n(E) and the features of the wave functions explicitly for each class of materials or models. Second, in real materials a fairly large number of matrix elements is required for a tight-binding model which accurately represents the electronic structure. The various kinds of disorder present affect these matrix elements and the states near the different band edges in quite different ways. These two circumstances lead to the approximate validity of two simplifying assumptions which we shall explore in the following:

1) We can use a tight-binding model without orbital multiplicity to represent the conduction or valence band of real materials, but we must suppose the disorder potential to have a Gaussian probability distribution. That is, the complexity of the real problem allows us to invoke the central limit theorem.

2) The valence and conduction bands can be taken as statistically independent because individual elements of the disorder affect the valence and conduction band edges quite differently.

III. THE MOBILITY EDGE CONCEPT

We now suppose that either there is no electron-phonon interaction or T = 0. In a crystal, there are extended states within the energy bands and no states within the gaps. The extended states are Bloch states which have equal amplitudes in every unit cell and complete phase coherence. If an isolated defect or impurity is introduced into the material, localized states can appear within the energy gaps. By localized states, we mean states which decay exponentially in amplitude asymptotically far away from the center of the region of localization, in this case the impurity or defect responsible for this existence of the state. Even though the band edge is smeared out in a disordered material, localized defect or impurity states can still occur deeper in the gap. These will no longer be sharp in energy, as in a crystal, but will be broadened by the disorder. These, however, are not the only localized states. Even in the absence of such defects or impurities, as in an ideal CRN, there is an energy E_c, called the mobility edge for reasons which will emerge below, which divides localized states in the tail of the energy band from extended states within the band. These localized states also decay exponentially, but the decay length, the so-called localization length, diverges as the mobility edge is approached. The extended states are also quite different from those in a crystal. The phase becomes incoherent over distances greater than the mean-free-path ℓ, and violent amplitude fluctuations set in on distance scales between ℓ and the amplitude coherence length ξ, which also diverges at the mobility edge. In fact, the wave functions are fractal between those limits.

In the absence of the electron-phonon interaction or when the important phonon energies are substantially less than k_BT, we can write the Kubo-Greenwood formulas for the dc conductivity in the following form,

$$\sigma = \int dE\ \sigma(E)\ \left(-\frac{df(E)}{dE}\right), \qquad (6)$$

$$\sigma = -\int dE n(E) e\mu(E) f(E), \qquad (7)$$

$$\frac{d\sigma(E)}{dE} = -n(E)e\mu(E) \quad . \qquad (8)$$

In Eq. (6), $\sigma(E)$ is the microscopic conductivity, and $f(E)$ is the electron occupation number for a state of energy E. In Eq. (7), $\mu(E)$ is the microscopic mobility; Eq. (8) relates the two. It has been shown that in the absence of the electron-phonon interaction,

$$\sigma(E) \propto (E-E_c)^s,\ E \to E_c^+\ , \qquad (9a)$$

$$= 0 \qquad ,\ E < E_c\ . \qquad (9b)$$

Eq. (9b) follows from the fact that electrons in localized states ($E < E_c$) are immobile, whereas electrons in extended states ($E > E_c$) can move macroscopic distances. Eq. (8) implies that $s \geq 1$, and the best current theories give $s = 1$. In that case Eq. (8) and Eq. (9a,b) imply that there is a step in the mobility from zero to a finite value at E_c. Thus E_c is termed a mobility edge. Eq. (9a) holds only near the mobility edge, where amplitude fluctuations suppress the conductivity. A major task of the theory is to attach an energy scale to the variation of $\sigma(E)$ and other quantities with energy. However, supposing that the characteristic energy scale is larger than k_BT leads immediately to the observed activated temperature dependence of the conductivity. Thus with the two concepts of somewhat smeared band edges but sharp mobility edges in disordered bands, we can understand the most characteristic properties of amorphous semiconductors, the apparent optical absorption edge and the activated dc conductivity.

IV. THE SIMPLEST BAND MODEL OF AN AMORPHOUS SEMICONDUCTOR

I shall use as an updated version of the original Mott-CFO model of the energy bands of an amorphous semiconductor a sketch of the density of states in a-SiH_x as it is now emerging from a wide variety of experiments. In Fig. 1a we show n(E) vs E. There is a valence band and within it a mobility edge E_v, and a conduction band and within it a mobility edge E_c. Below E_v and above E_c the states are extended. Between E_v and E_c the states are localized. Both bands have an exponential tail, the width being about 500K for the valence band and 300K for the conduction band.

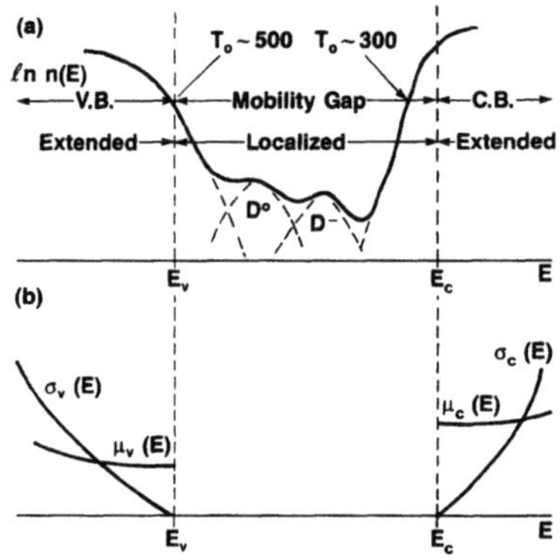

Figure 1. (a) The density of states of a-SiH_x after Fritzsche's synthesis of the experimental data. (b) A sketch of the microscopic conductivities and mobilities for the conduction and valence bands.

Near the center of the gap there is a bump associated with neutral 3-fold coordinated Si atoms, the dangling bond state $D°$, and above that a bump corresponding to the same state doubly occupied, D^-, the shift being due to the repulsion between the two electrons of opposite spin occupying the dangling bond state in D^-. The region between the two mobility edges is called the mobility gap E_c-E_v because the mobility vanishes there even though the density of states remains finite (at T = 0 or neglecting the electron-phonon interaction).

In Fig. (1b) are shown the corresponding microscopic conductivities and mobilities. However, to complete the correspondence between

Figs. (1a) and (1b) it is necessary to establish the energy scale for Fig. (1b), that for Fig. (1a) being known in most respects. Moreover, the effects of the electron-phonon interaction must be included as well. Thus, Figs. (1a) and (1b) enable us to understand the gross features of the optical absorption and the dc conductivity, but they are incomplete.

V. UNIVERSALITY, QUASIUNIVERSALITY AND PSEUDOUNIVERSALITY IN THE ELECTRONIC STRUCTURE AT BAND EDGES

Electronic properties arising from states near band edges appear to be quite similar in very different amorphous semiconductors. For example, the optical absorption α always displays a Tauc region in which

$$\{\alpha/\hbar\omega\}^{1/2} = B[\hbar\omega - E_{Go}] \tag{10}$$

and at lower photon energies $\hbar\omega$ an Urbach region in which

$$\alpha = \alpha_0 e^{\hbar\omega/E_{oo}}, \tag{11}$$

where B and α_0 are constants, E_{Go} is the apparent optical or Tauc gap, and E_{oo} is the width of the Urbach tail. Another example is provided by the dc transport in which similar patterns of behavior of σ and the thermopower S are found in dissimilar materials. These features of the data lead one to infer that there are universal or nearly universal features in the electronic structure near band edges, independent of structure, composition and the details of disorder, or nearly so.

It is a prime task for the theorist to establish whether and when such strict or approximate universality can exist. At present this task has been accomplished only for a specific class of models under the following conditions:

1. There exists a reference model from which the disorder can be measured.

2. The reference model need not be ordered but must have a normal band edge.

3. There is no orbital multiplicity in a tight-binding representation of the Hamiltonian.

4. The disorder is measured quantitatively via the variance w^2 of the random potential.

5. w must be substantially less than the unperturbed band width B.

6. Only states of energies E deviating from the unperturbed band edge by much less than the band width are considered.

7. Only distance scales much greater than λ, the larger of the interatomic separation a or the correlation length ℓ_c of the random potential, are considered.

8. The continuum limit of the tight-binding model must have no ultraviolet catastrophe, that is, no sensitivity to the short-wavelength cutoff λ or the high-energy cutoff B. The continuum limit is taken in such a way that $w \to \infty$, $B \to \infty$, $a \to 0$, $\lambda \to 0$ so that

$$Va^2 = \frac{\hbar^2}{2m^*} = \text{const}, \ w^2\lambda^d = \gamma = \text{const}. \tag{12}$$

In Eq. (12) V is an effective nearest neighbor electron transfer matrix element ($V \to \infty$ as $B \to \infty$), m* is the effective mass of the reference model, and d is the dimension of the model. No ultraviolet catastrophe occurs for $d < 2$, but it does occur for $d > 2$ leading to a shift of the unperturbed band edge to E_{Bd}. When energies are measured relative to E_{Bd}, the possibility of universality is restored.

The proof of universality follows from the observation that the model contains ony 2 parameters, $\hbar^2/2m^*$ and V. A natural unit of energy E_{od} and one of length L_{od} can be defined from these two parameters. In the simpler case that $\lambda = a$, E_{od} and L_{od} have the particularly simple forms

$$E_{od} = W^{\frac{4}{4-d}} \ V^{-\frac{d}{4-d}}, \tag{12}$$

$$L_{od} = a(V/w)^{\frac{2}{4-d}}, \tag{13}$$

$d \neq 4$. When all energies are measured relative to E_{Bd} in units of E_{od}, lengths in units of L_{od}, and conditions (1.) to (7.) are met, then all physical properties can be expressed either as universal functions or universal numbers. As an example of the former, consider the density of states,

$$n(E) = L_{od}^{-d} \ E_{od}^{-1} \ f_d \ ((E-E_{Bd})/E_{od}), \tag{14}$$

where $f_d(x)$ is a universal function of its argument x for each d. As an example of the latter consider the position of the mobility edge E_c,

$$E_c - E_{Bd} = C_{1d}E_{od}, \quad 2 < d, \tag{15}$$

where $C_{13} = 4.5 \times 10^{-3}$ and $C_{14} = 0$, $d > 4$. There are two limitations to universality deriving from the energy dependence of conditions (6.) and (7.) even when all the conditions are met. Condition (6.) breaks down both in the continuum and the gap when $|E - E_{Bd}|$ approaches V. Condition (7.) breaks down in the gap when the scale L of the potential fluctuations important for localization becomes smaller than λ.

We have obtained a broad range of detailed quantitative results relating to universality in the absence of the electron-phonon interaction by various theoretical techniques. The methods include perturbation theory, the coherent potential approximation (CPA), field theory, path integral methods, numerical calculations and the potential well analogy. The results include the density of states, the nature of the wave functions, the mean free path, the energy dependent conductivity and mobility, and the frequency-dependent dielectric function. We know how to add the electron-phonon interaction for fast processes, and some progress has been made for slow processes.

Various aspects of the methods and the results are discussed in detail in the papers of Soukoulis, Economou, and Zdetsis in this volume. In the present paper, we are concerned primarily with the conceptual picture which emerges from these results.

In Fig. 2, we show the distinct energy regions into which the spectrum divides near the band edge and summarize the principal characteristics of the density of states and the corresponding features of the wave functions. There is of course a sharp transition from localized to extended states at the mobility edge E_c. However, there is far more structure in the spectrum than a simple division into localized and extended states. There are in fact seven distinguishable regions, four within the localized domain and three within the extended domain. Only the transition from localized to extended states at E_c is sharp. The remaining five transitions are in fact smooth crossovers.

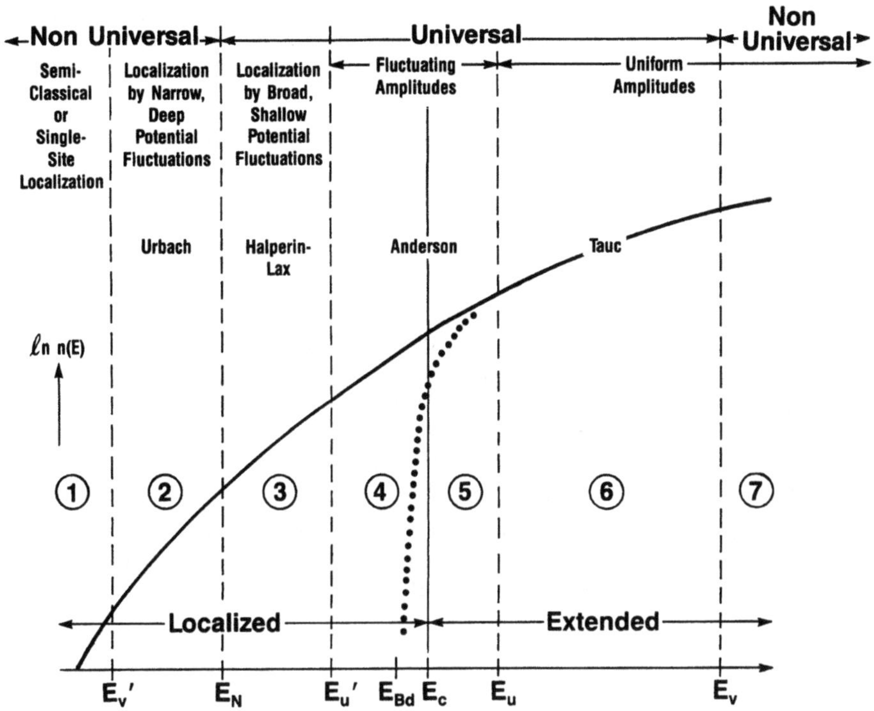

Figure 2. The seven distinguishable energy regions near the band edge of a simple model of a disordered material (no multiplicity of orbitals).

In addition to the localized/extended classification there are two other classifications used in Fig. 2: universal/nonuniversal and fluctuating/smooth. Universal behavior is found in regions 4, 5, and 6 of Fig. 1 (between E_N and E_V) including both localized states in 4, and extended states in 5 and 6. Nonuniversality occurs in region 7 because the condition $|E-E_{Bd}| \ll V$ is violated at E_V, and the electronic properties can no longer be represented by universal forms such as in Eqs. (14) and (15). Nonuniversality occurs also in regions (1.) and (2.) because the length scale L of the relevant potential fluctuations becomes smaller than λ at E_N. Amplitudes are smooth below E_u, and above $E_{u'}$ in regions (1.)-(3.) and (6.), (7.), respectively. Within regions (4.) and (5.) there are strong amplitude fluctuations on length scales lying between the mean free path ℓ (roughly) and the localization length L_{loc} in (4.) and the amplitude coherence length ξ in (5). This amplitude fluctuation occurs because of resonances in potential fluctuations not quite strong enough for localization. The crossover energy E_N occurs when ξ falls below ℓ, and the other one, $E_{u'}$, occurs when L_{loc} approaches within a factor of 2 of ℓ. We think it appropriate to term regions (4.) and (5.) the Anderson regime. In more quantitative terms, n(E) is accurately given by the CPA in regions (6.)

and (7.). In region (6.), which we term the Tauc region, the density of states has the normal form, Eq. (5), with E_N replaced by E_{Bd} and the coefficient fixed by universality. Transitions from region (6.) in the valence band to region (6.) in the conduction band are responsible for the Tauc region of the optical absorption (d = 3), Eq. (10), and give

$$E_{Go} = E_{B3c} - E_{B3v} \, . \tag{16}$$

Because C_{13} is so small, cf. Eq. (15), E_{Go} is nearly equal to the mobility gap $E_c - E_v$. These two statements must be understood with reference to optical absorption being a fast process, as discussed below.

In region (3.), the states are localized by broad, shallow potential fluctuations. This is the Halperin-Lax region, where the density of states is given by

$$\left.\begin{aligned} \ln n(E) &\propto -[|E-E_c|/E_{13}]^{1/2}, \\ E_{13} &= 7 \times 10^{-4}\, E_{03} \, . \end{aligned}\right\} \tag{17}$$

The interesting range between E_N and E_u comprising regions (3.), (4.), and (5.) is, if one chooses parameters for w and V appropriate to real materials for which $w/V \sim 1$, extremely narrow, and E_{13} is clearly very small. Accordingly, we have used a highly nonlinear energy scale in Fig. 2, expanding the range E_N to E_u greatly. The range itself and its characteristic phenomena, amplitude fluctuations and the Halperin-Lax tail, should be very hard to observe. Inelastic effects at finite temperature should wipe out or smear out these characteristic phenomena, which should clearly be observable only at low temperatures.

In region (2.), the states are localized by narrow, deep potential fluctuations, and the density of states becomes a simple exponential,

$$\left.\begin{aligned} \ln n(E) &\propto - |E|/E_{23} \, , \\ E_{23} &\simeq \tfrac{1}{8}\, w^2/V \, , \end{aligned}\right\} \tag{18}$$

but only for a Gaussian probability distribution, the one which should be used according to §2. We term this the Urbach region. The numerical coefficient in E_{23} is nonuniversal, and, in fact, is probably much smaller for real materials than in the simple models. We therefore,term the Urbach region pseudouniversal. We note that thermal disorder is also Gaussian and adds in the square to E_{23} in Eq. (18), giving E_{23} the observed T-dependence.

In new work with S. John, we have obtained an explicit analytic form for the density of states below E_u, including regions (1.), (2.), and (3.), which accurately fits our exact results.

VI. EFFECTS OF ELECTRON-PHONON INTERACTIONS

In amorphous semiconductors, electron-electron interactions are important only when two electrons of opposite spin occupy the same localized state or for Mott hopping at the Fermi level at very low temperature. Electron-phonon interactions are always present and more important in disordered than ordered semiconductors. The electron-phonon (e-p) interaction has two kinds of effects, elastic or static and inelastic. The elastic effects favor small polaron formation at $T = 0$ and lead to thermal disorder at finite T. Both increase localization. On the other hand, the inelastic effects destroy localization. The polaron effects shift E_c to higher energies and cause already localized states to be more tightly bound. Polaron resonances are formed in the range E_c to E_u with both energies shifted by the e-p interaction. As discussed above, this is a narrow range of energies, but within it we expect $\xi(E)$ to be substantially decreased by the presence of the small polaron resonances and possibly the exponent s in Eq. (9a) to be increased above unity.

Thermal disorder merely adds an additional Gaussian random potential whose variance adds linearly to w^2, as already discussed.

The inelastic processes lead to tunneling between localized states and between localized and extended and to scattering between extended states. This should substantially reduce the fractal behavior in regions (4.) and (5.) of Fig. 2, increasing the fractal dimension D towards d, reducing ξ, and decreasing $E_u - E_{u'}$. In other words, inelastic scattering tends to suppress amplitude fluctuations. Moreover, there is a profound modification of the transport processes. In particular, $\sigma(E)$ is decreased above E_u and increased everywhere below E_u, the increase growing rapidly with T.

The effect of the e-p interactions depends on the time scale of the physical process in question. In fast processes such as the optical absorption in which the frequency $\omega \gg \tau_{pol}^{-1}, \tau_{in}^{-1}$, where τ_{pol} is the time required for polaron effects to manifest themselves and τ_{in} is the inelastic scattering time, only thermal disorder plays a role, and it can be treated on the same basis as the static disorder previously described. On the other hand, in slow processes in which $\omega \ll \tau_{pol}^{-1}, \tau_{in}^{-1}$, all three effects play important roles, polaron effects, thermal disorder, and inelastic scattering.

VII. FAST PROCESSES: OPTICAL ABSORPTION

In fast processes, the effect of the electron-phonon interaction is solely to increase w^2 through the additional contribution of thermal disorder. Thus, up to some energy above E_V, one can write the imaginary part of the dielectric constant as

$$\varepsilon_2(\hbar\omega) = (2\pi e)[d(\hbar\omega)]^2 V_a \int dE\; n_v(E) n_c(E - \hbar\omega), \tag{19}$$

where V_a is the atomic volume and $[d(\hbar\omega)]^2$ is the mean squared matrix element of the coordinate operator. Both experimentally and theoretically, the latter has been shown to be constant, independent of ω, up to some value of $\hbar\omega$ above the upper limit of the Tauc regime, i.e. above $E_c - E_v$. This constancy of the matrix element is easily understood in terms of statistical independence of the valence and conduction band wave functions and their phase incoherence beyond the mean free path. In a-SiH_x, the observed value of d implies a mean free path on the atomic scale and incomplete randomness of phases. However, a detailed calculation of vertex corrections is still lacking because of the required multiplicity of the basis orbitals. The combined density of states, on the other hand, is understood quantitatively in simple models, and when multiplicity is included it is understood at the level of the CPA.

In summary, the optical absorption is the best understood of all physical properties.

VIII. SLOW PROCESSES: DC TRANSPORT

Without electron-phonon interactions or when $\hbar\omega_{ph} < k_BT$, the case we shall actually consider, one can write exact expressions for the dc conductivity and thermopower S

$$\sigma = \int dE\sigma(E)\ (-\frac{df(E)}{dE})\ , \tag{6 }$$

$$S = -\frac{k_B}{e}\frac{\beta}{\sigma}\int dE\ (E-E_p)\ \sigma(E)\ (-\frac{df}{dE})\ , \tag{20}$$

$$\int [n(E)f(E) - n_o(E)f_o(E)]dE = 0\ , \tag{21}$$

$$Q \equiv \ln \sigma/\sigma_o - (e/k_B)S\ . \tag{22}$$

In Eq. (21) a subscript o indicates T = 0 values, and σ_o is any convenient constant with dimensions of conductivity. Eq. (21) determines the temperature dependent Fermi level $E_F(T)$.

The experimental data fall into three categories: (1.) ln σ, S, and therefore Q are linear functions of $1/k_BT$, and the thermopower and the conductivity relate in the way expected if the dominant carriers are in extended states above the mobility edge and inelastic scattering is not important; (2.) ln σ, S and other dc transport properties are individually anomalous; (3.) lnσ and S show two distinct linear regions in k_BT joined by a kink, with the lower temperature region behaving as (1.) and the higher temperature region as (2.), while Q remains linear in $1/k_BT$.

One can understand the anomalous behavior in cases (2.) and (3.) in the following way. As shown by Thomas and coworkers, inelastic e-p effects wipe out the mobility edge, delocalizing the electron states. The mobility step at E_c for T = 0 is replaced by a mobility tail which falls off rapidly with E_c-E and increases rapidly with T. The integral in Eq. (6) or Eq. (20) therefore has a sharp maximum at E*. Exponentiating the integrand and expanding the result to lowest order in (E-E*) leads to

$$\sigma = \sqrt{2\pi}\ \frac{\delta E}{k_BT}\ \sigma(E^*)\ e^{-\beta(E^*-E_F)}\ , \tag{23}$$

$$S = \frac{-k_B}{e}\left[\frac{E^*-E_F}{k_BT} + \frac{a}{\sqrt{2\pi}}\ \frac{\delta E}{k_BT}\right]\ , \tag{24}$$

$$Q = \ln\left[\sqrt{2\pi}\ \frac{\delta E}{k_BT}\ \frac{\sigma(E^*)}{\sigma_o}\right] + \frac{a\,\delta E}{\sqrt{2\pi}\ k_BT}\ , \tag{25}$$

$$\delta E = \left\{\left[\frac{\Sigma'(E^*)}{\Sigma(E^*)}\right]^2 - \frac{\Sigma''(E^*)}{\Sigma(E^*)}\right\}^{-1/2}\ , \tag{26}$$

$$a = (\delta E)^{-2}\int (E-E^*)\ e^{f(E,E^*)}dE\ , \tag{27a}$$

$$f(E,E^*) = \ln[\Sigma(E)/\Sigma(E^*)] \approx -\beta(E-E^*)\ . \tag{27b}$$

Note that the temperature-dependent activation term $\beta(E^*-E_F)$ present in σ, Eq. (23), and S, Eq. (24), cancels out of Q, Eq. (25), as stressed by Overhof. If the complex and as yet uncertain temperature dependence of E* has the form

$$\sigma(E^*) = \sigma^* e^{-\beta\Delta E}, \tag{28}$$

then Q becomes

$$Q = \ln\left[\sqrt{2\pi}\, \beta\, \delta E\, \sigma^*/\sigma_0\right] - \beta\left[\Delta E - \frac{a}{\sqrt{2\pi}}\, \delta E\right], \tag{29}$$

consistent with the observations. The break in σ and S separately must therefore be due to variations of E^*-E_F with T. Overhof has proposed in fact that the transport anomalies arise from the T dependence of E_F. I find this less plausible than the variation of E* with T shown in Fig. 3. At low T, E* remains at E_c. Around a temperature T_0, it rapidly drops below the T = 0 mobility edge and continues to decrease for $T > T_0$ at a slower rate. This motion of E* and the associated transport via tail states, I propose, is responsible for the kinks in case (2.) and the anomalies in cases (2.) and (3.).

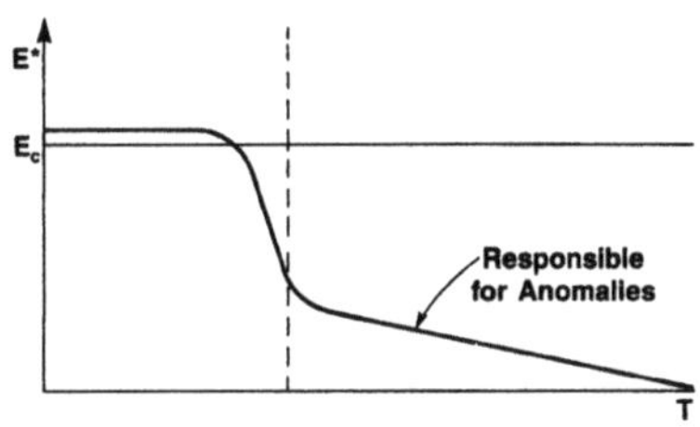

Figure 3. Variation of E* with T proposed to account for the transport anomalies.

IX. CONCLUSIONS

The density of states, the optical absorption and related quantities are now becoming well understood for simple models, and substantial progress has been made for more complex models containing multiplicity and actually applicable to real materials.

The transport properties will be well understood when there is a more complete treatment of the e-p interaction. Our degree of understanding of polaron efects is only partial, that of thermal disorder is good, and that of inelastic effects is partial. However, each of these aspects has been treated independently. The most pressing need is an integrated treatment of all three aspects of the e-p interaction.

After the e-p interaction is fully integrated into the theory of the simple models, the then pressing need will be to introduce more features of real materials into the simple models. In particular, our understanding of off-diagonal disorder, topological disorder, and multiplicity of the basis set and their interplay with diagonal disorder must be further developed.

Clearly, the outlook is optimistic, but, even though the rate of progress is accelerating, the time scale will be long.

ELECTRONIC STRUCTURE OF α-SiH

E.N. Economou

Department of Physics, University of Crete
and Research Center of Crete, Heraklio, Crete
Greece

ABSTRACT

Recent advances based on the coherent potential approximation and the potential well analogy suggest that, in spite of the great complexity of the potential felt by an electron moving in an amorphous semiconductor, the electronic structure of the latter possesses a certain universality. As a result (and to a first approximation) only a few parameters matter. The situation is analoguous to that of a crystalline semiconductor, where a single quantity - the effective mass - allows one to bypass the complexity of the crystalline potential.

INTRODUCTION

An electron moving in an amorphous semiconductor can be described by a Hamiltonian of the form

$$H = H_0 + H_1 \qquad (1)$$

where H_0 is a non-random part which is assumed soluble and H_1 is random with $\langle H_1 \rangle = 0$. For a typical semiconductor H_0 allows us to obtain the unperturbed density of states (DOS) $\rho_{0i}(E)$, where i stands for the s-like and p-like component, and the unperturbed Green's function $G_{0i}(E)$; $G_{0i}(E)$ can be expressed as an integral of $\rho_{0i}(E')/(E-E')$. The random part H_1 can be expressed in terms of some matrix elements ε_i, which are random variables possesing probability distributions $p_i(\varepsilon_i)$: To a first approximation all we need to know from $p_i(\varepsilon_i)$ is its variance w_i^2 and its form in the tail; in most cases, due to the many independent sources of disorder, $p_i(\varepsilon)$ possesses a Gaussian tail,

$$p_i(\varepsilon_i) \sim \exp[-\frac{\varepsilon_i^2}{2w_i^2}] \,. \qquad (2)$$

To summarize : To a first approximation the main inputs for constructing a calculational scheme for the electronic structure of α-semiconductors are the unperturbed DOS (and the unperturbed Green's function) near the band edges, and the variance w_i^2 of the random potential.

THE COHERENT POTENTIAL APPROXIMATION

The CPA is conceptually simple : it replaces the random matrix elements ε_i by appropriately choosen non-random, complex effective matrix elements $\Sigma_i(E)$. The criterion which allows us to choose Σ_i is that the fluctuation around Σ_i produces no scattering on the average. This conceptually simple condition leads to a rather complicated set of equations involving $G_{0i}(E)$, from which $\Sigma_i(E)$ are determined[1]. In the weak scattering case (w_i^2 small) the CPA reduces to second order perturbation theory. It is worthwhile to note that the CPA becomes exact in the limit where the eigenstates are bound around a single large isolated potential fluctuation; the CPA is also very satisfactory for states extending over a large number of sites. It is only in the intermediate case of eigenstates trapped in a cluster of a few sites where the CPA is inappropriate.

RESULTS FOR THE DENSITY OF STATES

The CPA for not-so-large disorder w_i^2 produces a DOS near the gap which can be deduced from the unperturbed $\rho_{0i}(E)$ by the following operations[2].

(a) Both the valence and the conduction bands are shifted towards the gap by $G_{0i}(E_{v0}).w_i^2$ and $G_{0i}(E_{c0}).w_i^2$ respectively, where E_{v0} (E_{c0}) is the unperturbed valence (conduction) band edge. Thus the gap is reduced by an amount equal to

$$\delta E_g = (|G_{0i}(E_{v0})| + |G_{0i}(E_{c0})|) w_i^2 . \qquad (3)$$

(b) Both the valence and the conduction bands develop tails towards the gap. The near tails are dominated by states trapped in clusters of atomic sites; these cluster trapped states (CTS) under normal conditions dominate over a very narrow range of energies. The deeper tails are usually dominated by single site bound (SSB) states. For a Gaussian probability distribution, the SSB states produce an exponential tail in the DOS,

$$\rho_i(E) \sim \exp[-|E|/E_i] . \qquad (4)$$

Equation (4) is an immediate consequence of the fact that the binding energy E in a single potential well of depth ε is approximately a linear function of ε^2 :

$$E \cong c_1\varepsilon^2 + b_2 . \qquad (5)$$

Equation (5) is valid[3] over a considerable range of energies E. Combining Eqs. (2) and (5) we obtain immediately Eq.(4) with

$$E_i = 2c_1 w_i^2 . \qquad (6)$$

It is worthwhile to note that Eqs. (3) and (6) imply a linear relation between δE_g and E_i as found by Cody[4].

CONDUCTIVITY AND LOCALIZATION

The CPA allows us to obtain not only the DOS but the ac conductivity $\sigma(\omega)$. From $\sigma(\omega)$ one can obtain the optical absorption $\alpha(\omega)$ with no additional assumptions about the dipole matrix elements involved. Our explicit calculations have shown that this matrix element is energy-independent over a rather wide range of energies (from 1.7 eV to 3.4 eV for α-Si:H) in agreement with the finding of Cody[4] and the analysis of Cohen et al[5]. The dc conductivity obtained by taking the limit $\omega \to 0$ will be denoted by σ_0. (The subscript is placed in order to distinguish the CPA result for the dc conductivity from the dc conductivity σ which includes localization effects, see below.) In the weak scattering limit the CPA result for σ_0 reduces to the usual second order perturbation formulae[1]

$$\sigma_0 = \frac{1}{12\pi^3} \frac{e^2}{\hbar} S_0 \ell , \tag{7}$$

$$\ell = v_0 \tau , \tag{8}$$

$$\frac{1}{\tau} = \frac{2\pi}{\hbar} \rho_0 w^2 , \tag{9}$$

where S_0 is the constant energy surface in k-space and v_0 is the average velocity.

The CPA result σ_0 is non zero even in the regions of the spectrum where the eigenstates are localized and the dc conductivity (at zero temperature) ought to be zero. Thus to obtain the various effects due to the localization of the states we need to go beyond the CPA. The simplest way to achieve that is by means of the potential well analogy (PWA). The PWA constructs at each energy E an effective potential well the depth and extent of which are given in terms of $\sigma_0 \ell^d$ and ℓ respectively. If this effective potential well can sustain a bound state then the states at E are localized; the decay length of the bound state gives the localization length. If the effective potential well is so shallow that only scattering states exist then the states at E are extended but fluctuating in general, with the largest fluctuation length ξ given by the scattering length of the effective potential well and the dc conductivity being proportional to[6] $1/\xi$. The mobility edge E_c, i.e. the critical energy separating the extended states region from the localized ones is obtained according to the PWA by the relation

$$S\ell^2 \simeq 9 , \tag{10}$$

which for small disorder and near the band edge, where $S = 4\pi k^2$, reduces to $k\ell = 0.84$ (Mott usually quotes $k\ell \simeq \pi$ as the criterion for localization).

OUR MODEL FOR α-Si:H

In recent years[7] we have introduced and studied a very simplistic model for α-Si:H, which nevertheless produced suprisingly good agreement with the experimental data. The explanation for this unexpected success has to do with the insensitivity of the results to the details of the Hamiltonian. As was pointed out in the introduction the results depend mainly on the unperturbed density of states (which is reproduced rather satisfactorily in our model) and on the variance of the disorder w_i^2.

The starting point for our model is a c-Si tight binding Hamiltonian the matrix elements of which have been obtained by fitting pseudopotential c-Si calculations. Although the fit is reasonable, there is room for improvement as recent unpublished work of ours indicates. Hydrogen is introduced by randomly replacing Si atoms either by clusters of 4H or by individual H atoms (in the second case it is formally assumed that the Si lattice heals immediately around the H atom).

Our model is definitely not realistic from a structural point of view : it possesses an underlying crystallinity and omits the topological disorder; it incorporates no reconstruction (which, however, for high H-concentration may not be so extensive); and it omits many alternative ways of H incorporation. Furthermore, it involves some further approximations in obtaining the H-H and Si-H matrix elements. In spite of the above shortcomings and the omission of the off-diagonal disorder, the results we obtain for the electronic structure seem to be very reasonable. More explicitly the widening of the gap upon hydrogenation is satisfactorily reproduced as are the H-induced peaks in the valence band. The optical absorption is obtained correctly in the range 1.7 eV to 3.4 eV both in shape and magnitude. The dipole matrix element turns out to be constant as obtained recently by Jackson[9]. The variation of the various optical features with temperature as well as their correlations seem to be in good agreement[10] with the analysis of Cody[4]. The prefactor in the dc conductivity as well as the mobility are within the experimental range of values[8,10]. The DOS at E_c is consistent with experimental estimates[8].

REFERENCES

1. See, e.g., E.N. Economou, "Green's Function in Quantum Physics", Springer-Verlag, Heidelberg, (1983).

2. E.N. Economou, C.M. Soukoulis, M.H. Cohen, and A.D. Zdetsis, Phys. Rev. B31, 6172(1985).

3. N. Bacalis and E.N. Economou, unpublished.

4. G.D. Cody in "Semiconductors and Semimetals", Vol. 21, J. Pankove, ed., Academic Press (1984).

5. M.H. Cohen, E.N. Economou and C.M. Soukoulis, submitted to Phys. Rev. B.

6. E.N. Economou, Phys. Rev. B31, 7710(85).

7. See, e.g., E.N. Economou, D. Papaconstantopoulos and A.D. Zdetsis, Phys. Rev. B28, 2232(83); ibid B31, 2410(85).

8. E.N. Economou, A.D. Zdetsis and D. Papaconstantopoulos in the Proceedings of the 11th International Conference on Amorphous and Liquids Semiconductors, to be published.

9. W.B. Jackson, S.M. Kelso, C.C. Tsai, J.W. Allen and S.J. Oh, Phys. Rev. B31, 5187(85).

10. E.N. Economou, A.D. Zdetsis, N. Bacalis and D. Papaconstantopoulos, unpublished.

ELECTRON DENSITY OF STATES IN RANDOM SYSTEMS

C. M. Soukoulis

Ames Laboratory-USDOE and Department of Physics
Iowa State University
Ames, Iowa 50011

ABSTRACT

The well-known result of Halperin and Lax for the tail of the density of an electron moving in a Gaussian random potential is reviewed. We then show that complete universality exists for the density of states near band edges for weak disorder in less than two dimensions and modified universality exists in more than two dimensions. Deep in the tail, non-universal behavior emerges as the localized states become sensitive to potential fluctuations on individual sites. This non-universal exponential behavior is responsible for the observed Urbach tails.

THE HALPERIN-LAX TAIL

Various attempts[1-11] have been made to calculate the low-energy tail of the density of states of a disordered system within the physical picture of independent electrons without mutual interactions, the electron-phonon interaction, or spin-flip scattering. In all these attempts it has been recognized that the density of localized states is several orders of magnitude smaller than the density of extended states, implying that the former arise from special atomic configurations. Consequently, one is forced to abandon mean-field-like or coherent-potential-approximation-like theories and even numerical simulations to obtain the density of localized states. They arise from potential fluctuations on wavelength scales of at least several atomic distances due to a variation in the physical parameters of the disordered system on the same scale. When the potential varies slowly enough, the fluctuations in the energy of states mirror the fluctuations in the potential energy. This approach has been discussed by Lifshitz[1] and by Bonch-Bruevich.[2] Kane[3] and Eggarter and Cohen[4] have combined the fluctuations with the semiclassical Thomas-Fermi method to calculate the density of states (DOS). Since the potential-energy fluctuations are Gaussian, the tail found by Kane and by Eggarter and Cohen is Gaussian. By including the kinetic energy of localization omitted in the Fermi-Thomas procedure, Halperin and Lax[5] obtained tails falling off less rapidly than Gaussian. Feynman path integrals,[6,7] field-theoretic treatments,[8] and integration in other function spaces[9] as well as variational calculations[10] yield essentially the same results as those of Halperin and Lax. It is clearly established by any of the above treatments[11] that wells of size λ in which the minimum kinetic energy of localization is proportional to χ_d/λ^2 gives a DOS $n(E)$ which behaves as

$$\ln n(E) = \ln(n_0) - \left(\frac{4}{4-d}\right)^2 \left(\frac{4-d}{d}\right)^{d/2} \frac{\chi_d^{d/2}}{2w^2 \ell_c^d} |E|^{2-(d/2)} \quad , \tag{1}$$

where n_0 is the preexponential, $\chi_d = d\pi^2\hbar^2/2m$, ℓ_c is the correlation length of the potential fluctuations and is of atomic size, d is the space dimensionality, and w^2 is the variance of the random potential. $|E|$ is measured from the bottom of the conduction band. The energy-dependence in Eq. (1) can be understood as follows: The factor $|E|^2$ comes from the amplitude of the potential fluctuation which becomes Gaussian on the length scale $\lambda >> \ell_c$, and the factor $|E|^{-d/2}$ comes from the spatial extent of the potential fluctuations λ_d, which scales as $|E|^{-d/2}$ when one assumes that the kinetic energy $|E|$ of confinement of the wave function for any dimension goes like $1/\lambda^2$.

The dominant feature of the Halperin-Lax (H-L) tail is an exponential factor given by

$$n(E) \sim \exp\left[-\left(\frac{|E|}{\varepsilon_{1d}}\right)^{2-(d/2)}\right] \tag{2}$$

where the energy E is measured from an appropriate unperturbed continuum edge, d is the dimension and the width ε_{1d} is given by

$$\varepsilon_{1dc} = c_d \gamma^{2/(4-d)} (2m^*/\hbar^2)^{d/(4-d)} \quad . \tag{3a}$$

The quantity c_d is a numerical factor having the values

$$c_1 = 5.200\times10^{-1}, \; c_2 = 8.547\times10^{-2}, \; c_3 = 6.999\times10^{-4} \tag{4}$$

in 1, 2, and 3 dimensions, respectively. The quantity γ is the volume integral of the autocorrelation function of the potential and may be written as $\gamma = w^2\ell_c^2$, where ℓ_c is the correlation length of the random potential. In the specific case of the continuum limit of a nondegenerate tight-binding model on a hypercubic lattice with only nearest-neighbor electron transfer and with ℓ_c reduced to its limiting value α, the atomic separation (3a) simplifies to

$$\varepsilon_{1d} = c_d \, w^{2/(4-d)} V^{-d(/4-d)} \quad . \tag{3b}$$

SHORT WAVELENGTH POTENTIAL FLUCTUATIONS

The Halperin-Lax result has been derived in so many ways that one might think it to be completely unchallengeable. Nevertheless, all derivations are approximate, and we have pointed out what appeared to be a flaw in the above physical interpretation of the H-L tail.[12] As mentioned above, Eq. (1) is based on the fact that the kinetic energy cost ΔE of localizing a state at the bottom of the unperturbed continuum into a smooth potential fluctuation of size L is

$$\Delta E = \frac{\hbar^2}{2m^* L^2} \quad . \tag{5}$$

We argued that such smooth, wide potential fluctuations appear superimposed on random, short wavelength potential fluctuations which cause violent amplitude fluctuations in states near the bottom of the unperturbed continuum. These states become fractal[13] on length scales L such that $L_l < L < L_u$. The upper limit to fractal behavior L_u equals the localization length for all states when $d \leq 2$ and equals the amplitude coherence length ξ for extended states when $d > 2$. Drawing upon the results of numerical studies carried out near the center of the band, we have concluded that the lower limit L_1 to fractal behavior is of atomic size, $L_1 \simeq a$. Using the Lloyd-Best variational principle,[10] we argued that the states in the tail are pulled down from a narrow energy range around the mobility edge for $d > 2$ and from the peak in the density of states for $d < 2$.

The energy of localization of such a fractal state in an additional potential fluctuation of size L was obtained from the Thouless scaling argument[14] and showed that the original H-L result was regained for $d \leq 2$. On the other hand, we found that for $d > 2$, Eq. (5) was to be replaced by

$$\Delta E \sim L^{-d} , \tag{6}$$

leading to an exponential tail in the density of states,

$$n(E) \sim \exp\,[-|E|/\varepsilon_{2d}] \tag{7}$$

with a width ε_{2d} not significally different from ε_{1d}, the width of the H-L tail.

The different length scaling relation, Eq. (6), of the kinetic energy can be also obtained from the following argument: For a random walker in a regular periodic system after time t the square of the average displacement $\langle r^2 \rangle$ from its starting point goes proportional to t. Therefore, by an uncertainty principle argument $((\Delta E)(\Delta t) \sim \hbar)$, one obtains the energy scale as $1/L^2$ in any dimension. Now if the random walker moves in a fractal space or in a highly non uniform space, after time t, $\langle r^2 \rangle \sim t^x$ with $x < 1$. This is so because the random walker spends most of its time wandering around its starting point. Using $(\Delta E)(\Delta t) \sim \hbar$, one obtains the energy scale as $1/L^{2/x}$, $x < 1$. There is numerical evidence that $x = 2/d$ for disordered systems.

The result of Eq. (7) was welcomed because the H-L variation $\exp\,[-(|E|/\varepsilon_{13})^{1/2}]$, had never been observed in the tails of disordered three-dimensional bands. Instead, exponential tails are observed everywhere in the optical absorption[15] (Urbach tails) and the density of states.[16]

We have undertaken a deeper analysis based on a field theoretic treatment in order to find Eq. (7) without resorting to scaling or similar heuristic arguments.

UNIVERSAL FEATURES OF THE DENSITY OF STATES

Those electronic properties which depend on electronic structure near band edges are quite similar in very different amorphous semiconductors, despite the complexity of those materials. This suggests the presence of universal features in the electronic structures of disordered materials, independent of structure, composition, and the detailed nature of the disorder.

We have indeed been able to show that complete universality exists[17,18] near band edges for weak disorder in less than two dimensions, and modified universality exists in more than two dimensions in the following broad class of models. First, there must be a suitable reference model from which the disorder can be measured. Second, this model need not be ordered but must possess a nondegenerate normal band edge. Third, a suitable measure w of the disorder must be much less than a suitable measure B of the band width. Fourth, the energy separations from the underlying band edge must be much less than B. Fifth, any characteristic lengths must be much larger than the larger of the interatomic spacing a and the correlation length ℓ_c of the random potential. Finally, the continuum limit of the model should contain no ultraviolet catastrophe, which implies no sensitivity to ℓ_c, a, or B.

Ultraviolet catastrophes do not occur for $d < 2$, and the desired universality is easily shown to exist.[17] In $d > 2$, the divergences exist and lead to a shift E_{Bd} of the original, underlying band edge. E_{Bd} is non-universal, depending on structure and probability distribution, but, once energies are measured relative to the shifted band edge, universality reemerges.

From the two parameters w and Va^2, where V is the nearest-neighbor transfer matrix, natural units of energy and length can be defined (for $d < 4$),

$$\begin{aligned} E_{od} &= w^{4/(4-d)} V^{-d/(4-d)} \\ L_{od} &= a(V/w)^{2/(4-d)} \end{aligned} \tag{8}$$

When energies are measured relative to the unperturbed band edge for $d < 2$ or E_{Bd} for $d > 2$ in units of E_{od} and lengths in units of L_{od}, all quantities can be expressed in universal form through scaling relations. For example, we have for the density of states

$$n(E) = L_{od}{}^{-d} E_{od}{}^{-1} f_d\big((E - E_{Bd})/E_{od}\big) \tag{9}$$

where $f_d(x)$ is a universal function of its argument varying only with dimension. As another example, we find[17]

$$E_c - E_{Bd} = b_{1d}\, E_{od} \tag{10}$$

for the mobility edge for $2 < d < 4$, where $b_{13} = 4.515 \times 10^{-3}$ and $b_{1d} = 0$, $d > 4$. This universality breaks down in the region of extended states when $E - E_{Bd}$ approaches V in violation of condition four above. It also breaks down deep in the tail when the potential fluctuations responsible for the localized states become of atomic size,[18] in violation of condition five above.

Abe and Toyozawa[19] have shown numerically that this non-universal form can be approximated by a simple exponential for the particular case of Gaussian single site disorder perturbing a semicircular density of states in the Coherent Potential Approximation (CPA).

We have shown[18] that the Abe-Toyozawa exponential tail can be obtained analytically from the CPA; that it is non-universal, depending in detail on atomic arrangement and nature of the disorder; and that it yields an exponential tail only for a Gaussian probability distribution of the disorder. Its physical interpretation is quite simple. In addition to the broad, relatively shallow, potential fluctuations considered by H-L, there are potential fluctuations on the scale of a single site or band which are deep enough to cause localization. Such fluctuations are accurately treated in the CPA. They become important deeper in the tail, where there is, in principle, a crossover from Halperin-Lax to Abe-Toyozawa. We have obtained the results

$$n(E) \sim \exp[-|E|/\varepsilon'_{23}\ ]; \quad \varepsilon'_{23} = 0.127 w^2/V \tag{11}$$

for the simple cubic structure and is non-universal.

CAN THE HALPERIN-LAX TAIL BE SEEN EXPERIMENTALLY?

As we have discussed above, there is some uncertainty about the exact form of n(E) in the H-L tail. We have argued[12] that if the wave functions possessed fractal character[13] in that energy range, the energy dependence of n(E) should be primarily a simple exponential. However, this distinction seems irrelevant because when values for w and V appropriate to real materials were inserted into ε_{13}, one finds that the Halperin-Lax region is squeezed between the continuum of extended states and the Abe-Toyozawa tail to such an extent that the preexponential and the corrections considered by Sa-Yakanit[7] become important. As ε_{13} is decreased the H-L region becomes so narrow that it is smeared out by inelastic scattering, i.e., $\varepsilon_{13} < \tau_{in}^{-1}$, where τ_{in} is the inelastic lifetime. The only way to observe it is to go to very low temperatures and study the band edges with methods capable of very high energy resolution.

The principal theoretical problem remaining in this area is to carry out formal or numerical analyses capable of covering the entire range from the interior of the band to the Abe-Toyozawa tail, which is beyond the reach of a continuum theory. The only relevant numerical work has been carried out in 2d by Thouless and Elzain, [20] who have shown that the density of the states in the interior of the band is well-fitted by the CPA while that in the tail is well-fitted by H-L theory. They did not go deeply enough into the tail to reach the Abe-Toyozawa region.

ACKNOWLEDGEMENTS

This research was done in collaboration with M. H. Cohen and E. N. Economou. This was partially supported by a North Atlantic Treaty Organization Travel Grant No. RG684/84. Ames Laboratory is operated for the U.S. Department of Energy by Iowa State University under Contract No. W-7405-Eng-82.

REFERENCES

1. I. M. Lifshitz, Adv. Phys. 13:483 (1964).
2. V. L. Bonch-Bruevich and A. G. Mironov, Fiz. Tverd. Tela. 3:3009 (1962) [Sov. Phys. Solid State 3:2194 (1962)].
3. E. O. Kane, Phys. Rev. 131:79 (1963).
4. T. P. Eggarter and M. H. Cohen, Phys. Rev. Lett. 25:807 (1970).
5. B. I. Halperin and M. Lax, Phys. Rev. 148:722 (1966) and 153:802 (1967).
6. S. F. Edwards, J. Phys. C3:L30 (1970), and J. Non-Cryst. Solids 4:417 (1970).
7. V. Sa-yakanit, J. Phys. C7:2849 (1947) and Phys. Rev. B19:2266 (1979).
8. J. L. Cardy, J. Phys. C11:L321 (1978).
9. J. Zittartz and J. S. Langer, Phys. Rev. 148:741 (1966).
10. P. Lloyd and P. R. Best, J. Phys. C8:3752 (1975).
11. J. Ziman, Models of Disorder Cambridge University Press, Cambridge, England, (1979).
12. C. M. Soukoulis, M. H. Cohen and E. N. Economou, Phys. Rev. Lett. 53:616 (1984).
13. H. Aoki, J. Phys. C16:L205 (1984); C. M. Soukoulis and E. N. Economou, Phys. Rev. Lett. 52:565 (1984).
14. D. J. Thouless, Phys. Rev. Lett. 39:1167 (1977).

15. G. D. Cody, T. Tiedje, B. Abeles, B. Brooks, and Y. Goldstein, Phys. Rev. Lett. 47:1480 (1981); G. D. Cody, "The Optical Absorption Edge of a-Si:H_x in Amorphous Silicon Hydride," in Semiconductors and Semimetals Vol. 21B, J. Pankove, ed., Academic, New York (1984) p. 11 and references therein.
16. T. Tiedje, J. M. Cebulka, D. L. Morel, and B. Abeles, Phys. Rev. Lett. 46:1425 (1981).
17. M. H. Cohen, E. N. Economou and C. M. Soukoulis, "Band Edge Features in Disordered Systems" Phys. Rev. B to be published.
18. E. N. Economou, C. M. Soukoulis, M. H. Cohen, and A. D. Zdetsis, Phys. Rev. B 31:6172 (1985).
19. S. Abe and Y. Toyozawa, J. Phys. Soc. Jpn. 50:2185 (1981).
20. D. J. Thouless and M. E. Elzain, J. Phys. C11:3425 (1978).

A CPA DESCRIPTION OF THE ELECTRONIC AND TRANSPORT PROPERTIES OF α-GeH, α-SiH and α-SiGeH

Aristides D. Zdetsis

Research Center of Crete and
Department of Physics
University of Crete, Heraklio
Crete, Greece

ABSTRACT

The coherent potential approximation (CPA) within the tight binding Slater-Koster scheme is briefly reviewed as a theoretical tool for the study of the electronic properties of amorphous semiconductor hydrides.

Basic quantities of interest include the density of electronic states, the optical absorption, and the DC and AC conductivities. The optical gap, as well as other quantities of interest are obtained as a function of hydrogen and semiconductor concentrations.

The overall agreement with experimental measurements is very good.

Recent theoretical developments allow the incorporation of corrections beyond the CPA.

I. INTRODUCTION

Over the last five years a rather simple model [1-4] based on the coherent potential approximation [5] (CPA) has emerged as a very successful scheme for the description of electronic and transport properties of hydrogenated amorphous semiconductors. The success of the model for quantities such as the density of states (DOS), the optical gap, the optical absorption, the AC and DC conductivities and, more recently [4], the mobilities, position of mobility edges, mean free path, localization length, etc, was very impressive. This is even more so, in view of the criticism [6] for the inadequacy of the CPA method and the underlying crystallinity of the model. This criticism, not totally unfounded, shall be considered below in the discussion of the drawbacks and the ways of improving and extending the model. However, the remaining fact still is the undeniable success of the model thus far in producing very reasonable results for many physical quantities of interest.

In the following section (II) an outline stressing the key features of the model, such as the tight binding formalism and the zeroth order Hamiltonian [7,1], the CPA method and the way in which disorder is introduced will be given. Based on this outline, ways of improving the model will be considered in section III, followed by several selective results in section IV. Some of the latest developments [4,8] allowing the calculation of several quantities of interest (such as the mobility at the critical energy E_c, the prefactor of the DC-conductivity, etc) including localization corrections well beyond the CPA regime, will be reviewed in section V. Results concerning several of these quantities will be also presented in the same section. It should be emphasized here that the localization corrections to the CPA results are fully determined by the CPA calculation and do not need any additional complicated computations of intermediate quantities. In section VI finally, the conclusions of this work are given summarizing also the success and the drawbacks of this CPA model. The concept of universality[8,9] in the electronic properties of disorder systems which we believe is of primary importance in understanding the success of such simple models as the present one is also discussed in this last section, to offer a final unified picture of this work.

II. THE MODEL

The theoretical framework of this model which is dominated by the CPA theory has been given in more detail elsewhere[1,2]. The CPA method starting from a zeroth order crystalline Hamiltonian treats the disorder by introducing an effective medium in which, at the first level of approximation, the on site matrix elements of the underlying tight-bidning Hamiltonian are replaced by a common complex self-energy Σ. To put it otherwise, an effective lattice is constructed whose sites may have probability c of being vacant, and probability 1-c of having a semiconductor atom. (assuming one semiconductor type). In addition it has been assumed that H atoms may be located along the lines connecting a vacant site with its nearest neighbors. Thus the model includes, at random, semiconductor sites, vacancy sites, and sites surrounded by one, two, three, or four H atoms which saturate the semiconductor dangling bonds. This model neglects topological and reconstruction disorder. In the case of two or more types of semiconductors these conditions are easily extended by the introduction of additional probabilities and concentrations having a total sum of unity (see equation 6). The resulting effective medium is a crystalline medium with complex (self) energies which are defined by the CPA condition of zero scattering on the average, when one replaces an effective atom by a semiconductor atom or by a hydrogenated vacancy. (A good account of the tight-binding form of the CPA method and the involved formalism is given in reference 10).

To determine the Green's function $\tilde{G}_e$ of the effective medium, we need the zeroth order crystalline semiconductor Hamiltonian. For this we use a Slater-Koster [7] (SK) Hamiltonian which includes up to third-neighbor interactions [1]. The basis used includes the four outer orbitals of Si, one s and three p orbitals, but sometimes it is more convenient (for instance in considering the Si-H matrix elements) to use the equivalent four sp^3 hybridized orbitals. The parameters for the SK Hamiltonian are chosen by fitting as accurately as possible the pseudopotential band structure of Si or Ge [11]. Although the fit is considered good it can be further improved. Already more satisfactory sets of parameters have been obtained and the question of uniqueness for such parameters has been examined [12]. The H-H and Si-H matrix elements are evaluated from small-molecule matrix elements[1]. Since we are using only diagonal disorder in the CPA, due to the complexity of the off-diagonal disorder in the CPA theory, with a unit cell of two atoms (or vacancies) we take an average of the first neighbor matrix elements for Si-Si and Si-H in order to form the corresponding SK parameters. Thus, in addition to the other approximations, the nearest-neighbor off-diagonal disorder is treated within the virtual crystal approximation (VCA) while the second and third neighbor interaction matrix elements are assumed unaffected by the disorder.

Having now the zeroth order Hamiltonian, and thus the corresponding

Green's function $\tilde{G}_0$ matrix, we can go to the next step (step 1 in the schematic representation of Fig.1) to calculate Itteratively the CPA condition. This condition for α-SiH (and similarly for α-GeH) and, for simplicity, only 1H and 4H hydrogenated vacancies, is:

$$(1-c)\tilde{t}_{Si} + \frac{x_1}{4} \sum_{i=1}^{4} \tilde{t}_{1i} + x_4 \tilde{t}_4 = 0 \quad (1)$$

$$\text{where } \tilde{t}_A = \tilde{U}_A (\tilde{1} - \tilde{G}_e \tilde{U}_A)^{-1}, \quad (2)$$

and A stands for the various types of sites, Si (or Ge), 1H, 4H, of the effective lattice. $\tilde{G}_e$ is the effective medium Green's function obtained from the corresponding crystalline semiconductor Green's function $\tilde{G}_0$ by replacing ε_s and ε_p with the CPA self energies Σ_s and Σ_p, respectively. The scattering matrices in the SK basis of the s and p orbitals are given by[1],

$$\tilde{U}_{Si} = \begin{bmatrix} \varepsilon_s-\Sigma_s & 0 & 0 & 0 \\ 0 & \varepsilon_p-\Sigma_p & 0 & 0 \\ 0 & 0 & \varepsilon_p-\Sigma_p & 0 \\ 0 & 0 & 0 & \varepsilon_p-\Sigma_p \end{bmatrix} \quad (3)$$

$\tilde{U}_{1i}$ is a matrix corresponding to the four equivalent configurations where only one hydrogen atom is present with probability of occurence $x_1/4$,

$$\tilde{U}_{11} = \tilde{S} \begin{bmatrix} \gamma_1' & 0 & 0 & 0 \\ 0 & \infty & 0 & 0 \\ 0 & 0 & \infty & 0 \\ 0 & 0 & 0 & \infty \end{bmatrix} \tilde{S} - \tilde{\Sigma} \quad (4)$$

and similarly for $\tilde{U}_{12}$, $\tilde{U}_{13}$ and $\tilde{U}_{14}$. The matrix corresponding to the case of four hydrogen atoms present with probability x_4 is given by

$$\tilde{U}_4 = \tilde{S} \begin{bmatrix} \gamma_1' & \gamma_2' & \gamma_2' & \gamma_2' \\ \gamma_2' & \gamma_1' & \gamma_2' & \gamma_2' \\ \gamma_2' & \gamma_2' & \gamma_1' & \gamma_2' \\ \gamma_2' & \gamma_2' & \gamma_2' & \gamma_1' \end{bmatrix} \tilde{S} - \tilde{\Sigma} \quad (5)$$

In the above relations (4) and (5) $\tilde{S}$ is the 4x4 transformation matrix[1] between the SK (s and p) and the sp^3 basis. The matrix $-\tilde{\Sigma}$ is a diagonal matrix like (3) but without ε_s and ε_p. The elements γ_1' and γ_2' are the hydrogen matrix elements with values -3.38 eV and -1.78 respectively.

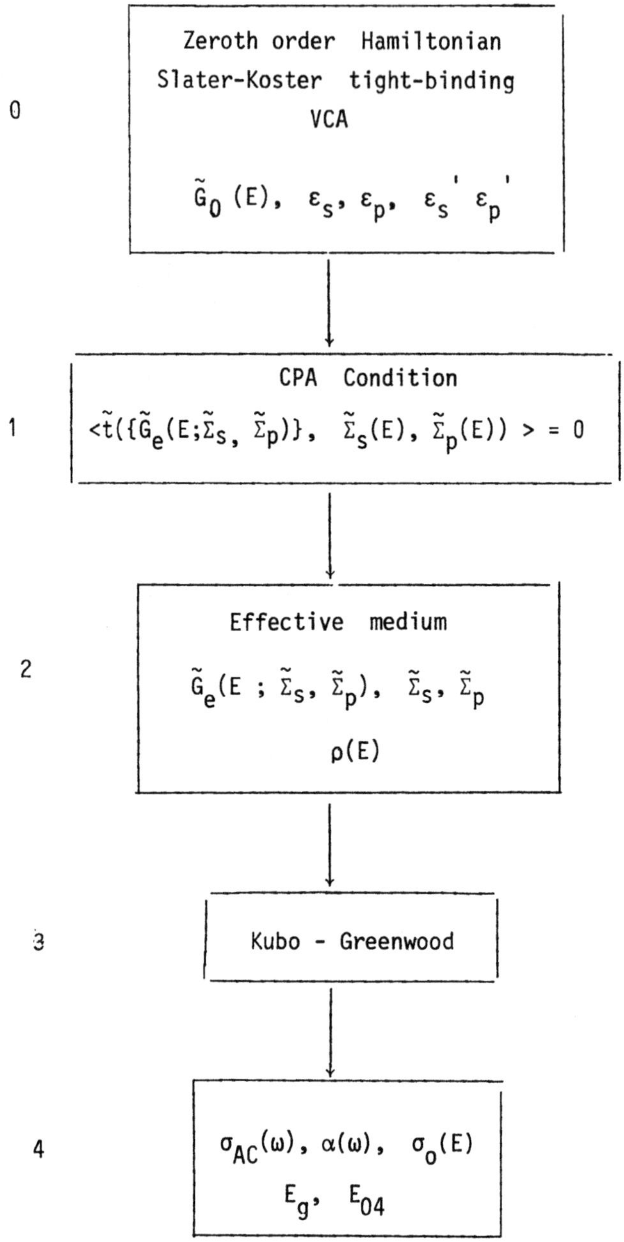

Fig.1. Diagramatic representation of a full CPA calculation.

In the case that we have Si and Ge present together with concentrations x and y respectively and 4H sites only, for simplicity, the corresponding CPA equation is given by [3],

$$x\tilde{t}_{Si} + y\tilde{t}_{Ge} + c\,\frac{x}{x+y}\,\tilde{t}_{HSi} + c\,\frac{y}{x+y}\,\tilde{t}_{HGe} = 0 \qquad (6)$$

where the notation is obvious, and $x+y+c=1$.

The self-consistent solution of equations (1) or (6) in a second step yields Σ_S, Σ_p and the effective medium Green's function $\tilde{G}_e$ from the imaginary part of which the total (as well as partial [1,3]) CPA density of states (DOS) results. Having $\tilde{G}_e(E,\vec{k})$ the next 2 steps involve the setting up of the necessary matrix elements, such as the momentum matrix elements $p_{x,ij}$ in the basis of SK orbitals, and evaluation of standard Kubo-Greenwood-type formulas[13]

$$\mathrm{Re}\,\sigma_{AC}(\omega) = \frac{2e^2}{\pi\Omega m^2\omega} \times \sum_k \int_{E_F-\hbar\omega}^{E_F} dE\ \mathrm{Tr}\Big[\tilde{p}_x \mathrm{Im}\,\tilde{G}_e(\vec{k}, E + i0)$$

$$\times\,\tilde{p}_x \mathrm{Im}\tilde{G}_e(\vec{k}, E + h\omega + i0)\Big] \qquad (7)$$

In the above formula (7) for the real part of the AC conductivity at frequeny ω we have set $\langle G\rangle = \tilde{G}_e$ and $\langle GPG\rangle \simeq \langle G\rangle p \langle G\rangle$ within the CPA spirit, ignoring vertex corrections arising from correlations in intermediate scattering. In the limit $\omega \to 0$ equation (7) gives the CPA energy-dependent conductivity $\sigma_0(E)$. The absorption coefficient $\alpha(\omega)$ and the optical gaps E_g and E_{04} can be obtained in a straightforward way [1,3] from the AC conductivity $\sigma_{AC}(\omega)$.

Having finished now the plan outlined in figure 1, we can get into the next section to examine possible ways of improving the model.

III. IMPROVING THE MODEL

Aside from all the particular approximations, serious or non-serious, mentioned in the previous section, the general feeling one gets is that "the model does not have enough disorder" (underlying crystallinity). The only source of disorder is the hydrogen induced disorder. In this respect the concentration x of the hydrogen should be considered as a disposable parameter having no exact relation with the actual hydrogen concentration of the sample. It is almost obvious that to improve the model we must introduce somehow more disorder. However if we restrict ourselves to diagonal CPA theory, for several reasons (to mention only the mathematical simplicity, the considerable success of the model in its present form and the easiness to accommodate corrections and extention, as will be seen below) then we have a limited number of choices to increase the disorder. Besides the obvious, but not unimportant, choice of increasing drastically the effective concentration of hydrogen we will consider two additional choices.

(a) <u>To increase the configurational disorder induced by hydrogen</u>

To do this we have introduced[2] the 1H-model, characterized by the four different configurations corresponding to the scattering matrices $\tilde{U}_{1i}(i=1,4)$ of equation (4). However to avoid the dangling bonds and the zero bandgaps associated with the infinite matrix elements of relations (4), we have replaced these elements with finite values. These values are eventually chosen[2] intermediate between those of the semiconductor and those of hydrogen. So in this extreme single hydrogen case it is assumed that the lattice reconstructs itself around each isolated H atom so that no dangling bonds survive. Although the success of this 1H model (or mixture of 1H and 4H) for reasonable values of hydrogen concentration is considered extremely good, it has the drawback of introducing the finite value of the matrix element in (3) as an additional parameter.

(b) To increase the reconstruction disorder

We have attempted to evaluate and incorporate into the model the effect of disorder induced by reconstruction, by modeling the latter through a continuous distribution of the diagonal matrix elements ε_s and ε_p [2,14]. The variation, introduced by a rectangular distribution of width W, is performed in such a way as to leave the bond length constant while changing the bond angle. However, for widths up to W=0.25 the effects were negligible in comparison with the hydrogen-induced disorder (especially for the 1H model). These effects could become larger if together with the semiconductor matrix elements the hydrogen matrix elements were varied.

Although, as will be seen in the next section, the results obtained by the improved version of the model were significantly better than before, the inclusion of off-diagonal disorder in the CPA theory should be pursued seriously in the future.

IV. RESULTS

According to the plan outlined in figure 1, first we show results for the DOS of α-SiH. The DOS around the gap displayed in figures 2 and 3 correspond to the 4H and 1H models respectively for two values of W(W=0 and W=0.60) each. The total hydrogen concentration in both cases is 0.10. At W=0 (no reconstruction disorder) the 1H model produces a DOS gap, E_{CV}=1.25 eV, whereas the 4H model for W=0, gives E_{CV}=1.15 eV. At W=0.60 the corresponding quantities are E_{CV}=1.12 for the 1H case and E_{CV}=1.05 for the 4H one. The difference between the two models is due to the larger recession of the top of the valence band in the 1H model. This, in turn is related to the higher hydrogen-induced disorder in the 1H model which is also reflected in a much smoother DOS in that case. The hydrogen-induced disorder, contrary to the disorder due to reconstruction (W>0) tends to open the gap due to its reduction of the effective ppπ interaction, and to the passivation of dangling bonds. Thus because of the fact that the two types of disorder we have considered tend to counterbalance each other, the DOS features of the 1H model with W=0.60 are almost identical to those of the 4H model with W=0.0, as can be seen in figures 2 and 3. Similar results are obtained for α-GeH with the corresponding band gaps drastically reduced in comparison to α-SiH. For instance, for x=0.20 the 1H model gives $E_{CV}\simeq$0.95 eV, whereas the same quantity for α-SiH is 1.4 eV. Also, the variation of the gap with hydrogen content is slightly slower in α-GeH. Besides total DOS, partial DOS can be obtained by properly projecting out of the G_e the desired components[1,2]. Figure 2 of ref.2, which is not reproduced here for space economy, shows a quite good agreement between the calculated H-site partial DOS of α-SiH and the experimentally observed peaks from photoemission measurements.

Having the DOS and before resorting to the Kubo-Greenwood equation (7) one can calculate $\sigma_{AC}(\omega)$ and in turn $\alpha(\omega)$ simply from the convolution of the DOS in the valence and the conduction bands after making some simplifying assumptions for the matrix element $p_{x,ij}$. The usual convenient assumption is that this momentum matrix element on the average is constant. Another assumption that Cody [15] and others have proposed is that the dipole matrix

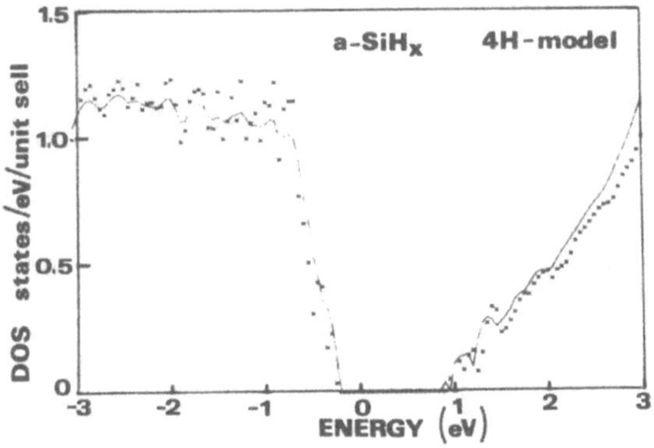

Figure 2
DOS for α-SiH_x in the 4H model, for two values of W. W=0.0 (points) and W=0.60 (line)

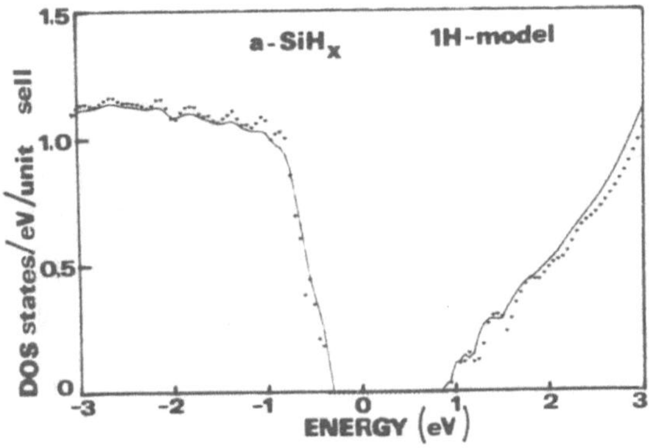

Figure 3
DOS for α-SiH_x in the 1H model, for two values of W. W=0.0 points and W=0.60 (line)

element (related to the momentum matrix element by a factor of $i\omega m$) is a constant rather than the momentum matrix element. It has been shown [2] that indeed this is the case. This can be also seen in figure 4 where the two theoretical curves for the absorption coefficient versus $E=\hbar\omega$ of α-SiH literally coincide with each other and with the experimental data [16] up to 3 eV. Both theoretical curves (solid and dashed) have been obtained with the 1H model for x=0.25. The solid curve is based on equation 7 without any simplifying assumption; the dashed curve is obtained by convoluting the CPA DOS and assuming that the dipole matrix element is on the average constant. This constant average value of the dipole matrix element (DME) is estimated here to be

$$\overline{\text{DME}} \simeq 3.6 \text{ Å}$$

in good agreement with the value obtained by Jackson et al [17]. This good agreement with experiment in view of the simplicity of the model is indeed impressive. For plots of the CPA DC-conductivity $\sigma_0 = \sigma_0(E)$ the reader is refered to figure 3 of ref.2. Here, as a final plot the optical gap E_{04} of α -SiGeH versus Ge-concentration is displayed in figure 5 together with experimental data [18]. The optical gap E_{04} is defined as the value of E at which the absorption coefficient α is ~ 10^4 cm^{-1}, in addition to the optical

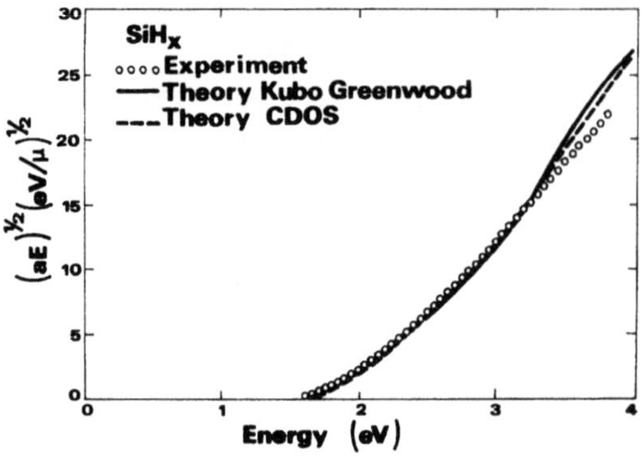

Figure 4
Optical absorption coefficient α vs $E=h\omega$. For α-SiH_x, x=0.25. The exp. data are from ref.16. The solid line is the full Kubo-Greenwood calculation. The dashed line is for constant dipole matrix element.

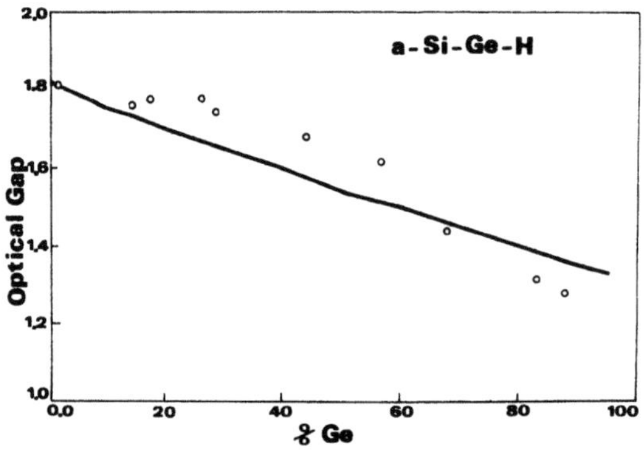

Figure 5
Optical gap of $Si_{1-x}Ge_xH_{0.20}$ as a function of Ge content. The dots correspond to the experimental data of ref.18.

gap E_g which is usually determined by an extrapolation of the $(\alpha E)^{\frac{1}{2}}$ versus E graph. The results in figure 5 are consistent with the experimental data especially in view of the uncertainties in the operational definition of optical gap and in the experimental determination of the hydrogen content which is varying, whereas the calculation was performed for a constant hydrogen concentration of 0.20 and was based on the 4H model. In ref.3 there are further results on the α-Si-Ge-H alloys.

V. EXTENDING THE MODEL

Besides the attempts in section III to improve the model within the CPA program of figure 1, one can go far beyond the CPA regime and include localization effects into the model[4]. These effects can be accounted for by the potential well analogy (PWA) method[8] which uses the CPA results as input. In figure 6 there is a schematic illustration of this combination of CPA + PWA. More details can be found in refs 4 and 8. Here the discussion will be restricted to a brief explanation of the plan outlined in figure 6 (which can be thought as a continuation of figure 1). The equation in the box of step 5 gives the CPA conductivity $\sigma_0(E)$ in the Born approximation

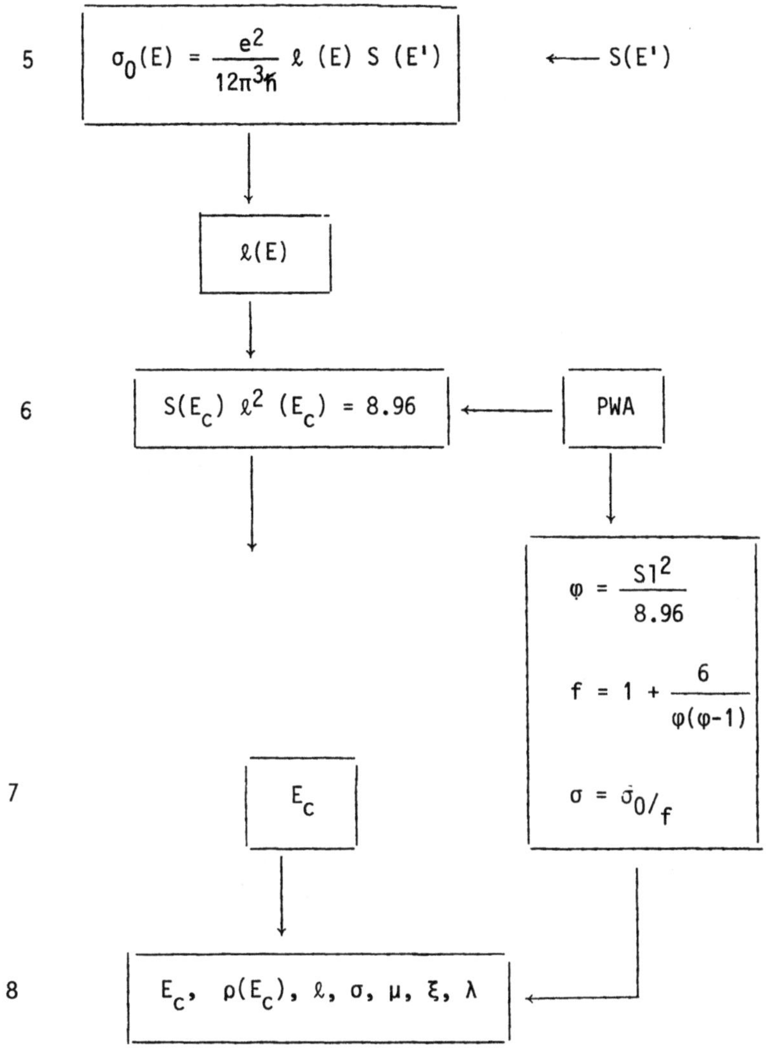

Fig. 6 Schematic representation of extending the CPA calculation of Fig.1 with the help of the PWA method.

in terms of the Born mean free path $\ell(E)$, and the surface of constant energy $S(E)$ for the disordered system. This last quantity, defined [4,8] for the disordered system as in the ordered one but with E measured from the CPA band edge, is easy to calculate. From the CPA conductivity and $S(E')$ the mean free path $\ell(E)$ can be calculated. Then, the critical energy E_c separating extended from localized states can be calculated in step 7 using the PWA localization criterion (6). The conductivity σ including the localization correction, which near the band edge is very important, but without phonon corrections, can be calculated from σ_0 using the simple function $f(\varphi)$ as shown in the figure. Thus in step 8 all energy dependent interesting quantities can be calculated exactly at the mobility edge E_c. Such important quantities which are directly accessible to measurements are; the DOS, the conductivity σ, the mobility μ, and the mean free path ℓ, slope of the Urbach tail etc. The localization ξ has been also included in the list and can be easily calculated [8].

Some of the results of the calculation for α-SiH outlined in figure 6 are listed below. These results were obtained for x=0.30 using the 1H model. The mobility edge measured from the CPA band edge is

$$\underline{E_c \approx 0.005 \text{ eV}}$$

The prefactor σ_0 of the T-dependent conductivity, including localization effects, but without phonons, is

$$\underline{\sigma_0 = 137\ (\Omega\text{cm}^{-1})}$$

The slope of the Urbach tail E_{0c} is approximately $\underline{E_{0c} \simeq 20 \text{ meV}}$. The calculation gives also

$$\underline{\rho(E_c) \simeq 10^{21}\ \text{eV}^{-1}\ \text{cm}^{-3}}, \qquad \underline{\ell(E_c) \simeq 17\ \text{Å}},$$

$$\underline{\mu(E_c) \simeq 28\ \text{cm}^2/\text{V sec}}$$

It is very important to mention that the experimental measurements of all these key quantities give values which are between the quoted values and values obtained by increasing the disorder (W) of the model by a factor of 2 (see ref.4).

VI. CONCLUSIONS

As has been explained so far, the rather crude and simple model described in this article with the often criticized "underlying crystallinity" is based on a series of non-serious approximations within the general diagonal SK tight binding CPA scheme. Many of the less significant approximations can be improved in a simple and rather straightforward way, for instance, by a better fit of the SK parameters, or by introducing clusters of 2 and 3 hydrogens in the hydrogenated vacancy sites, or even by better calculating the hydrogen related matrix elements by first principles Hartree-Fock cluster techniques [19]. However, the real improvement of the general scheme would definitely be the inclusion of d-basis in the SK Hamiltonian and the introduction of off-diagonal disorder, neither of which is simple or straightforward. And yet, as was illustrated, this simple model, especially after its extension by the PWA theory, is capable of producing good and realistic results for so many physical quantities of direct scientific and technological interest.

This overall good agreement with experiment seems to indicate that

the hydrogen induced disorder (the only disorder included explicitly in this model) could dominate around the gap making other important features of the disordered system less relevant.Thus, the electronic properties seem to be rather insensitive to the structural details, while they depend strongly on local chemical bonding, the size of the unperturbed gap and in general the amount of disorder. Although this is true the deepest reasons for the apparent success of this rather crude model should be traced to more profound concepts such as the concept of universality, introduced in refs 4 and 8, for the electrical and transport properties of disordered systems.
It has been shown [4,8] that in a simple band resulting from one orbital per unit cell, several quantities near the band edge exhibit universal behavior, while others depend on the details of the band. This universality for more complex systems as the ones examined here is not fully understood.

The success of the present model is suggestive of the existence of such universality which makes possible to draw correct results from simplified models, as the present one, which retain only the truly significant features of the disordered systems. However, even if only to check this universality, future work should include off-diagonal CPA theory (with d-orbitals in the basis set) for the electronic properties and static and dynamic effects of the electron phonon interaction for the DC conductivity and the other transport properties.

REFERENCES

1) D.A. Papaconstantopoulos and E.N. Economou, Phys.Rev. B22, 2903 (1980); Phys.Rev. B24, 7233 (1981); W. Pickett, D.A. Papaconstantopoulos, E.N. Economou, Phys.Rev. B28, 2232 (1983)

2) A.D. Zdetsis, E.N. Economou, D.A. Papaconstantopoulos, and N. Flytzanis, Phys.Rev. B31, 2410 (1985)

3) D.A. Papaconstantopoulos, E.N. Economou and A.D. Zdetsis, "Proceedings of the 17th International Conference on the physics of semiconductors", edited by J.D. Chadi and W.A. Harrison, 795 (Springer-Verlag, New York (1985)

4) E.N. Economou, A.D. Zdetsis and D.A. Papaconstantopoulos, "11th International Conference on amorphous and liquid semiconductors", Journal of non-crystalline solids, in press; C.M. Soukoulis, M.H. Cohen, E.N. Economou and A.D. Zdetsis ibid

5) P. Soven, Phys.Rev. B2, 4715 (1970); J.S. Faulkner, ibid 13, 2391 (1976)

6) See for instance the article of M.H. Brodsky in these proceedings

7) J.C. Slater and G.F. Koster, Phys.Rev. 94, 1498 (1954)

8) E.N. Economou and C.M. Soukoulis, Phys.Rev. B28, (1983); E.N. Economou, C.M. Soukoulis and A.D. Zdetsis, ibid 30 1686 (1984); E.N. Economou, C.M. Soukoulis, A.D. Zdetsis ibid B31, 6483 (1985); E.N. Economou, C.M. Soukoulis, M.H. Cohen, A.D. Zdetsis, ibid B31, 6172 (1985); E.N. Economou, ibid, B31, 7710 (1985)

9) See articles of E.N. Economou, C.M. Soukoulis and M.H. Cohen, in these proceedings

10) E.N. Economou, "Green's functions in quantum physics", 2nd ed.(Springer-Heidelberg, 1983)

11) W.E. Pickett (unpublished). This calculation utilizes the local pseudopotential form factors of Chelikowsky and Cohen [Phys.Rev. B10, 5095 (1974)]

12) A.D. Zdetsis (unpublished)

13) R. Kubo, J.Phys.Soc. Japan 12, 570 (1957); D.A. Greenwood, Proc.Roy.Soc. (London) 71, 585 (1958)

14) A.D. Zdetsis, N. Flytzanis, E.N. Economou, D.A. Papaconstantopoulos (unpublished)

15) G.D. Cody, C.R. Wronski, B. Abels, R.B. Stephens, and B. Brooks, Sol. Cells, 2, 227 (1980)

16) G.D. Cody in "Semiconductors and semimetals", Vol.21 B, J. Pankove, ed. Academic Press (1984), (Fig.14) and private communication

17) W.B. Jackson, C.C. Tsai, and S.M. Kelso, "11th International Conference on amorphous and liquid semiconductors", Journal of Non-Crystalline Solids, in press.

18) B. von Roeden, D.K. Paul, J. Blake, R.W. Collins, G. Moddel and W. Paul, Phys.Rev. B25, 7678 (1982)

19) A.D. Zdetsis (unpublished)

FLUCTUATION INDUCED GAP STATES IN AMORPHOUS HYDROGENATED SILICON

B.K. Chakraverty

Laboratoire d'Etudes des Propriétés Electroniques des

Solides - BP 166, 38042 Grenoble Cedex, France

ABSTRACT

It is shown that in amorphous hydrogenated Silicon, self trapped localised electronic states can occur in the band gap due to intense local fluctuation of the hydrogen concentration. The trapping is determined by the mean hydrogen concentration of the system and by the strength of the electron interaction with the fluctuation potential. The line of stability between localised and delocalised electron state is calculated. The fluctuation induced gap states (FIGS) could be deep enough to trap one or two electrons, are metastable and are probably the reason of the Staebler-Wronski effect and other numerous light-induced effects observed in amorphous hydrogenated Silicon.

I - THEORY

A. The theoretical considerations leading to fluctuation induced gap states (henceforth called FIGS, for simplicity) go back to Lifshitz (1967). Considerable literature exists of electron trapping in dense Helium, e.g. : (Levine and Sanders, 1966) (Springett, Jortner and Cohen, 1968) (Kukushkin and Shikin, 1973) Krivoglaz, 1974) Friedberg and Luttinger, 1975) Moore, Cleveland and Gersch, 1978) where more exhaustive references can be obtained. In Hydrogenated amorphous Silicon and electron in the conduction band can lead to local migration of hydrogen in such a way that a sphere is formed in which the concentration of hydrogen becomes small, so that the electron in turn is self-trapped in this sphere. This is a central assumption of this paper. We shall calculate the self-trapping process of electrons in amorphous Silicon when the local concentration c of hydrogen deviates significantly from the mean hydrogen concentration c_o, assumed evenly distributed in the systems. The details of the calculations are published elsewhere (Chakraverty, 1985).

The extra local potential that the electron will see due to this fluctuation is written as

$$\Delta V = G\,(c - c_o) \qquad \text{(A-1)}$$

where G is assumed positive or repulsive so that an attractive local potential well results whenever $c \ll c_0$. We shall give in the later part of this section a realistic estimate for G for amorphous hydrogenated Silicon.

If the system is assumed both incompressible and isotropic, the Helmholtz free energy change due to fluctuation of the local concentration will be given by

$$W = \int_\tau [f(c) - f(c_0) + g(\nabla c)^2] d\tau \tag{A-2}$$

where f(c) is the Helmholtz free energy per unit volume of a homogeneous system of concentration c per atom, and $g(\Delta c)^2$ is a gradient energy that hinders spontaneous concentration fluctuation, akin to surface energy. The fluctuation must conserve average concentration, hence

$$\int_\tau (c - c_0) d\tau = 0 \tag{A-2a}$$

using (2 a), we get

$$W = \int_\tau \left[f(c) - f(c_0) - \left(\frac{\partial f}{\partial c}\right)_{c=c_0} (c - c_0) + g(\nabla c)^2 \right] d\tau \tag{A-3}$$

If the electron localises itself in the potential well created by the fluctuation the total free energy change will be given by

$$\Delta F = \frac{\hbar^2}{2m} \int |\nabla \Psi|^2 d\tau + \int \Delta V(r) \Psi^2 d\tau + W \tag{A-4}$$

The first two terms represent the kinetic energy and the attractive potential energy of localisation of the electron with wave-function Ψ using the bottom of the conduction band as the reference zero of energy, while the third term gives the thermodynamic energy expense to create the fluctuation. We need to minimize $\Delta F(\Psi, c)$ with respect to c and Ψ, to obtain the ground-state energy of the FIGS. In the evaluation of W, the gradient energy term will be neglected in this paper.

Although the Schrödinger equation resulting from (A-4) is highly nonlinear, we shall proceed with a trial wave-function $\Psi(r)$ that reflects the assumed spherical symmetry of the problem. Let the normalised wave-function be given by

$$\Psi(r) = (2\pi R_0)^{-\frac{1}{2}} \sin\left(\frac{\pi r}{R_0}\right) \Big/ r \tag{A-5}$$

where R_0 represents a localisation radius.

In order to keep the calculation tractable, let us further assume, that the work W of the fluctuation is uniquely due to entropy decrease from the uniform composition state and we write, neglecting the gradient energy term

$$W = \frac{kT}{\Omega} \int_\tau \left[\sigma(c) - \sigma(c_0) - \left(\frac{d\sigma}{dc}\right)_{c_0} (c - c_0)\right] d\tau \tag{A-6}$$

where Ω is the atomic volume of Silicon, $d\tau$ a volume element, $\sigma(c)$ is the entropy term at concentration c or C_0 given by

$$\sigma(c) = c \ln c + (1-c) \ln (1-c) \tag{A-6a}$$

Omitting the details, we get

$$c/1-c = c_0/1-c_0 \; \exp - \frac{G\Omega |\Psi^2|}{kT} \tag{A-7}$$

Substituting the expression for c, equation (A-7) into equation (A-4) and using equation (A-6), we get

$$\Delta F = \frac{\hbar^2}{2m} \int |\nabla \Psi|^2 d\tau - G c_0 \int \Psi^2 d\tau + \frac{kT}{\Omega} \int \ln \frac{1-c}{1-c_0} \, d\tau \tag{A-8}$$

Let us make the assumption

$$\Psi(r) = (2\pi R_o)^{-\frac{1}{2}} \frac{\sin\left(\frac{\pi r}{R_o}\right)}{r} \qquad \text{for } r < R_o \qquad \text{(A-9a)}$$

$$= 0 \qquad \text{for } r > R_o$$

and

$$c(r) = c \qquad \text{for } r < R_o \qquad \text{(A-9b)}$$

$$= c_o \qquad \text{for } r > R_o$$

Equations (A-9a) and A-9b) are used in the rest of this paper and only a single fluctuation is considered.

Now the volume integration in equation (A-8) can be performed to give us

$$\Delta F = \frac{\hbar^2 \pi^2}{2m R_o^2} - G c_o - \frac{kT}{\Omega} \frac{4}{3}\pi R_o^3 \left\{ \ln\left[(1-c_o) + c_o \exp \frac{-G|\Psi|^2 \Omega}{kT}\right]\right\} \qquad \text{(A-10)}$$

The assumption (A-9a) gives us $|\Psi|^2 = \frac{1}{4/3\pi R_o^3}$ which is an average over R_o and allows us to express ΔF as a function of localisation radius R_o, a parameter of the problem to be determined by minimisation of ΔF. We use dimensionless variables for energy and length, in terms of Hartree E_H and Bohr radius a_B, of Silicon given by

$$\mathcal{F} = \frac{\Delta F}{E_H} \; ; \; l = \frac{R_o}{a_B} \qquad \text{(A-11)}$$

where

$$E_H = \frac{E_H^o}{\epsilon^2} \; ; \; E_H^o = \frac{me^4}{\hbar^2} = 27.2 \text{ e.V} \qquad \text{(A-11a)}$$

$$a_B = a_B^o \epsilon \; ; \; a_B^o = \frac{\hbar^2}{me^2} = 0.53 \text{ Å}$$

Here ε is the static dielectric constant of Si, taken as equal to 12.

Equation (A-10) in terms of the dimensionless variables becomes

$$\mathcal{F}(l) = \frac{\pi^2}{2l^2} - \frac{G c_o}{E_H} - \gamma_o l^3 \left\{ \ln\left[(1-c_o) + c_o \exp \frac{-\eta}{l^3}\right]\right\} \qquad \text{(A-12)}$$

where

$$\gamma_o = \frac{4/3\,\pi a_B^3}{\Omega} \frac{kT}{E_H} \qquad \text{(A-12a)}$$

$$\eta = \frac{G}{kT} \Omega \Big/ \frac{4}{3}\pi a_B^3$$

Making the substitution $\frac{1}{y^3} = \frac{\eta}{l^3}$, we obtain

$$\mathcal{F}(y) = \frac{\pi^2}{2\eta^{2/3} y^2} - \frac{G c_o}{E_H} - \frac{G y^3}{E_H} \left\{ \ln\left[(1-c_o) + c_o \exp \frac{-1}{y^3}\right]\right\} \qquad \text{(A-13)}$$

where

$$l = \sqrt[3]{\eta}\, y \qquad \text{(A-13a)}$$

Examination of equation (A-13) reveals that depending on η, it may have two values of y, for which $\mathcal{F}(y)$ is either a minimum or a maximum. This is seen from the derivative $\mathcal{F}(y)$ with respect to y and setting it equal to zero, which gives the condition

$$\frac{E_H \pi^2}{3G\eta^{2/3}} = -y^5 \left\{ \ln\left[(1-c_o) + c_o \exp \frac{-1}{y^3}\right]\right\} - c_o y^2 \left[\frac{\exp \frac{-1}{y^3}}{(1-c_o) + c_o \exp \frac{-1}{y^3}} \right] \qquad \text{(A-14)}$$

The right hand-side of equation (A-14) depends uniquely on y and is traced called g(y) in figure 1 (the algebra is similar to Moore et al). For every value of the left hand-side of equation (A-14), the curve g(y) is intersected at two points, until at a critical value of η, we reach the maximum of the curve (at 0.027) beyond which no solution is possible. The lower value of the y corresponds to the minimum of the free energy $\mathcal{F}(y)$ while the larger value of y is a maximum. We are now in a position to trace $\mathcal{F}(y)$ of equation (A-13) numerically, for each C_o, at different values of η. Figure 2 gives the curve for $C_o = 0.2$ corresponding to average hydrogen concentration in Silicon of $1.2 \times 10^{22}/cm^3$. The curves are drawn for the room temperature and we have used

$$kT = 0.025 \text{ e.v.}$$

$$E_H = 0.2 \text{ e.v.}$$

$$\frac{\frac{4}{3}\pi a_B^3}{\Omega} = 27, \text{ for an } a_B \sim 6\,\text{Å}$$

and G = 2, 4, 6, 7, 8, 10 e.v.

We see, typically for $C_o = 0.2$ and $G = 6$ e.v., a FIG state exists at $y = 0.6$ (i.e. $\ell = 1.2$, giving a localisation radius of ~ 7.2 Å) with the state localised about 0.3 e.v. below the zero which corresponds to completely delocalised electron. We also note from the curves that a slight activation barrier exists to come in from $y \simeq \infty$ and that below a certain critical η no localisation is possible. The critical η reflects the three dimensional nature of the problem.

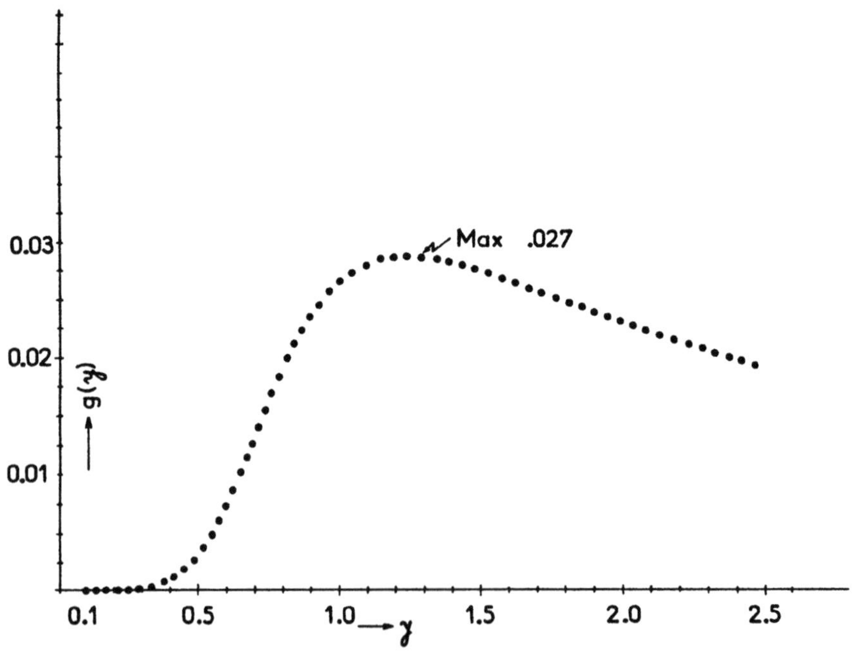

Fig. 1. Free energy derivative function g(y) against y.

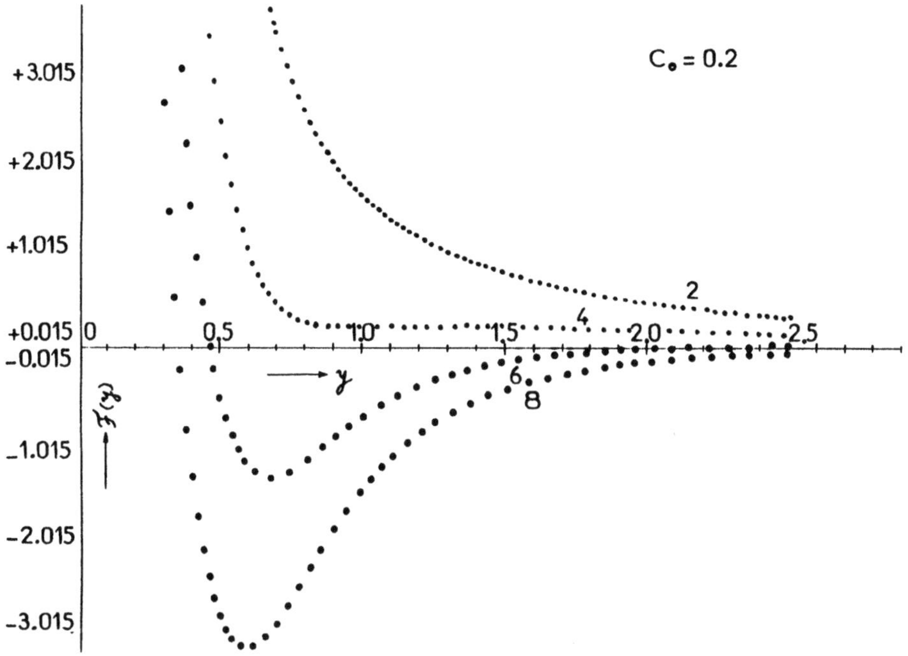

Fig. 2. Reduced free energy (y) against dimensionless reduced length y for average hydrogen concentration $c_o = 0.2$. Energy unit, E_H. Different G-values, see text.

We now want to find out if the FIG state can accomodate two electrons. The analysis consists in comparing energy of two infinitely separated electrons localised in two fluctuation wells with that of two electrons localised in the same potential well. These are given by, for two separated electrons

$$\mathcal{F}_2^{\circ}(y) = \frac{\pi^2}{\eta^{2/3} y^2} - \frac{2GC_0}{E_H} - \frac{2G}{E_H} y^3 \left\{ \ln\left[(1-C_0) + C_0 \exp\frac{-1}{y^3}\right]\right\} \qquad \text{(A-15a)}$$

Similarly it can be easily shown, for the two electrons in the same well, we have

$$\tilde{\mathcal{F}}_2(y) = \frac{\pi^2}{\eta^{2/3} y^2} + \frac{1}{\eta^{1/3} y} - \frac{2GC_0}{E_H} - \frac{Gy^3}{E_H} \left\{ \ln\left[(1-C_0) + C_0 \exp\frac{-2}{y^3}\right]\right\} \qquad \text{(A-15b)}$$

The second term in (A-15b) represents the Coulomb repulsion of the two electrons $\sim e^2/R_o$, while the crucial difference in the exponential of the last terms comes from the fact that $|\Psi|^2$ is now $= \frac{2}{4/3\,\pi R^3}$, where Ψ is a two-electron wave-function. We have plotted in figures 3 and 4, the function $_2(y) - {}_2^{\circ}(y)$ as function of y, at $C_o = 0.1$ and 0.2, for different values of G and we observe that beyond a critical value of G, the two electron states are always favorable energetically. The localisation radius of the two electron FIG states, tend to be higher than its 1-electron counter-part, and has a typical value of $y \sim 1$, corresponding to (for $G = 6$ e.v.) a radius ~ 12 Å. We note finally that the curves 2 and 3 are in reduced energy units and must be multiplied by $E_H = 0.2$ e.v., to obtain the energy in electron volt. We also note that these curves represent the total free energy change due to localisation and is not the energy of the localised electron-state, which is given by only the electronic terms of equations (A-10) and (A-15b), to be computed with R_o at the minimum. See

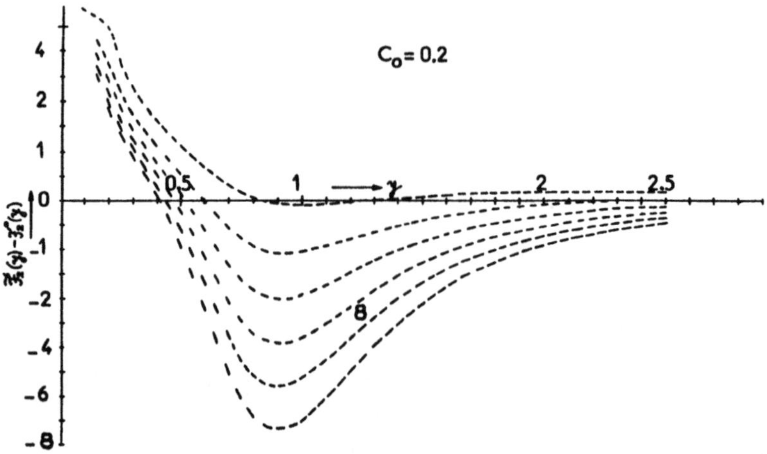

Fig. 3. Two electron state free energy with respect to two separated localised electron, against y. Same energy unit as before, average hydrogen concentration $c_o = 0.2$ Different G-values.

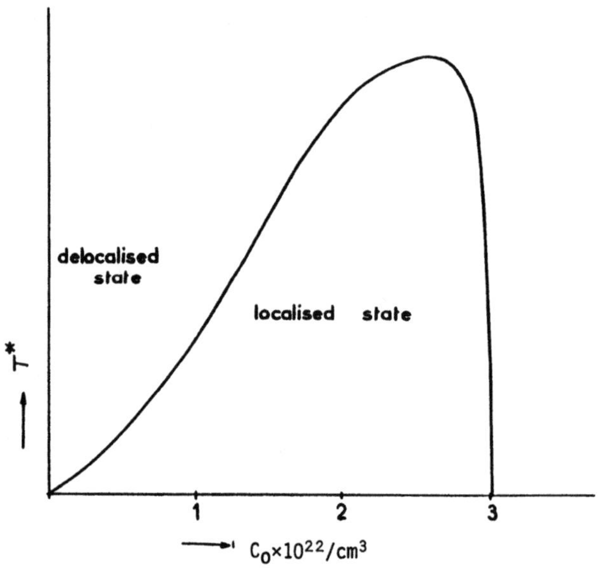

Fig. 4. Critical temperature versus hydrogen concentration as boundary between the localised state and localised state.

ref. Chakraverty 1985. One notes that for each average concentration of hydrogen, the localisation radius shifts very little. The analysis leading to the two electron FIG states is analogous to that of the bipolaronic states derived by Cohen, Economou and Soukoulis, (1984) although in our paper we make no attempt to improve the estimates of the kinetic energy of localisation due to the amorphous nature of the medium. The details of the microscopic calculation for G is to be found in Chakraverty 1985.

B. Stability of the localised state. In this section, we shall calculate the boundary between the localised and the delocalised state, in an approach similar to Yabukov et al (19) and to Cleveland and Gersch (1981) although calculations of the later are full of algebraic errors. The analysis consists in abandonning the infinite square-well potential assumed in the section A (equation (A-5) for the wave-function) and we start from the continuity of Ψ and Ψ' at $r = R$ given by

$$\frac{\hbar^2 k^2}{2m} + cG = \frac{-\hbar^2 k'^2}{2m} + C_0 G \tag{B-1}$$

where k is the wave-vector inside the well and k' outside the well.

k and k' are connected by

$$k \cot kR = -k' \tag{B-1a}$$

or

$$\tan kR = \left[\frac{\hbar^2 k^2}{2mG(C_0-c) - \hbar^2 k^2}\right]^{\frac{1}{2}} \tag{B-2}$$

This gives

$$\sin kR = \frac{\hbar k}{\left[2mG(c-C_0)\right]^{\frac{1}{2}}} \tag{B-3}$$

Let us write

$\alpha = kR$ and we have

$$\frac{\sin\alpha}{\alpha} = \frac{1}{\pi}\left[\frac{\frac{\hbar^2}{2m}\left(\frac{\pi}{R}\right)^2}{G(C_0-C}\right]^{\frac{1}{2}} \tag{B-4}$$

α measures the extent to which the electron state is bound. For an infinitely deep potential well (section A) $\frac{\sin\alpha}{\alpha} = 0$, or $\alpha = \pi$, corresponding to an wave-function with no over-lap at the wall-edges. The smallest value α can have is $\frac{\pi}{2}$; the kinetic energy of the infinite square well being 4 times the depth of the actual well.

The electronic energy of localisation is then

$$E = \frac{\hbar^2 k^2}{2m} + G\int (c-c_0)\Psi^2 d\tau = \frac{\hbar^2}{2m}\left(\frac{\alpha}{R}\right)^2 + G\,(c-c_0) \tag{B-5}$$

The work of creating the fluctuation well is (rewriting equation (A-6) section A

$$W = \frac{kTC_0}{\Omega}\int\left[z \ln z + (z^* - z)\ln\left(\frac{z^*-z}{z^*-1}\right)\right] d\tau \tag{B-6}$$

where $Z = \frac{C}{C_0}$, $Z^* = \frac{C^*}{C_0}$ where C^* is the limiting concentration of hydrogen beyond which the amorphous state is not mechanically stable (~ 50 atomic per cent).

(B5) and B6), in the same dimensionless units as before gives

$$\Delta F = \frac{\alpha^2}{2l^2} + \gamma_1 (z-1) + \gamma_0 C_0 f(z)\, l^3 \tag{B-7}$$

where $\gamma_1 = \frac{G C_0}{\varepsilon_H}$; $f(z) = z \ln z + (z^* - z) \ln\left(\frac{z^* - z}{z^* - 1}\right)$

From $\frac{\partial(\Delta F)}{\partial l} = 0$, and $\frac{\partial(\Delta F)}{\partial Z} = 0$, we get respectively

$$-2\alpha\left(\alpha + \frac{y}{y'}\right) + 6\gamma_0 l^5 f(z) = 0 \quad \text{(B-8)}$$

and

$$\frac{\gamma_1 \alpha y^3}{y'} + \gamma_1 + \gamma_0 C_0 l^3 f'(z) = 0 \quad \text{(B-9)}$$

where we have written

$$y = \frac{\sin\alpha}{\alpha} \; ; \; \frac{d\alpha}{dl} = \frac{-1}{l} \frac{y(\alpha)}{y'(\alpha)} \quad \text{(B-9a)}$$

The sensibility between the localised and the delocalised state is determined by setting equation (B-7) equal to zero, to give

$$\alpha^2 + 2\gamma^0 f(z) l^5 = \frac{1}{y^2(\alpha)} \quad \text{(B-10)}$$

Substracting from (B-8), we obtain

$$-2\alpha\left(\alpha + \frac{y}{y'}\right) - 3\alpha^2 = \frac{-3}{y^2(\alpha)}$$

This gives us the constitutive equation for α at the stable-unstable boundary

$$\frac{5}{3}\sin^2\alpha + \frac{2}{3}\frac{\sin^3\alpha}{\alpha\cos\alpha - \sin\alpha} = 1 \quad \text{(B-11)}$$

The value of $\alpha = 2.12$ constitutes an universal value and thus along the stable-unstable boundary $\alpha = kR$ is a constant irrespective of the temperature or G.

Rewriting (B-8) and (B-9), we have

$$\left.\begin{aligned} -2\alpha\left(\alpha + \frac{y}{y'}\right) &= -6\gamma^0 l^5 f(z) \\ \frac{\gamma_1 \alpha y^3}{y'} + \gamma_1 &= -\gamma^0 l^3 f'(z) \end{aligned}\right\}$$

Using the universal value of α, we now get the equation for the stability, giving Z from

$$\frac{z \ln z + (z^* - z)\ln\left(\frac{z^* - z}{z^* - 1}\right)}{(1 - z)\ln\left[\frac{z(z^* - 1)}{z^* - z}\right]} = -0.4 \quad \text{(B-12)}$$

where $z = \frac{C}{C_0}$, $z^* = \frac{C^*}{C_0}$

The temperature associated, with a choosen C_0 and C^* is given by

$$T^* = \frac{G\Omega_{Si}}{4/3\,\pi a_B^3 k} \; \frac{-1 - \frac{\alpha y^3}{y'}}{l^3 f'(z)} \quad \text{(B-13)}$$

where T^* represents the critical temperature beyond which the electron state is delocalised. Equation (B-13) is plotted in figure 4 for an arbitrary value of G.

II - DISCUSSION AND CONCLUSION

Literature is rich in experimental evidence of numerous light induced effects in hydrogenated amorphous Silicon (see the recent review by Guha 1985, as well as the two recent international conference proceedings, Chakraverty and Kaplan, 1981 and Tanaka and Shimizu, 1983), the most striking of which is the Staebler-Wronski effect (Staebler and Wronski 1977). In the Staebler-Wronski effect, prolonged light-exposure of amorphous hydrogenated Silicon leads to decrease of dark and photoconductivity by orders of magnitude making all device-application problematic but the normal state before the light exposure can be recovered by a thermal anneal at 150-200° C. More recently, based on electron spin resonance measurements, one kind of defect caused by light exposure, the dangling bond has been identified (Dersch, Stuke and Beichler 1981). Creation of dangling bonds implies that prolonged light exposure can snap weak silicon-silicon bonds. In the absence of any subsequent movement of hydrogen atoms such dangling bonds will certainly cause deterioration of photoconductivity but can hardly be expected to be the cause of a reversible, metastable effect. Dersch et al (1981) pointed out that hydrogen movement occurs to give metastability. It is highly speculative at this stage to affirm what exactly light does, but one can safely assume that there are at least two principal effects or processes : (a) In most regions of the amorphous hydrogenated silicon, photo electrons and holes are created if $h\nu > E_g$ (b). But there must also exist regions, so called 'weak regions' of weakly-bonded silicon-silicon atoms and/or $(Si\text{-}H_2)_n$ chains where light may not only break a bond but can cause significant local redistribution of hydrogen atoms through the reaction of type $2\,(Si\text{-}H) \rightarrow (Si\text{-}Si) + H_2$ where local weak-bond reconstruction and hydrogen formation can proceed, light providing the activation energy for the reaction. In such an eventuality, the photo-electrons created by the first process (i.e. (a)) will be trapped in its turn due to second where a dense mobile hydrogen will now exist ; the electron-hydrogen interaction invoked in this paper describes the trapping process and the metastable state. We are led to believe that key to the deterioration of photo-voltaic properties of amorphous hydrogenated silicon lies in the process (b), where either photoexcitation or photodecomposition of weak silicon-hydrogen bonds occurs locally (in regions of $(Si\text{-}H_2)_n$ chains or Si-H-H-Si bonds) so that mobile or excited hydrogen atoms are available which can then readily interact with an excess electron or hole. The mobile hydrogen atom migration needs to take place only on a scale ~ 10-50 Å, to pull an electron state out of the continuum. In pure non-hydrogenated amorphous silicon, one should not see these light-induced effects. As the electron (or hole) becomes self-trapped due to hydrogen atoms both the fermi and the quasi-fermi level will shift (downwards for electrons and upwards for holes). New tail states would appear which are quite different from those due to dangling bonds. Although the chemical reaction described in the step (b) may even be exothermic, it would not occur thermally not at room temperature anyway, as the activation barrier for the reaction is probably $\sim h\nu$, and hence the deleterious effect of light as observed. This is why, once the fluctuation well is created, it cannot go away at room temperature, and a higher temperature thermal anneal is needed.

In our analysis, we have neglected the gradient energy term associated with fluctuation, as well as possibility of creation of dangling bonds at the internal surface between the potential well and the medium. In fact for a region of 10 Å radius, the total entropy increase, from equation (A-8) is 1 e.v., for $C_0 = 0.2$, while the relevant energy increase of newly created dangling bonds in this region <<electron-volt. In a forthcoming paper, the total change of free energy that includes both internal energy as well as the gradient term will be evaluated. As a matter of fact we have intentionally neglected in this paper the spatially fluctuating eigen states near the mobility edge, whose inclusion is known to give (Cohen et al, 1984) an easier localisation. We have kept the physics simple so that several straightforward conclusions can be drawn about the FIGS :

a) The potential well (1) will create both electron and hole trapping. The presence of hydrogen is attractive for electron in the valence band and repulsive for the electron in the conduction band bonding and antibonding states respectively. (Johnson et al, 1980), hence holes in the valence band and electrons in the conduction band will be trapped although the value of G may be different.

b) The creation of such potential wells imply movement of hydrogen atoms. How exposure of light >1.6 e.v. or double injection or electron bombardment can create this movement at room temperature remains to be elucidated, but very recently careful neutron-measurements of light exposed α-hydrogenated Silicon give the first direct evidence, (Chenavas-Paule et al, 1985), that massive hydrogen motion does occur.

c) The fact that the 2-electron FIGS readily occur show that the FIGS will be as efficient recombination center (the repulsive energy ~ 0.1 e.v. for a 6 Å region) as the dangling bonds (Street, 1981).

d) The self-trapped FIGS are metastable. Increasing temperature will destroy them, either by back-diffusion of hydrogen or due to critical $\frac{G}{KT_c}$, at $T \sim T_c$; whichever effect predominates.

ACKNOWLEDGEMENT

The author is indebted to Drs. R. Cinti, E. Al-Khoury and M. Avignon of this laboratory for help in the numerical computation. He is appreciative of Dr. R. Bellissent of Léon Brillouin Laboratory, Saclay for communicating some of his experimental results before publication. Discussion with Dr. J.C. Bruyere and Prof. A. Deneuville of this laboratory is acknowledged.

REFERENCES

A. Chenavas-Paule, R. Bellissent, M. Roth and I. Pankove - 1985 ; to be presented at 11th International Conference on "Amorphous and Liquid Semiconductors" Rome, Sept. 1985. Private Communication, R. Bellissent.

B.K. Chakraverty - 1985, Phil Mag. B, to be published.

B.K. Chakraverty, Kaplan D. - 1981, Proceedings of the Ninth International Conference on "Amorphous & Liquid Semiconductors" 'Editions de Physique, Paris) C4-371 to C4-407.

S. Chandrasekhar - 1943, Rev. Modern Phys. 15, 1, 1.

C.L. Cleveland and H.A. Gersch - 1981, Phys. Rev. A 23, 261.

M.H. Cohen, E.N. Economou, C.M. Soukoulis - 1984, Physical Review B 29, 4496.

H. Dersch, J. Stuke and J. Beichler - 1981, Appl. Phys. Letter 38, 456.

T.M. Eggarter - 1972, Phys. Rev. A, 5 2496.

R. Friedberg, J.M. Luttinger - 1975, Phys. Rev. B 12, 4460.

S. Guha - 1985, "Physical Properties of Amorphous Materials" (Plenum Press, N.Y. and London) Edited by : D. Adler, B.B. Schwartz, M.C. Steele, 423.
K.H. Johnson, H.J. Kolair, J.P. De Neufville, D.L. Morel - 1980, Phys. Rev. B 21 , 643.
M.A. Krivoglaz - 1974, Sov. Phys. Usp 16, 856.
L.S. Kukushkin, V.B. Shikin - 1973, Sov. Phys. JETP, 969.
J.L. Levine, T.M. Sanders - 1967, Phys. Rev. 154, 138.
I.M. Lifshitz - 1968, Sov. Phys. JETP 26, 462.
A. Messiah - 1962 "Quantum Mechanics" (John Wiley & Sons, New York) Vol. 2, 806.
R.L. Moore, C.L. Cleveland, H.A. Gersch - 1978, Phys. Rev. B 18 , 1183.
N.F. Mott, H.S.W. Massey - 1965, "The Theory of Atomic Collisions" (Oxford University Press) Chapter V, 86.
B.E. Springett, J. Jortner, M.H. Cohen - 1968, Journal Chem. Phys. 48, 2720.
D.L. Staebler, C.R. Wronski - 1980, Jl. Appl. Phys. 51, 3262.
R. Street - 1981, Advances in Phys. 30, 593.
K. Tanaka, T. Shimizu - 1983, Proceedings of the Tenth International Conference on "Amorphous and Liquid Semiconductors" (North Holland) 393 - 437.

HYDROGEN DISTRIBUTION IN AMORPHOUS SILICON AND SILICON BASED ALLOYS

B. L. Jones

GEC Research Limited
Hirst Research Centre
Wembley, Middlesex UK

ABSTRACT

The results of hydrogen evolution experiments on amorphous silicon alloys prepared by high frequency PECVD of gas mixtures containing SiH_4, NH_3, PH_3, B_2H_6 are compared. Using a very low heating rate of 5°/min it is possible to resolve fine structure on the exodiffusion spectra. Three evolution processes are observed:

(a) low temperature effusion due to included gas
(b) mid temperature effusion due to 'clustered' hydrogen bonds
(c) high temperature effusion due to 'isolated' hydrogen bonds

In addition it is possible to oberve very fine structure 'puffing' due to the release of molecular hydrogen at mid to high temperature.

Silicon and silicon nitride films have been annealed at low temperatures before the exodiffusion experiments and changes in the evolution spectra are observed, dependent on the annealing process.

A scanning electron microscope study of the effect of high temperature heat treatment has also been undertaken. These results are correlated with infra-red absorption measurements and the influence of doping concentration and substrate character discussed. Under certain preparation conditions the films blister on heating and finally burst forming circular craters, and these effects are shown to be dependent on substrate material and intrinsic stress of the as-grown films.

An understanding of these effects is an essential item for multilayer thin film devices e.g. VLSI where any post-deposition heat treatment can adversely/beneficially change the properties of neighbouring layers.

INTRODUCTION

Two techniques of hydrogen characterisation of amorphous silicon based alloys readily lend themselves to industrial use by virtue of their cost and ease of execution and interpretation.

(a) Infra-red vibrational spectroscopy - which provides information on bonding configuration and total content.

(b) Thermal evolution - which provides information on atomic distribution and total content.

With so much recent interest in NMR techniques[1,2,3] to elucidate the H distribution it is pertinent to question the place of thermal evolution as an analytical study. Apart from the obvious choice of simplicity and lack of expense in carrying out such a study, which is of course essential to consisently and efficiently fabricate device worthy material there is another major consideration. In nearly every IC manufacturing process any thin film deposition (insulator or semiconductor) is likely to undergo further heat treatment - annealing, diffusion drive in, or deposition. In order to use these amorphous materials in 3D VLSI it is necessary to predict what will occur to the material during these treatments.

In a typical multilayer structure for example (Figure 1) it is possible, by this process, to promote desirable or undesirable effects in neighbouring structures. Hydrogen and in some cases impurity atoms can diffuse through and become trapped in neighbouring layers and interfaces. It is also possible in extreme cases to cause the film to breakup on heating, caused by occluded hydrogen at certain interfaces.

The results in this paper represent the first stage in identifying some aspects of the hydrogen distribution in the alloys that are likely to be used in novel 3D VLSI thin film structures.

AMORPHOUS SILICON

The hydrogen bonding configuration of hydrogenated amorphous silicon i.e. monohydride ($Si\equiv H$), dihydride ($Si=H_2$), or trihydride ($Si-H_3$) can be determined by infra-red vibrational spectroscopy[4] and is a strong function of deposition temperature. For high substrate temperatures $>200°C$ the hydrogen present is predominantly ($>90\%$) of the monohydride form (w stretch = 2000 cm^{-1}). Low substrate temperature films $<100°C$ contain significant quantities ($>50\%$) of polysilane chains $(SiH_2)_n$ (w stretch = 2100 cm^{-1}). The total hydrogen content in the film can be determined by integrating the area under the 640 cm^{-1} peak (SiH_x rocks), using an appropriate value of material oscillator strength. (This is disadvantaged because of the necessary assumption of a particular well defined background atomic lattice).

Alternatively the total hydrogen content in the film can be determined from the integral of hydrogen exodiffuison spectra (executed at low heating rates $<5°/min$)[5] where also the rate of release of hydrogen molecules as a function of temperature produces fine structure which determines whether the hydrogen is bonded in the 'clustered' phase (a group of $Si\equiv H$ bonds in close proximity similar to a silicon vacancy with the four dangling bonds satisfied by four hydrogen atoms), the 'isolated' phase (isolated $Si=H$ bonds) or exists in the form of gas 'inclusions' (Figure 2a). On a microscopic scale very fine structure ('puffing') is observed (Figure 2b) due to repeated bursts of small quantities of molecular hydrogen which are thought to arise as microvoids begin to form. The position of the high temperature 'isolated' peak ~450°C is a function of film thickness (Figure 3b) as the hydrogen diffuses to the surface of the film, by the formation and dissociation of $Si\equiv H$ bonds, where it combines molecularly. The 'clustered' peak ~300°C occurs due to the

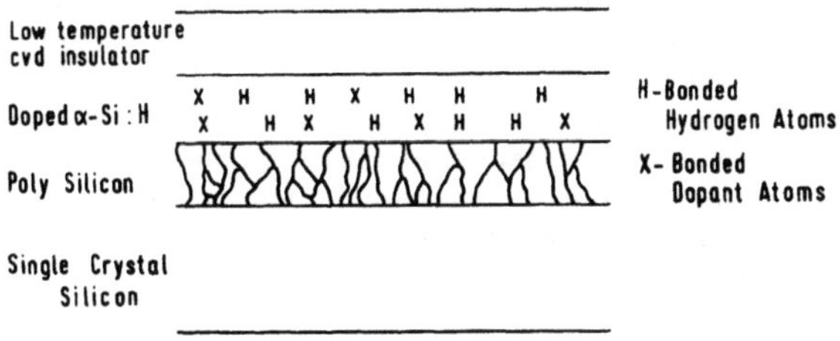

Figure 1: Multilayer structure

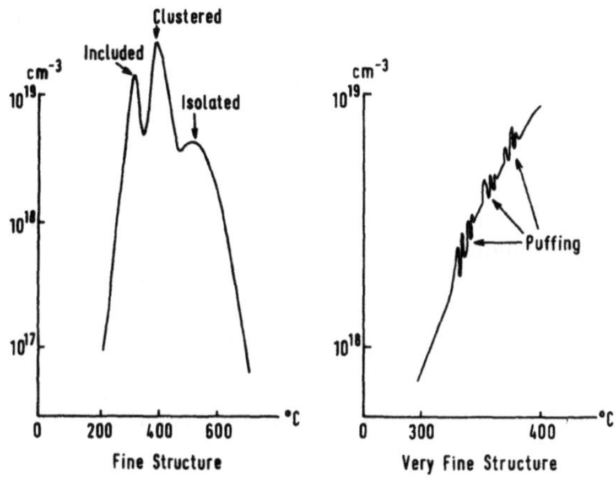

Figure 2: Thermal evolution model

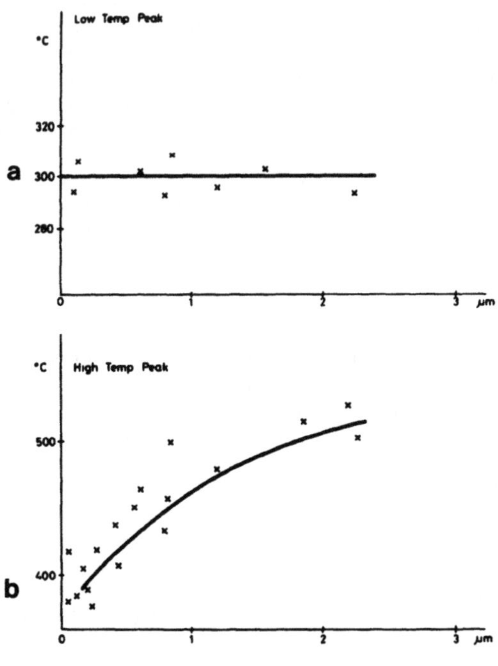

Figure 3: Variation of peak temperature with film thickness

direct diffusion of H_2 molecules formed by simultaneous dissociation of neighbouring Si≡H bonds, and this energy required is partially regained by the formation of H-H and Si-Si bonds. The position of the peak is independent of the film thickness (Figure 3a). Any included gas in the film is evolved at a lower temperature (below 250°C). Previous theoretical calculations have indicated that a single bonding configuration provides better film characteristics and thus it is assumed that the highest quality 'trap-free' material consists of Si≡H only in the 'dilute' phase.

It is also possible to modify the hydrogen content and bonding phase by annealing the sample at the 'clustered' peak temperature ~350°C for one hour (Figure 4). The unwanted 'clustered' and 'included' peaks can be removed and the hydrogen content reduced. Annealing below this temperature shows no effect. The difference observed between the annealing atmosphere (H_2 or vacuum) shows that the presence of molecular hydrogen at the surface of the material inhibits the exodiffusion process.

Boron doped (Figure 5) and phosphorous doped (Figure 6) samples have also been investigated using this technique. In the former case the spectra appear to broaden (progressively with dopant incorporation) in shape indicating an enhanced hydrogen diffusion coefficient whilst the phosphorus spectra increase in complexity with dopant content. Not surprisingly the total hydrogen content increases with the amount of dopant present and by implication there must be a corresponding increasing amount of dopant passivation. This is to some extent reflected in the poor dopant incorporation efficiency and increased number of defect states introduced during in situ doping amorphous silicon[6].

For undoped amorphous silicon films on c-Si substrates, the αSi:H dehydrogenates microbubbles grow and burst to form circular craters[7,8,9] when the yield strength of the material $4x10^9$ dyn cm^{-2} is exceeded by the high pressure of occluded hydrogen at the α-Si:H/c-Si interface due to the difference in diffusion rates between the amorphous layer and substrate. For oxidised Si substrates this effect does not occur (Figure 7) indicating that lateral diffusion of H atoms or H_2 is relatively fast at elevated temperatures, compared to occlusion at the SiO_2 layer[10]. This effect has been utilised to form an archival optical storage technique.

A similar effect is not observed for doped (B or P) materials and craters occur on oxidised surfaces (Figure 8) illustrating that the presence of dopant atoms also affects the lateral diffusion mechanism.

SILICON NITRIDE

The absorbance values from IR spectra at the stretching vibrational frequencies SiH_x (2160 cm^{-1}) and NH (3350 cm^{-1}) groups can be used[11] to determine the distribution of hydrogen between the silicon and nitrogen sites, as well as estimating the total content of the film. For NH_3 and SiH_4 prepared films there is very little change in this hydrogen content when altering the molar gas ratio (CH% ~25%). The ratio of the H located at the nitrogen sites compared to that located at the silicon sites shows a marked dependence on the molar gas ratio (Figure 9). The SiH_x stretch mode appears at 2160 cm^{-1} rather than at 2100 cm^{-1} which is explicable in terms of the average coordination number of N atoms contiguously bound to SiH_x sites and the concomitant inductive shift in the wavenumbers, although the choice between a NSi_2H_2 or N_2Si_2H assignment

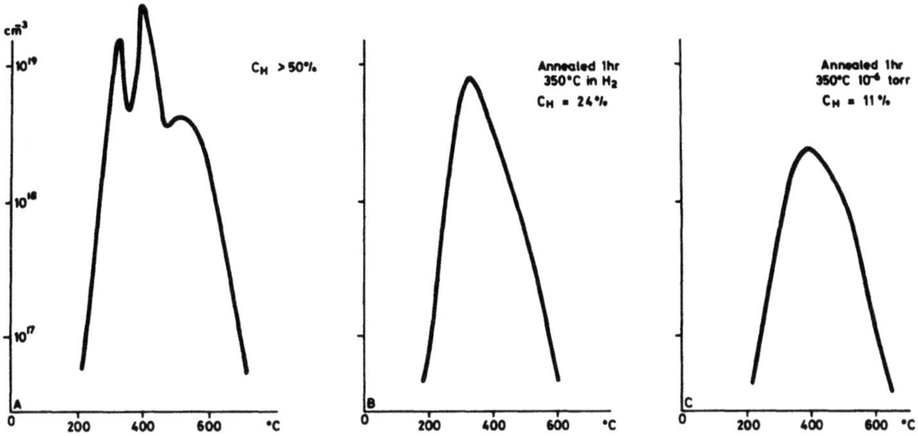

Figure 4: Room temperature amorphous silicon, annealed

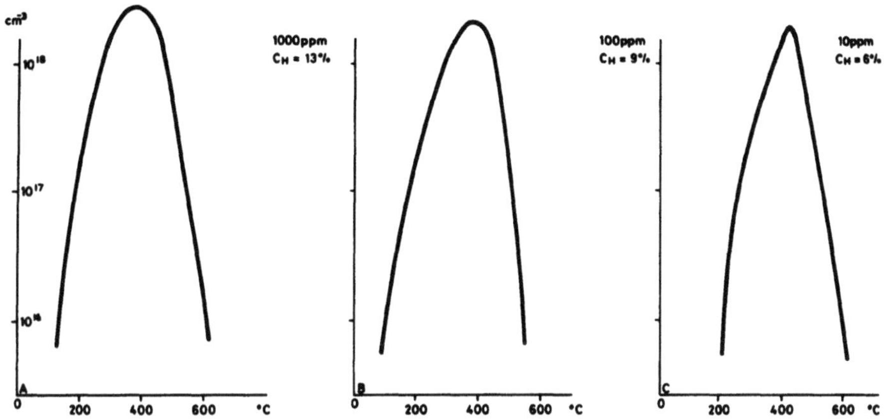

Figure 5: Boron doped amorphous silicon

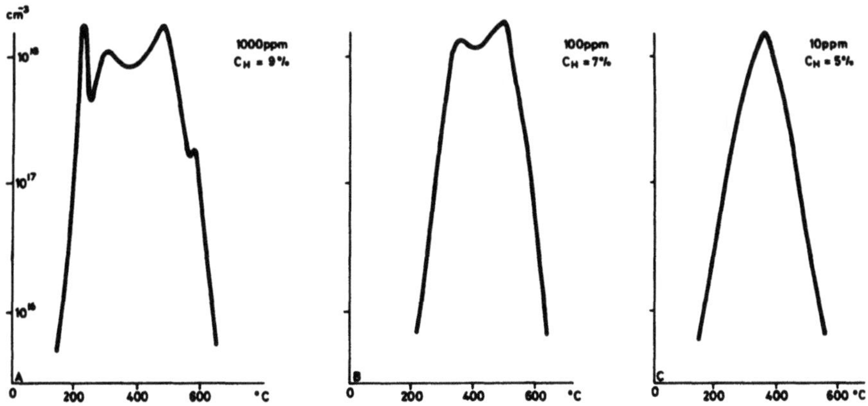

Figure 6: Phosphorous doped amorphous silicon

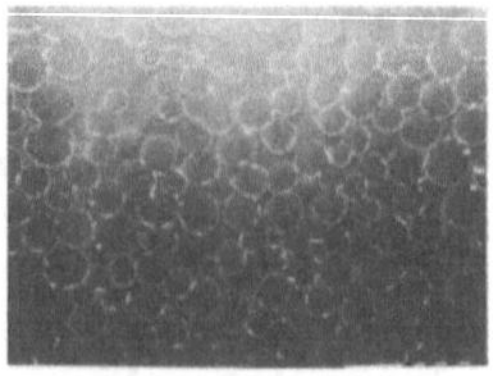

(a) Room temperature deposited α-Si:H on c-Si annealed to 750°C

(b) Room temperature deposited α-Si:H on oxidised Si annealed to 750°C

Figure 7: SEM of annealed amorphous silicon on oxidised and unoxidised c-Si

(a) 10 ppm P_2H_6 : SiH_4 - annealed to 750°C

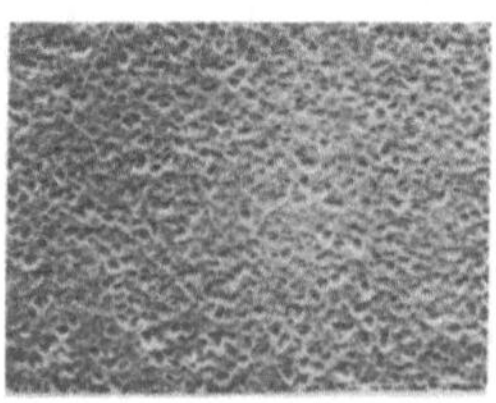

100 μm

(b) 1000 ppm P_2H_6 : SiH_4 - annealed to 750°C

Figure 8: SEM of annealed doped amorphous silicon on oxidised c-Si is ambiguous (Figure 10)[12]. The value of the Si/N atomic ratio, determined from XPS measurements is found to obey a linear relationship with refractive index and thus it is possible to express the fraction of NH bonds as a function of the stoichiometry of the film (Figure 11). It can be seen from this relationship that it is possible to 'tune' the amorphous silicon nitride to the required refractive index, e.g. for optical waveguide applications, within the range 1.6 to 2.1.

I-V and C-V analyses using MNS capacitor structures on n^+ Si yield information on dielectric performance including stability which can be correlated to structural information i.e. location of H atoms[13]. A voltage ramp of 1 V/sec has been applied to the MNS structures to determine the non-destructive breakdown field. Preliminary capacitance voltage measurements indicate that the initial flatband voltage (V_{fb}) depends on the proportion of NH groupings (Figure 12). The highest quality nitride suitable for gate dielectric applications shows a flatband voltage of +1V, non-destructive breakdown field of 8 MV/cm and a surface state value of 10^{12} cm^{-2} eV^{-1} [14]. The trapped charge in the nitride has been further investigated using a V_{fb} hysteresis measurement where the voltage ramp is applied in forward and reverse diretions. The amount of hysteresis shows the same dependence on the NH groupings as the initial flatband voltage, with no hysteresis for NH >85%[12].

From thermal evolution investigations on Si rich (NH_3/SiH_4 = 4/1) nitride films it can be seen that the main thermal evolution peak occurs at ~450°C with a secondary peak at ~650°C[15] (Figure 13). The former temperature is similar to that observed in α-Si:H films and may be ascribed to the release of hydrogen via Si-H bond fission. By analogy the secondary peak at 650°C can be associated with the release of H from NH sites. Films treated for 1 hour at 350°C have not significantly changed the spectrum integral i.e. total hydrogen content, although the H appears to have changed sites from Si to N. This is reflected in the improved dielectric performance of annealed films. Under high temperature anneals ~650°C there is evidence that the total amount of hydrogen is reduced and almost totally distributed to the Si sites[16].

As with the α-Si:H films microbubbles form on annealing and erupt with sufficient ferocity to reveal the underlying substrate. Films grown with a high degree of intrinsic stress appear to experience the most degradation and the manner in which the films tear off the substrate depends on the crystallographic orientation of the substrate (Figure 14). Films on <111> Si replicate the trigonal symmetry of the underlying crystallographic planes compare to the 90° fracture surfaces observed on <100> Si [17]. Films prepared on oxidised C-Si substrates do not rupture; a feature previously described in this paper.

CONCLUSIONS

These results show that we have gone some way down the line in estabilishing what form hydrogen distribution occurs in amorphous silicon and silicon based alloys. Using the basic techniques of vibration spectroscopy and thermal evolution we can differentiate between mono-, di- and tri-hydride formations in α-Si:H as well as between 'dilute' and 'clustered' formations. With dopant impurity atoms the picture becomes more complex where defects and included gas formations become important. Clearly this is an area for further work which would complement the recent models on the doping mechanisms. The distribution of the H atoms between the silicon and nitrogen sites in α-Si_xN_y:H determines the optical and

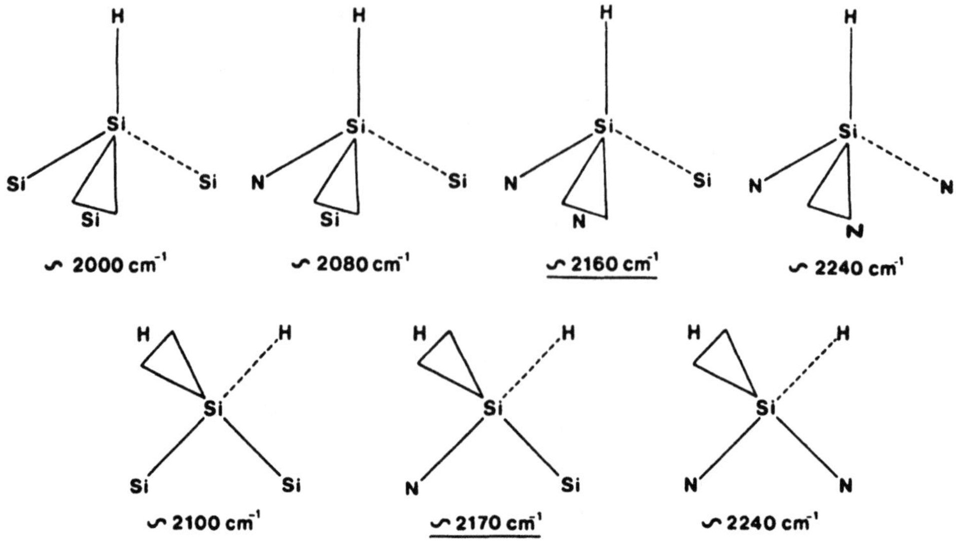

Figure 9: Variation of N-H% with molar ratio

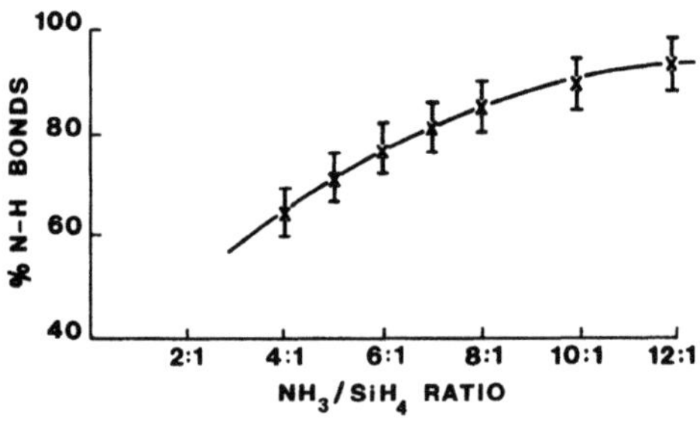

Figure 10: Possible vibration groupings for ~2100 cm^{-1}

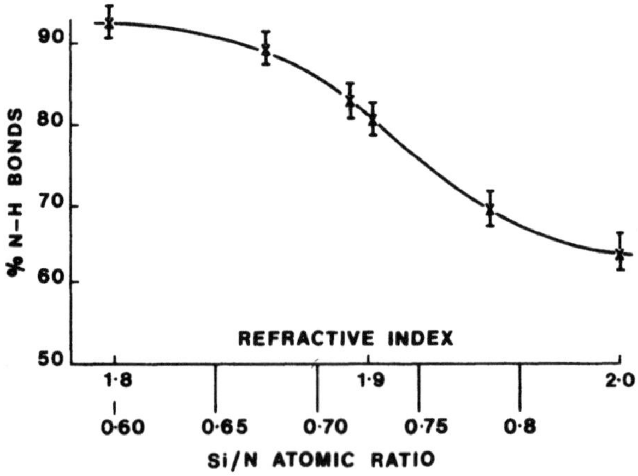

Figure 11: Variation of N-H% with refractive index

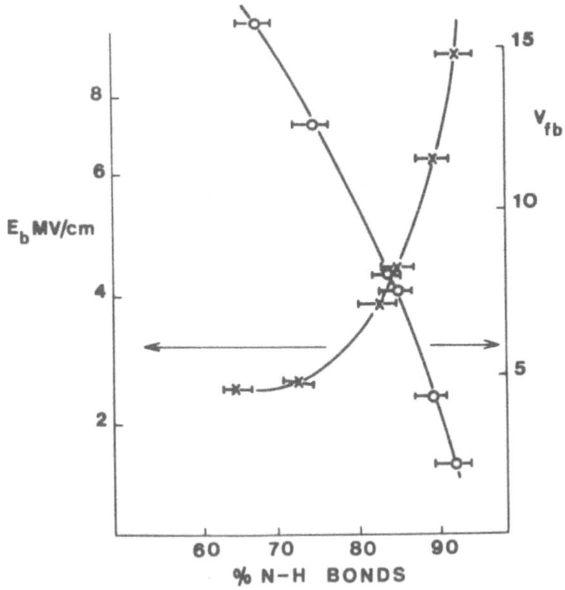

Figure 12: Variation of N-H% with V_{fb} and E_b

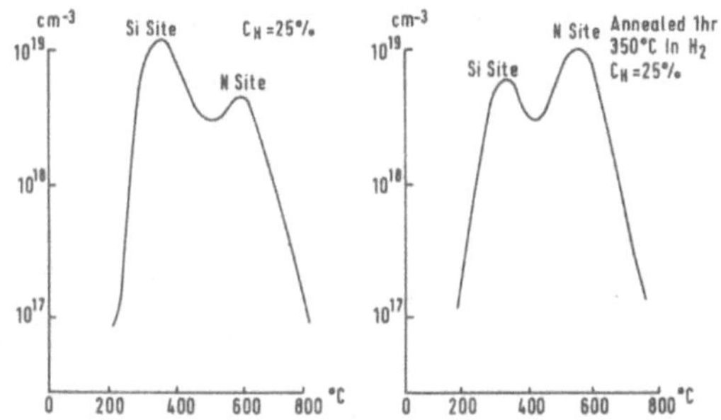

Figure 13: Amorphous silicon nitride, annealed

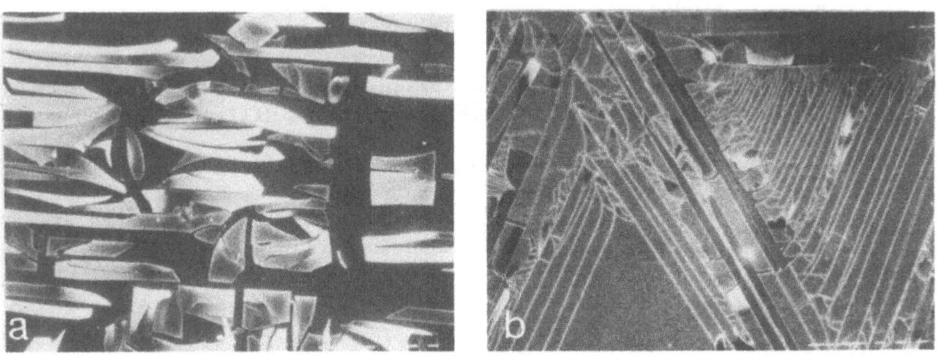

Figure 14: Si-N annealed to 750°C on Si substrates (a) <100>, (b) <111>
—— 100 μm

dielectric properties of the material which allows us to 'tune' the material to suit the application. Mid- and high-temperature anneals on this material serve to adjust this distribution according to our requirements.

For all the amorphous silicon based materials the adhesion to the substrate (under anneal conditions) is dependent on the nature of the substrate. With this in mind the information that we are steadily occurring on the hydrogen diffuison and distribution allows us to establish which combination of multilayer films are likely to be of use in novel VLSI development.

ACKNOWLEDGEMENTS

The author would like to thank Dr P John (Heriot-Watt University) and Dr. P.LeCamber(Dundee University) for helpful discussions. This work has been supported by the Procurement Executive, Ministry of Defence (Directorate of Components, Valves and Devices) and sponsored from the RSRE Establishment.

REFERENCES

1 A. Reimer, R.W. Vaughan and J.C. Knights, Phys. Rev. B, 24:3360 (1981)
2 W.E. Carlos and P.C. Taylor, Phys. Rev. B, 26:3605 (1982)
3 D.J. Leopold, J.B. Boyce, P.A. Fedders and R.E. Norberg, Phys. Rev. B, 26:6053 (1982)
4 M.H. Brodsky, M. Cardona and J.J. Cuomo, Phys. Rev. B, 16:3556 (1977)
5 B.L. Jones, J. Burrage and R. Holtom, Advances in Electronics and Electron Phyics, ed Morgan Vol 64 (1985)
6 R.A. Street, Proc. XIth International Conference on Amorphous and Liquid Semiconductors, Rome (1985)
7 H.R. Shanks and L. Ley, J. Appl. Phys., 52:811 (1981)
8 P. John, I.M. Odeh, M.K. Thomas, M.J. Tricker, J.I.B. Wilson and R.Dhariwal, J. Mat. Sci., 16:1305 (1981)
9 H.R. Shanks, C.J. Fang, L. Ley and M. Cardona, Phys. Status Solidi B, 100:43 (1980)
10 P. John and B.L. Jones, Appl. Phys. Lett., 45(1):39 (1984)
11 W.A. Lanford and M.J. Rand, J. Appl. Phys., 49:2473 (1978)
12 B.L. Jones, Proc. XIth International Conference on Amorphous and Liquid Semiconductors, Rome, (1985)
13 B.L. Jones and D. B. Meakin, To be presented at MRS Fall Meeting, Boston, (1985)
14 B.L. Jones and P. John, Proc. 6th International Conference on Thin Films, Stockholm, (1984)
15 B.L. Jones, P. John and K. Welham, Presented at Inst. of Phys. mtg 'Physics of Amorphous and Microcrystalline Silicon', London (1984)
16 I.D. Frech, B.L. Jones, Presented at Chelsea Conference on Amorphous and Liquid Semiconductors, London, (1984)
17 B.L. Jones, P. John, L. McConachie and D. Gold, To be presented at Inst. of Physics mtg on Amorphous Thin Films, London, (1985)

HYDROGEN ON SEMICONDUCTOR SURFACES

James W. Corbett[1], D. Peak[2], S.J. Pearton[3], and A.G. Sganga[1]

1 Physics Department, SUNY/Albany, Albany, NY 12222
2 Physics Department, Union College, Schenectady, NY 12308
3 AT&T-Bell Laboratories, Murray Hill, NJ 07974

ABSTRACT

The state of understanding concerning hydrogen at semiconductor surfaces (external and internal) will be reviewed, the emphasis being on silicon and germanium. Topics will include hydrogen at crystalline-vacuum interfaces in ultra-high-vacuum systems, at grain boundaries in polycrystalline material, and at internal defect surfaces as arise in irradiated or ion-implanted material. The atomic configurations of hydrogen in semiconductors will be surveyed, including both monoatomic hydrogen [H] and molecular hydrogen [H_2]. The diffusion coefficients and profiles associated with free [H] diffusion, trapping of [H] at defects, and formation of [H_2] will be covered. The role of hydrogen in passivating defects will be discussed, as will the passivation mechanisms (bonding and non-bonding), and defect/surface reconstruction and chemically driven reconstruction.

INTRODUCTION

Hydrogen plays an important technological and interesting scientific role in semiconductors. In this brief review, we will survey these results, including what is known about the atomic and electronic configurations of hydrogen and hydrogen-related defects, and about their electrical and other properties. Predominantly hydrogen occurs at the dangling bonds at surfaces, and we will consider external surfaces and internal surfaces, e.g., grain boundaries and defects. As we will see, there remain exciting new developments and open questions.

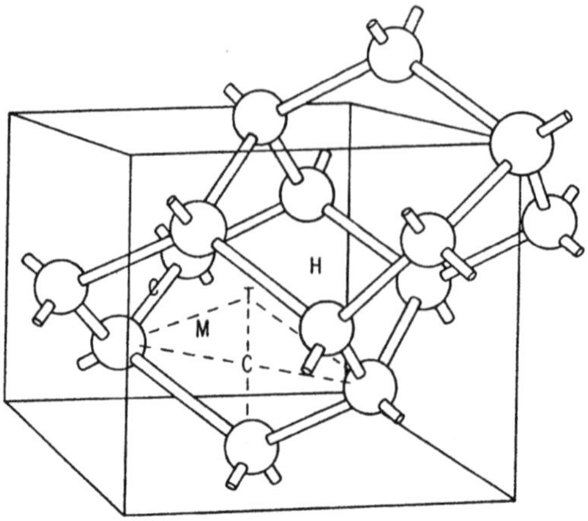

Fig. 1 Portion of the silicon lattice showing the tetrahedral (T_d), hexagonal (H), C- and M-interstitial site. The C-site is at the center of the rhombus formed by three adjacent substitutional sites and the nearest T_d site. An M-site is the mid-point between two adjacent C-sites.

DEFECT CONFIGURATIONS: SILICON

There was at one time a controversy concerning the states of hydrogen in crystalline silicon; some arguments persist, but a consensus is emerging, and we will not present the full chronology. Suffice it to say that the early permeation/diffusion data of Van Wieringen and Warmoltz [1], made it clear that hydrogen dissolved and diffused in silicon as atomic hydrogen, presumably on the tetrahedral (T_d) interstitial site. Picraux et al. [2-4] found in channeling studies that the hydrogen was not on the T_d site; they favored what is called the anti-bonding site, i.e., a site along the <111> direction away from the neighboring atom, a site very much like the back-bond site found by Appelbaum et al. [5] in their theoretical studies of hydrogen at silicon surfaces. Infrared studies on silicon implanted with hydrogen [6-16] and on silicon grown in a hydrogen atmosphere [17] suggested that Si-H bonds occurred, presumably at defects sites, but these sites were not consistent with the channeling studies.

Singh et al. [18] used the semi-empirical Extended Hückel Theory (EHT) [19] to explore the states of atomic hydrogen in the crystalline silicon lattice. They concluded that the minimum energy site for atomic hydrogen was at the tetrahedral (T_d) interstitial site (see Fig. 1). They also examined the electronic states associated with atomic hydrogen (see also Wang & Kittel [20]), and concluded that the hydrogen ground state was a resonance ca. 1 eV below the top of the valence band, but that as the hydrogen was moved away from the T_d site this level moved into the gap, i.e., the state would be electrically active (a point to which we will have to return). They also considered hydrogen bound at defect sites, specifically a vacancy with one, two, three, or four hydrogens, and a divacancy with six hydrogens. They found the energy of the bound hydrogen to be substantially lower than that of the free (T_d) atomic hydrogen, i.e., given an available dangling bond (d) the following reaction will occur:

$$[H] + [d] \underset{\leftarrow \Delta_d}{\overset{K_d \rightarrow}{\Leftrightarrow}} [H \cdot d]. \tag{1}$$

The electrical levels associated with these defects they found to be those of the vacancy as perturbed by the presence of the one, two or three hydrogens; there is no electrically active level for the fully saturated defects $[V \cdot H_4]$ and $[V_2 \cdot H_6]$. (See also, DeLeo et al. [21].) Singh et al. [18] also calculated the vibrational energies associated with the hydrogen in these defects and found frequencies comparable to those observed in the infrared studies. Later Singh, et al. [22] used EHT to consider the anti-bonding site. They concluded that that site had a metastable energy minimum, the T_d site being lower in energy.

Subsequently [23,24] ab-initio and semi-ab-initio quantum techniques were used to address some of these questions, the bulk of the work being done with the MNDO (modified neglect of diatomic overlap) method [25]. On the basis of these more sophisticated techniques, it was concluded that the hydrogen bound to a defect was still the lowest energy site, but that for the free hydrogen atom the T_d interstitial site was not the lowest energy site, rather it was a site called the M-site. The M-site can be visualized using another site called the C-site (see Fig. 1.) The C-site is at the center of the rhombus formed by three connected substitutional atoms (S) and the nearest T_d interstitial site; each triad of lattice atoms has a C-site. The S-, C- and T_d-sites form a simple cubic lattice. The M-sites, which are mid-way between adjacent C-sites, is the lowest energy site for the unbound hydrogen, although one may view the hydrogen in this site as forming a three-center bond with the adjacent substitutional atoms, as found by Applebaum et al. [5] in their pseudopotential calculations and by Snyder et al. [26,27] in ab-initio calculations for hydrogen near a tetra-silyl-silane cluster; this multi-center bonding does not occur in diamond [24], where the T_d site is the lowest energy site. Diffusion occurs from M- to C- to M-site with a theoretical diffusion energy of ca. 0.32 eV, to be compared to the Van Wieringen-Warmoltz [1] experimental value of 0.48 eV. The calculated heat of solution of 1.65 eV also agrees roughly with the experimental value [1] of 1.87 eV.

The MNDO studies also concluded that molecular hydrogen could form in silicon. This conclusion was reached independently by Pearton et al. [28] and by Hall [29,30] in considering experimental data, and in comparable calculations (using CNDO) by Mainwood and Stoneham [31]; clearly the idea of molecular hydrogen in silicon was an idea whose time had come. The molecule has its lowest energy site with the molecule centered at the T_d interstitial site. This configuration fits the channeling results better than does the anti-bonding site [24], and is consistent with the channeling results being carried out with a high concentration of hydrogen, in which it would be expected that molecular hydrogen would predominate. The diffusion of the molecule occurs with an activation energy of ca. 2.7 eV [24] along the path from T_d- to H- to T_d-site, i.e., the standard route for interstitials, the H-site being the so-called hexagonal site (see Fig. 1). Continued MNDO studies [32] have concluded that the presence of a hydrogen molecule in a T_d site increases the barrier by ca. 0.3 eV (i.e., almost doubles the barrier) for diffusion of atomic hydrogen in the M-C-M network around that T_d site. Further, since the atomic hydrogen is on the M sites in the lattice, two hydrogens would have to leave that network to form a molecule situated on the T_d site; there is therefore a small (ca. 0.3 eV) barrier to the formation of the molecule, although the overall reaction is favored:

$$[H_2]_{Si} \underset{\leftarrow \Delta_m}{\overset{K_m \rightarrow}{\Leftrightarrow}} 2*[H]_{Si} + 1.6\ \text{eV}, \qquad (2)$$

as in the gas phase, where the energy is 1.7 eV, i.e., the molecule is the lower energy state. No information concerning electrical levels was obtained in these MNDO calculations. We will return to the influence of molecules when we discuss diffusion below.

In summary, hydrogen can exist in silicon as free atomic hydrogen, as free molecular hydrogen, and as bound hydrogen. But hydrogen may be bound in a variety of configurations. It may be bound at a dangling bond, or trapped at a site where nominally a dangling bond does not exist. This may be seen by considering the "passivation," i.e., electrical inactivation, of electrical levels associated with metallic impurities in silicon.

The metal-related defects in silicon are almost ubiquitous. Due to the high diffusivities of the metals at moderate temperatures they are easily introduced into silicon wafers during processing, or they may be present as metallic clusters or microprecipitates in as-received wafer; normal processing steps involving elevated temperatures act to redissolve the impurities into solution, making them electrically active and creating havoc with device yields. A number of the metallic impurities are intentionally used as lifetime controllers in silicon (eg. Au, Pt, Pd, Mo [33]), and many can act as generation-recombination centers because of their high capture cross-sections for both electrons and holes. Precipitates of these impurities, particularly Fe, may bridge across the emitter-collector region of transistors, creating a short circuit. Even though the solubilities of these impurities are relatively low in silicon at typical processing temperatures, the concentrations are easily high enough to degrade device performance because of the high demands placed on modern circuits and the increasing degree of integration, where device degradation is additive.

Many of these metal-related deep levels in silicon can be neutralized by reaction with atomic hydrogen. Hydrogen passivation of defects at interfaces on silicon surfaces had been observed as far back as 1968 by Brown and Gray [34], but it is only in recent years that passivation of bulk defects has been pursued. Typically this passivation is achieved by placing the sample containing the electrical level in a hydrogen plasma; the plasma creates the atomic hydrogen which diffuses into the silicon and passivates the electrical level. (The diffusion often requires heating the sample in the plasma to > 300°C for several hours; we return to questions of diffusion below.) The Au donor level (E_V +0.35 eV) in p-type silicon may be neutralized by atomic hydrogen [35-37], as can a number of Pd-, Pt- [38], Cu-, Ni- [39], Ag-, Fe- [40] and quenched-in metal-related [41] centers in bulk silicon. The Fe interstitial is mobile even at 300°K, but plasma exposures at this temperature have determined that it is not affected by atomic hydrogen. Insufficient data is available to reveal any trends in susceptibility to hydrogenation with trends in chemical nature of the impurities, and one problem as mentioned previously is that many of the metal-related centers are complexes. Indeed the Cu- and Ni-related levels neutralized by hydrogen are certainly impurity complexes, as Cu and Ni are extremely rapid diffusers and cannot be retained in electrically active form even by rapid quenching of diffused samples [42]. Even in complexes, however, the defects need not have "dangling bonds," in the usual sense. But even lacking "broken bonds," as Baraff has conjectured [43], the presence of a nearby hydrogen may change the nature of the electrical activity. For example, recent MNDO calculations by Singh et al. [44] have found that the last-filled energy level associated with a substitutional sulfur in silicon is lowered substantially in energy with a hydrogen at an adjacent C-site, the lowest-energy configuration for the hydrogen.

It is also clear that hydrogen can bond to a dangling bond that is created "chemically" by the hydrogen, what we will call a chemically-induced reconstruction. This can be seen in the great deal of recent work [45-62] which has shown that the electrical activity of the shallow acceptors (B, Al, Ga, In, Tl) can be neutralized by the association with atomic hydrogen, while the shallow donors (P, As, Sb) are unaffected by hydrogenation [45,50]. The hydrogen has been introduced in these experiments by releasing it from a water-related defect in the oxide layer of MOS capacitors subjected to avalanche injection or low-energy electron irradiation [45-47,55], injected into bulk silicon from a plasma [48-52,54,57-60], by electrolysis [36, 61], or from boiling water [62]! Of course, these shallow acceptors are tetrahedrally coordinated at substitutional sites, but the model that has emerged from these studies has the hydrogen interrupting one of the bonds between the acceptor and its neighboring atoms, so that the hydrogen forms an Si-H bond with that atom, leaving the acceptor bonded to three neighboring atoms, the model receiving confirmation from the infra-red absorption [54,57] associated with this Si-H bond, and the observation of its isotope effect [57]. But these studies have opened up a several questions. Johnson [57] and Pankove et al. [58] have carried out experiments which indicate that a positively-charged free hole is required to incorporate hydrogen into silicon (at least at low temperatures, e.g., 120°C). And Tavendale and co-workers [59,60,62] have shown that the active species (the one which neutralizes B acceptors in p-type silicon on its exposure to atomic hydrogen) drifts as a positively charged ion in the electric field of a depletion region of a Schottky diode, indicating that the species has a donor ionization level in the upper half of the band gap, not what is expected of atomic hydrogen, as discussed above. As Pearton [62] has noted, the state of the species in the depletion region is not an equilibrium state, and the diffusing species may involve hydrogen bound to another defect (such as suggested by Ma et al. [63,64] who have argued that a radiation-defect is involved in the neutralization mechanism). One must conclude that the understanding of the exact nature of the reaction between hydrogen and the acceptor atoms is incomplete.

A number of the well-known radiation-induced centers in silicon, which are known to involve broken bonds, are also neutralized by atomic hydrogen. For example, the A-center (oxygen+vacancy complex, E_c -0.18 eV), and divacancy level (E_c -0.23 eV) may be passivated [65]. Point defects produced by the Q-switched ruby-laser-annealing of both n-type [66] and p-type [67] Si surfaces are neutralized to the melt depth of ~ 1 μm by plasma exposures of just 10 minutes at 100°C. And there has been a lot of work on the passivation of the electrical activity associated with dislocations (and their attendant point defects) and grain boundaries [66-76], which is so helpful in making photovoltaic solar cells from polycrystalline materials. While these defects may involve broken bonds, it should be remembered that these same defects also exhibit reconstruction, i.e., the lattice can distort to permit (at least a partial) bonding to occur; that is, some electrical activity may persist (i.e., have a level in the forbidden gap), but some of the reconstructed bonds may be in the valence band. Passivation implies that the hydrogen interacts with the level in the forbidden gap rendering that level inactive. But it should be realized that hydrogen may also interact with the inactive reconstructed bonds, altering the structure, i.e., we may have a chemically driven reconstruction, besides having hydrogen interact with a dangling bond. (Such a reconstruction is presumably what is associated with hydrogen-induced interfacial trapping states in diodes [77-79].) This type reconstruction may be seen by considering the infra-red absorption associated with hydrogen in silicon.

Hydrogen introduces into silicon a number of infra-red bands, the stretching, wagging and bending modes [6-16] of silicon-hydrogen bonds. We will be concerned only with the stretching modes. Comparison of these bands to those of the silanes [80,81] indicate that Si-H, $Si\text{-}H_2$, and $Si\text{-}H_3$ configurations exist in crystalline silicon, i.e., a single silicon with one, two, and three hydrogens. But there is considerable knowledge available concerning multi-vacancy defects [82-83], and hydrogenation of these defects yields only Si-H configurations. This can be seen in Fig. 2 where we show the divacancy saturated with hydrogens $(V_2 \cdot H_6)$. But an $Si\text{-}H_2$ configuration can be obtained [84] by partially dissociating the divacancy, as shown in Fig. 3, yielding V_2^* which takes eight hydrogens to saturate, $(V_2^* \cdot H_8)$. Since the partial dissociated configuration is the saddle-point for divacancy motion, we know that partial dissociation costs ~ 1.3 eV in energy [85], but the Si-H bond energy is ~ 3 eV, so that

$$(V_2 \cdot H_6) + 2 \cdot H \Leftrightarrow (V_2^* \cdot H_8) + 4.7 \text{ eV}. \qquad (3)$$

A similar partial dissociation of the trivacancy can yield [84] an $Si\text{-}H_3$ configuration, which is also the smallest of the {111} micro-cracks which may occur in hydrogenated silicon, and may be the micro-cleavage cracks needed by Wolf [86] to explain the properties of hydrogen-doped silicon. Lu et al. [87] found, however, that neutron transmutation doping of floating zone silicon grown in a hydrogen atmosphere did not change the micro-hardness. But there are more stretching mode IR bands than can be accounted for by these simple Si-H, $Si\text{-}H_2$, and $Si\text{-}H_3$ configurations. Lucovsky [88,89] has used the electro-negativity of atoms neighboring the silicon of the vibrating unit to explain the differences in vibrating units in amorphous silicon, and this interpretation has been applied to crystalline silicon and extended by quantum calculations [90], which show that the nature of the bond of the silicon of the vibrating unit (i.e., its s- or p-character) determines the

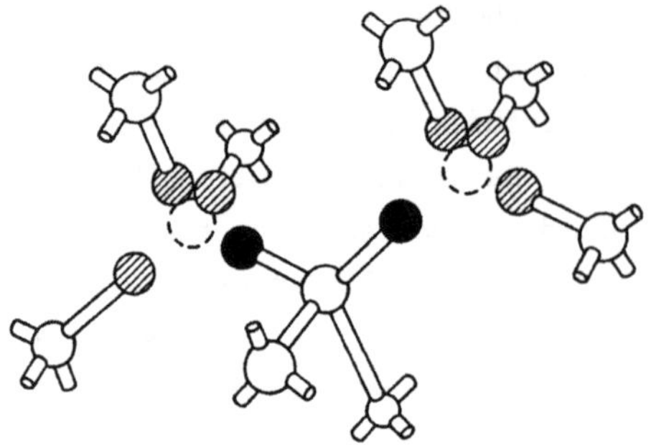

Fig. 2 The divacancy saturated with hydrogens.

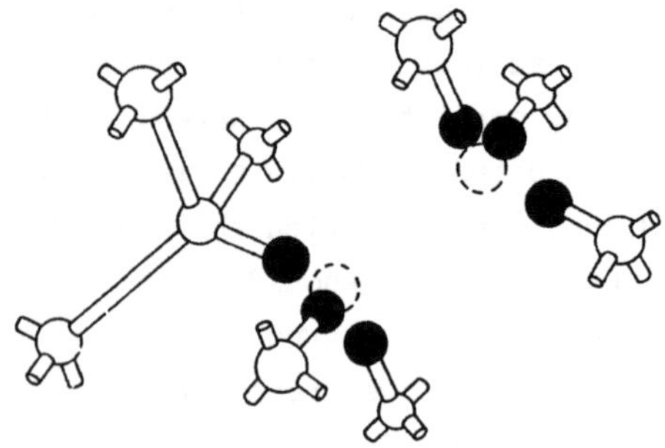

Fig. 3 A partially dissociated divacancy saturated with hydrogens.

frequency of the stretching vibration, i.e., the nature of the chemical environment, and the structural (stress) environment can alter the frequency. In fact, anything that influences the nature of the bonds to that silicon, influences the frequency of the vibrating unit. This recognition provides an abundance of models for vibrating units, and is being confirmed by emerging results [13,14,16]. The use of hydrogen to decorate defects then proves to be a powerful tool in the study of defects, but further work is required before a full understanding exists of the influence of hydrogen on silicon [90]. For example, there are the so-called "hydrogen donors" which are observed in hydrogen implanted silicon [91-94]. How can hydrogen give rise to donors? A possible explanation for the appearance of such donors is that they are donors due to oxygen precipitation which has occured due to the radiation-enhanced diffusion of the oxygen, such as was observed by Pflueger et al. [95] in neutron-transmutation-doped silicon, i.e., the donors may be only indirectly due to hydrogen; but since hydrogen can passivate oxygen donors [96], this explanation needs more study.

Clearly hydrogen is useful in decorating surfaces within crystals, and has been used in amorphous silicon in comparable ways [88,89,97,98] (albeit with less structural resolution), including exciting recent work on multi-layered amorphous superlattices [99]; the prospect of comparable work on crystalline superlattices we also find exciting.

This brings us to consider external surfaces. There has been an enormous amount of experimental work done studying the complicated reconstructed surface structures which occur under ultra-high vacuum conditions on the silicon {100}, {110} and, in particular, {111} surfaces, and these structures have spawned a myriad of models. This work has been reviewed extensively [100-105]. The properties of these surfaces have been studied with an extraordinary variety of techniques: Auger electron spectroscopy, low-energy electron diffraction [106], electron energy loss spectroscopy [107,108], low energy ion scattering [109], ion channeling [110,111], nuclear reaction analysis [112], atom scattering [113], photoemission [114], noble gas physisorption [115], electron tunneling spectroscopy [116,117], infra-red [118-123], photo-stimulated desorption [124], ion neutralization and ultraviolet photoemission spectroscopy [125-129], and field-ion microscopy [130]. It is not clear, to us at least, that there is a consensus on structure of all these surfaces, but our purpose is to discuss the role of hydrogen in these studies. It has been found that Si-H, $Si-H_2$, and $Si-H_3$ can exist on hydrogenated surfaces, and that there can be sites with varying degrees of binding [114, 124]. Infra-red studies can clearly give information on the structure of sites which bind hydrogen, including revealing special sites [123]. It is also clear that hydrogenation can alter the structure of the surface, once again revealing chemically-driven reconstruction. And finally we should note that such surface experiments (and therefore the structure that they reveal) are very sensitive to impurities which may deposit on the surface even under ultra-high vacuum conditions (for example, Culbertson et al. [112] found even their "clean" surfaces may have 0.2 monolayer of hydrogen), or which may come from the sample (for example, Binnig et al. [117] found boron segregation to the surface).

DEFECT CONFIGURATIONS: OTHER MATERIALS

As has been reviewed recently by Haller [131], a variety of hydrogen-related shallow- and deep-level defects has been observed [132-134] in Ge. It is beyond the scope of this review to detail the experiments leading to the identification and modeling of the properties of these centers, but for the shallow level acceptors and donors, a model of the light interstitial hydrogen trapped at and tunnelling around a heavier substitutional impurity (such as Cu, C, Si or O) is fully consistent with experimental results. The incorporation of hydrogen in these shallow centers can be monitored directly by observing an isotope shift in the ground state energy of the center when growing a crystal under a deuterium atmosphere [135]. Uniaxial stress splitting of the ground state of the defect gives information on the symmetry of the centers when combined with the high-resolution Fourier-transform infrared spectroscopy technique [136,137] on high-purity material. This method is capable of great sensitivity, and combined with its high resolution may prove valuable in the search for hydrogen-related electrically active centers in Si and other materials.

A large number of defects and impurities in Ge have been observed to be neutralized by reaction with atomic hydrogen. The most troublesome impurity in Ge is undoubtedly the triple acceptor Cu, which is an extremely rapid diffuser; the passivation of the various Cu centers after exposure to a hydrogen plasma has been found [138] - the passivation mechanism is assumed to be conversion of all the defects to Cu-H , which is electrically neutral. For a 3 hr, 300°C treatment the neutralization depth in Ge is ~80 μm, compared to ~15 μm for a comparable treatment of defects in Si. Similar results are obtained for another rapidly diffusing acceptor defect, Ni [139], which has the property of gettering oxygen in Ge.

An examination of the effects of hydrogen on a variety of other defects in Ge revealed that they could be divided into three classes - those which were either weakly or strongly neutralized to the depth of the hydrogen incorporation, and those which were unaffected by the hydrogen plasma heat treatment [140]. This behaviour had immediate benefits in differentiating between defects with similar activation energies and electron capture cross-sections [140]. It was also speculated that the incorporation of the mono-atomic hydrogen near metal impurities in Ge may act to relieve some of the lattice strain induced by these impurities, which is then observed as a reduction in electrical activity. Two hydrogen-related electron traps were created in Te-doped p-type Ge after plasma exposure, which is not surprising as hydrogen is known to be involved in a number of Cu-related centers in p-type Ge [135].

The passivation of two deep-level electron traps induced in n-type Ge by a Q-switched ruby laser anneal was found [141]. As with the corresponding case in Si, plasma exposures of 10 minutes at 100°C were sufficient to neutralize the electrical activity of such centers. In Ge most radiation damage or quenched-in centers that have been found are vacancy-related, and again the propensity of hydrogen to neutralize this type of defect is seen by the effectiveness of hydrogen plasma exposures in passivating Co-60 γ-induced hole traps [142], quenched-in acceptors [143], and ion-implanted oxygen-related deep levels [144].

As with Si, the electrical activity of dislocations in Ge can be neutralized by reaction with hydrogen [145]. The difference in Ge is that the electrical levels associated with dislocations are hole traps, rather than electron traps as in Si. Interesting topological features are observed after preferential etching of hydrogen plasma heated Ge surfaces. The chemically etched surface of a dislocated wafer from a crystal grown in a molecular hydrogen atmosphere displays only large conical etch pits related to rapid etching of the strained region around the dislocations [135]. It is assumed that these dislocations act as precipitation sites for the hydrogen during cooling of the crystal after growth. In dislocation-free Ge, however, (a potentially useful substrate for epitaxial GaAs growth) there are no sinks for the hydrogen, and it precipitates uniformly throughout the wafer. Preferential etching of the surface of such a wafer reveals a high concentration of small shallow pits ascribed to hydrogen precipitates [135]. The contrast between a dislocated and a dislocation free Ge surface is thus quite stark. If one exposes a dislocated Ge wafer to a hydrogen plasma at < 400°C, and then uses a preferential etchant, one observes both the smooth shallow pits and the large conical pits which are reduced in size by up to a factor of two [146]. This latter phenomenon is ascribed to hydrogen relieving the strain around dislocations and hence reducing the etch rate. The smooth etch pits are, however, direct evidence for a tremendous amount of hydrogen precipitation during the plasma exposure, and this has important consequences for the diffusivity of hydrogen at low temperatures.

The relatively high concentrations of defects in even the purest GaAs available makes the promise of a low temperature neutralization method even more important for this material. The initial experiments were performed on bulk n-type GaAs, containing two prominent donor levels (E_c-0.36 eV), (E_c-0.70 eV) [147]. Only the (E_c-0.36 eV) level reacted with atomic hydrogen. At low temperatures (< 200°C), the diffusion coefficient of hydrogen does not appear to be strongly temperature dependent in GaAs, however experiments utilizing high purity n-type LPE material showed [148] that a 3 hour exposure at 300°C neutralized electron traps to a depth of ~9 μm. [148]. In this material all five of the deep level centers present were neutralized. Pearton et al. [149] found that neutralization of defects was slightly more pronounced for defects in GaAs than in GaP.

In typical undoped, LEC grown, bulk GaAs the main defect is the so-called EL2 center, which has been tentatively identified as consisting of a family of defects related to antisite defects (As on a Ga site). The EL2 center is typically present at high concentrations, and, as a deep double donor, it compensates the shallow level impurities present, creating semi-insulating material provided the crystal is not intentionally doped with, say, silicon. It has recently been shown that EL2 can be neutralized with atomic hydrogen [150], though it is not clear that this will be useful from a device point of view, as the presence of EL2 negates the need for the unreliable Cr doping process in producing semi-insulating GaAs.

Infrared studies have been begun in crystalline germanium and GaAs [151], and the characteristic stretching bands observed. And infrared studies have been done in amorphous and recrystallized germanium [152] and in amorphous-hydrogenated GaAs [153]. These results are promising.

Indeed all the work on other systems must be viewed as promising, but even more incomplete than that in silicon.

DIFFUSION

There has been considerable confusion concerning hydrogen diffusion in semiconductors, and it is now clear that the presence of traps, and the formation of hydrogen molecules (self-trapping) greatly complicates the problem. We do not have the space to discuss all the data and the issues, but we can make some observations. The coupled set of equations for the problem with immobile $[H_2]$ and $[H\cdot d]$ follows:

$$\partial_t [H] = D_H \partial_x^2 [H] - 2 K_m [H]^2 + 2 \Delta_m [H_2] - K_d [d][H] + \Delta_d [H\cdot d], \quad (4)$$

$$\partial_t [H_2] = K_m [H]^2 - \Delta_m [H_2], \quad (5)$$

$$\partial_t [H\cdot d] = K_d [d][H] - \Delta_d [H\cdot d], \quad (6)$$

where we have used $\partial_t = \partial/\partial t$, $\partial_x = \partial/\partial x$. One can sum these equations and get

$$\partial_t [H_T] = D_H \partial_x^2 [H], \quad (7)$$

with

$$[H_T] = [H] + 2[H_2] + [H\cdot d]. \quad (8)$$

But Eq. (7) is not a proper diffusion equation. What is required is the solution of the coupled Eqs. (4)-(6), a project that we have not completed.

Following Shi et al. [154], we can, however, get some insight by some approximations. Consider first the case with a large concentration of traps [d], and no molecule formation. Assume an approximate local equilibrium,

$$[H\cdot d] \Delta_d \simeq K_d [d] [H]. \quad (9)$$

Let $\eta = K_d [d]/ \Delta_d$. Then because of Eq. (9)

$$[H_T] = (1 + \eta) [H], \quad (10)$$

which may be differentiated yielding with Eq.(7)

$$(1 + \eta) \partial_t [H] = D_H \partial_x^2 [H], \quad (11)$$

a proper diffusion equation with an effective diffusion coefficient, D_{eff},

$$D_{eff} = \{1/(1+\eta)\} D_H = \{[H]/([H]+[H\cdot d])\} D_H, \quad (12)$$

i.e., a normal diffusion equation with D_{eff} reflecting the fraction of hydrogen free to diffuse. Solutions to Eq. (11) are standard error functions; the only steady-state is equilibrium. On the other hand suppose that the dissociation of [H·d] is ignorable. Then Eq. (4) can have a (quasi) steady state satisfying

$$D_H \partial_x^2 [H] = K_d [d][H] \quad (13)$$

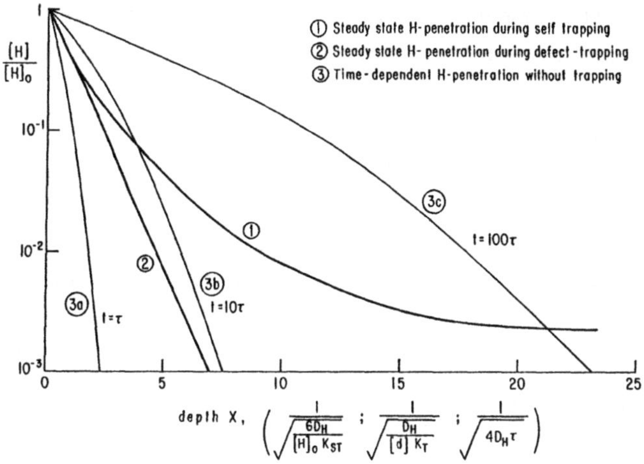

Fig. 4 Diffusion profiles in c-Si for the steady-state approximation for (1) molecule formation, and (2) defect-trapping. Case (3) shows the time-dependence of normal diffusion. The depth is X and diffusion time τ.

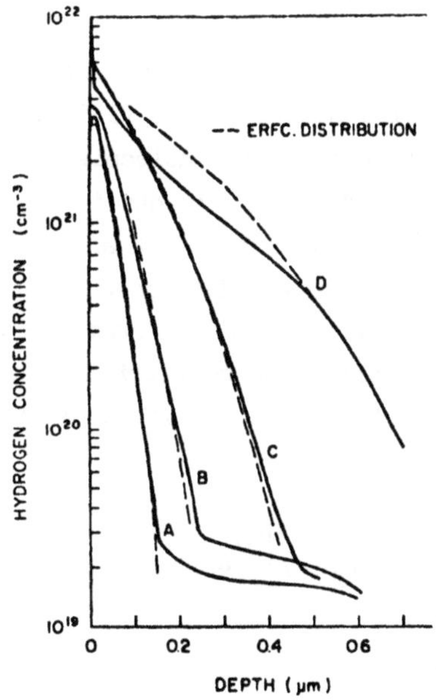

Fig. 5 Hydrogen profiles after hydrogenation of a-Si in a plasma. [155]. Note the drop below the erfc distribution in curve D.

with the solution (see Fig. 4)

$$[H] = [H_o] \exp [-(K_d[d]/D_H)^{\frac{1}{2}} x], \tag{14}$$

with $[H_o]$ the surface concentration of hydrogen (assumed constant).

We can follow the comparable analysis [154] for the case for the formation of molecules and no trapping. Again, assume local equilibrium,

$$[H_2]\ \Delta_m \simeq K_m\ [H]^2. \tag{15}$$

Let $\xi = K_m / \Delta_m$. Then Eq. (8) becomes

$$[H_T] = [H] + 2\ [H_2] = [H] + 2\ \xi\ [H]^2. \tag{16}$$

and

$$\partial_t\ [H_T] = \partial_t\ [H] + 4\xi\ [H]\ \partial_t\ [H] = (1 + 4\xi\ [H]\ )\ \partial_t\ [H] = D_H\ \partial_x^{\ 2}\ [H]. \tag{17}$$

Now if one casts this in the form of a diffusion equation, then

$$D_{eff} = \{1+/(1+4\xi\ [H])\}\ D_H, \tag{18}$$

and the diffusion coefficient explicitly contains [H]. And further,

$$D_{eff} = \{[H]/([H]+4[H_2]\}\ D_H, \tag{19}$$

i.e, this differs from the intuitive result by an extra factor of two. Solutions to Eq. (17) are not simple error functions. The only steady state is equilibrium. If we assume that molecule break-up is ignorable, then Eq. (4) can have a (quasi) steady state satisfying

$$D_H\ \partial_x^{\ 2}\ [H] = 2\ K_m\ [H]^2 \tag{20}$$

which has the solution (see Fig. 4)

$$\frac{1}{[H]} - \frac{1}{[H_o]} = \left(\frac{1}{3}\right)\left(\frac{K_m}{D_H}\right)\ x^2. \tag{21}$$

We can compare the steady-state profiles discussed above to experiments. In Fig. 5 we see a profile from Widmer et al. [155] in amorphous silicon, suggestive of the steady-state profile for molecule formation. Molecule formation has also been indicated in experiments in crystalline silicon where near surface concentrations much higher than the solubility have been found [51,57,156, 157]. We can also see data on the diffusion coefficient in Fig. 6, where the diffusion coefficient drops from the value for atomic hydrogen much as an effective diffusion coefficient would, but then does not continue to drop; the interpretation of the latter trend is not clear.

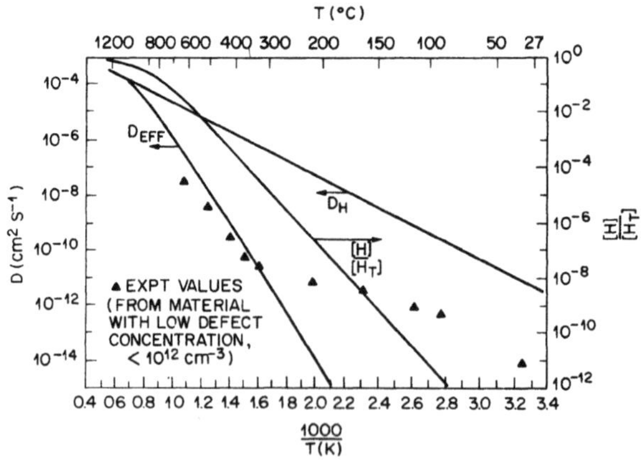

Fig. 6 Diffusion coefficients observed for hydrogen in c-Si. [62].

The point we would like to emphasize on diffusion, is that the details of the profiles must be studied, and more data obtained before we fully understand the diffusion of hydrogen in silicon. And what data there is on diffusion in other semiconductors, suggests that there is comparable complexity.

Acknowledgement: This research was supported in part by the Jet Propulsion Laboratory, California Institute of Technology (Flat Panel Solar Array Project) and the Mobil Foundation.

REFERENCES

1. A. Van Wieringen and N. Warmoltz, Physica 22:849 (1956).
2. S.T. Picraux and F.L. Vook, Phys. Rev. B14:1593 (1976).
3. S.T. Picraux and F.L. Vook, Phys. Rev. B18:2066 (1978).
4. S.T. Picraux, F.L. Vook, and H.J. Stein in :"Defects and Radiation Effects in Semiconductors, 1978," ed. J.H. Albany (Inst. Phys. Bristol, 1979) p. 31.
5. J.A. Appelbaum, D.R. Hamann, and K.H. Tasso, Phys. Rev. Lett. 38:1478 (1977).
6. H.J. Stein, J. Electron. Mater. 4:159 (1975).
7. N.N. Gerasimenko, M. Rollé, L.-J. Cheng, Y.H. Lee, J.C. Corelli, and J.W. Corbett, Phys Stat. Sol. (b) 90::689 (1978).
8. H.J. Stein, Phys. Rev. Lett. 43:1030 (1979).
9. B.N. Mukashev, K.H. Nussupov, and M.F. Tamendarov, Phys. Lett. 72A:381 (1979).
10. J. Tatarkiewicz, and K. Wieteska, Phys. Stat. Sol. (a) 66:K101 (1981).
11. S.-F. Cui, Z.-H. Mai, P.-W. Ge, and D.-Y. Sheng, Kexue Tongbao, 27:382 (1982).
12. M. Cardona, Phys. Stat. Sol. (b) 118:463 (1983).
13. G.-R. Bai, M.-W. Qi, L.-M. Xie, and T.-S. Xie, Sol. State. Comm. in press (1985).
14. T.-S. Shi, L.-M. Xie, G.-R. Bai, and M.-W. Qi, Phys. Stat. Sol., in press (1985).
15. B.-Y. Gu, Z.-Y. Xu, and P.-W. Ge, to be published.
16. M.-W. Qi, G.-R. Bai, T.-S. Shi, and L.-M. Xie, Mater. Lett. in press.
17. S.-F. Cui, P.-W. Ge, Y.-Q. Zhao, L.-S. Wu, Acta Phys. Sinica, 28:791 (1979).
18. V.A. Singh, C. Weigel, J.W. Corbett, and L.M. Roth, Phys. Stat. Sol. (b) 81:637 (1977).
19. R. Hoffman, J. Chem. Phys. 39:1397 (1963).
20. J. Wang and C. Kittel, Phys. Rev. B7:713 (1973).
21. V.A. Singh, J.W. Corbett, C. Weigel, and L.M. Roth, Phys. Lett. 65A:261 (1978).

22. G.G. DeLeo, W.B. Fowler, and G.D. Watkins, Phys. Rev. B29:1819 (1984).
23. J.W. Corbett, S.N. Sahu, T.S. Shi, and L.C. Snyder, Phys. Lett. 93A:303 (1983).
24. T.-S. Shi, S.N. Sahu, J.W. Corbett, and L.C Snyder, Scientia Sinica (A) 27:98 (1984).
25. M.J.S. Dewar and W. Thiel, J. Amer. Chem. Soc. 99:4899 (1977).
26. L.C. Snyder, J.W. Moskowitz, and S. Topiol, Int. J. Quantum Chem. 21:565 (1982).
27. L.C. Snyder, J.W. Moskowitz, and S. Topiol, Phys. Rev. B26:6727 (1982).
28. S.J. Pearton, J.M. Kahn, W.L. Hansen, and E.E. Haller, J. Appl. Phys. 55:1461 (1984).
29. R.N. Hall, IEEE Trans. Nucl. Sci. NS-31:320 (1984).
30. R.N. Hall in "Thirteenth Int'l. Conf. on Defects in Semiconductors," eds. L.C. Kimerling and J.M. Parsey, Jr. (AIME, NY, 1985) p.759.
31. A. Mainwood and A.M. Stoneham, Physica 116B:101 (1983).
32. T.-S. Shi, S.N. Sahu, A.G. Sganga, J.W. Corbett and L.C. Snyder, to be published.
33. A. Rohatgi, R.H. Hopkins, J.R. Davis and R.B. Campbell, Solid State Electron. 23:1185 (1980).
34. D. Brown and P. U. Gray, J. Electrochem. Soc. 115:670 (1968).
35. S.J. Pearton and A.J. Tavendale, Phys. Rev. B26:1105 (1982).
36. S.J. Pearton, W.L. Hansen, E.E. Haller, and J.W. Kahn, J. Appl. Phys. 55:1221 (1984).
37. A. Mogro-Campero, R.P. Love and R. Schubert, J. Electrochem. Soc., (1985) in press.
38. S.J. Pearton and E.E. Haller, J. Appl. Phys. 54:3613 (1983).
39. S.J. Pearton and A.J. Tavendale, J. Appl. Phys. 54:1375 (1983).
40. S.J. Pearton and A.J. Tavendale, J. Phys. C (in press).
41. A.J. Tavendale and S.J. Pearton, J. Phys. C16:1665 (1983).
42. See the review paper by E.R. Weber, Appl. Phys. A 30, 1 (1983).
43. G.A. Baraff, private communication.
44. R.K. Singh, S.N. Sahu, J.W. Corbett, and L.C. Snyder, to be published.
45. C.T. Sah, J.Y.C. Sun, and J.J.T. Tzou, J. Appl. Phys. 54:4378 (1983).
46. C.T. Sah, J.Y.C. Sun, and J.J.T. Tzou, Appl. Phys. Lett. 43:204 (1983).
47. C.T. Sah, J.Y.C. Sun, J.J.T. Tzou, and S.C.S. Pan, Appl. Phys. Lett. 43:962 (1983).
48. J.I. Pankove, D.E. Carlson, J.E. Berkeyheiser, and R.O. Wance, Phys. Rev. Lett. 51:2224 (1983).
49. W.L. Hansen, S.J. Pearton, and E.E. Haller, Appl. Phys. Lett. 44:608 (1984).
50. J.I. Pankove, R.O. Wance, and J.E. Berkeyheiser, Appl. Phys. Lett. 45:1100 (1984).
51. N.M. Johnson and N.D. Moyer, Appl. Phys. Lett. 46:787 (1985).
52. J.C. Mikkelsen, Jr., Appl. Phys. 46:882 (1985).
53. G.G DeLeo and W.B. Fowler, Phys. Rev. B31:6861 (1985).
54. J.I. Pankove, P.J. Zanzucchi, C.W. Magee, and G. Lucovsky, Appl. Phys. Lett. 46:421 (1985).
55. C.-T. Sah, S.C.-S. Pan, and C.C.-H. Hsu, J. Appl. Phys. 57: 5148 (1985).
56. G.G. DeLeo and W.B. Fowler, J. Electron. Mater. 14a:745 (1985).
57. N.M. Johnson, Phys. Rev. B 31:5525 (1985).
58. J.I. Pankove, C.W. Magee, and R.O. Wance, to be published (1985).
59. A.J. Tavendale, A.A. Williams, and D. Alexiev, Appl. Phys. Lett, in press.
60. A.J. Tavendale, A.A. Williams, D. Alexiev, and S.J. Pearton, to be published.
61. G.S. Oehrlein, J.L. Lindström, and J.W. Corbett, Phys. Lett. 81A:246 (1981).
62. S.J. Pearton, private communication.
63. M.R. Chin and T.P. Ma, Appl. Phys. Lett. 42:883 (1983).
64. C.C. Wei and T.P. Ma, Appl. Phys. Lett. 45:900 (1984).
65. S.J. Pearton, Phys. Stat. Sol. A72:K73 (1982).
66. J.L. Benton, C.J. Doherty, S.D. Ferriss, D.L. Flamm, L.C. Kimerling and H.J. Leamy, Appl. Phys. Lett. 36:670 (1980).
67. E.M. Lawson and S.J. Pearton, Phys. Stat. Sol. A 72:K155 (1982).

68. B. Pohoryles, Phys. Stat. Sol. A67:K75 (1981).
69. C.H. Seager, D.J. Sharp, J.K.G. Panitz, and J.I. Hanoka, J. de Phys. Colloq. C1 43:C1 (1982).
70. K. Noguchi, T. Saito, A. Shimura, and S. Ichikawa, Jpn. J. Appl. Phys. Suppl. 2 21:39 (1982).
71. W. Schmidt, K.D. Rasch, and K. Roy, in "16th IEEE Photovoltaic Specialists Conf." (IEEE, NY 1982) 357.
72. C.H. Seager, D.J. Sharp, J.K.G. Panitz, and R.V. D'Aiello, J. Vac. Sci. Technol. 20:430 (1982).
73. J.I. Hanoka, C.H. Seager, D.J. Sharp, and J.K.G. Panitz, Appl. Phys. Lett. 42:618 (1983).
74. S.J. Pearton and E.E. Haller, J. Appl. Phys. 54:3613 (1983).
75. C. Dubé, J.I. Hanoka, and D.B. Sandstrom, Appl. Phys. Lett. 44:425 (1984).
76. C. Dubé and J.I. Hanoka, Appl. Phys. Lett. 45:1135 (1984).
77. B. Keramati and J.N. Zemel, J. Appl. Phys. 53:1091 (1982).
78. B. Keramati and J.N. Zemel, J. Appl. Phys. 53:1100 (1982).
79. M.C. Petty, Electronics Lett. 18:314 (1982).
80. H.S. Gutowsky and E.O. Stejskal, J. Chem. Phys. 22:939 (1954).
81. G.J. Janz and Y. Mikawa, Nippon Kagakukai, 34:1405 (1961).
82. J.W. Corbett, R.L. Kleinhenz, E. Wu, and Z.P. You, J. Nucl. Mater. 108-109:617 (1982).
83. J.W. Corbett, J.C. Corelli, U. Desnica, And L.C. Snyder in "Microscopic Identification of Electronic Defects in Semiconductors," eds. N. Johnson, S. Bishop, and G.D. Watkins (Mater. Res. Soc. NY 1985) in press.
84. T.S. Shi, S.N. Sahu, G.S. Oehrlein, A. Hiraki, and J.W. Corbett, Phys. Stat. Sol. A74:329 (1982).
85. L.-J. Cheng, J.C. Corelli, J.W. Corbett, and G.D. Watkins, Phys. Rev. 152:761 (1966).
86. E. Wolf, Phys. Stat. Sol. (a) 70:K59 (1982).
87. C.-G. Lu, Y.-X. Li, C.-T. Sun, and J.-H. Yin in "Neutron Transmutation Doping of Semiconductor Materials," ed. R.D. Larrabee (Plenum Press, NY, 1984) p. 193.
88. G. Lucovsky, G., Sol. State Commun. 29:571 (1979).
89. G. Lucovsky, R.J. Nemanich, and J.C. Knights, Phys. Rev. 19:2064 (1979).
90. S.N. Sahu, T.S. Shi, P.W. Ge, A. Hiraki, T. Imura, M. Tashiro, V.A. Singh, and J.W. Corbett, J. Chem. Phys. 77:4330 (1982).
91. Y. Zohta, Y. Ohmura, and M. Kanagawa, Jpn J. Appl. Phys. 10:532 (1971).
92. Y. Ohmura, Y. Zohta, and M. Kanagawa, Sol. state. Commun. 11:263 (1972).
93. Y. Gorelkinskii and N.N. Nevinnyi, Nucl. Inst. Meth. 209:677 (1983).
94. Z.Y. Wang and L.Y. Lin in "Neutron Transmutation Doping of Semiconductor Materials," ed. R.D. Larrabee (Plenum, NY, 1983) p. 311.
95. R.J. Pflueger, J.C. Corelli, and J.W. Corbett, Phys. Stat. Sol. (a) (1985) in press.
96. S.J. Pearton, A.M. Chantre, L.C. Kimerling, K.D. Cummings, and W.C. Dautremont-Smith, to be published.
97. H. Wieder, M. Cardona, and C.R. Guarnieri, Phys. Stat. Sol. (b) 92:99 (1979).
98. A. Hiraki, T. Imura, K. Mogi, and M. Tashiro, J. de Phys. Suppl. C4 42:C4-277 (1981).
99. B. Abeles, L. Yang, P.D. Persans, H.S. Stasiewski, and W. Lanford, to be published.
100. W. Mönch, Surf. Sci. 86:672 (1979).
101. D.E. Eastman, J. Vac. Sci. Technol. 17:492 (1980).
102. D.J. Chadi, Surf. Sci. 99:1 (1980).
103. D.J. Chadi, J. Phys. Soc. Jpn. Suppl. A 49:1035 (1980).
104. D.J. Miller and D. Haneman, Surf. Sci. 104:L237 (1980).
105. D. Haneman, Adv. in Phys. 31:165 (1982).
106. E.G. McRae, Surf. Sci. 124:106 (1983).
107. J.E. Rowe and H. Ibach, Surf. Sci. 48:44 (1975).
108. H. Wagner, R. Butz, U. Backes, and U. Bruckmann, Sol. State, Comm. 38:1155 (1981).
109. M. Aono, R. Souda, C. Oshima, and Y. Ishizawa, Phys. Rev. Lett. 51:801 (1983).

110. R.J. Culbertson, L.C. Feldman and P.J. Silverman, Phys. Rev. Lett. 45:2043 (1980).
111. R.M. Tromp, E.J. van Loenen, M. Iwami, and F.W. Saris, Sol. State Comm. 44:971 (1981).
112. R.J. Culbertson, L.C. Feldman, P.J. Silverman, and R. Haight, J.. Vac. Sci. Technol. 20:868 (1982).
113. M.J. Cardillo, Phys. Rev. B23:4279 (1981).
114. K. Fujiwara, Phys. Rev. B24:2240 (1981).
115. E.H. Conrad and M.B. Webb, Surf. Sci. 129:37 (1983).
116. G. Binnig, H. Rohrer, Ch. Gerber, and E. Weibel, Phys. Rev. Let.. 50:120 (1983).
117. G. Binnig, H. Rohrer, F. Salvan, Ch. Gerber, and A. Baro, to be published.
118. G.E. Becker and G.W. Gobeli, J. Phys. Chem. 38:1 (1963).
119. J.A. Applebaum and D.R. Hamann, Phys. Rev. Lett. 31:806 (1975).
120. J.A. Applebaum and D.R. Hamann, Phys. Rev. B15:2006 (1977).
121. G. Lucovsky, J. Vac. Sci. Technol. 16:1225 (1979).
122. T.-S. Shi, S.N. Sahu and J.W. Corbett, Surf. Sci. Lett. 130:L289 (1983).
123. Y.J. Chabal, Phys. Rev. Lett. 50:1850 (1983).
124. M.L. Knotek, G.M. Loubriel, R.H. Stulen, C.E. Parks, B.E. Koel, Z. Hussain, Phys. Rev. B26:2292 (1982).
125. T. Sakurai and H. D. Hagstrum, Phys. Rev. B12:5349 (1975).
126. K.C. Pandey, T. Sakurai, and H.D. Hagstrum, Phys. Rev. Lett. 35:1728 91975).
127. T. Sakurai and H.D. Hagstrum, J. Vac. Sci. Technol. 13:807 (1976).
128. T. Sakurai, M.J. Cardillo, and H.D. Hagstrum, J. Vac. Sci. Technol. 14:397 (1976).
129. T. Sakurai and H.D. Hagstrum, Phys. Rev. B14:1593 (1976).
130. T. Sakurai, E.W. Müller, R.J. Culbertson, and A.J. Melmed, Phys. Rev. Lett. 39:578 (1977).
131. E.E. Haller in "Microscopic Identification of Electronic Defects in Semiconductors," eds., N. Johnson, S. Bishop, and G.D. Watkins (Mat. Res. Soc., NY 1985) in press.
132. R.N. Hall in " Lattice Defects in Semiconductors-1974," ed. F.A. Huntley (Inst. Phys., London, 1975) p. 190.
133. E.E. Haller, R. Joós, and L.M. Falicov, Phys. Rev. B21:4729 (1980).
134. B. Joós, E.E. Haller, and L.M. Falicov, Phys. Rev. B22:832 (1980).
135. E.E. Haller, W.L. Hansen and F.S. Goulding, Adv. Phys. 30:193 (1981).
136. Sh.M. Kogan, Sov. Phys. Semicond. 1:828 (1973).
137. R.J. Bell, "Introductory Fourier Transform Spectroscopy" (Academic Press, N.Y., 1972).
138. S.J. Pearton, Appl. Phys. Lett. 40:253 (1982).
139. A.J. Tavendale and S.J. Pearton, J. Appl. Phys. 54:1156 (1983).
140. S.J. Pearton and A.J. Tavendale, J. Appl. Phys. 54:823 (1983).
141. S.J. Pearton and A.J. Tavendale, J. Appl. Phys. 54:440 (1983).
142. S.J. Pearton, AAEC E-Report 523 (1981).
143. A.J. Tavendale and S.J. Pearton, Rad. Effects 69:39 (1983).
144. A.J. Tavendale and S.J. Pearton, J. Appl. Phys. 54:3213 (1983).
145. S.J. Pearton and J.M. Kahn, Phys. Stat. Sol. A78:K65 (1983).
146. A.J. Tavendale and S.J. Pearton (unpublished).
147. S.J. Pearton, J. Appl. Phys. 53, 4509 (1982).
148. S.J. Pearton and A.J. Tavendale, Electronics Lett. 18:715 (1982).
149. S.J. Pearton, E.E. Haller, and A.G. Elliott, Electron. Lett. 19:1052 (1983).
150. J. Lagowski, M. Kaminska, J.M. Parsey, H.C. Gatos and M. Lichtensteiger, Appl. Phys. Lett. 41:1078 (1982).
151. P.W. Wang and J.W. Corbett, to be published.
152. D. Bermejo and M. Cardona, J. Non-Cryst. Sol. 32:421 (1979).
153. Z.P. Wang, L. Ley, and M. Cardona, Phys. Rev. B26:3249 (1982).
154. T.-S. Shi, A.G. Sganga, S.N. Sahu, D. Peak, and J.W. Corbett, to be published.

155. A.E. Widmer, R. Fehlmann, and C.W. Magee, J. Non-Cryst. Sol. 34:199 (1983).
156. A.E. Jaworoski and J.W. Corbett in "Proc. 13th Int'l. Conf. on Defects in Semiconductors," eds. L.C. Kimerling and J.W. Parsey, Jr. (AIME, NY, 1985) p. 767.
157. A.E. Jaworowski, L.S. Wielunski, and T.W. Listerman in "Microscopic Identification of Electronic Defects in Semiconductors," eds. N. Johnson, S. Bishop, and G.D. Watkins (Mater. Res. Soc. NY 1985) in press.

HYDROGEN PASSIVATION OF POLYCRYSTALLINE SILICON

Jack I. Hanoka

Mobil Solar Energy Corporation
16 Hickory Drive
Waltham, Massachusetts 02254

ABSTRACT

Defects in polycrystalline silicon sheet which are associated with high angle grain boundaries, twin boundaries, and intragranular dislocation arrays have all been shown to be amenable to hydrogen passivation. Using the EBIC technique, it has been shown that nearly all the recombination producing defects can be passivated down to depths of 10–20 µm and that passivation depths of several hundred microns can sometimes occur. A model is advanced whereby the key defects being passivated are dislocations and where deep passivation is effected through rapid thermal diffusion down dislocation arrays, either in intragranular regions or in the grain boundaries themselves. Hydrogen diffusivity down grain boundaries and twin boundaries has been measured to be 10^{-8} to 10^{-10} cm^2/sec. Diffusivity down intragranular dislocation arrays is $> 10^{-8}$ cm^2/sec. Surface recombination velocities, S, of the grain boundaries have been measured as a function of the passivation depth, X; and for $X < 35$ µm, $\ell n\ S \propto -X$. Enhanced EBIC contrast at T = 100°K has been interpreted as due to shallow electron traps within ~0.1 eV of the conduction band edge. Such shallow traps do not seem to be subject to passivation while presumed deeper lying recombination levels at the same spatial location do respond to hydrogen passivation.

INTRODUCTION

A feature common to all silicon sheet growth techniques is that the resultant material contains considerably more defects than does single-crystal Czochralski silicon. Such a higher concentration of defects is usually an inevitable outcome of the constraints imposed in growing sheets or ribbons [1,2]. As a result of such higher defect concentrations, solar cell efficiencies in sheet silicon materials are lower than that of single-crystal silicon. It has been recognized for several years that hydrogen passivation of defects in silicon may be the way to overcome this efficiency shortfall for polycrystalline sheet silicon [3,4]. In this paper, we will describe (1) defect properties of one form of silicon sheet, Edge-defined Film-fed Growth (EFG) ribbon [5], (2) passivation of many of these defects using atomic hydrogen, (3) characterization of the hydrogen passivation process using the electron beam induced current (EBIC) technique, and (4) a model for passivation in polycrystalline silicon.

EXPERIMENTAL

The passivation was done using a Kaufman ion source [6]. Such ion sources provide the ability to vary the ion beam energy (up to ~2 keV) and ion beam current density (up to several mA/cm^2) somewhat independently. Ultra high purity H_2 was used as the feed gas for the ion source. The ion beam energies used here were 1500-1700 eV, beam current densities were 1-2 mA/cm^2. Sample temperatures were difficult to measure accurately, but they were estimated to be 400-500°C during passivation. Passivation times were generally one to four minutes in duration. The silicon ribbon studied was p-type, $\rho = 5\ \Omega$-cm and had either a PH_3 diffused junction ~0.5 μm deep or a Schottky barrier formed by ~125 Å of evaporated Al. The EBIC work was performed using a Cambridge S250 SEM. For the low temperature EBIC studies, a cold stage with an LN_2 feed was employed. The EBIC work was done at 40 keV unless otherwise noted.

DEFECT PROPERTIES OF EFG RIBBON

In polycrystalline silicon, one can have point defects such as vacancies, interstitials, and impurities; line defects or dislocation; and finally, planar defects--twin boundaries or more generally, grain boundaries. Hydrogen has been shown to be capable of passivating defects in virtually all of these categories in silicon but here we will confine ourselves to the latter two categories. Pearton has recently summarized passivation of point defects [7]. We have good evidence to believe that the minority carrier diffusion length is generally determined by the dislocation content in many varieties of polycrystalline silicon sheet. Thus, emphasis will be given to dislocation passivation in this paper. Also, in the model which we will advance further on in this paper, grain boundary passivation is viewed as just a special case of dislocation passivation.

Figure 1 is an illustration of both the general phenomenon of hydrogen passivation in silicon ribbon and also of the utility of the EBIC technique in studying passivation. In this paper, we will not describe the general EBIC technique in detail; the reader is referred to recent review articles for that [8,9]. Suffice it to say that EBIC is an electron beam analogue of the photovoltaic effect and provides us with a recombination map or current collection map. A later section will describe, however, several specific applications of EBIC. In Fig. 1, the right-hand part of the sample has been masked and the unmasked portion exposed to the hydrogen plasma in the Kaufman ion source. The rather abrupt termination of the electrical activity of the dislocation arrays in the passivated portion dramatically shows that hydrogen passivation can be quite effective. In this particular instance, near-total passivation down to a depth of ~15 μm (the extent of the electron beam penetration) can be seen.

Figure 2(a) is an EBIC micrograph which shows more clearly the two prototypical defects in silicon ribbon, dislocations and grain boundaries. The darker, mostly linear boundaries are twin boundaries and grain boundaries and the less poorly resolved contrast between the boundaries is due to dislocations. The evidence that this intraboundary contrast is primarily due to dislocations is shown in Fig. 2(b) which is the same region as seen in Fig. 2(a) but has been etched to reveal dislocation etch pits. There are several important features to be noted in Figs. 2(a) and 2(b) which are relevant for some of the subsequent discussion. First of all, there is a rather clear correlation between the spatial distribution of the EBIC contrast and distribution of the dislocation etch pits. Also note that on closer examination, there are differences in the shapes of some of the etch pits, presumably indicative of underlying differences in dislocations, and there are differences in degree of EBIC contrast for different dislocations.

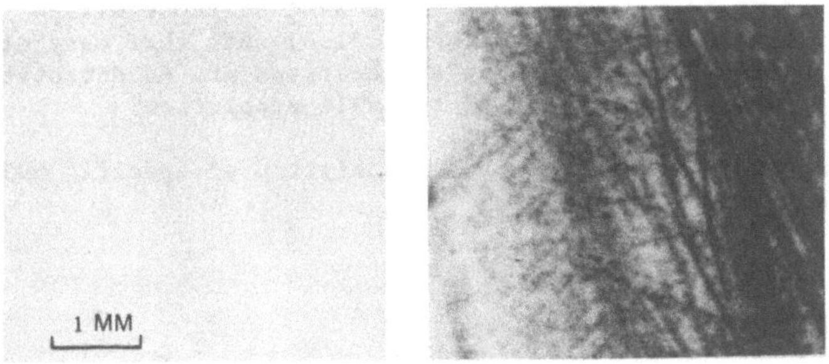

Fig. 1. EBIC micrograph showing the effects of hydrogen passivation. The left half of the sample was subjected to the H ion beam; the right half was masked from the beam.

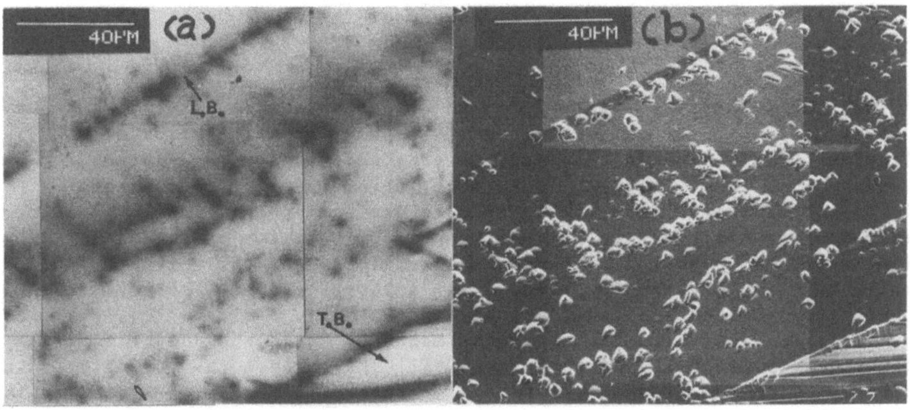

Fig. 2. (a) EBIC micrograph of ribbon silicon showing dislocation and grain boundary recombination; (b) the same region after etching to reveal dislocation etch pits.

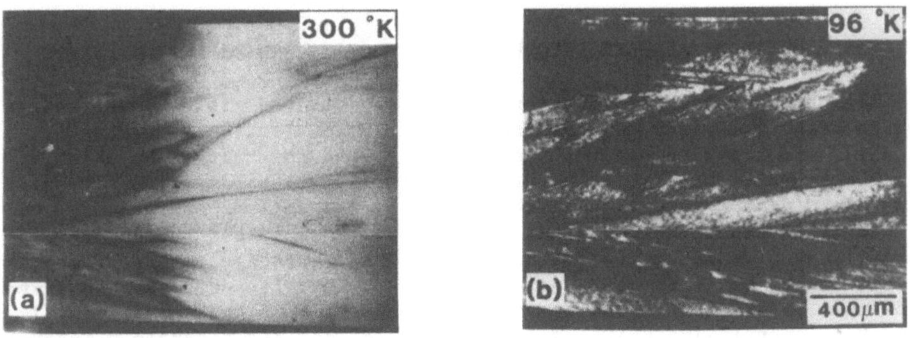

MASKED ↑ PASSIVATED

MASKED ↑ PASSIVATED

Fig. 3. EBIC micrographs showing both enhanced contrast and lack of passivation effects at low temperatures.

Note that what we term linear boundaries such as those marked "ℓ.b." in the figure generally contain dislocations. Also, note that many of the twin boundaries (marked "t.b.") show no dislocations and no detectable electrical activity, at least on the scale of the EBIC resolution.

We turn now to a more detailed description of specific EBIC applications.

LOW TEMPERATURE EBIC

Several years ago, it was noted that there is a significant amount of EBIC recombination contrast enhancement in EFG silicon in going from room temperature to, say, 96°K [10]. Figures 3(a) and (b) illustrate the phenomenon. Note that both the linear boundaries and the dislocation arrays between the boundaries exhibit this contrast enhancement. A trapping model was used to explain this enhanced contrast. It assumed that at shallow levels just below the conduction band edge, electrons thermally trap and de-trap. For a level at $E_c - \Delta E$, it was possible to show that

$$L^2 T^{1/2} \propto e^{-\Delta E/kT}$$

Thus, measuring the value of L, the diffusion length, at a particular defect as a function of temperature would permit the calculation of the trap activation energy, ΔE. In this way, three shallow traps were found, E_c − 0.04 eV, E_c − 0.10 eV, and E_c − 0.13 eV. The behavior of these traps indicates considerable sensitivity to both the particular growth conditions (quartz or graphite crucibles), addition or non-addition of CO_2 to the growth ambient used, and also to the temperature employed for the n-type diffusion. In spite of this sensitivity, these traps apparently do not respond to hydrogen passivation [11] as is illustrated in Figs. 3(a) and (b), where it can clearly be seen that EBIC contrast imaged at 300°K and which is eliminated by hydrogen is present at 100°K. Thus, given the validity of the shallow trapping model, one is led to a picture of two sets of states in the band gap which correspond to the dislocation-related defects in EFG silicon ribbon: a deeper, mid-gap state or set of states--nothing is yet known about these--which account for recombination imaged with room temperature EBIC and which are amenable to hydrogen passivation and a second set of shallow states near the conduction band edge which are sensitive to oxygen and insensitive to hydrogen. At present, this remains a qualitative and somewhat speculative picture in what is probably a complex situation, but it is a start and may reveal an interesting property of hydrogen passivated defects.

EBIC MEASUREMENTS OF SURFACE RECOMBINATION VELOCITY AND DIFFUSION LENGTH

The inhomogeneous distribution of dislocations and grain boundaries as shown in Figs. 1, 2, and 3 means that the diffusion length, L, varies spatially. In order to use EBIC to measure L, one must decide on which of several approaches to take to handle the spatial variability of L. We have employed two different methods which represent two extreme approaches. One approach is to treat the defect as a perturbation such that L changes abruptly when a defect is encountered. With this approach to the problem, an analysis can be built up using point defects so that line defects and planar defects can be studied. Donolato [12] has used this method extensively in his work on quantifying EBIC. With this method, one can accurately measure the surface recombination velocity, S, of a grain boundary simply by measuring geometric properties of an EBIC line scan across the grain boundary. Values of L on either side of the boundary can also be determined. The only problem with this method is that it requires a boundary which is not bordered by dense dislocation arrays, a condition which is not always easy to satisfy in silicon ribbon.

The second technique to measure L takes a different tack and assumes that the defect fills the hole-electron pair generation volume and that L varies in a smoother and less abrupt fashion in moving from a defect to a defect-free region. The low temperature shallow trapping model uses this approach to calculate L [10]. This method allows one to measure L at any point, irrespective of the presence of surrounding defects [13]. We will illustrate the use of both techniques for hydrogen passivation.

To assess the depth of passivation, and also to measure S as a function of depth of passivation, Chris Dubé of our laboratory developed a method to passivate the sample edge and at the same time mask the front surface from the hydrogen beam [14]. Figure 4 shows EBIC micrographs using this method. Figure 4 (a) shows recombination due to a grain boundary which has not undergone passivation. Figures 4(b) and (c) show similar micrographs following passivation. Note that there is a finite distance x from the sample edge to the beginning of significant recombination along the grain boundary. The measurement of this distance is used as a measure of both the passivation depth and the hydrogen diffusion depth. In this way, passivation depths varying from ~5 to ~200 μm have been found. Since differing depths are found in the same sample, this is interpreted to mean that there are large differences in the number of defects in the grain boundaries themselves. Work is now in progress to further quantify this.

Differences in the shape of the end (arrows in Figs. 4(b) and 4(c)) of the remaining recombination pattern and the passivation depth are found for boundaries with differing degrees of recombination. Using the method of Donolato [12] for measuring surface recombination velocities S of the grain boundary, we find that boundaries with high values of S ($S = 2.5 \times 10^5$ cm/s for the boundary in Fig. 4(b)) show relatively shallow passivation ($X = 7$ μm) with a tapered end going towards the sample edge. On the other hand, for lower values of S ($S = 7.0 \times 10^4$ cm/s for the boundary in Fig. 4(c)), the passivation is deeper ($X = 25$ μm), with the end of the recombination pattern exhibiting a rounded or more flattened appearance.

In this way, we find that S can be reduced several orders of magnitude following passivation, in fact to values so low that it cannot be measured by this method. A relationship between ℓn S and the passivation depth, X, has been found [14]. For $2 \times 10^4 \lesssim S \lesssim 6 \times 10^5$ cm/s, ℓn S decreases linearly with increasing X up to X = 35 μm. We do not know how to explain this result as of yet, but suspect it may be connected with the kinetics of the diffusion process.

If we assume that the degree of passivation is also a measure of the diffusion of the hydrogen down the grain boundaries, then we can measure X as a function of time, t. A linear relation between X and $\sqrt{t}$ at 400°C has been found for a number of boundaries and so grain boundary diffusivities of 10^{-8} to 10^{-10} cm^2/sec have been determined [14,15]. These values are in good agreement with measurements of Ginley and Hellmer [16] using a totally different technique on quite different silicon material. They measured grain boundary conductance as a function of silicon removal by etching of neutron transmutation-doped chemical vapor-deposited (CVD) silicon. Their diffusivities ranged from $1.7\text{-}7.0 \times 10^{-9}$. CVD silicon is very likely to be much purer than EFG silicon and so this suggests that both the passivation and diffusive process for grain boundaries in silicon is not particularly sensitive to impurities, at least under the passivation conditions we have employed where $T \gtrsim 400$°C and where passivation of point defects is unlikely to be stable [7].

STABILITY

Is there a crystalline counterpart to the light-enhanced instability effect seen in a-Si? Such a question has important scientific and technological implications, the former because it could indicate different bonding states for H in crystalline as opposed to a-Si and the latter because of the obvious possible benefits hydrogen could produce in making low-cost polycrystalline solar cells more efficient. Most of the evidence in the literature and our own work, as well, indicate there is no such effect. For example, Schmidt et al. [17] report that a sample solar cell made of cast polycrystalline silicon, "Silso", kept in air at 200°C for 700 days showed no significant change in photovoltaic parameters. Seager and Ginley [3] reported at the very outset of hydrogen passivation work that they found no decay in the grain barrier conductance for samples of NTD-doped, FZ quality polycrystalline Si with grain sizes of 200-500 μm and annealed in vacuum for 50 hours at 320°C. In our own laboratory we have conducted stability tests by doing before and after EBIC on vacuum-annealed solar cells. We find that no change in the recombination behavior or the degree of passivation occur until $T > 325°C$ under long anneals, on the order of tens of hours. At $T = 400°C$ out-diffusion of the hydrogen and bond breaking is observed to be more rapid and at $T = 450°C$, it becomes quite rapid but still requiring about an hour at this temperature before total bond breaking is observed.

The lone report of an instability effect in polycrystalline silicon comes from a French group. Mautref et al. [18] report that a hydrogenated solar cell made on ribbon against drop (RAD) polycrystalline silicon shows significant degradation in both the open circuit voltage and the short circuit current. They distinguish between a highly mobile hydrogen species near the surface and a more tightly bound, more immobile hydrogen species in the grain boundaries and dislocations. The instability they observe is attributed to the former species. It is possible that what they are observing is a point defect-H pairing of some sort which is unstable under light. The H compensation effect on boron is known to exhibit such behavior [19].

DISLOCATION PASSIVATION

Continued EBIC studies on silicon ribbon led to the conclusion that the dominant defects in this material are dislocations or at least are centered on dislocations. Through the use of low electron beam currents (10^{-11} A) and a high gain EBIC amplifier, a great deal of the background or "noise" seen in earlier EBIC micrographs was shown in fact to be due to recombination at dislocations. In addition, the low temperature EBIC work further supported this idea of the rather ubiquituous nature of these dislocations in this material, as can be seen in Figs. 1, 2, and 3.

Thus, with sensitive EBIC techniques, it is possible to demonstrate that the vast majority of the bulk passivation seen in silicon ribbon is, in fact, dislocation passivation. Figure 5 shows this. All the gray areas between the linear boundaries are dislocation arrays. The hydrogen passivation can be seen to proceed in a "front" down these intragranular dislocation arrays. Note also that the diffusion of the hydrogen down the dislocation arrays is invariably deeper than that for the adjacent linear boundaries. Since the passivated/non-passivated demarcation is somewhat diffuse for these dislocation arrays, it is not possible to measure a diffusivity for them as was done for the linear boundaries. But we can assert that the diffusivity is greater than that of even the highest diffusivity boundaries; i.e., it is $> 10^{-8}$ cm²/sec.

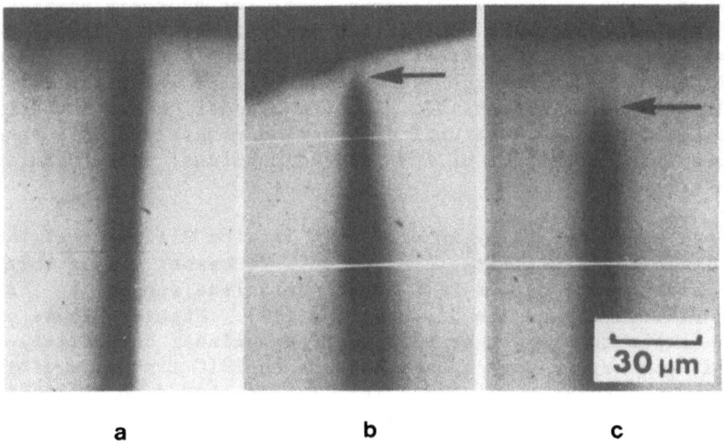

Fig. 4. EBIC micrographs illustrating depth of grain boundary passivation. Hydrogen ions were incident from the top. (a) Shows a typical boundary before any passivation; (b) and c) show other boundaries after hydrogen passivation.

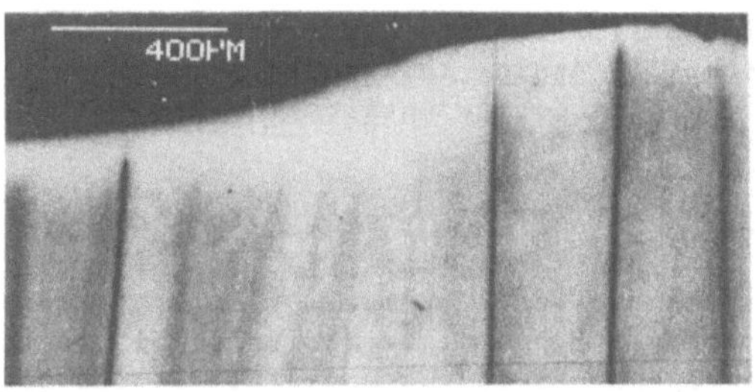

Fig. 5. EBIC micrograph of recombination in an EFG ribbon sample showing passivation of grain boundaries and intragranular dislocation arrays.

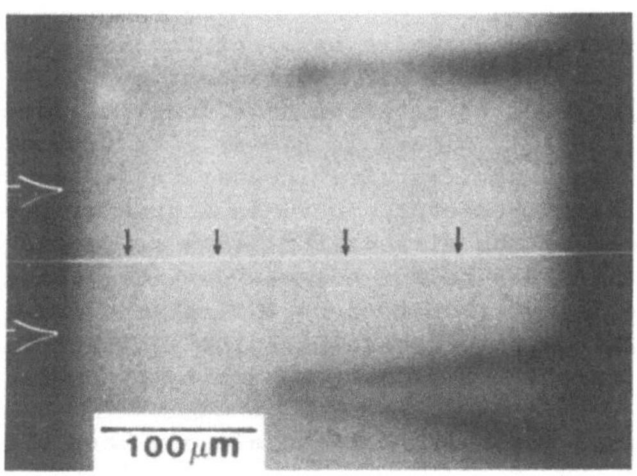

Fig. 6. Cross-sectional EBIC micrograph showing passivation of several grain boundaries to ~150 µm. Arrows show side exposed to hydrogen. Diffusion length values for various points along the line scan (small arrows) between the boundaries were measured and all were L = 117 µm ± 6 µm.

This then leads us to the following model for hydrogen passivation of polycrystalline silicon. The dislocation arrays occur throughout the bulk of the material and provide, in effect, a three-dimensional interconnected set of rapid diffusion paths for the hydrogen as well as being the sites of the principal defects being passivated. In such a model, grain boundaries are viewed as basically two-dimensional dislocation arrays and as a special case of the more general situation of the three-dimensional intragranular dislocation arrays.

The above model basically ignores bulk lattice diffusion of the hydrogen. Is this a legitimate thing to do? We cannot answer this question completely yet but some recent EBIC experiments lend support to the approach of focusing exclusively on the dislocations [11]. Figure 6 shows passivation of a ribbon sample to a depth of ~150 μm of two linear boundaries, indicating that the hydrogen has reached such a depth. An EBIC line scan of the region between these boundaries was taken. This region showed no dislocations and so any hydrogen which might be diffusing down this region would have to do so by bulk lattice diffusion. If we use the diffusion length, L, as a measure of any passivation down this region, we find that L is virtually unchanged throughout the thickness of the sample, using the point-wise EBIC measurement technique of L discussed above. $L = 116\ \mu m \pm 6\ \mu m$ at all four of the points indicated on the line scan. This result shows that whatever defects are limiting the diffusion length in this region are not being passivated to any significant degree. Thus, a region such as this, where dislocations are not found, does not seem to be responsive to passivation and therefore can probably be ignored in considering the general picture of passivation in such a polycrystalline material. Again, this would be so when the passivation is done at $T > 400°C$ where point defect passivation becomes unstable [7].

SUMMARY AND CONCLUSIONS

In this paper, we have given a summary of some of the aspects of the work done in our laboratory on hydrogen passivation in polycrystalline silicon ribbon. We have shown that the dominant defects being passivated are dislocations, that many of the dislocation-related defects can be passivated quite readily using a Kaufman ion source, and that passivation can proceed to depths > 200 μm in some cases with diffusivities down dislocation arrays > 10^{-8} cm^2/sec. Several examples have been chosen to demonstrate the enormous utility of the EBIC technique in studying passivation.

An important question which we have not touched upon is just what is the hydrogen doing at the dislocations? A frequent approach taken by many authors is to assign the rectification of a dangling bond to the hydrogen (see, for example [20,21]). There is ESR data to support such a view [22]. However, there is also experimental evidence that reconstructed dislocations can be passivated [23]. Moreover, there is a general viewpoint among many TEM workers that most of the dislocations (95% or so) in silicon are reconstructed and therefore do not possess dangling bonds. In such a case, the unreconstructed dislocation core or the interaction of the dislocation with various possible point defects [24,25] may result in recombination. The resolution of this question will have to await more work.

In the literature recently [7,26] there has been significant progress in understanding lattice diffusion of hydrogen in silicon. The general view [19,27] seems to be that below temperatures somewhere between 300 and 400°C, it is energetically favorable to form H_2 which is barely mobile and which would not be likely to contribute to the passivation process. Diffusion of hydrogen in silicon can then be modeled [26] as an effective diffusion, taking into account this strong tendency to form a barely mobile H_2. The values of diffusivity predicted by such a model and experimentally determined by DLTS measurements using high reverse biasing to profile hydrogenation effects, are in close agreement at 400°C and give a value of ~6 x 10^{-10} cm^2/sec [7]. This is in rather close agreement with the grain boundary diffusivities we have measured. Since from our work dislocation diffusivities are believed to be > 10^{-8} cm^2/sec, this would indicate that bulk lattice diffusion is still slower than diffusion down bulk dislocation arrays and could lend further support to our model of the dislocation arrays acting as rapid diffusion paths.

Two important areas which we have not touched upon because of space limitations are solar cell device efficiencies after passivation and the question of surface damage introduced as a result of the low energy hydrogen ion implant performed with the Kaufman ion source. Briefly, solar cell efficiency improvements are quite in consonance with the EBIC results shown here and conversion efficiency improvements of anywhere from 10 to 40% can be obtained [4]. The best small area EFG ribbon cell made this way was 14.5% efficient [4], a value approaching single-crystal solar cells. Surface damage is a significant problem and not as readily solved as one would like, and work is ongoing in this area [28,29,30].

ACKNOWLEDGEMENTS

This work has been made possible through the efforts of Chris Dubé, Jim Gregory, Ron Micheels, Don Sandstrom, and Zeke Vayman. I am indebted to Fritz Wald for helpful comments concerning the manuscript and to Dot Bergin and Sigrid Wile for help in preparing the manuscript.

REFERENCES

1. F.V. Wald, Poly-Micro-Crystalline and Amorphous Semiconductors (Proc. of the European Materials Research Soc.), ed. by P. Pinard and S. Kalbitzer.

2. B. Chalmers, J. Crystal Growth, 70, 3-10 (1984).

3. C.H. Seager and D.S. Ginley, Appl. Phys. Lett., 34, 537 (1979).

4. J.I. Hanoka, C.H. Seager, D.J. Sharp, and J.K.G. Panitz, Appl. Phys. Lett., 42(7), 618 (1983).

5. F.V. Wald, in: Crystals: Growth Properties and Applications 5, edited by J. Grabmaier (Springer, Berlin, 1981), pp. 147-198.

6. J.M.E. Harper, J.J. Cuomo, and H.R. Kaufman, Ann. Rev. Mater. Sci., 13, 413 (1983).

7. S.J. Pearton, Thirteenth International Conf. on Defects in Semiconductors, ed. by L.C. Kimerling and J.M. Parsey, Jr.; Metallurgical Society of AIME, Warrendale, PA, 1985, p. 737.

8. J.I. Hanoka and R.O. Bell, Ann. Rev. Mater. Sci., 11, 353 (1981).

9. H.J. Leamy, J. Appl. Phys., 53, R51 (1982).

10. J.I. Hanoka, R.O. Bell, and B. Bathey, in: Symposium on Electronic Optical Properties in Polycrystalline or Impure Semiconductors, Novel Crystal Growth Techniques, edited by K.V. Ravi and B. O'Mara (The Electrochemical Society, Princeton, New Jersey, 1980), pp. 76-86.

11. J.I. Hanoka, C.E. Dubé, and D.B. Sandstrom, to be published in Mat. Res. Soc. Symposium on "Microscopic Identification of Electron Defects in Semiconductors", San Francisco Meeting, Spring 1985.

12. C. Donolato, Appl. Phys. Lett., 34, 80 (1979), and C. Donolato, Scanning Electron Microsc., Part I (1979), p. 257; also, C. Donolato and R.O. Bell, Rev. Sci. Instrum., 54, 1005 (1983).

13. R.O. Bell and J.I. Hanoka, J. Appl. Phys., 53, 1741 (1982).

14. C. Dubé, J.I. Hanoka, and D.B. Sandstrom, Appl. Phys. Lett., 44, 425 (1984).

15. C. Dubé and J.I. Hanoka, Appl. Phys. Lett., 45, 1135 (1984).

16. David S. Ginley and R.P. Hellmer, 17th IEEE Photovoltaic Specialists Conference (IEEE: New York; 1984), p. 1213.

17. W. Schmidt, K.D. Rasch, and K. Roy, 16th IEEE Photovoltaic Specialists Conference (IEEE: New York; 1982, pp. 537-54.

18. M. Mautref, C. Belouet, A. Buenas, M. Aucouturier, and M. Groos, in Ref. 1, pp. 129-36.

19. S.J. Pearton and A.J. Tavendale, private communication.

20. J.I. Pankove, M.A. Lampert, and M.L. Tarng, Appl. Phys. Lett., 32(7), 439 (1978).

21. C. Belouet, Ref. 1, pp. 53-66.

22. N.M. Johnson, D.K. Biegelson, and M.C. Moyer, Appl. Phys. Lett., 40, 882 (1982).

23. M.N. Zolotukhin, V.V. Kveder, and Yu. A. Osip'yan, Sov. Phys., JETP 55(b), 1189 (1982).

24. S. Marklund, J. de Physique, Colloque C4, 44, 25 (1983).

25. M.I. Heggie and R. Jones, ibid., 44, 43 (1983).

26. R.N. Hall, Ref. 7, p. 759.

27. J.W. Corbett, S.N. Sahu, T.S. Shi, and L.C. Snyder, Phys. Lett., 93A 303 (1983).

28. J.W. Wang, S.T. Fonash, and S. Ashok, IEEE Electron Device Lett., EDL-4, No. 12, 432 (1983).

29. J.K.G. Panitz, D.J. Sharp, and C.H. Seager, Thin Solid Films, 111, 277 (1984).

30. A. Barhdadi, A. Mesli, E. Courcelle, D. Salles, and P. Siffert, in Ref. 1, pp. 373-377.

THE INFLUENCE OF HYDROGEN ON THE DEFECTS AND INSTABILITIES IN HYDROGENATED AMORPHOUS SILICON

P.C. Taylor, W.D. Ohlsen, C. Lee, and E.D. VanderHeiden

Department of Physics
University of Utah
Salt Lake City, UT 84112

ABSTRACT

The presence of hydrogen in amorphous silicon alloys affects both the defect structure and the instabilities. Specific examples where the presence of hydrogen is either directly or indirectly important include (1) the elimination of silicon dangling bonds, (2) the trapping of molecular hydrogen in voids and (3) the presence of optically-induced, reversible metastabilities in the optical and electronic properties.

INTRODUCTION

In tetrahedrally coordinated amorphous alloys the theraputic role of hydrogen in passivating "dangling bonds" is well known. In amorphous silicon without hydrogen (a-Si) the densities of silicon "dangling bonds" as measured by electron spin resonance (ESR) are almost always greater than 10^{18} spins cm^{-3}. With the addition of hydrogen the spin densities in a-Si:H can be as low as $\sim 10^{15}$ spins cm^{-3}.

Although the passivation of silicon dangling bonds is extremely important, especially for electronic applications, the presence of hydrogen has a much greater influence on the structure and the defects in a-Si:H than merely satisfying these dangling bonds. There may be 10^{18}-10^{19} spins cm^{-3} in a-Si, but typical samples of a-Si:H contain over 10^{21} hydrogen atoms cm^{-3}. Therefore, the vast majority of hydrogen atoms in a-Si:H serves to reduce the number of silicon-silicon bonds in the alloy. The common perception is that hydrogen atoms replace "strained" silicon-silicon bonds with silicon-hydrogen bonds. This process is thought to be especially prevalent on the surfaces of internal voids which remain even in the best device-quality alloys.

In addition to dramatically altering the local structural order and to passivating dangling bonds, the presence of hydrogen in amorphous silicon also affects the defect structure in more subtle ways. For example, several experiments[1-5] have detected the presence of trapped molecular hydrogen (H_2 molecules) in a-Si:H. These molecules are probably trapped in small internal voids where they are stable up to fairly high temperatures (> 500°C in many cases). Even though most a-Si:H alloys have considerable oxygen contamination ($\geq 10^{18}$ cm^{-3}), the presence of OH groups in these alloys is very rare. The vast majority of the hydrogen is found to bond preferentially to silicon. In alloys with oxygen purposely added at levels of 1-2 at.% one can trap atomic hydrogen, probably on oxygen-rich internal surfaces.[6] An indirect influence of hydrogen on the defect structure in a-Si:H alloys involves the role of hydrogen in eliminating nonradiative processes such as those dominated by silicon dangling bonds. The presence of hydrogen allows efficient photoluminescence (PL) to occur, and the PL processes are strongly influenced by the hydrogen concentration in the alloys.

In addition to these examples of the influence of hydrogen on the formation of defects in a-Si:H, hydrogen may play at least an indirect role in the occurrence of electronic and optical metastabilities in the "best" a-Si:H alloys.* The first observation of such an effect was a decrease in the photoconductivity after optical excitation as reported by Staebler and Wronski.[7] Since this work, metastable changes in many other transport and optical properties have been observed. Unfortunately, all of these changes, many of which may not be directly related, are often grouped together under one generic title as "the Staebler-Wronski effect." We shall discuss only one aspect of this effect, the occurrence of a metastable, optically-induced paramagnetism in a-Si:H.

DEFECTS

Although the dangling bond on the group IV atom is commonly considered to be the primary paramagnetic defect with an energy near the center of the gap, there are several other defects which are potentially important in determining the electronic properties of these alloys. In hydrogenated amorphous silicon the dangling bond defect yields an ESR response at $g = 2.0055$ which is always present on a level of at least 10^{15} spins cm^{-3}. In silicon-germanium alloys two ESR signals are generally observed, one attributed to a dangling bond on Si and the other to a dangling bond on Ge. The most common interpretation of these defects is that they are paramagnetic and neutral in the ground state,[8] but it has also been suggested[9] that at least some of the Si and Ge dangling-bond defects might exist positively and negatively charged (negative U_{eff} system) in the ground state.

Several shallow defects near the band edges have been postulated in these alloy systems, but none of them directly involve the presence of hydrogen. The most accepted interpretation of these defects is in terms of strained bonds on the group IV atoms.[10] It has also been suggested that neutral two-fold coordinated Si and Ge atoms[11] may be responsible for these shallow electronic states.

The occupancy of many of these defects can be altered optically by the application of band-gap light. Three transient optically-induced ESR

---

*In the common perception, the "best" a-Si:H alloys are those whose electronic properties, such as photoconductivity, electronic mobility and so forth, yield the highest quality electronic devices. Since the technology is currently driven almost exclusively by photovoltaic or thin film transistor applications, the best usually translates into something like the highest solar cell efficiency or the fastest switching speed. In terms of the basic properties of the material, the best refers in some qualitative fashion to the lowest density of electronically-active defects. Perhaps the most accurate definition of the best a-Si:H alloys is those which exhibit the greatest instabilities in their electronic and optical properties.

signals are observed in a-Si:H at g-values of 2.004, 2.013 and 2.0055. These three ESR signals are attributed to electrons trapped in localized electronic states below the conduction band edge, holes trapped in localized states above the valence band edge and silicon dangling bonds, respectively.

Hydrogen plays a direct role in at least two defects in a-Si:H and related alloys. Molecular hydrogen (H_2) exists in most films at levels on the order of 10^{-2} to 10^{-1} at.%.[3,4] Under the appropriate conditions, such as in oxygen-doped samples, after x-irradiation at low temperatures, atomic hydrogen can also be trapped in a-Si:H.[6]

In the case of molecular hydrogen, some of the H_2 molecules can be probed by examining the spin lattice relaxation rates T_1^{-1} of hydrogen bonded to the silicon network.[12] It has been established[1,13] that a characteristic minimum in T_1 near 40K is due to the relaxation of the bonded hydrogen atoms via some of the H_2 molecules trapped in the films. A typical example of this T_1 minimum is shown in Fig. 1. It can be shown[1,13] that the magnitude of the minimum value of T_1 is proportional to that concentration of H_2 molecules which is effective in relaxing the bonded hydrogen.

The data of Fig. 1 show that there are fewer H_2 molecules effective in relaxing the bonded hydrogen in flakes of a-Si:H which have been removed from the substrate than in similar films which remain on the substrate. It has been postulated[14] that removal of the films from the substrate provides a strain relief mechanism which allows some of the trapped H_2 to diffuse out of the films. This speculation is supported by the fact that one can also achieve the same decrease in trapped H_2

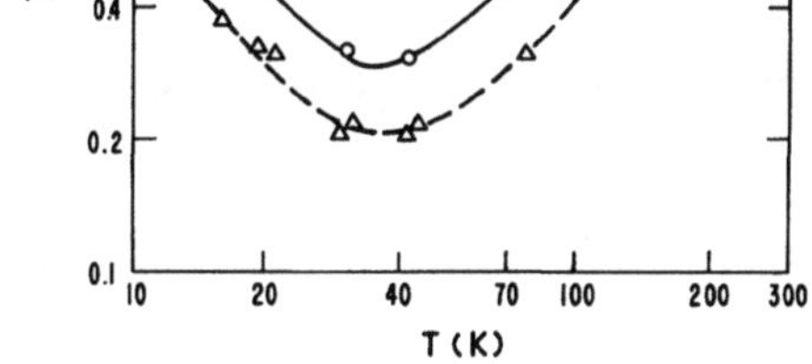

Fig. 1. ^1H spin lattice relaxation time in a-si:H as a function of temperature. The solid line represents data on flakes which have been removed from the substrate. Triangles indicate data taken on films on quartz substrates within two weeks of deposition. Circles denote data taken on the same films after ten months. The magnitude of the minimum value of T_1 (near 40K) is inversely proportional to the number of H_2 molecules contributing to the relaxation process.

concentration by allowing the films on the substrate to "age" for ten months (circles in Fig. 1).

After the films have aged, one can reinduce H_2 by immersing the sample in an atmosphere of H_2 or by heating the samples and creating H_2 from bonded hydrogen.[14] This "diffusion" process takes only about three weeks at room temperature with approximately one atmosphere pressure of H_2. Molecular hydrogen can be reintroduced into the film by either of these two processes only up to the level observed in the virgin films. Furthermore, the reintroduced hydrogen is not stable and diffuses out at room temperature on a time scale of a few weeks. This process is much faster than the initial aging process.

It has been speculated[14] that the stress relief involved in the initial aging process results in a relaxation of the initial void structure such that future reinducing can proceed more rapidly but that the molecular hydrogen is no longer as strongly trapped. Because the reinducing process appears to saturate and because the temperature dependence doesn't yield parameters consistent with a simple diffusion process,[14] a more complicated mechanism will be necessary to explain the effect.

While trapped molecular hydrogen is a ubiquitous "impurity" in a-Si:H, trapped atomic hydrogen is only rarely observed. When films which contain ≥ 2 at.% oxygen are electron-irradiated at low temperatures ($T \leq 77K$), atomic hydrogen is observed in electron spin resonance (ESR) experiments.[6] The characteristic ESR spectrum for atomic hydrogen is shown in Fig. 2. The derivative features near 3200 G (center of magnetic field scan) are due to defects associated with oxygen or silicon, but the doublet structure near 3000 and 3500 G is due to atomic hydrogen. Once created, these defects are stable at temperatures up to ~ 300K in a-Si:H. Above 300K the atomic hydrogen defects are unstable and they recombine in a bimolecular fashion to form H_2 which, as mentioned above, probably diffuses out of the film.[6]

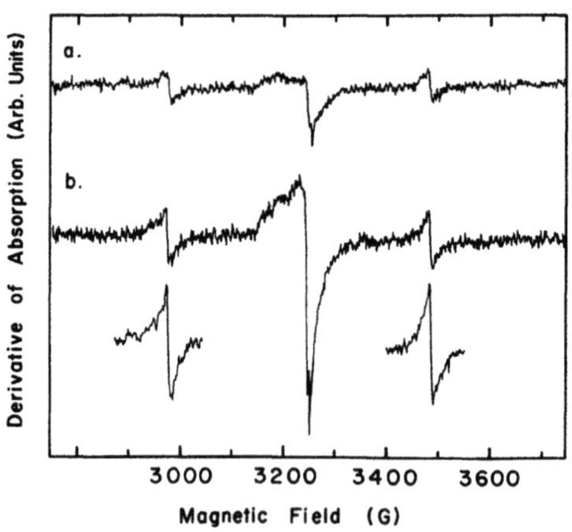

Fig. 2. ESR in x-irradiated flakes of a-Si:H doped with approximately 2 at. % oxygen. The lines near 3000 and 3500 G are due to atomic hydrogen. The features near 3200 G are due to other centers.

The occurrence of optically-induced, metastable changes in the paramagnetism in films of a-Si:H is well known.[15-17] The effect was first observed with white light in powdered samples,[15] but recent experiments with monochromatic light incident on films on substrates have observed a similar effect.[17] This optically-induced ESR appears to be stable at temperatures ≤ 300K.

The kinetics of the inducing and annealing of the optically-induced ESR have been studied by several groups.[16-19] Typical inducing curves are shown in Fig. 3 for a sample where the dark spin density is $> 10^{16}$ spins cm^{-3} (triangles) and for a sample where the dark spin density is $< 10^{16}$ spins cm^{-3} (circles). The sample with the larger spin density is representative of those which show a $t^{1/3}$ behavior for the spin density n_s at long times. This behavior has been interpreted as due to the optically-induced creation of additional dangling bonds by electronic transitions between localized states in the band tails.[17,19] This inefficient process is considered to be limited by the presence of existing dangling bonds which provide an effective competing recombination path.

The sample with the lower spin density (circles in Fig. 3) is representative of those which show a continuously decreasing slope with time which does not exhibit a power law behavior over any extended range of time. In samples of this type the growth rate appears to decrease continuously with increasing time. In these samples it has been suggested that a second type of inducing process may be important.[18] Several other experiments have also suggested that more than one center may be involved in these metastable effects.[17,20,21]

The dependence of the metastable, optically-induced ESR on the power density of the inducing light is shown for a representative sample with low dark spin density ($< 10^{16}$ spins cm^{-3}) in Fig. 4. In the samples with high initial spin densities the inducing curves appear to scale[17,19] with the intensity I of the exciting light as $I^{2/3}$. For samples with low initial spin densities such simple scaling does not appear to hold (see Fig. 4).

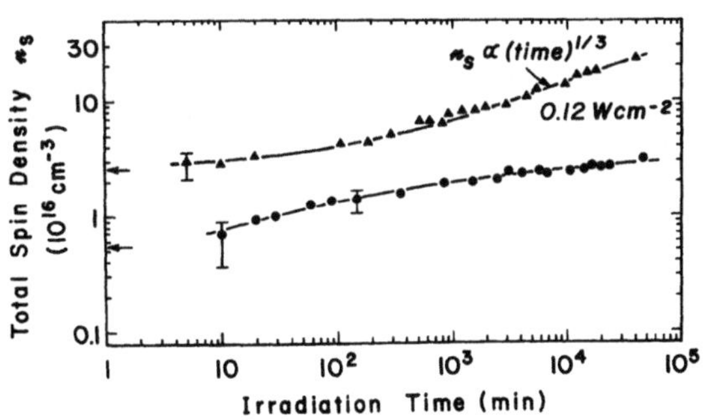

Fig. 3. Total ESR density in two films of a-Si:H on quartz substrates. The triangles represent data on a film with an initial spin density of ~ 3×10^{16} cm^{-3}. The circles represent data on a film with an initial spin density of ~ 5×10^{15} cm^{-3}. The two samples were both irradiated with 0.12 W cm^{-2} of light from a tungsten source.

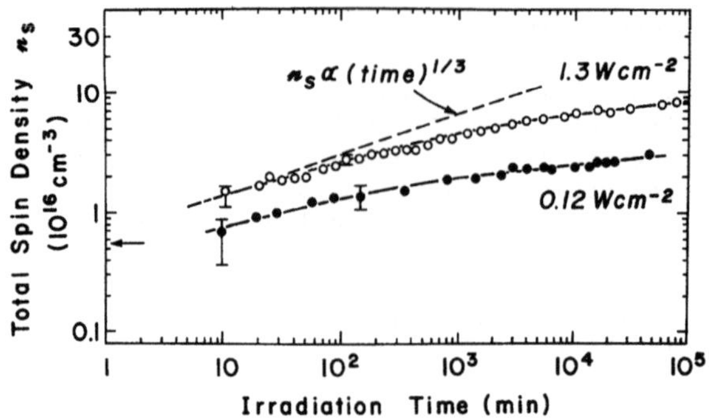

Fig. 4. Total ESR spin density in a film of a-Si:H with ~ 3×10^{15} spins cm^{-3} initially. Open circles are data after irradiation with 1.3 W cm^{-2} and filled circles are data after irradiation with 0.12 W cm^{-2}.

The connection of these rather complicated metastable effects with the presence of hydrogen is not at all clear. What is clear is that most of the effects are only observed in films of relatively low intitial (dark) ESR spin densities. Although these films all contain approximately 10 at.% hydrogen a direct role of hydrogen in the instabilities has yet to be shown. If there is any direct role of hydrogen in these processes, it must be very special because these metastable changes are reversible upon annealing around 200°C.

RADIATIVE PROCESSES

In amorphous silicon without hydrogen (a-Si) radiative recombination processes are greatly suppressed and not measurable experimentally due to the presence of large densities of silicon dangling bonds which serve as effective non-radiative recombination centers. When the density of Si dangling bonds is greatly reduced in a-Si:H, an efficient photoluminescence (PL) which peaks near 1.2-1.4 eV is observed. The most commonly accepted interpretation of this PL process is in terms of recombination between electrons trapped in localized band-tail states below the conduction band with holes trapped in localized states above the valence band.[22] It should be noted that this explanation has been very successful in explaining large quantities of data taken on a-Si:H and, hence, appears at present to rest on rather firm ground. With this explanation the role of hydrogen is the indirect one of removing the dominant competing non-radiative channel by bonding to silicon atoms which would otherwise have at least one unsatisfied bond.

There is some evidence, albeit preliminary, that this explanation may not be the entire story and that hydrogen may perhaps play a more direct role in the recombination processes.[23] In amorphous alloys containing both silicon and germanium (a-$Si_{1-x}Ge_x$:H), the peaks of the PL spectra shift to lower energy with increasing germanium content (increasing x) in a manner which parallels the decrease in the energy gap with x. In addition, these spectra exhibit a low energy exponential region in which the PL efficiency is independent of germanium content as shown in Fig. 5.

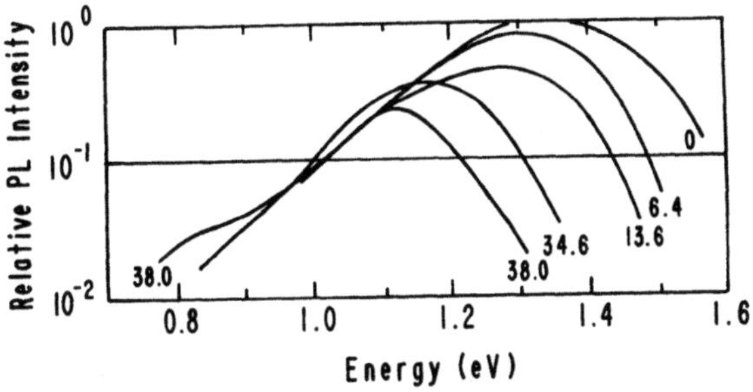

Fig. 5. PL intensitites for several samples in the system a-$Si_{1-x}Ge_x$:H where x is indicated in percent on the curves. Data were taken at 77K and are all scaled by the same factor such that the peak intensity for a-Si:H (x = 0) is unity.

The shift of the PL peak with the band gap is easy to explain with the model described above because the band tail states become closer together in energy as the band gap decreases. However, the constant magnitude of the low energy portion of the PL can only be a fortuitous coincidence in the model--a coincidence which is achieved if the recombination rate at a given energy decreases exponentially with the band gap.

In a recent publication Gal et al.[23] have suggested that an alternative explanation is that the PL process consists of recombination between the excited and ground states of an ensemble of specific defects. With this assumption both the constant magnitude of the low energy portion of the PL and the shift of the peak with decreasing band gap arise naturally as, for example, the higher energy portion of the density of excited states for this defect becomes resonant with extended states in the conduction band and no longer contributes to the PL. In this interpretation of the PL process it is possible that hydrogen plays a more direct role in the radiative recombination in a-Si:H.

It remains to be seen whether the attribution of the PL in a-Si:H and related alloys to an ensemble of specific defects will be as successful in explaining the great accumulation of related data as the attribution of this process to transitions between localized electronic states within the band tails has been.

SUMMARY

Although the dangling bond on the group IV atoms is perhaps the most important defect deep in the gap in the tetrahedrally-bonded amorphous semiconductors, there are several other important defects. Hydrogen plays a direct role in at least two of these defects. Molecular hydrogen (H_2) is trapped in most films of a-Si:H and related alloys. In films of a-Si:H which are alloyed with oxygen ($\geq$ 2 at. % oxygen) one can also observe trapped atomic hydrogen after irradiation at low temperature ($T < 300$ K).

The application of light to films of a-Si:H and related alloys produces several metastable changes in the optical and electronic properties. The changes are accompanied by an increase in the paramagnetism of the films as measured by ESR. An understanding of the kinetics of formation and annealing of the ESR is of great importance for an understanding of the microscopic mechanisms responsible for these changes. The connection of these rather complicated metastable effects with the presence of hydrogen is not clear.

Efficient radiative recombination in tetrahedrally-bonded amorphous films depends critically on the presence of hydrogen which removes non-radiative recombination centers from the middle of the gap. Recent PL experiments on amorphous silicon-germanium alloys show that the intensity of the PL at low energies is independent of germanium content. This observation may require a reinterpretation of the commonly accepted explanation for the dominant PL process in these alloys.

This research was supported by the National Science Foundation under grant number DMR-83-04971 and by the Solar Energy Research Institute under subcontract number XB-5-05009-2.

REFERENCES

1. W.E. Carlos and P.C. Taylor, Molecular Hydrogen in a-Si:H, Phys. Rev. B25:1435 (1982).
2. J.E. Graebner, B. Golding and L.C. Allen, Solid Hydrogen in Hydrogenated Amorphous Silicon, Phys. Rev. Lett. 52:553 (1984); J.E. Graebner, L.C. Allen and B. Golding, Solid H_2 in a-SI:H at Low Temperatures, Phys. Rev. B31:904 (1985).
3. H.V. Löhneysen, H.J. Schink and W. Beyer, Direct Experimental Evidence for Molecular Hydrogen in Amorphous Si:H, Phys. Rev. Lett. 52:549 (1984).
4. J.B. Boyce and M. Stutzmann, Orientational Ordering and Melting of Molecular H_2 and in an a-Si Matrix: NMR Studies, Phys. Rev. Lett. 54:562 (1985).
5. P.A. Fedders, R. Fisch and R.E. Norberg, Dense H_2 and Proton NMR in a-Si:H, Phys. Rev. B31:6887 (1985).
6. W.M. Pontuschka, W.E. Carlos and P.C. Taylor, Radiation-induced Paramagnetism in a-Si:H, Phys. Rev. B25:4362 (1982).
7. D.L. Staebler and C.R. Wronski, Reversible Conductivity Changes in Discharge-Produced Amorphous Si, Appl. Phys. Lett. 31:292 (1977).
8. D.K. Biegelsen, Electron Spin Resonance Studies of Amorphous Silicon, Proc. Electron Resonance Symp. 3:85 (1981).
9. D. Adler and and F.R. Shapiro, Effective Correlation Energy of the Dangling Bond in Amorphous Silicon, Physica 117B and 118B:932 (1983).
10. R.A. Street, D.K. Biegelsen and J.C. Knights, Defect States in Doped and Compensated a-Si:H, Phys. Rev. B24:969 (1981).
11. D. Adler, Electronic Properties of Amorphous Silicon Alloys, Kinam 4C: 225 (1982).
12. W.E. Carlos and P.C. Taylor, ^1H NMR in a-Si, Phys. Rev. B26:3605 (1982).
13. M.S. Conradi and R.E. Norberg, Molecular H_2: Nuclear Spin Relaxation Centers for Protons in a-Si:H, Phys. Rev. B24:2285 (1981).
14. E.D. VanderHeiden, W.D. Ohlsen and P.C. Taylor, NMR Studies of H_2 in a-Si:H, Bull. Am. Phys. Soc. 30:354 (1985).
15. H. Dersch, J. Stuke and J. Beichler, Electron Spin Resonance of Doped Glow-Discharge Amorphous Silicon, Phys. Status Solidi B105:265 (1981); Temperature Dependence of ESR Spectra of Doped a-Si:H, Phys. Status Solidi B107:307 (1981).

16. C. Lee, W.D. Ohlsen and P.C. Taylor, Kinetics of the Metastable Optically Induced ESR in a-Si:H, Phys. Rev. B31:100 (1985).
17. M. Stutzmann, W.B. Jackson and C.C. Tsai, Light-Induced Metastable Defects in Hydrogenated Amorphous Silicon: A Systematic Study, Phys. Rev. B32:23 (1985).
18. C. Lee, W.D. Ohlsen, P.C. Taylor, H.S. Ullal and G.P. Ceasar, Dependence of the Metastable Optically-Induced ESR in a-Si:H on Temperature and Power, in AIP Conf. Proc. 120:205 (1984).
19. M. Stutzmann, W.B. Jackson and C.C. Tsai, The Kinetics of Formation and Annealing of Light Induced Defects in Hydrogenated Amorphous Silicon, in AIP Conf. Proc. 120:213 (1984).
20. D. Han and H. Fritzsche, Study of Light-Induced Creation of Defects in a-Si:H by Means of Single and Dual-Beam Photoconductivity, J. Non-Cryst. Solids 59+60:397 (1983).
21. S. Guha, C.-Y. Huang, S.J. Hudgens and J.S. Payson, Effects of Light Soaking at Different Temperatures on the Properties of Hydrogenated Amorphous Silicon Alloys, J. Non-Cryst. Solids 66:65 (1984).
22. R.A. Street, Luminescence in Amorphous Silicon, Adv. Phys. 30:593 (1981).
23. M. Gal, J.M. Viner, P.C. Taylor and R.D. Wieting, Existence of a Universal Low-Energy Tail in the Photoluminescence of a-$Si_{1-x}Ge_x$:H Alloys, Phys. Rev. B31:4060 (1985).

NMR INVESTIGATION OF PAIRED HYDROGEN ATOMS IN PLASMA-DEPOSITED AMORPHOUS SILICON

J. B. Boyce

Xerox Palo Alto Research Center

Palo Alto, CA 94304

ABSTRACT

When two identical spins have separations which are smaller than those between the other spins in the solid, their NMR spectrum consists of a characteristic doublet, the Pake doublet. This doublet provides information on the bonding of the atoms and their separations. Two different Pake doublets have been observed in the hyrdogen NMR spectrum of plasma-deposited amorphous silicon. The first is due to SiH_2, part of the H that is bonded to the a-Si network. The second is due to H_2, molecular hydrogen which is trapped in voids in the a-Si:H matrix.

INTRODUCTION

When amorphous silicon (a–Si) is prepared by the plasma decomposition of silane gas, about 10 atomic % H is incorporated in the a–Si matrix. This hydrogen plays the important role of removing a majority of the defect states from the band gap, thereby allowing hydrogenated a–Si (a–Si:H) to be doped either p– or n–type.[1] Such is not the case for unhydrogenated a–Si which has a high density of dangling–bond defects which pin the Fermi level in the gap. Typical dangling–bond defect densities are $5\times10^{19}/cm^3$ for a–Si, and this is reduced to $10^{16}/cm^3$ for a–Si:H.[2] Considerable effect has gone into the study of H in a–Si:H due to its key role in making amorphous silicon an electronically useful semiconductor. NMR has provided a significant portion of this knowledge.

In this paper, a specific subset of the information that NMR can provide on the role of H in a–Si:H is discussed. This subset is the information contained in the NMR spectrum due to paired H atoms, those which, due to their bonding, have a characteristic H–H distance which is shorter than the distances between the pair and the other H atoms in the solid. This configuration gives rise to the well–known Pake doublet[3] in the NMR spectrum. The paired H in a–Si:H that give rise to a Pake doublet occur in two forms: (1) as SiH_2, part of the H that is bonded to the a–Si network, and (2) as H_2, the quantum rotor, molecular hydrogen, which is trapped in voids in the a–Si:H matrix. Before discussing the results on these two structures, the Pake doublet spectrum is reviewed.

PAKE DOUBLET

When two identical spins (two protons) have separations which are smaller than those between other spins in the solid, a splitting of the resonance into a doublet occurs. A requirement for a resolved doublet is that the dipolar interaction between the two spins be larger than that with the other spins. Since the dipolar interaction energy goes as $\gamma^2\hbar^2/r^3$ (γ is the proton gyromagnetic ratio, $\hbar$ is Planck's constant and r is the spacing), this requires that the spacing between the protons of the pair be smaller than that between the pair and the other protons in the sample. Due to the rapid $1/r^3$ variation, this difference in spacing need not be too large.

Since the Pake doublet has been much discussed in the literature,[4] only the pertinent facts will be given here. The NMR spectrum of an isolated pair of spins is

$$\omega = \omega_0 \pm \alpha\,(3\cos^2\Theta - 1)\,, \qquad (1)$$

where $\omega_0 = \gamma H_0$, the NMR resonance frequency, H_0 is the applied external magnetic field, Θ is the angle the vector r joining the two spins makes with the external magnetic field, and $\alpha = 3\gamma^2\hbar/4r^3$, the dipolar coupling constant. Eq. (1) describes a doublet spectrum with splitting $2\alpha(3\cos^2\Theta-1)$. In a single crystal, Θ has a well-defined value, but in a-Si:H there is a random distribution of all angles Θ. Such a distribution smears out the doublet into a powder-averaged doublet,[3] the powder pattern $S(\omega)$ shown in Fig. 1a. This pattern has a square-root singularity at $\pm\alpha$ and a cutoff at $\pm 2\alpha$, relative to ω_0. In fact, the paired spins are never completely isolated and so the pattern is broadened by the dipolar interaction of the paired protons with other protons in a-Si:H. This broadening is accurately represented by the convolution of the powder pattern with a Gaussian of width β. The resulting spectrum is also shown in Fig. 1a. The second moment of this line is

$$\Delta\omega^2 = 4\alpha^2/5 + \beta^2\,. \qquad (2)$$

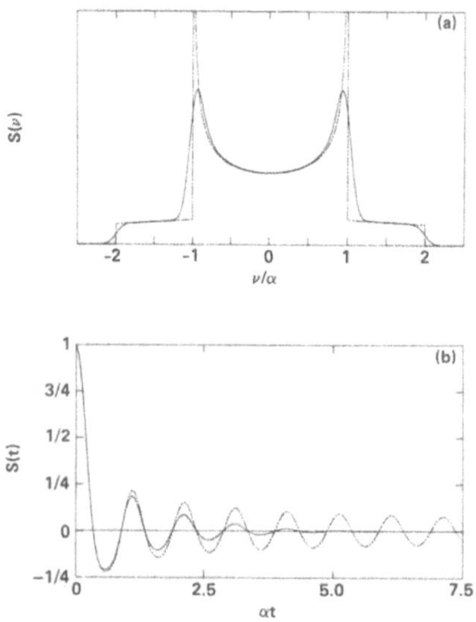

Fig. 1. (a) The Pake doublet powder pattern for an isolated pair of spins (dashed curve) and with a Gaussian broadening of width β (solid curve) with $\beta/\alpha = 0.08$, in the frequency domain (a) and time domain (b).

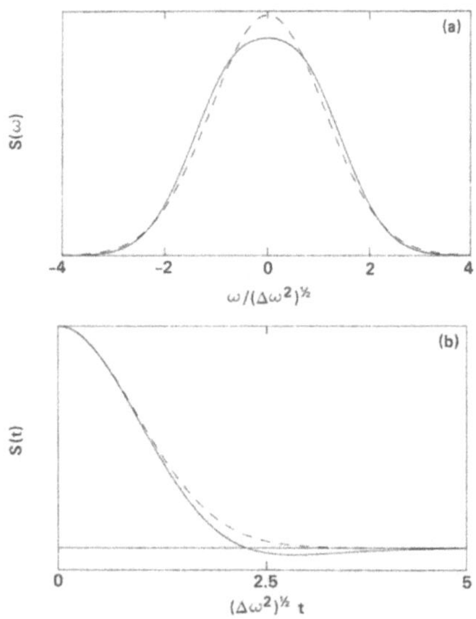

Fig. 2. Comparison between a Gaussian (dashed line) and a Pake "doublet" (solid line) for the broadening comparable to the dipolar splitting, $\beta/\alpha = 2/3$. The second moments, $\Delta\omega^2$, are the same in each case.

Often the NMR spectrum is observed in the time domain rather than in the frequency domain. The time domain signal is the response to a 90° radio frequency pulse, the free induction decay (FID). The two domains are related by a Fourier transform.[5] The free induction decay, S(t), is given by the sum of Fresnel integrals times $\cos\alpha t$ and $\sin\alpha t$,[6] and is shown in Fig. 1b. The Gaussian broadening of the spectrum due to the dipolar interaction with neighbors occurs in S(t) as a Gaussian damping, $\exp(-\beta^2 t^2/2)$, of the oscillatory decay, also shown in Fig. 1b. From either S(t) or $S(\omega)$, α, and thereby r, can be determined. Also β can be extracted, giving additional structural information.

When the Gaussian broadening due to the neighbors becomes comparable to the Pake splitting, then the powder-pattern spectrum approaches a Gaussian spectrum, which is the line shape for a dense distribution of interacting spins. This is seen in Fig. 2 where $\beta = 2\alpha/3$ and the second moments, $\Delta\omega^2$, are the same. The second moment of the Gaussian is the Gaussian width squared, σ^2, whereas that of the Pake doublet is given by Eq. (2). Despite the fact that $\beta \sim \alpha$, the "doublet" spectrum can still be discerned in this figure as a flat-topped line in $S(\omega)$ or as a beat in S(t). Given an adequate signal-to-noise ratio, the relevant structural parameters, α and β, can still be determined in this case, although not with the same degree of precision as for the case shown in Fig 1.

SPECTRUM OF SiH_2

The NMR spectrum of H was measured on a number of a-Si:H samples made from the plasma decomposition of silane under different preparation conditions, as listed in Table 1. For samples prepared on heated substrates (#5-#8) the spectrum

Table 1. The total hydrogen content, n(H), and the line shape parameters for the broad and the narrow components of the NMR line for eight plasma-deposited a-Si:H samples prepared under different conditions as listed in column 1 (RF power density, substrate temperature, gas composition and deposition electrode). For the first four deposited on room-temperature substrates, the broad line is a Pake doublet due to SiH_2. For the second four, deposited on heated substrates, the broad line is a Gaussian.

Sample	n(H) at%	Broad				Narrow	
		Gaussian		Powder Pattern		n(H)	FWHM
		n(H) at%	σ kHz	n(H) at%	$(\Delta\omega^2)^{1/2}$ kHz	at%	kHz
#1: 0.25 W/cm², RT, 5% SiH_4 in Ar, Anode	32.4			27.9	8.7	4.5	2.4
#2: 0.05 W/cm², RT, 100% SiH_4, Cathode	21.3			17.6	9.4	3.6	3.8
#3: 0.41 W/cm², RT, 5% SiH_4 in He, Cathode	19.3			13.0	8.6	6.2	3.3
#4: 0.045 W/cm², RT, 100% SiH_4, Anode	22.5			19.3	9.1	3.2	3.1
#5: 0.025 W/cm², 230°C, 100% SiH_4, Anode	11.5	6.8	10.2			4.6	2.6
#6: 0.025 W/cm², 250°C, 100% SiH_4, Anode	10.7	6.5	9.8			4.2	3.1
#7: 0.025 W/cm², 230°C, 100% SiH_4, Anode	8.3	4.8	9.5			3.5	2.9
#8: 0.025 W/cm², 230°C, 100% SiH_4 -10^{-6} PH_3	8.6	4.5	9.2			4.1	2.7

is as shown in Fig. 3 (for sample #5). It consists of a broad line and a narrow line and has been much discussed in the literature.[7-9] The total H content is typically about 10 at.%, with about 4 at. % in the narrow line and the remaining 6 at. % in the broad line. The narrow line is Lorentzian in shape and has a full width at half maximum (FWHM) of about 3 kHz. The broad line is Gaussian in shape with a Gaussian width, σ, of about 10 kHz, equivalent to a FWHM $\simeq$ 25 kHz. These parameters for samples #5-#8 are given in Table 1 and are evident in the spectrum of Fig. 3 (for sample #5). The narrow line corresponds to H that is randomly distributed throughout the Si marix and is bonded to the Si atoms. The broad line consists of H atoms that are clustered together, such as, the H on the hydrogenated surfaces of voids in the material. It is suspected that part of the broad line consists of SiH_2 units since these are observed in the infrared spectrum. Nonetheless, no direct evidence has been obtained for SiH_2 units in the NMR spectrum.

For an isolated SiH_2 unit bonded to the Si matrix and far from other H atoms, the NMR spectrum is a Pake doublet. The H-H spacing for SiH_2 is 2.33A, so that α = 7 kHz and the Pake splitting, ν_P, is $\nu_P = 2\alpha = 14$ kHz. This splitting is quite small so that, if the other H atoms are near to the SiH_2, i.e., within about 2.33Å, then the doublet may not be distinguishable from a Gaussian line shape. Add to this the fact that the spectrum may also contain a broad and a narrow line and one sees that it is not unreasonable that the observation of the doublet spectrum due to SiH_2 has not been reported.

Despite these facts, a beat in the FID due to the Pake doublet from SiH_2 has

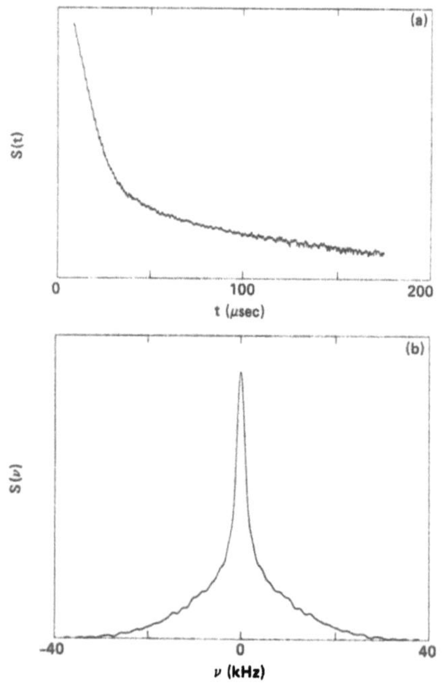

Fig. 3. (a) The free induction decay and (b) its Fourier transform for the H signal in Sample #5, an electonically good a-Si:H sample prepared on a heated substrate.

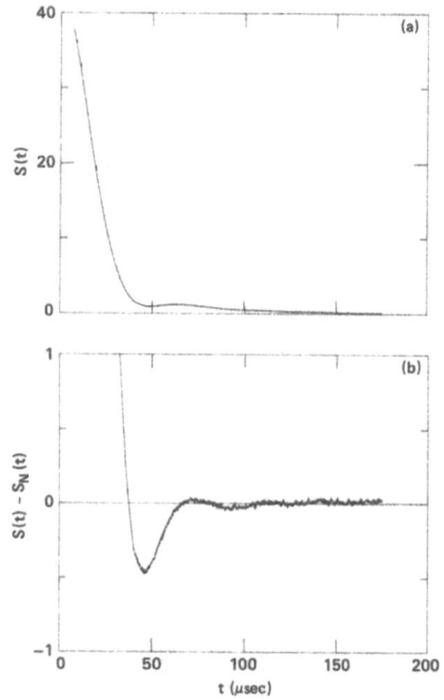

Fig. 4. (a) The free induction decay for Sample #1, an electronically poor a-Si:H sample prepared on a RT substrate. (b) The FID minus the slow component, the narrow line, on an expanded scale, showing the beat due to SiH_2.

been observed in a class of a-Si:H samples that had been prepared on room temperature substrates (samples #1-#4). A typical spectrum is shown in Fig. 4 (for sample #1). Fig. 4a is the total signal and Fig. 4b is the total signal minus the narrow line (the slow decay in the FID). This time-domain signal is very similar to the calculated spectrum of Fig. 2b, solid curve. It is, however, quite different from the signal due to sample #5 shown in Fig. 3. A contrast of the broad components of the lines, that is, the total signal minus the narrow line, is similar to that shown for the calculation of Fig. 2. The fact that there is a beat in the FID for sample #1 is strong evidence for a doublet spectrum and rules out a Gaussian line shape. The fact that only one beat is observed indicates that the extra broadening, β, due to other H atoms, is comparable to α. This is born out in a detailed fit to the data. A fit to a Gaussian gave a poor fit, a factor of two larger quality of fit parameter, R-value, than for a doublet line shape. A fit to a Pake doublet gave a good fit and yielded $\beta = 6\pm1$ kHz and $\alpha = 7.5\pm1$ kHz. Since $\beta < \alpha$, a beat could be observed. Also since $\alpha = 7.5\pm1$ kHz agrees with the value of 7 kHz calculated for SiH_2, this spectrum is due to SiH_2 units.

Similar spectra were observed on several samples (#1-#4) and the results are listed in Table 1. All the samples have a narrow NMR line which contains about 4 at.% of the H and has a FWHM$\simeq$ 3 kHz, similar to the narrow line of the heated-substrate samples, #5-#8. They also have a very high H-content, 20-30 at.% versus 10 at. % for the heated-substrate samples. This extra H goes into the broad line which is due almost entirely to SiH_2 units with $\alpha \simeq 7$ kHz and $\beta \simeq 6$ kHz. The resulting second moment, given by Eq.(2), is $(\Delta\omega^2)^{1/2} \simeq 9$ kHz. This is comparable to the Gaussian width of heated-substrate samples (#5-#8), namely, $\sigma \simeq 10$ kHz.

These results indicate that the H in a-Si:H which is prepared on room temperature substrates is predominately bonded as SiH_2. For samples prepared on heated substrates ($\simeq$230°C) and with low RF power, no doublet is observed, indicating that there is little SiH_2. Another possibility is that there is some SiH_2 but that it is clustered with other Si-H units so that $\beta > 7$ kHz, giving rise to a Gaussian line. In either case the heated-substrate samples contain significantly less SiH_2 than the room-temperature substrate samples. These conclusions are in agreement with infrared and Raman results,[10-11] where it was observed that the absorption constant is larger for SiH_2 than for SiH in the low-temperature-substrate material but that the reverse is true for substrate temperatures above 200°C.[11] Due to the fact that the dipole matrix elements for SiH and SiH_2 are not well known,[12] it is difficult to obtain the precise number of H atoms in each of the two bonding configurations from the optical absorption coefficient. Since NMR measures the quantities of H directly, these results can help to quantify the fraction of H in each of the bonding configurations as a function of the sample preparation conditions.

SPECTRUM OF MOLECULAR H_2

In addition to the approximately 10 at. % H that is incorporated in plasma deposited a–Si:H and which is bonded to the Si, there is a small amount of molecular H_2 that is trapped in voids in the a–Si matrix.[13-15] The H_2 is inert so that it does not affect the electronic properties of the a–Si; however, it exhibits some interesting properties since the molecules are quantum rotors interacting with each other and with their a–Si container.

The NMR spectrum of molecular H_2 differs from that of SiH_2 in two significant respects. First, the H-H spacing for H_2 is significantly smaller, 0.74Å versus 2.33A. As a result, for a rigid H_2 molecule, the dipolar coupling constant $\alpha_R = 222$ kHz, and the difficulties encountered with SiH_2 where $\beta \simeq \alpha$ should not occur. Secondly, since H_2

is a quantum rotor, there is a partial averaging of the dipole-dipole interaction due to the rapid rotational motion of the molecule.[16] This reduces α_K by a factor of 2/5 so that $\alpha(H_2)$ = 88 kHz. Third, since H_2 is not rigidly fixed in the lattice but rather is a rotor interacting weakly with the lattice, there can be a total averaging to zero of the H-H dipolar interaction when $k_BT > E_Q$, where E_Q is the quadrupole interaction of the molecule with the lattice. This interaction tends to orient the angular momentum, **J** , of the molecule in a specific direction relative to the a-Si matrix. So at low temperatures ($k_BT < E_Q$), one should observe a Pake doublet, and for high temperatures ($k_BT > E_Q$), the rapid tumbling of the molecule will average the doublet splitting the zero -- only an unsplit central line will be observed.

Four a-Si:H samples were prepared by plasma decomposition of pure SiH_4 gas, using low power density (0.025 W/cm^2) and Al substrates heated 230 C. Two of these samples were then annealed at 500°C for ½ hour to increase the molecular H_2 content, and are similar to the samples used for calorimetry experiments.[17] The results are very similar for the two annealed samples, so the specific results are quoted only for one sample. The case is similar for the two unannealed samples.

The NMR spectral results at room temperature show the standard unsplit line with two width components (a broad line and a narrow line) for all the samples studied, similar to that of Fig. 3. When the samples are cooled to low temperatures, a third component appears in the NMR spectrum,[18] a powder-averaged Pake doublet shown in Fig. 5. The observed doublet is due to orientationally ordered molecular H_2 and is very similar to that seen in bulk, solid normal-H_2.[16] When placed in a crystal field, the degeneracy of the three m_J levels of o-H_2 (J = 1; m_J = 0, ±1) can be lifted; i.e., the rapid tumbling of the H_2 molecule is frozen out. For H_2 in a-Si, the orientational part of the crystalline potential, V_c, consists of two parts: the

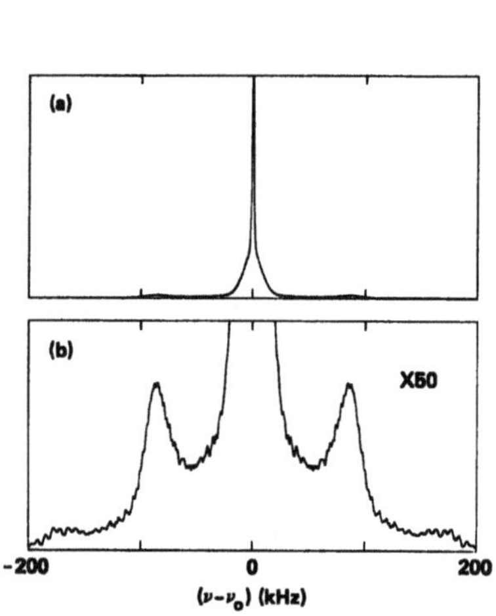

Fig. 5. H NMR spectrum from a Fourier transform of the free induction decay at 92 MHz and 1.43 K in one of the annealed samples showing (a) the broad and narrow central lines and (b) the molecular H_2 powder pattern with broadened singularities at ± (88 ± 5) kHz.

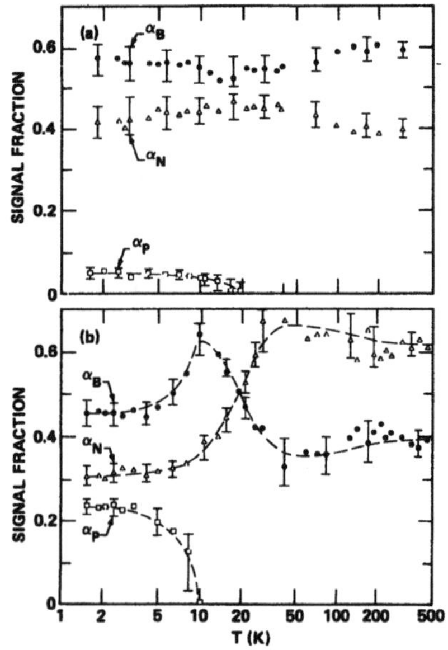

Fig. 6. Fraction of the total H NMR signal in the three components of the line versus temperature for (a) an unannealed sample and (b) an annealed sample.

electrostatic quadrupole–quadrupole (EQQ) interaction between the o–H_2 molecules and the electric field gradients (EFG) due to the electronic charge distribution of the a-Si matrix interacting with the o–H_2 quadrupole moment. Both contribute to the crystal-field splitting, Δ, which for an axially symmetric field gradient is the splitting between the $m_J = 0$ ground state and the doubly degenerate $m_J = \pm 1$ states. The EQQ part of V_c is responsible for the order–disorder transition in solid normal–H_2 (75% o–H_2) at $T_c \simeq 1.6$ K. For $T \ll \Delta$, the probability that the $m_J = 0$ state is occupied, p_0, is 1 and the H_2 molecules are *locally* ordered in their respective crystal fields. For $T \gg \Delta$, $p(m_J) = 1/3$ for all three m_J and the H_2 molecules are orientationally disordered with respect to their local environment. For such an axially symmetric potential, the resonance due to the orientationally ordered H_2 in a–Si is a powder averaged Pake doublet with a splitting of $\nu_P = 176$ kHz $(3p_0-1)/2$. The splitting 176 kHz is merely 2α. So for $T \ll \Delta$, $\nu_P = 176$ kHz, and for $T \gg \Delta$, $\nu_P = 0$; i.e., the doublet has collapsed. In the intermediate temperature region, $T \sim \Delta$, the lineshape can be approximated by a broadened powder pattern, with a reduction in ν_P, plus a broad central component due to the molecules for which $\nu_P = 0$.

This general lineshape behavior was observed for both the annealed and unannealed samples. The Pake doublet splitting is $\nu_P = 180 \pm 10$ kHz for both sets of samples and for all temperatures below T_C, except possibly just below T_C where the doublet signal is disappearing. This splitting is consistent with that observed in solid H_2. The different samples have the same ν_P but differ in two other important aspects, namely, in the concentration of molecular H_2, $n(H_2)$, and in T_C.

First, $n(H_2)$ is larger in the annealed samples (A) than the unannealed ones (U), as expected. From the spectral area under the Pake doublet, we obtain $n(H_2)|_A \approx 1$ at.% and $n(H_2)|_U \approx 0.25$ at.%. These values of $n(H_2)$ are larger than that inferred from T_1 measurements. An analysis of T_1 yields $n(H_2)|_U \simeq 0.1$ at.%[19] and $n(H_2)|_A \approx 0.2$ at.%. But the T_1 is affected only by molecules that serve as relaxation centers for the H bonded to the Si. This is the H_2 on the surface of a void, not that in the void interior since they do not couple to the H in the bulk.[20] Thus the different determinations of $n(H_2)$ can give a rough estimate of the mean surface-to-volume ratio of the void (assuming they are completely filled with H_2). This is about 0.4 for U and 0.2 for A. A rough estimate of the mean void radius, assuming solid H_2 density and spherical shape, is of order 25 Å and 50 Å for the unannealed and annealed material, respectively. The void size is larger for the annealed sample, as is to be expected.

Secondly, the value of T_C for sample U is ≈ 20 K, whereas for sample A, $T_C \approx 10$ K. This is seen in Fig. 6 where the fraction of the total signal in the various lines – Pake doublet (α_P); broad, unsplit line (α_B); and narrow, unsplit line (α_N) – is plotted against temperature. Well below T_C, α_P is constant as observed in solid H_2, indicating that all the H_2 molecules are orientationally ordered and are contributing to the Pake doublet. This implies that our value of $n(H_2)$ includes all the H_2 molecules. As $T \to T_C$, the value of α_P drops to zero.

The different T_C's for these samples can be understood as follows. For bulk, solid n–H_2, $T_C \simeq 1.6$ K and Δ is due to the EQQ interaction. For H_2 in large voids in a–Si, this same interaction is operative, but there is also the added a-Si matrix EFG, which increases the average value of Δ. As a result, the annealed material has a larger T_C of ≈ 10 K. For the unannealed sample, the amount of H_2 in the voids is smaller and the void size itself is smaller. Thus most H_2 experience the larger EFG from the a-Si and T_C is increased to $\approx$ 20K.

Above T_C, the line shape for the unannealed samples are about the same as that at room temperature. This is due to the fact that the H_2 is only 4% of the total H signal and its contribution to the lineshape parameters is within the uncertainties, except when split off as a doublet below T_C. For the annealed samples, this is not the case since the H_2 contributes 25% of the total signal. For these samples, the lineshape parameters are seen to change above T_C, as seen in Fig. 6b. Above T_C, the doublet has collapsed onto the unsplit central line due to the averaging of the spectral splitting by the rapid tumbling of the molecules; all the m_J states are equally occupied. This component contributes to the broad line since the *inter*molecular dipole interaction is not averaged out by the tumbling. This causes an increase in α_B. When the H_2 molecules begin to diffuse, with their hopping rate comparable to the linewidth, this *inter*molecular interaction will be averaged and the line will narrow. This happens between 15 K and 40 K, above which the H_2 molecules contribute only to the narrow line. Their width has been motionally narrowed. This causes an increase in α_N with a corresponding decrease in α_B. This happens in pure solid H_2 where above T_C only a broad line exists, and it narrows as the melting temperature of 14 K is approached.[21] It is fully narrowed within a few degrees of the melting temperature, T_m. As the pressure is increased, both transition temperatures move up. Assuming that the T_m-versus-pressure curve for pure H_2 applies,[22] our value of $T_m \approx$ 15–40 K gives p ~ 0–1.5 kbar. This result is comparable to the 2 kbar result from infrared measurements,[23] but also indicates that there is a rather broad range of pressures.

Acknowledgment: I gratefully acknowledge the help of M. Stutzmann, who annealed the a-Si:H samples and provided significant physical insight, and the technical help of S. E. Ready.

REFERENCES

1. For a review, see, for example, "Fundamental Physics of Amorphous Semiconductors", edited by F. Yonezawa, Springer-Verlag, Berlin, 1981.

2. Ref. 1, M. Brodsky, p. 56.

3. G. E. Pake, J. Chem Phys. 16, 327 (1948).

4. A. Abragam, "The Principles of Nuclear Magnetism", Oxford University Press, London (1961), p. 216-223.

5. I. J. Lowe and R. E. Norberg, Phys. Rev. 107, 46 (1957).

6. D. C. Look, I. J. Lowe and J. A. Northby, J. Chem. Phys. 44, 3441 (1966).

7. J. A. Reimer, R. W. Vaughan and J. C. Knights, Phy Rev. B 24, 3360 (1981).

8. W. E. Carlos and P. C. Taylor, Phys. Rev. B 26, 3605 (1982).

9. M. Lowry, F. R. Jeffrey, R. G. Barnes and D. R. Torgeson, Solid State Comm. 38, 113 (1981).

10. M. H. Brodsky, M. Cardona and J. J. Cuomo, Phys. Rev. B 16, 3556 (1977).

11. G. Lucovsky, R. J. Nemanich and J. C. Knights, Phys. Rev. B 19, 2064 (1979).

12. M. Cardona, Phys. Stat. Sol. (b) 118, 463 (1983).

13. M. S. Conradi and R. E. Norberg, Phys. Rev. B 24, 2285 (1981).

14. W. E. Carlos and P. C. Taylor, Phys. Rev. B 25, 1435 (1982).

15. D. J. Leopold, J. B. Boyce, P. A. Fedders and R. E. Norberg, Phys. Rev. B 26, 6053 (1982).

16. F. Reif and E. M. Purcell, Phys. Rev. 91, 631 (1953).

17. J. E. Graebner, B. Golding, L. C. Allen, D. K. Biegelsen and M. Stutzmann, Phys. Rev. Lett. 52, 553 (1984).

18. J. B. Boyce and M. Stutzmann, Phys. Rev. Lett. 54, 562 (1985).

19. J. B. Boyce and M. J. Thompson, J. Non-Cryst. Solids 66, 129 (1984).

20. P. A. Fedders, R. Fisch and R. E. Norberg, Phys. Rev. B 31, 6887 (1985).

21. G. W. Smith and C. F. Squire, Phys. Rev. 111, 188 (1958).

22, V. Diatschenko and C. W. Chu, Science 212, 1393 (1981).

23. Y. J. Chabal and C. K. N. Patel, Phys. Rev. Lett. 53, 210 and 1771 (1984).

DEUTERON MAGNETIC RESONANCE IN SOME AMORPHOUS SEMICONDUCTORS*

V. P. Bork, P. A. Fedders, and R. E. Norberg
Washington University, St. Louis, Missouri 63130

D. J. Leopold,† K. D. Mackenzie, and W. Paul
Harvard University, Cambridge, Massachusetts 02138

INTRODUCTION

Deuteron magnetic resonance (DMR) measurements have been made on plasma-deposited samples of a-Ge:D,H; a-Si:D,F; and a-SiGe:D,F. The experiments were performed at 30.7 MHz with a Bruker CXP 200 spectrometer and a probe constructed to permit operation at high rf power down to temperatures near 4 K. The 50 to 100 mg samples comprise less than 1 mm flakes in Kel-F containers of 5 mm diameter.

Spin-lattice relaxation times (T_1) were obtained by the saturation-recovery method. Transverse relaxation times (T_2) for the narrow central component of the line shapes were measured directly from the Lorentzian line shapes. The line shapes are Fourier transforms (FT) of quadrupole echo (QE) or free induction decay (FID) transients. Complete line shapes and long T_1 measurements were obtained with a composite pulse QE sequence with a 3.5 μsec $\pi/2$ pulse length. Long T_1 and weak signal strength made data accumulation tedious: A single T_1 measurement or spectral line shape often required over 12 hours of data acquisition.

Samples were prepared by rf glow discharge decomposition of gas mixtures in a system designed[1] to reduce contamination by water vapor or oxygen. Some details of the sample preparations and the DMR spin counts are listed in Table I. Also included in the table for comparison with present results are two Xerox PARC a-Si:D,H samples (designated I and V), for which DMR results have been reported previously.[2,3] The three-digit numbers for the other materials are Harvard sample designators.

DMR LINE SHAPES

The DMR line shapes show three spectral components: a resolved quadrupolar doublet with splittings in good agreement[3] with bonded-D infrared stretching mode absorptions; a broad central component associated with D

*Supported in part by NSF Grants DMR 82-04166, 83-04473, and 85-03083 and by SERI under subcontract XB-2-02144-1 of prime contract EG-77-C-01-4042 with the Department of Energy.

†Now at McDonnell Douglas Research Laboratories, St. Louis, Missouri.

quadrupolar couplings in more disordered and perhaps more weakly-bound configurations; and a narrow central component associated with D_2 in micro-voids. The relative populations vary with sample composition and preparation conditions.

Table I. Sample Preparation and Deuterium Concentrations

Sample	Si:D,H #I	Si:D,H #V	Ge:D,H #258	Si:D,F #325	SiGe:D,F #357
Gas Mixture (%)	SiH_4 (5) D_2 (95)	SiD_4 (5) Ar (95)	GeH_4 (9) D_2 (91)	SiF_4 (70) D_2 (30)	SiF_4 (49) GeF_4 (1) D_2 (50)
Substrate Temperature (C)	25	25	230	300	200
RF Power (W)	18	2	10	35	50
n(D) (at.%)	24	10.5	0.7	3.9	2.4
Resolved Doublet (D)	20	7	0.1	2.8	1.5
Broad Central (D)	1	3.5	0.3	0.4	0.8
Narrow Central (D_2)	3	---	0.3	0.7	0.1

Figure 1 shows FTQE DMR spectra between 4 and 14K for a-Si:D,H (I and V); a-Si:D,F (325); and a-SiGe:D,F (357). Figure 2 shows similar spectra for a-Si:D,H (I and V); a-Ge:D,H (258); and a-SiGe:D,F (357) at warmer temperatures. All of the a-Si and a-SiGe spectra show the 66 kHz zero asymmetry Pake quadrupolar doublet characteristic[2] of SiD and SiD_2 bonded configurations. The anticipated[2] corresponding doublets (61.5 and 54 kHz) in the a-Ge sample are not resolved, and only a much broadened spectral feature appears in Fig. 2 near the vertical lines drawn to indicate 58 kHz. This component, which is also unresolved in the a-SiGe sample, indicates that the bonded D configurations are more widely distributed in a-Ge than in a-Si. However, well defined IR features associated with bonded D were obtained for all a-Ge samples, including two a-SiGe:D,H and a-SiGe:D,F samples for which no useful DMR signals were observed. A relatively large fraction of the DMR signal for these samples may correspond to distributed asymmetry parameters.

Sample I is unusual among a-Si samples for which DMR has been reported because of its large D content and large narrow central D_2 component. The sample was deposited rapidly on a cold substrate at high rf power. Sample V is a more typical high quality hydrogenated a-Si material. There is much less D_2, but about the same relative amounts of broad central and resolved doublet D as seen in Sample I. The a-Si:D,F and a-Ge:D,H samples both show small total D content and a relatively larger D_2 content, which presumably reflects the fractional accessible void volume. The a-SiGe:D,F alloy sample also shows a significant narrow central D_2 component.

Deuteron nuclear spin relaxation times T_1 and T_2 for these spectral components show remarkable features. Figure 3 compares relaxation times observed for a-Ge;D,H (258) (solid symbols) with those previously reported[2] for a-Si:D,H (I) (open symbols), the a-Si sample which contained an unusually

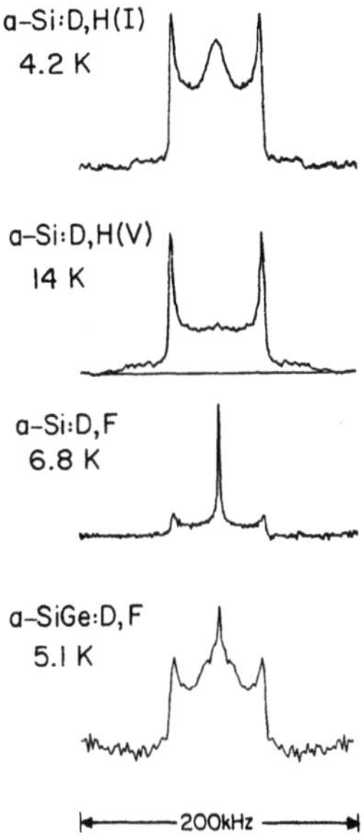

Figure 1. DMR spectra between 4 and 14 K.

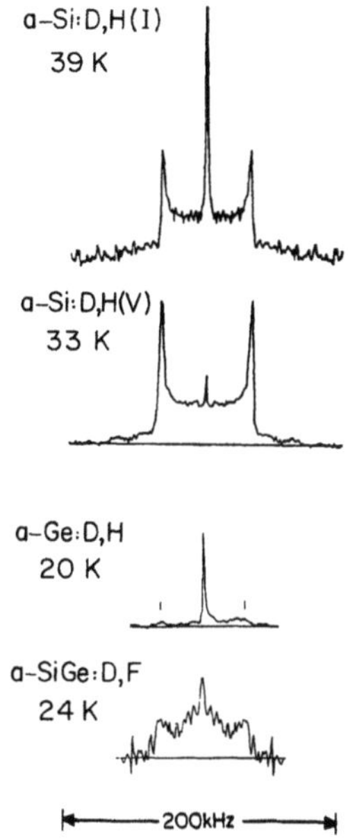

Figure 2. DMR spectra between 20 and 39 K.

large amount of D_2. The T_1 results for the narrow central D_2 component in the two samples are nearly identical at 30 MHz and reflect relaxation by effectively-dilute relaxation center[4] p-D_2 located at void surfaces. The $T_1(D_2)$ predicted[2,5] for decoupled relaxation centers is shown by the curved line. T_1 for the broad central components in a-Si:D,H (I and V) has been reported to be nearly as rapid as that for the narrow central component. The broad central lines presumably reflect relatively weakly bound deuterons (WBD) located in the vicinity of the voids.

The circles and solid triangles at the bottom of Fig. 3 show that the temperature dependent T_2 for the narrow central components are the same in the a-Si and a-Ge samples. The data can be well-approximated by a power law, $T_2 = aT^{0.6}$, whose temperature variation may reflect transverse relaxation by local disorder modes associated with migration on fractal surfaces of microvoids. The diamonds indicate temperature independent relaxation times T_{2Q} associated with the resolved doublet component in a-Si:D,H (I). Three narrow central T_2 values observed between 14 and 33 K for the small D_2 components in a-Si:D,H (V) (Figs. 1 and 2) agree with the data in Fig. 3.

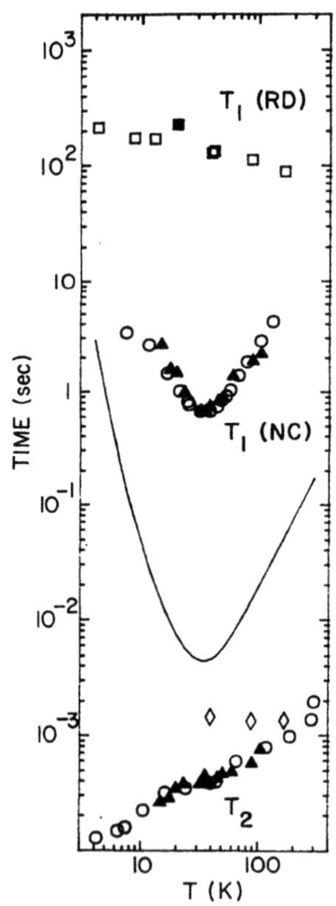

Figure 3. Comparison of a-Ge:D,H DMR relaxation times (solid symbols) with a-Si:D,H (open symbols).

The open squares at the top of Fig. 3 indicate T_1 for the resolved doublet in a-Si:D,H (I). The solid square at 20 K represents a 230 sec $T_1(D)$ measured for a-Ge:D,H. Closer inspection revealed that there is in fact an exceedingly wide distribution of T_1 associated with this component in the a-Ge sample. $\pi/2$ pulses with repetition intervals as short as 5 msec did not completely saturate out the smeared doublet feature visible in Fig. 2. The spin-lattice relaxation rate of the bonded D in the a-Ge sample ranges over five orders of magnitude and presumably reflects the geometrical distribution of bonded D sites with respect to the microvoid D_2 relaxation centers. Because spin diffusion of bonded D is reduced considerably by the doublet quadrupolar splitting, the degree to which bonded D will relax via D_2 relaxation centers will depend on their proximity to the relaxation centers.

Figure 4 compares the a-Si:D,H (I) reference data (open symbols) with the relaxation times now observed in a-Si:D,F (325), The striking new results indicate the presence of bulk solid and liquid D_2 in the a-Si:D,F microvoids. The solid squares at the top of Fig. 4 indicate the very long relaxation times (≈ 500 sec) observed for the resolved doublet component in the a-Si:D,F. The a-Si:D,F narrow central component exhibits two relaxation times, designated by solid triangles and dots. The dots in the middle of Fig. 4 show a D_2-related T_1 minimum somewhat slower than that reported in a-Si:D,H (I). The triangles show a more rapid narrow central relaxation component which appears to be temperature independent for $T \geq 14$ K. This rapid relaxation component was not observed for $T < 10$ K.

Narrow central component magnet-corrected transverse relaxation times T_2 (solid dots) are shown near the bottom of Fig. 4. There is a rapid T_2 increase between 10 K and 16 K followed by a slower increase to 20 K and a leveling off at higher temperatures.

The sloping solid line shows the reported[6] temperature variation of T_2 between 13 K and 17 K for unconstrained solid D_2 with an $x = 0.33$ p-D_2 fraction. The dashed curve shows the coefficient of self-diffusion (on the right hand scale) reported[7] for liquid n-H_2 at SVP. Liquid D_2 diffusion must follow a similar curve. It is probable that the observed temperature variation of T_2 for the narrow central DMR component in a-Si:D,F (325) reflects the melting of bulk solid and diffusion in dense fluid D_2 in microvoids. Either the presence of F produces unusually large voids (which does not seem likely) or else void surfaces are rendered less effective in controlling the relaxation properties of the contained D_2 than was the case in a-Si:D,H (circles).

The same hypothesis about the presence of bulk void D_2 explains the unusual rapid T_1 (solid triangles) observed for the narrow central DMR component in a-Si:D,F (325). The curved line near 10 sec shows $T_1(D_2)$ reported[8] at 4.7 MHz for solid hcp D_2 with $x = 0.33$. The two crosses at 19.9 K indicate the corresponding[8] $T_1(D_2)$ for liquid o-D_2 (1200 sec) and p-D_2 (1.2 sec) in $x = 0.33$ liquid D_2. It is evident that the faster narrow central T_1 fraction observed above 19 K in a-Si:D,F probably reflects the intrinsic relaxation of bulk dense fluid p-D_2 in the voids. The disappearance of the fast T_1 component below 14 K then corresponds to the freezing of the bulk D_2, in agreement with the T_2 decrease.

Both the a-Si:D,F and the alloy a-SiGe:D,F show a small (≈5%) component characterized by a broad line shape and a very short T_1. The preliminary T_1 for the a-Si:D,F sample (indicated by solid diamonds in Fig. 4) suggest that this component is closely related to the p-D_2 relaxation centers, whose theoretical T_1 are given by the dashed line. This component probably arises from dilute or effectively-dilute p-D_2 relaxation centers decoupled from most bonded D. The broad lines may reflect a distribution of adsorption sites or may reflect the short intrinsic T_2 for relaxation center p-D_2 on the cold side of the $T_1(D_2)$ minimum.

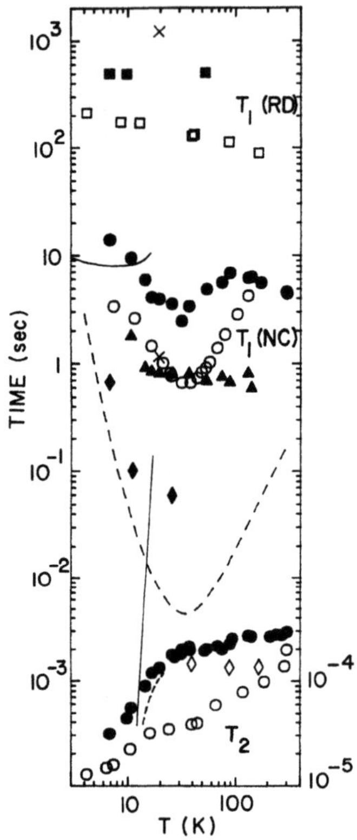

Figure 4. Comparison of a-Si:D,F DMR relaxation times (solid symbols) with a-Si:D,H (open symbols) and with solid and liquid D_2.

CONCLUSIONS

Deuterium tightly bonded in a-Ge has a wider distribution of configurations than does D in a-Si. The fractional void volume in a-Ge:D is larger than in a-Si:D samples prepared under similar conditions. The introduction of F reduces the fraction of bonded D in well defined doublet configurations and produces significant amounts of D_2 in voids in which the void surfaces have little effect on the D_2 relaxation.

REFERENCES

1. K. D. Mackenzie, J. R. Eggert, D. J. Leopold, Y. M. Li, S. Lin, and W. Paul, Phys. Rev. B 31, 2198 (1985).
2. D. J. Leopold, J. B. Boyce, P. A. Fedders, and R. E. Norberg, Phys. Rev. B 26, 6053 (1982).
3. D. J. Leopold, P. A. Fedders, R. E. Norberg, J. B. Boyce, and J. C. Knights, Phys. Rev. B 31, 5642 (1985).
4. P. A. Fedders, R. Fisch, and R. E. Norberg, Phys. Rev. B 31, 6887 (1985).
5. P. A. Fedders, Phys. Rev. B 20, 2588 (1979).
6. F. Weinhaus, H. Meyer, S. M. Myers, and A. B. Harris, Phys. Rev. B 7, 2960 (1973).
7. W. P. A. Hass, G. Seidel, and N. J. Poulis, Physica 26, 834 (1960).
8. B. Maraviglia, F. Weinhaus, S. M. Myers, and H. Meyer, Proc. 12th International Conference on Low Temperature Physics, 847 (1970), Kyoto.

CHARGE TRANSFER ELECTRON-EXCITON COMPLEXES

IN INSULATORS AND SEMICONDUCTORS

Constantine Mavroyannis

Division of Chemistry
National Research Council of Canada
Ottawa, Ontario, Canada K1A 0R6

ABSTRACT. We have considered the excitation spectrum arising from the coherent electron-exciton pairing in crystals at low temperatures. For the pairing processes, three different types of excitons have been considered: Frenkel excitons (tightly bound), Wannier-Mott excitons (loosely bound) and excitons of the intermediate binding. Expressions for the gap functions and transition temperatures at which metal to nonmetal phase transitions occur have been derived and discussed for each type of pairing. In the electron pairing with the intermediate exciton, the dispersion relations which determine the exciton and the electron-exciton modes are solved numerically and the derived results are graphically presented. The physical picture for the electron-exciton coupled modes is analogous to that for polaritons and magnon-phonon modes in crystals.

INTRODUCTION

Considerable interest has been recently given to the existence of charged excitons in insulating and semiconducting crystals. A charged exciton is a quasiparticle consisting of a bound state of an exciton and either an electron in the conduction band or a hole in the valence band, which moves with definite energy and wave vector through the crystal.[1-3] At the present time, experimental evidence for the existence of such conducting states is limited[4,5] but these types of excitations may play an important role in the metal-insulator transitions at high excitation intensities.[6] Theoretical calculations have been done so far when conduction electrons interact with Frenkel excitons[1,2,7,8] and Wannier-Mott excitons,[3] which are the appropriate collective excitations for molecular crystals and semiconductors, respectively. However, excitons with intermediate binding have also been found to exist in rare-gas solids.[9,10] We shall describe here our recent work[11,12] concerning the formation of a bound state consisting of an exciton of the intermediate binding and an electron which is excited into the conduction band. The derived results will be compared with those obtained when the pairing process takes place with a Frenkel exciton[1,7,8] and a Wannier-Mott exciton,[13] respectively.

INTERMEDIATE EXCITON-ELECTRON PAIRING

We consider a simple model for the crystal consisting of N identical atoms on a rigid cubic lattice with one atom per unit cell and lattice spacing a_0. Each atom is assumed to have two non-degenerate levels corresponding to the valence and conduction bands and all effects due to electron spin are discarded. At T=0K, the valence band is full of electrons while the conduction band is empty. The exciton (electron-hole pair) under consideration is formed when an electron from the valence band is excited into the conduction band leaving behind a hole in the valence band at the same lattice site. In addition, the usual band motions of electrons and holes from one lattice site to another without creating excitons are included. The Hamiltonian of such a system may be taken as[11-14]

$$
\begin{aligned}
H = & \sum_{\langle \ell\ell' \rangle} T_\alpha(\ell\ell')\alpha^+_{\ell'}\alpha_\ell - \sum_{\langle \ell\ell' \rangle} T_\beta(\ell\ell')\beta^+_{\ell'}\beta_\ell + E_\alpha \sum_\ell \alpha^+_\ell\alpha_\ell \\
& - E_\beta \sum_\ell \beta^+_\ell\beta_\ell + u_0 \sum_\ell \beta^+_\ell\beta_\ell\alpha^+_\ell\alpha_\ell - \sum_{\langle \ell\ell' \rangle} u(\ell\ell')\beta^+_\ell\alpha_\ell\alpha^+_{\ell'}\beta_{\ell'} \\
& + \sum_{\langle \ell\ell' \rangle} v(\ell\ell')\alpha^+_\ell\alpha_\ell\beta^+_{\ell'}\beta_{\ell'} ,
\end{aligned}
\tag{1}
$$

where $\alpha^+_\ell, \alpha_\ell$ are the creation and destruction operators for electrons at the lattice site $\vec{R}^0_\ell$ in the conduction band, and β^+_ℓ, β_ℓ are the corresponding operators for electrons in the valence band. E_α, E_β denote the on-site energy of the electron in the conduction and valence bands, respectively, and $\langle \ell\ell' \rangle$ denotes that the sum is to be done over nearest-neighbor pairs $\vec{R}^0_\ell$ and $\vec{R}^0_{\ell'}$. T_α and T_β are the hopping energies for electrons in the respective bands, u_0 is the on-site Coulomb interaction, and $u(\ell\ell')$, $v(\ell\ell')$ are the exchange and direct Coulomb interactions between electrons on neighboring lattices sites.

Let us introduce the exciton operator $b_\ell \equiv \beta^+_\ell\alpha_\ell$. For the two level system, we have $\beta^+_\ell\beta_\ell + \alpha^+_\ell\alpha_\ell = 1$. Using this and the usual anticommutation relations for the electron operators, we have $\beta^+_\ell\beta_\ell = b_\ell b^+_\ell$ and $\alpha^+_\ell\alpha_\ell = b^+_\ell b_\ell$. Making use of these results in equation (1), we find that the exchange Coulomb term is quadratic in the exciton operators. We also find that the direct Coulomb term is quartic in (a) the exciton operators and (b) the exciton and electron operators. When written in the form (a), we find that the direct Coulomb term describes an exciton density interacting with an exciton density at a neighbouring lattice site. The form (b) describes an electron density interacting with an exciton density on a neighbouring lattice site.

The excitation energy of the conduction electron-exciton complex is obtained by making use of the Green's function method. The two Green's functions which we are going to evaluate are defined as $G(\ell\ell';t) = \langle\langle b_\ell(t);b^+_{\ell'}(0)\rangle\rangle$ and $F(\ell\ell';t) = \langle\langle \alpha_\ell(t);b^+_{\ell'}(0)\rangle\rangle$, where the operators are in the Heisenberg reprsentation and use has been made of the notation, $\langle\langle A(t);B(0)\rangle\rangle = -i\theta(t)\langle [A(t),B(0)]_+\rangle$ for the operators A and B. The diamond brackets stand for an ensemble average. The properties of the Green's functions can be found elsewhere.[15]

Commuting the operators b_ℓ and α_ℓ in turn with the Hamiltonian in equation (1) and then decoupling the resulting equations of motion, we find that the Green's functions $G(\ell\ell';t)$ and $F(\ell\ell';t)$ are solutions of the two sets of coupled equations, whose Fourier transforms with respect to $(\ell-\ell')$ and the time, are given by $(\hbar=1)$

$$
G(k,\omega) = G^{HF}(k,\omega) - G^{HF}(k,\omega)\Delta(k)F(k,\omega) , \tag{2}
$$

$$[\omega-E_\alpha(k)-u_0-v(k=0)]F(k,\omega) = -\Delta^*(k)G(k,\omega) \, , \tag{3}$$

where

$$\Delta(k) = \frac{1}{N}\sum_{k'} v(k-k')\langle b_{k'}\alpha_k^+\rangle \, , \tag{4}$$

$$G^{HF}(k,\omega) = \frac{G_0(k,\omega)}{\varepsilon_+(k,\omega)} \tag{5}$$

with the single-particle Green's function given by

$$G_0(k,\omega) = \frac{1}{N}\sum_q \frac{1}{\omega-E_\beta^0-E_\alpha(q-k)-E_\beta(q)} \tag{6}$$

and $\varepsilon_+(k,\omega)$ is defined by

$$\varepsilon_+(k,\omega) = 1+[u(k)G_0(k,\omega)-v(k=0)G_0(k=0,\omega)] \, . \tag{7}$$

$E_\alpha(k)$ and $E_\beta(k)$ are the energy bands[14] and we have chosen $E_\beta=E_\beta^0+6T_\beta$ and $E_\alpha=6T_\alpha$ where $T_{\alpha,\beta}(\ell\ell')=-T_{\alpha,\beta}$ for nearest-neighbour lattice sites. u(k) and v(k) are the Fourier transforms for the nearest-neighbour exchange and direct Coulomb interactions of strength u and v, respectively. In eqs. (2)-(7), $G^{HF}(k,\omega)$ is the exciton-exciton Green's function which is correct in the Hartree-Fock (HF) approximation, $G_0(k,\omega)$ is the non-interacting Green's function[14] while $\Delta(k)$ is the coupling parameter describing the pairing of an exciton at the lattice site $\vec{R}_\ell$ with an electron, which is excited into the conduction band, at the nearest-neighbour lattice site $\vec{R}_{\ell'}$.

Solving equations (2) and (3), we obtain

$$G(k,\omega) = [\omega-E_\alpha(k)-u_0-v(k=0)]\,\frac{G_0(k,\omega)}{D^B(k,\omega)} \, , \tag{8}$$

$$F(k,\omega) = -\Delta^*(k)\,\frac{G_0(k,\omega)}{D^B(k,\omega)} \, , \tag{9}$$

where

$$D^B(k,\omega) = [\omega-E_\alpha(k)-u-v(k=0)]\varepsilon_+(k,\omega)-|\Delta(k)|^2G_0(k,\omega) \, . \tag{10}$$

Singularities of $G(k,\omega)$ and $F(k,\omega)$ are located at the solutions of $D^B(k,\omega)=0$. These correspond to the bound states of the electron-exciton complex, provided the roots of $D^B(k,\omega)=0$ are separated from the solution of the equation $\varepsilon_+(k,\omega)=0$ which determine the frequencies of the exciton band. They must also be different from $\omega=E_\alpha(k)+u_0+v(k=0)$ which corresponds to the band energies of the conduction electron renormalized by the screened Coulomb interactions. Referring to equation (10), we see that a finite value of $\Delta(k)$ may produce a bound state for the electron-exciton complex.

In the denominator of equation (6), E_β^0 is the separation between the valence and conduction bands for zero wave vector and it also corresponds to the value of the lower bound of the electron-hole excitations for k=0. For a stable electron-exciton complex, we therefore require that E_β^0 should satisfy the relation $E_\beta^0>u_0+6v$. For a simple cubic lattice, since we have $\Delta(k)=[v(k)/6v]\Delta_0$, where $\Delta_0=\Delta(k=0)$, it is straightforward to calculate the

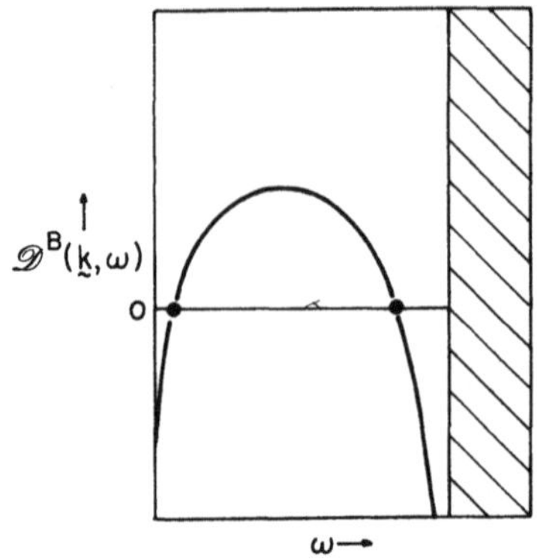

Fig. 1. Schematic of $D^B(k,\omega)$ as a function of frequency, for a fixed wave vector. The two solutions of $D^B(k,\omega)=0$ correspond to the coupled conduction electron-exciton modes. The hatched area corresponds to the single-particle excitons of electrons and holes.

frequency of the (electron-exciton) mode in the long wavelength limit. In particular, the frequency of the mode for k=0 is

$$\omega_0 = E^0_\beta - 2x_0(T_\alpha + T_\beta) \ . \tag{11}$$

where x_0 is the solution of the nonlinear equation

$$[E^0_\beta - u_0 - 6v - 2x_0(T_\alpha + T_\beta)][3(u-v)J(x_0) - (T_\alpha + T_\beta)] = \tfrac{1}{2}\Delta_0^2 J(x_0) \ . \tag{12}$$

Here, $J(x_0)$ is a one-dimensional integral:

$$J(x_0) = \int_0^\infty ds\ e^{-s(3+x_0)}\ I_0^3(S) \ , \tag{13}$$

where $I_0(S)$ is the modified Bessel function of the first kind, of order zero.

We have evaluated $D^D(k,\omega)$ numerically, assuming for simplicity that the coupling parameter $\Delta(k)$ is a constant Δ. We have chosen the bandwidth of the conduction band ($12T_\alpha$) and that of the valence band ($12T_\beta$) to be 0.6 eV. The nearest-neighbour exchange interaction (u) and the nearest-neighbour direct interaction (v) are set equal to 0.3 and 0.01 eV, respectively. The energy gap between the valence and conduction bands at zero wave vector is E^0_β=1.5 eV and the coupling parameter Δ is chosen as 0.1 eV. A schematic plot of $D^B(k,\omega)$ is shown in Fig. 1. There we see that, for a given value of k, there are two coupled mode frequencies below the electron-hole continuum so that there are two branches to the dispersion relation. In Fig. 2, we have plotted the dispersion relation for the electron-exciton complex in the (100) direction. The dispersion relation for the coupled mode has a form similar to that for polaritons and magnon-phonon modes in ferromagnets. Referring to Fig. 2, we find that as $k\to\pi/a_0$, the upper branch approaches the exciton band and the lower branch approaches the conduction band. However, as k decreases the separation between these modes becomes larger.

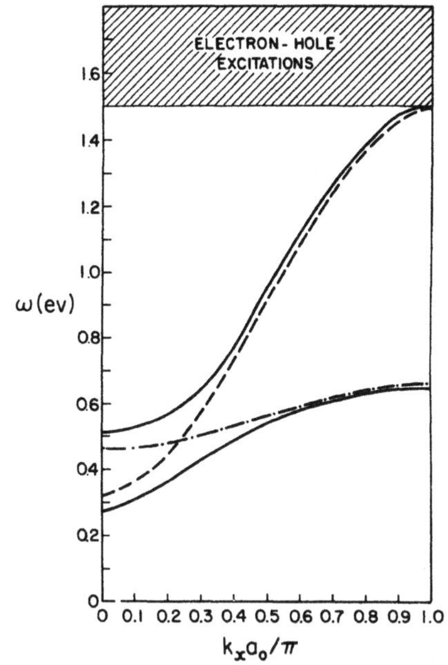

Fig. 2. Dispersion relation for coupled electron-exciton modes (——) in an insulator, with intermediate binding for the excitons. Also shown are the exciton band (----) and the conduction electron band (-•-•-) from which the coupled mode is formed. The parameters used in the calculation are given in the text.

FRENKEL EXCITON-ELECTRON PAIRING

For the process of coherent pairing of a Frenkel exciton with an excited electron, the fifth term in the Hamiltonian (1), which describes the Coulomb interaction u_0 between the electron and hole densities located on the same lattice site in the crystal, does not make any contribution and, therefore, is discarded. Such a model is appropriate for excitations occurring in molecular crystals where van der Waals interactions between the atoms are prevailing. Then using the Hamiltonian (1) and the Green function formalism, which is analogous to that used in the theory of superconductivity,[16] the gap functions Δ(k) describing the Frenkel exciton-electron pairing[1,8] is found to be

$$\Delta(k) = \frac{\bar{\omega}}{\sinh(1/\rho)} \approx 2\bar{\omega}e^{-1/\rho} , \tag{14}$$

where

$$\bar{\omega} = \xi^2\sigma/4m , \quad \xi = (6\pi^2 n)^{1/3} , \tag{15}$$

$$\rho = N(0)v(k) = 3V(k)/2\bar{\omega} , \quad \sigma = m\left|\frac{1}{m_e} - \frac{1}{m_{exc}}\right| , \tag{16}$$

with n=N/V is the exciton (electron) density, N(0) is the density of states, m_e and m_{ex} are the effective masses for the electron and exciton, respectively; m is the mass of the free electron and v(k) is the Fourier transform of the interaction energy $v(\ell\ell')$. In deriving eq. (14), use has been made of the effective mass approximation for the Frenkel exciton and

Table 1. Computed values for the gap function Δ/σ, transition temperature T_c, binding energy W_0/σ, and average kinetic energy $\bar{\omega}/\sigma$.

n (cm^{-3})	ρ	$v(k)/\bar{\omega}$	Δ/σ (cm^{-1})	T_c/σ (°K)	$-W_0/\sigma$ (cm^{-1})	ω/σ (cm^{-1})
2×10^{18}	1.5	1.0	52	43	96	37
	2.0	1.3	74	61	121	37
	2.5	1.6	93	76	146	37
	3.0	2.0	111	91	174	37
4×10^{18}	1.5	1.0	83	68	152	59
	2.0	1.3	117	96	191	59
	2.5	1.6	146	120	232	59
	3.0	2.0	167	137	277	59
6×10^{18}	1.5	1.0	109	89	200	77
	2.0	1.3	154	126	250	77
	2.5	1.6	193	158	304	77
	3.0	2.0	231	189	363	77
8×10^{18}	1.5	1.0	132	108	242	93
	2.0	1.3	187	153	303	93
	2.5	1.6	234	192	369	93
	3.0	2.0	281	230	440	93
1×10^{19}	1.5	1.0	153	125	281	108
	2.0	1.3	217	188	352	108
	2.5	1.6	271	222	428	108
	3.0	2.0	326	267	511	108

excited electron spectra, respectively. The transition temperature T_c may be found by the relation

$$T_c = 0.57\ \Delta(k)/k_B\ , \tag{17}$$

where k_B is Boltzmann's constant. The coherent energy W_0, which describes the binding energy corresponding to the electron-exciton pairing, has been calculated by averaging the Hamiltonian (1) to obtain the ground-state energy of the system[1,7,8] and it is found to be

$$W_0 = -\frac{N(0)\bar{\omega}^2}{\tanh(1/\rho)} = -\frac{3\bar{\omega}}{2\tanh(1/\rho)}\ . \tag{18}$$

Estimates for the gap function $\Delta(k)$, transition temperature T_c, binding energy W_0 divided by σ are given in Table 1 as a function of the exciton (electron) density n. The data in Table 1 indicate that for exciton concentrations $n > 2\times10^{18}$ cm^{-3}, the energy-gap function is a measurable quantity provided that we are dealing with the strong coupling limit which occurs when the average interaction, energy is equal or greater than the average kinetic energy, $v(k) > \bar{\omega}$. The last column in Table 1 gives the values $\bar{\omega}/\sigma$ for different exciton densities which indicate the condition of strong coupling limit, $v(k) > \bar{\omega}$ can be easily fulfilled with a reasonable strength of the required interaction v(k). The results of the present study imply that at exciton densities $n > 2\times10^{18}$ cm^{-3}, the observation of the charge-carrying electron-exciton states is feasible in molecular organic solids.

For the process under consideration, the sixth term in the Hamiltonian (1), which describes the interaction between two Frenkel excitons, does not contribute and, hence, it is neglected. Then the Hamiltonian (1) is appropriate to describe excitations occurring in semiconducting crystals which are characterized by small energy gaps between the valence and conduction bands and large dielectric constants. In this case, a similar calculation[3] to that used before gives the following expression for the gap function

$$\Delta(k) = \frac{\bar{\omega}}{\sinh(1/\rho)} \left[1 - \frac{2d}{\bar{\omega}}\left(1 + \frac{d}{\bar{\omega}}\right)\left(\cosh\frac{1}{\rho} - 1\right)\right]^{\frac{1}{2}} \tag{19}$$

$$\approx \frac{\bar{\omega}}{\sinh(1/\rho)}, \quad \text{for either } \frac{d}{\bar{\omega}} << 1 \text{ or } \frac{d}{\bar{\omega}} \approx 1, \tag{20}$$

where $d=\mu e^4/4$ is one-half of the energy of the exciton Rydberg and μ is the reduced mass of the electron-hole pair. The binding energy W_0 is found to be

$$W_0 = \frac{N(0)\bar{\omega}^2}{\tanh(1/\rho)} \left[\left(1 + \frac{d}{\bar{\omega}}\right)^2 + \left(\frac{d}{\bar{\omega}}\right)^2 - \frac{2d}{\bar{\omega}}\left(1 + \frac{d}{\bar{\omega}}\right)/\cosh(1/\rho)\right] \tag{21}$$

$$\approx \frac{-N(0)\bar{\omega}^2}{\tanh(1/\rho)} = \frac{-3\bar{\omega}}{2\tanh(1/\rho)}, \quad \text{for either } \frac{d}{\bar{\omega}} << 1 \text{ or } \frac{d}{\bar{\omega}} \approx 1. \tag{22}$$

For most semiconducting solids $v(k)$ is expected to be of the order of the energy of the exciton Rydberg. Then considering eq. (16) for ρ, the limit $d<<\bar{\omega}$ also implies that $\rho<<1$ provided that $v(k)\approx 2d$, which is the weak coupling limit. In the limiting case when $d\approx\bar{\omega}$ then $\rho\gtrsim 1$, which may be considered as the intermediate or strong coupling limit. In both cases, $\Delta(k)$ and W_0 are determined by eqs. (20) and (22), respectively. Finally, when $d>>\bar{\omega}$, $\Delta(k)$ is given by eq. (19) provided that the relation

$$\frac{2d}{\bar{\omega}}\left(1 + \frac{d}{\bar{\omega}}\right)\left(\cosh\frac{1}{\rho} - 1\right) < 1$$

is satisfied. This corresponds to the extreme strong coupling limit where $\rho>>1$ provided that $v(k)\approx d$. In this case an enhancement of the binding energy of the complex is anticipated to occur.

In conclusion, we have discussed the excitation spectra arising from the interaction of Frenkel, Wannier-Mott, and intermediate type excitons with conduction electrons to form bound states at low temperatures; such conducting states are appropriate to occur in molecular crystals, semiconductors and rare-gas solids, respectively. We hope that the present discussions will stimulate experimental interest on the spectroscopic properties of electron-exciton complexes and their transport properties as well.

REFERENCES

1. C. Mavroyannis, Physica 77, 373 (1974).
2. E.L. Nagaev and E.B. Sokolova, Phys. Status Solidi (b) 64, 441 (1974).
3. C. Mavroyannis, J. Low Temp. Phys. 20, 285 (1975).
4. B. Stebe, T. Sauder, M. Certier and C. Comte, Solid State Commun. 26, 637 (1978).
5. G. Bader, L. Caron and L. Sanche, Solid State Commun. 38, 849 (1981).
6. G.A. Thomas and T.M. Rice, Solid State Commun. 23, 359 (1977).
7. C. Mavroyannis, Phys. Rev. B15, 1906 (1977).

8. C. Mavroyannis, Phys. Rev. B16, 2863 (1977).
9. J. Hermanson, Phys. Rev. 150, 660 (1966).
10. M. Altarelli and F. Bassani, J. Phys. C4, L328 (1971).
11. G. Gumbs and C. Mavroyannis, Solid State Commun. 41, 237 (1982).
12. C. Mavroyannis, Can. J. Chem. in press.
13. G. Gumbs and C. Mavroyannis, J. Phys. C: Solid State 14, 2199 (1981).
14. G. Gumbs and C. Mavroyannis, Phys. Rev. B24, 7258 (1981).
15. E.N. Economou, Green's Functions in Quantum Physics, Vol. 7, Springer-Verlag, New York, 1979.
16. J.R. Schrieffer, Theory of Superconductivity, Benjamin, Reading, 1964.

PREPARATION, STRUCTURE AND PROPERTIES OF GLASSY METAL HYDRIDES

A. J. Maeland

Allied/Signal Corporation

Morristown, NJ 07960

ABSTRACT

Large quantities of hydrogen can be dissolved in some metallic glasses without causing crystallization. Hydrogen may be introduced from the gas phase, electrochemically and by ion implantation. While long-range order is characteristically absent in metallic glasses, the presence of short-range order is well established. The structure can be conveniently explored by neutron scattering methods. Selected properties of metallic glasses containing hydrogen such as absorption isotherms, hysteresis, hydrogen capacity, volume expansion, hydrogen diffusion and embrittlement are discussed and compared to the corresponding properties in crystalline materials.

INTRODUCTION

The atoms in crystalline materials are arranged in a regular three dimensional pattern. The pattern repeats and extends over distances which are large compared to the interatomic distances and this leads to long-range order in the solid. The atomic arrangement in amorphous solids is, by contrast, nonperiodic and the long-range order is thus absent. It has long been known that amorphous (glassy) structures are readily formed in certain solids on cooling from the melt. Such glass-forming solids include the oxide glasses and organic polymers. By contrast solidification of metals and alloys generally produces crystalline structures and prior to 1960 amorphous structures in these materials were rarely obtained. This all changed in 1960 when Duwez, Willens, and Klement developed a technique for obtaining cooling rates in excess of 10°C/sec. and were able to produce an amorphous alloy of gold and silicon. Subsequent investigations and improvements in rapid cooling techniques have produced a large number of metallic glasses, some of which are now commercially available. The amorphous state is no longer believed to be a property of a few glass-forming solids, but is quite universal; according to Turnbull[3], "Nearly all materials can, if cooled fast enough and far enough, be prepared as amorphous solids".

Hydrogen absorption in crystalline metals and alloys has been studied extensively for more than a hundred years[4]. Much of the recent interest has been motivated by possible uses in energy storage systems. Both the electronic structure and the crystal structure, i.e. the type and size of the interstitial sites in the lattice, are major factors in hydrogen absorption. The relative importance of these two factors may be assessed in appropriately selected systems by comparing hydrogen absorption in metallic glasses with crystalline phases of the same compositions[5,6]. However, metallic glass compositions which do not have single phase crystalline counter parts can also be prepared in many cases and their hydrogen absorption characteristics can be studied. It is also possible to use hydrogen as a probe to explore the local environment in metallic glasses by using such techniques as inelastic neutron scattering[7] and nuclear magnetic resonance[8]. A number of metallic glass-hydrogen systems have been investigated to date. This review is focused on the preparation and structure of metallic glasses and on selected properties of metallic glass-hydrogen systems such as absorption isotherms, hysteresis, hydrogen absorption capacity, volume expansion, hydrogen diffusion and embrittlement. Whenever possible these properties are compared to the corresponding properties in crystalline materials.

PREPARATION

Most metallic glasses are produced by rapid quenching techniques. Figure 1 illustrates how this may conveniently be done in a continuous process using the melt spinning method. A stream of molten alloy is driven against the surface of a rapidly rotating, copper cyclinder kept at room temperature or below. Thin ribbon or foil, 0.01-0.05 mm thick, can be produced at a rate of 1800 m/min in this process[6]. Techniques other than rapid quenching of the liquid may also be used in preparing metallic glasses. These include thermal evaporation, sputtering, electrodeposition and ion implantation[9].

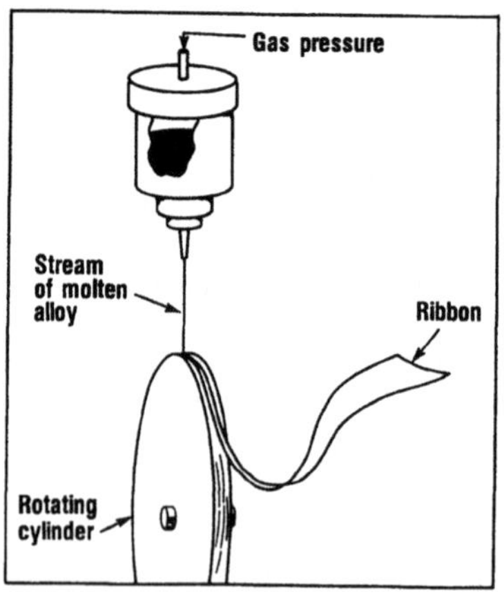

Figure 1. Melt spinning method for producing metallic glass

Hydrogen Absorption from the Gas Phase

Metallic glasses can be charged with hydrogen from the gas phase under appropriate conditions of temperature and pressure[5,10-13]. The reaction is exothermic for the metallic glasses considered here and care must be taken to prevent the heat evolved during the reaction from increasing the sample temperature to the point of crystallization of the glass. Sufficient time must be allowed to ensure that equilibrium is reached and that a homogeneous distribution of hydrogen in the alloy is obtained. Surface preparation to remove oxide layers is usually required; this may simply consist of abrading the ribbon with emory paper and ultrasonically cleaning in acetone followed by ether[5]. Coating with an overlayer of Pd has been shown to greatly improve the kinetics of the absorption.[13]

Electrochemical Charging with Hydrogen

Electrolytic charging is a convenient way of introducing hydrogen into metallic glasses provided no reaction with the electrolyte takes place[10,12,14,15]. A highly viscous electrolyte such as phosphoric acid in glycerine, is often used to minimize the loss of molecular hydrogen from recombination of atomic hydrogen in the electrolyte[15]. Non-homogeneous distribution may result at high hydrogen concentrations because the metallic ribbon becomes brittle upon hydrogen absorption and may rupture during electrolysis before reaching the maximum hydrogen content[12]. Surface preparation as indicated above may be necessary.

Ion Implantation of Hydrogen

Proton implantation of amorphous Pd/Si alloys at low temperatures has been done.[16] Hydrogen concentrations corresponding to a hydrogen to metal ratio, H/M, of 1.50 have been reported.

Formation of an Amorphous Metallic Hydride by Reaction of Hydrogen with Crystalline Intermetallic Compounds

Certain metallic glass hydrides can be formed from crystalline intermetallic phases by reaction with hydrogen at low temperature[17]. Amorphous $Zr_3RhH_{5.5}$, for example, was recently reported[17] to form when crystalline Zr_3Rh was reacted with hydrogen at temperatures below 200°C.

STRUCTURE

X-ray and neutron diffraction measurements on metallic glasses provide structural information which, due to the absence of long-range order, is limited to radial distributions of neighbors. Modelling of the structure is most frequently done using the concept of dense random packing of hard spheres (DRPHS)[18-23]. Recent diffraction experiments have established that chemical short-range order is present in many amorphous alloys.[24] Hydrogen is useful as a probe in exploring this short-range order using inelastic neutron scattering techniques. This is due to the fact that hydrogen has a very large incoherent scattering cross section for neutrons, and its mass is very small in comparison with most metals. Inelastic neutron scattering measurements provide a direct measure of the hydrogen vibrations both in-phase (acoustic) and out-of-phase (optical) with the alloy metal atoms in the material. Figure 2 shows the results obtained recently[7] on amorphous $TiCuH_{1.3}$. The optical phonon energy distribution for the metallic glass $TiCuH_{1.3}$ is seen to be very broad in contrast to that of the corresponding crystalline compound ($TiCuH_{0.93}$). However, both distributions are peaked at nearly the same phonon energy which is also very close to the peak obtained for γ-TiH_2.

Similar results have been obtained in other metallic glass-hydrogen systems.[25-28]

PROPERTIES

Pressure-Composition Isotherms

One of the most informative representations of the reaction between metals and alloys with hydrogen is the pressure-composition isotherm which is obtained by measuring the equilibrium hydrogen pressure at constant temperature as a function of hydrogen concentration, expressed as hydrogen to metal ratio (H/M). Typical isotherms for a crystalline material are shown in Figure 3. Hydrogen first dissolves in the metal according to equation (1);

$$M + y/2\ H_2 \rightleftarrows MH_y \text{ (solution phase)}\ , \qquad (1)$$

to form a solid solution whose composition depends on the hydrogen pressure as shown by the steeply, rising portion on the left hand side of the isotherm in Figure 3. The solid solution is saturated with hydrogen at point y and the nonstoichiometric hydride, MH_x, now begins to form. As hydrogen is continually absorbed, more of the saturated solid solution is converted to hydride according to reaction (2):

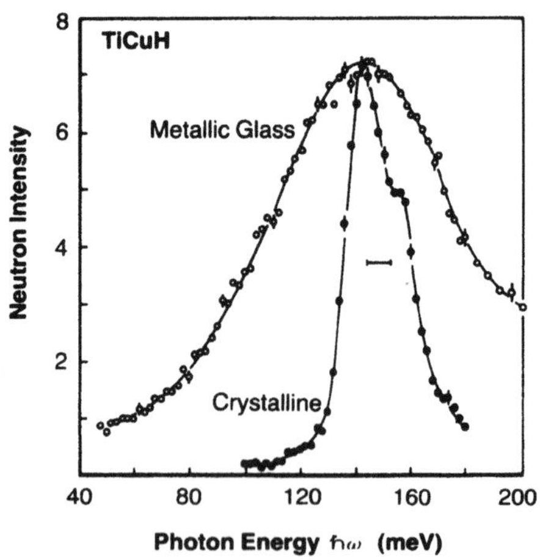

Figure 2. Neutron spectra[7] for $TiCuH_{0.93}$ and the metallic glass $TiCuH_{1.3}$

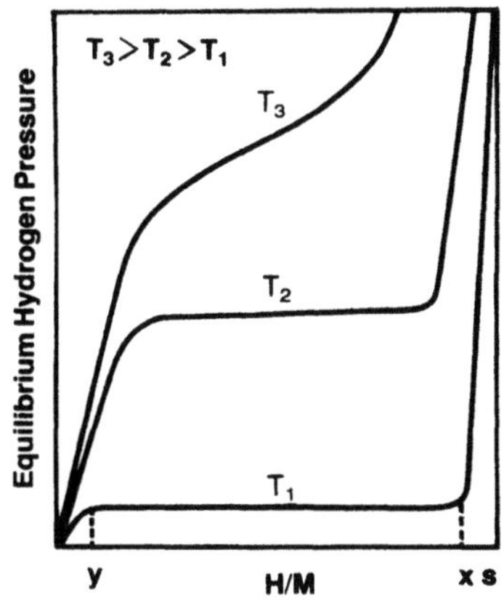

Figure 3. Pressure-Composition Isotherms (Schematic)

$$MH_y + (x-y)/2\ H_2 \rightleftharpoons MH_x\ , \qquad (2)$$

while the pressure remains constant, as required by the Phase rule, across the two-phase region indicated by the horizontal portion of the isotherm in Figure 3.

The invariant plateau pressure is the equilibrium absorption (dissociation) pressure of the hydride at the temperature of the isotherm. Reaction (2) is complete at point x and the hydrogen pressure again increases as the nonstoichiometric hydride absorbs hydrogen according to equation (3):

$$MH_x + (s-x)/2\ H_2 \rightleftharpoons MH_s\ . \qquad (3)$$

The plateau narrows with increasing temperature and disappears at a critical temperature, Tc; above this temperature the system exhibits one-phase behavior throughout the entire composition range. Pressure-composition isotherms obtained for metallic glasses to date do not exhibit plateaus[13,29,30,31] although corresponding crystalline alloys do.

Hysteresis

The equilibrium hydrogen pressure during hydride formation in crystalline metals and alloys, i.e. the forward direction of reaction (2), is often observed to be higher than during the dissociation (reverse) reaction. This well known hysteresis effect in the plateau region has long been recognized and discussed in the literature.[32] Since there have been no reports of plateaus in metallic glass-hydrogen systems, hysteresis in the pressure-composition isotherms are not expected. However, hysteresis has been reported in the Ni/Zr metallic glass-hydrogen system,[29,30] while no hysteresis was observed in the amorphous Fe/Ti-hydrogen system.[13]

Hydrogen Absorption Capacity

The maximum hydrogen contents in metallic Zr/Ni glass alloys are the same or less than in the corresponding crystalline alloys[30]. Similar results have been reported for rare earth-transition metal glasses.[13] By contrast hydrogen absorption capacities at room temperature and near atmospheric pressure for Ti/Cu metallic glasses are greater than for the corresponding intermetallic compounds.[5,12] In the case of TiCu the difference amounts to more than 36%. Very high hydrogen contents have been reported for the glass $Zr_{0.76}$ $Fe_{0.24}$; H/M ratios as high as 2.46 were obtained by electrolytic charging of the glass.[33] Since Zr forms a dihydride and Fe is not a hydride former, one would expect H/M to be less than 2. We have charged the same glass composition with hydrogen from the gas phase and obtained H/M= 1.76 at atmospheric hydrogen pressure and room temperature. There is apparently no data available for crystalline Zr_3Fe so no comparison can be made at this time. The very high hydrogen concentration reached by ion implantation of amorphous $Pd_{0.80}Si_{0.20}$, (H/M=1.50), is also quite remarkable[16].

Volume Expansion

It has been noted that the solution of hydrogen in a large variety of crystalline metals and alloys causes a volume expansion, ΔV, which on the average is $2.9x10^{-3}nm^3$ per hydrogen atom[33,34]. The expansion is independent of hydrogen concentration[34] and corresponds to a molar volume for hydrogen, V_H, of $1.7cm^3$. Similar expansion has been observed in metallic glasses.[4]

Hydrogen Diffusion

Hydrogen diffusion in the metallic glass $TiCuH_{1.3}$ is significantly greater than in the corresponding crystalline hydride ($TiCuH_{0.94}$) according to NMR studies.[8,36] Non-Arrhenius behavior has been observed in the temperature range 110-420K with four different diffusion activation energies ranging from 0.05eV to 0.40eV while the crystalline compound reportedly has a single activation energy of 0.79eV for hydrogen diffusion.[36] A strong dependence of the hydrogen diffusion coefficient on hydrogen concentration has been observed in amorphous $Ti_{0.52}Cu_{0.48}$ and $Ti_{0.30}Cu_{0.70}$ using an electrochemical permeation technique.[31] The effective hydrogen diffusivity was found to increase with increasing hydrogen concentration while the activation energy decreased with an increase in hydrogen concentration.[31]

Enhanced proton mobility has also been observed in the $Zr_2PdH_{2.9}$ metallic glass relative to the crystalline compound of the same composition.[37] Again non-Arrhenius temperature dependence of the activation energy was observed for the glass with three distinct activation energies in the temperature range 140-420K.[39]

The diffusion of hydrogen in Pd/Si and related metallic glass alloys is less than in crystalline palladium.[6,14,31,38-41] Data on single phase crystalline compositions,[42] $Pd_{5.25}Si$, Pd_5Si, Pd_9Si_2 and/or Pd_4Si, which are near the metallic glass compositions studied are not available for comparison.

Hydrogen Embrittlement

Absorption of hydrogen into crystalline metals and alloys generally leads to severe embrittlement. Metallic glass alloys also become embrittled when charged with hydrogen.[43] The embrittlement is less severe in the metallic glasses and hydrogen may be absorbed and desorbed many times without spontaneous disintegration of the material as often occurs with crystalline alloys.[44] The original ductility of the glass returns when hydrogen is removed.[43,44]

DISCUSSION

The inelastic neutron scattering data in Figure 2 suggest that in the metallic glass, hydrogen basically occupies sites which are similar in geometry to those in the corresponding crystalline intermetallic phase. In the case of crystalline TiCuH it has been determined by neutron diffraction that hydrogen is tetrahedrally surrounded by four titanium atoms.[45] The Ti-H distances are quite similar to those found in TiH_2 and the narrow density-of-states distribution for crystalline TiCuH with maximum around 142 meV is, as might be expected, close to that observed in TiH_2.[7] The shoulder at 157 meV may be caused by H-H interactions due to the tetragonal TiCuH structure. The fact that the peak in the distribution for the glass $TiCuH_{1.3}$ is roughly at the same energy as in the crystalline alloy, indicates that on the average the hydrogen atoms occupy tetrahedral type holes similar to those of the crystalline alloy (short range order). However, the much broader distribution of vibrational states indicates a wider distribution of local environments. Studies of hydrogen in Zr-Pd metallic glasses,[28] Zr-Ni glasses,[46,47] and Ni-Ti glasses[26] have yielded similar conclusions with respect to hydrogen site occupancy in the metallic glass vs. the crystalline phases. However, the situation in amorphous FeTi appears to be quite different. Nakamura[13] recently reported that hydrogen in amorphous FeTi had a local environment quite different from that in crystalline FeTi. His conclusion was based on thermodynamic measurements and it would be interesting to check its validity by comparing the inelastic neutron spectra of crystalline and amorphous FeTi hydrides.

The phase-change behavior illustrated in Figure 3 has been described with some success by statistical mechanical models.[48] The central concept in these models is that of a lattice-gas in which particles, in this case hydrogen atoms, are confined to lattice sites, i.e. the sublattice formed by the interstitical sites in the metal, but are free to jump from site to site. Each site can accomodate one particle and interactions between the particles vary with the distance between them. The interaction energy, which may also involve an elastic interaction energy between hydrogen and metal atoms, is responsible for the phase change. The configurational disorder in a metallic glasses leads to interstitial site energies which vary from site to site. Griessen[49] has shown, using a mean field treatment, that the absence of phase separation in metallic glass-hydrogen systems is related to the width of the site energy distribution, Δ. No separation takes place when Δ is larger than the mean field interaction energy. Richards,[50], using an analytical approximation for a lattice gas with many interactions and random site energies,[51] obtained results which are qualitatively different from those of Griessen; in this model a finite critical temperature $T_c \propto 1/\Delta$ is predicted for large Δ. Both calculations account for the absence of phase separation in metallic glass-hydrogen systems, but do not exclude the possibility. In view of the absence of plateaus in the pressure-composition isotherms of metallic-glass-hydrogen systems studied to date, the term "metallic glass hydride" is not strictly correct since it is a solid solution of hydrogen in the metallic glass.

A number of models have been proposed to explain plateau pressure hystersis. A central idea to many of these models is the concept of lattice strain. During hydride formation plastic deformation takes place leading to strain, while no deformation occurs during the decomposition of the hydride phase.[32] A more recent proposal requires the production of dislocations both during hydride formation and decomposition.[52] Regardless of the explanation of this type of hysteresis, suffice it to say that it is directly associated with the phase change. Hysteresis in one-phase regions of metal-hydrogen systems is not observed except after cycling through a two-phase region in which case it is possible to observe solvus hysteresis.[32,52] The reported hysteresis in the Ni/Zr metallic glass-hydrogen system[29,30] is not associated with a phase change and is difficult to explain. However, hydrogen is clearly more stable in the matrix after cycling, but it is not obvious if this increase in stability continues with each cycle. More work is needed to resolve the problem.

If the maximum hydrogen absorption capacity is determined by the electronic structure, we expect similar hydrogen absorption capacities in the amorphous and crystalline states since it has been shown that the electronic structures[53] are to a first approximation quite similar. However, the type and size of the interstitial sites are also important. In the case of TiCuH, for example, the hydrogen atoms are located at the centers of slightly distorted tetrahedra of titanium atoms[45]; crystallographically the limiting composition is TiCuH. Absorption of hydrogen beyond TiCuH would involve placing hydrogen atoms in higher energy sites. In the metallic glass structure, however, there is wide distribution of local environments as shown by the neutron scattering studies[7] discussed above. Such fluctuation in local symmetry may increase the number of sites available for hydrogen occupation (with respect to the crystalline state) by lowering the energy. It may, therefore, be possible to find a higher hydrogen absorption capacity in certain metallic glass structures than in their crystalline counterparts.

The approximately constant volume expansion per hydrogen atom, ΔV, observed in crystalline metals and alloys on hydrogen absorption is apparently also characteristic of metallic glasses at least for moderate to high hydrogen contents.[4] There is, however, a dramatic difference between metallic glasses and crystalline solids at very low hydrogen concentrations.[54] A negative volume change was reported[54] for the metallic glass $Pd_{0.80}Si_{0.20}$ for hydrogen concentrations up to 5×10^{-5} H/Pd. This negative partial volume of hydrogen was attributed to the presence of large interstices (excess volume) where hydrogen would attract neighboring host atoms. Above H/Pd=5×10^{-5} the partial molar volume was positive and increased gradually with concentration to a value near 1.5cm^3/mole H. The concentration dependence is related to the distribution of the volume of interstices which also determines the energy distribution in the metallic glass.[54] By contrast the molar volume of hydrogen in crystalline palladium at these low concentration was determined to be 1.60 cm^3/mole and independent of the hydrogen concentration as expected.[54]

The volume expansion accompanying hydrogen uptake in metals and alloys has recently been calculated[55] using a semi-empirical model which relates the heat of formation, ΔH, of a metal hydride to the difference of the Fermi energy, E_F, and the center, E_s, of the s band of the host metal.[56] The remarkable constancy of ΔV was shown to be in agreement with the model and due mainly to the volume dependence of the width, W, of the electronic bands.[55]

The increase in proton mobility in the metallic glass $TiCuH_{1.3}$ relative to the crystalline form may be understood by referring to the crystal structure of TiCuH. The structure consists of alternate double layers of Ti and Cu atoms with hydrogen between Ti planes, tetrahedrally surrounded by four Ti atoms.[45] Diffusion is thus restricted to jumps between nearest neighbor tetrahedral sites within the Ti planes.[36] This restriction is removed in the metallic glass which has a distribution of local environments leading to alternative diffusion paths and greatly enhanced diffusion. The concentration dependence of hydrogen diffusion in metallic glasses is related to the broad spectrum of energy states for hydrogen in the glass. Hydrogen atoms occupy the lowest energy states at low concentrations with an expected higher diffusion barrier height and lower diffusion coefficient. This was shown recently for the $Pd_{0.80}Si_{0.20}$ glass by Kirchheim et al.[14] and for Cu/Ti and Pd/Si glasses by Lee and Stevenson.[31] As the hydrogen content increases, the hydrogen atoms occupy sites with lower binding energy and the energy to move hydrogen through the structure decreases, i.e. lower activation energy and increased diffusion constant.

A distribution of activation energies for hydrogen diffusion also explains the increase in activation energy with temperature observed in $TiCuH_{1.3}$.[8,36] Hydrogen atoms in sites where they are weakly bound will have a low energy of activation (0.05eV); at higher temperatures more tightly bound hydrogens have a high activation energy (0.40eV). The fact that four distinct activation energies were observed in the range 110-420K rather than a continuous change, suggests that in this case relatively few (four) jump paths contribute to hydrogen diffusion. Similar arguments apply to the Zr_2Pd-H_2 system.[37]

Hydrogen embrittlement is a complex phenomenon and despite numerous studies, none of the theories to date appears to provide a general explanation of the process. Metallic glasses can also be embrittled thermally. However, this process is irreversible. The reversible nature of hydrogen embrittlement in metallic glasses has been used as a basis for preparing powders.[43] While in the embrittled state, the glass is ground to a powder; when hydrogen is removed the original ductility returns.

REFERENCES

1. P. Duwez, R.H. Willens and W. Klement, J. Appl. Phys., 31: 1136(1960).

2. W. Klement, R.H. Willens and P. Duwez, Nature, 187: 869(1960).

3. D. Turnbull, Contemporary Phys., 10: 473(1969).

4. A.J. Maeland, Hydrogen in Crystalline and Non-Crystalline Metals and Alloys: Similarities and Diferences, in: "Rapidly Quenched Metals," S. Steeb and H. Warlimont, eds., Elsevier Science Publishers B.V., Amsterdam(1985).

5. A.J. Maeland, Hydrogen in Metallic Glasses, in: "Metal Hydrides", G. Bambakidis, ed., Plenum Publishing Corporation, New York(1981).

6. G.G. Libowitz and A.J. Maeland, J. Less-Common Met., 101:131(1984).

7. J.J. Rush, J.M. Row and A.J. Maeland, J. Phys. F: Metal Physics, 10:283(1980).

8. R.C. Bowman, Jr. and A.J. Maeland, Phys. Rev. B, 24:2328(1981).

9. H.S. Chen, Rep. Prog. Phys., 43:353(1980).

10. A.J. Maeland, Comparison of Hydrogen Absorption in Glassy and Crystalline Structures, in: "Hydrides for Energy Storage", A.F. Andresen and A.J. Maeland, eds., Pergamon Press, Oxford(1978).

11. F.H.M. Spit, J.W. Drijver and S. Radelaar, Z. Phys. Chem. N.F., 116:225(1979).

12. A.J. Maeland, L.E. Tanner and G.G. Libowitz, J. Less-Common Met., 74:279(1980).

13. K. Nakamura, Scripta Met., 18:793(1984); Ber. Bunsenges. Phys. Chem., 89:191(1985); D.W. Forester, P. Lubitz, J.H. Scheller and C. Vittoria, J. Non-Cryst. Solids, 61 & 62:685(1984).

14. R. Kirchheim, F. Sommer and G. Schluckebier, Acta Metall., 30:1059(1982).

15. B. Chelluri and R. Kirchheim, J. Non-Cryst. Solids, 54:107(1983).

16. A. Traverse, H. Bernas, J. Chaumont, Xiang-Jun Fan and L. Mendoza-Zelis, Phys. Rev. B., 30:6413(1984).

17. X.L. Yeh, K. Samwer and W.L. Johnson, Appl. Phys. Lett., 42:242 (1983).

18. J.D. Bernal, Nature, 185:68(1960).

19. J.L. Finney, Proc. Roy. Soc., A319:479(1970).

20. D.E. Polk, Acta Metall., 20:485(1972).

21. C.S. Cargill II, Solid State Phys., 30:227(1975).

22. J.L. Finney, Nature, 266:309(1977).

23. D.S. Boudreaux, Phys. Rev. B., 18:4039(1978).

24. J.F. Sadoc and C.N.J. Wagner, in: "Glassy Metals II", H.J. Güntherodt and H. Beck, eds., Springer Verlag, Berlin(1983).

25. H. Kaneko, T. Kajitani, M. Hirabayashi, M. Ueno and K. Suzuki, in: "Proc. 4th. Int. Conf. on Rapidly Quenched Metals", T. Masumoto and K. Suzuki, eds., Japan Inst. Metals, Sendai, Japan (1982).

26. K. Kai, S. Ikeda, T. Funkunaga, N. Watanabe and K. Suzuki, Physica, 120B:342(1983).

27. K. Suzuki, J. Less-Common Met., 89:183(1983).

28. A. Williams, J. Eckert, X.L. Yeh, M. Atzmon and K. Samwer, J. Non-Cryst. Solids, 61&62: 643(1984).

29. F.H.M. Spit, J.W. Drijver and S. Radelaar, Scripta Met., 14:1071 (1980).

30. K. Aoki, A. Horata and T. Masumoto, Hydrogen Absorption and Desorption Properties of the Amorphous Zr-Ni Alloys, in: "Proc. 4th. Int. Conf. on Rapidly Quenched Metals," T. Masumoto and K. Suzuki, eds., Japan Inst. Metals, Sendai, Japan (1982); K. Aoki, M. Kamachi and T. Masumoto, J. Non-Cryst. Solids, 61 & 62: 679(1984).

31. Y.S. Lee and D.A. Stevenson, J. Non-Cryst.Solids, 72:249(1985).

32. T.B. Flanagan and J.D. Clewley, J. Less-Common Met., 83:127(1982) (review paper).

33. S.M. Fries, H.G. Wagner, S.J. Campbell, U. Gonser, N. Blaes and P. Steiner, J. Phys. F:Metal Physics, 15:1179(1985).

34. H. Peisl, Lattice Strains due to Hydrogen in Metals, in: "Hydrogen in Metals I", Top. Appl. Phys., 28, G. Alefeld and J. Völkl, eds., Springer-Verlag, Berlin(1978).

35. D.G. Westlake, J. Less-Common Met., 90:251(1983).

36. R.C. Bowman, Jr., A.J. Maeland and W.-K. Rhim, Phys. Rev. B, 26:6362(1982).

37. R.C. Bowman, Jr., A. Attalla, A.J. Maeland and W.L. Johnson, Solid State Commun., 47:779(1983).

38. A.J. Maeland, Diffusion of Hydrogen in the Metallic Glass $Pd_{0.80}Si_{0.20}$, in "Proc. Int. Meeting on Hydrogen in Metals", E. Wicke and H. Züchner, eds., Akademische Verlagsgesellschaft, Wiesbaden (1979).

39. B.S. Berry and W.C. Pritchet, Phys. Rev. B, 24:2299(1981).

40. Y. Tagagi and K. Kawamura, Trans. Japan Inst. Metals 22:677(1981).

41. Y. Sakamoto, H. Miyamoto, P.H. Shingu, R. Suzuki, and S. Takayami, in "Proc. 4th Int. Conf. on Rapidly Quenched Metals," T. Masumoto and K. Suzuki, eds., Japan Inst. Metals, Sendai, Japan(1982).

42. H. Langer and E. Wachtel, Z. Metallk., 74:535(1983).

43. A.J. Maeland and G.G. Libowitz, Mat. Lett., 1:3(1982).

44. F.H.M. Spit, J.W. Drijver, W.C. Turkenburg and S. Radelaar, J. de Physique, Colloque C8, Suppl. au no. 8, Tome 41:C8-890(1980).

45. A. Santoro, A.J. Maeland and J.J. Rush, Acta Cryst. B34(1978) 3059.

46. H. Kaneko, T. Kajitani, M. Hirabayashi, M. Ueno and K. Suzuki, J. Less-Common Met., 89:237(1983).

47. K. Suzuki, N. Hayashi, Y. Tomizuka, T. Fukunaga, K. Kai and N. Watanabe, J. Non-Cryst. Solids, 61 & 62:637(1984).

48. M. Futran, S.G. Coats and C.K. Hall, J. Chem. Phys., 77:6223 (1982) (review paper).

49. R. Griessen, Phys. Rev. B, 27:7575(1983).

50. P.M. Richards, Phys. Rev. B, 30:5183(1984).

51. P.M. Richards, Phys. Rev. B, 28:300(1983).

52. T.B. Flanagan, B.S. Bowerman and G.E. Biehl, Scripta Met., 14: 443(1980).

53. A. Amamou, Solid State Commun., 33:1029(1980); P. Steiner, M. Schmidt, and S. Hüffner, ibid., 35:493(1980); J. Kübler, K.H. Bennemann, R. Lapka, F. Rösel, P. Oelhafen, and H.J. Guntherodt, Phys. Rev. B23:5176(1981); S. Frota-Pessoa, Phys. Rev. B28:3753 (1983).

54. U. Stolz, V. Nagorny and R. Kirchheim, Scripta Met., 18:347(1984).

55. R. Griessen and R. Feenstra, J. Phys. F:Metal Physics, 15:1013 (1985).

56. R. Griessen and A. Griessen, Phys. Rev. B, 30:4372(1984).

THEORY OF ELECTRONIC STATES IN DISORDERED ALLOY HYDRIDES

D.A. Papaconstantopoulos, P.M. Laufer and A.C. Switendick*

Naval Research Laboratory
Washington, DC 20375-5000
*Sandia Laboratories
Albuquerque, NM 87185

ABSTRACT

Band structure calculations of metal hydrides have provided understanding of the bonding characteristics of these materials and clarified the mechanisms involved in various physical properties such as superconductivity. Since the electronic structure of stoichiometric hydrides has been discussed extensively in the literature, in this article we will focus on the methodology and results of disordered hydrides. This disorder can occur on the metal site of the hydride by considering random substitutions of the host metal or on the hydrogen sublattice where vacancies appear.

The theory that we use is based on the coherent potential approximation and proceeds as follows. Self-consistent band structure calculations of typical metal hydrides are performed by the augmented plane wave method. The results of these calculations are fitted to a tight-binding Hamiltonian using a Slater-Koster type of procedure. This Hamiltonian is then used in a tight-binding version of the coherent potential approximation which gives us the electronic densities of states for different metal and hydrogen compositions.

INTRODUCTION

In a previous volume[1] of this series one of us described the methodology of performing band structure calculations for metal hydrides and discussed the electronic properties of such systems. In that article, although non-periodic systems were reported, the emphasis was on periodic materials. This paper deals exclusively with disordered materials.

A description of the basic equations involved in the coherent potential approximation is given and the discussion of results concentrates on the following three types of non-stoichiometric hydrides: 1) the titanium-iron-hydrogen system, 2) transition-metal dihydrides, and 3) palladium-noble metal hydrides.

COHERENT POTENTIAL APPROXIMATION

The coherent potential approximation (CPA) is a mean field theory which is now accepted as the state-of-the-art method of calculating the electronic energy states of disordered materials. The CPA assumes that the electrons in a solid move in an effective medium described by a Hamiltonian that is determined self-consistently from the condition of zero average scattering. The basic paper on the CPA is that of Soven[2] published in 1967. The first numerical implementations of the CPA were presented by Kirkpatrick et al.[3] and by Stocks et al[4] using a tight-binding (TB) formalism to study Cu-Ni alloys. From the point of view of hydrides with vacancy disorder on the hydrogen sites the basic paper is that of Faulkner[5] published in 1976. In later years Stocks and co-workers[6] championed a muffin-tin-potential version of the CPA known as the KKR-CPA due to its resemblance with the Korringa-Kohn-Rostocker band theory method. The latest review article on the CPA is that of Faulkner[7] which discusses many formal aspects of the CPA method as well as detailed calculations and comparisons with experiment.

In this article we discuss TB-CPA calculations for $TiFeH_x$ and for dihydrides such as TiH_x. For these calculations we follow closely Faulkner's formalism,[5] and make the assumption that the metal sublattice is perfectly periodic, while the non-metal sublattice is randomly occupied by hydrogen or vacancies.

We also present calculations of Pd-noble metal hydrides such as $Pd_{1-y}Ag_yH_x$ where we have included, in addition to the vacancy disorder on the H sites, a substitutional disorder between, for example, Pd and Ag. In this case the CPA condition for zero average scattering is given by the following equation:

$$\sum_{i=i}^{4} c_i\tilde{t}_i = 0 \qquad (1)$$

where c_i are the concentrations for Pd, noble metal, hydrogen and vacancy, and t_i are the corresponding scattering matrices which have the following form:

$$\tilde{t}_i = (\tilde{\varepsilon}_i-\tilde{\Sigma})\,[1-(\tilde{\varepsilon}_i-\tilde{\Sigma})\tilde{G}]^{-1} \qquad (2)$$

where ε_i is the matrix of the on-site Slater-Koster (SK) parameters of the corresponding periodic Hamiltonian, $\tilde{\Sigma}$ is the CPA self-energy diagonal matrix including ℓ = s, p, t_{2g}, e_g, and s_H symmetries; and $\tilde{G}$ is the CPA Green's function found from the following integral over the irreducible wedge of the Brillouin zone (IBZ):

$$\tilde{G}(z,\Sigma) = \int_{IBZ} \frac{d^3k}{z-H(\Sigma,k)} \qquad (3)$$

where $z=E+i\delta$ is the complex energy, and $H(\Sigma\ ,k)$ is the 10x10 periodic SK Hamiltonian where the on-site parameters have been replaced by the self-energies Σ and the rest of the parameters are a weighted average of the parameters of the constituent hydrides i.e. PdH and AgH. Equations (1) and (3) are solved simultaneously, by a fairly elaborate iterative procedure, to determine Σ and thus $G(z,\Sigma)$. Having determined $G(z,\Sigma)$ we proceed to calculate the angular momentum decomposed density of states (DOS) $N_\ell(E)$ using the standard formula:

$$N_\ell(E) = -\frac{1}{\pi} \lim_{z\to E^+} \mathrm{ImTr}G_\ell(z,\Sigma) \qquad (4)$$

The procedure that we have described above differs from the early[3,4] TB CPA calculations in two respects. First, our SK Hamiltonians which are fit to self-consistent relativistic augmented plane wave (APW) calculations of the stoichiometric hydrides are much more accurate than those used by other workers. Our rms error is typically less than 5 mRy for the first 7 bands. Second, previous[3,4] TB-CPA calculations freeze the s- and p-bands in transition metals and apply a so-called one-level TB model which ignores the coupling of t_{2g} and e_g orbitals. In our calculations we use five coupled CPA conditions corresponding to the self-energies of s, p, t_{2g}, e_g, and s_H symmetry.

There are two possible shortcomings of our calculations. One is the neglect of off-diagonal disorder and the other is setting the on-site parameter for a vacancy equal to infinity. The off-diagonal matrix elements in our Hamiltonian are treated in a virtual crystal sense which is adequate only when the parameters involved are similar. On the other hand we have found that the "infinity" assumption is reliable for alloys with high hydrogen content.

THE TITANIUM-IRON-HYDROGEN SYSTEM

We performed[8] CPA calculations for the system $TiFeH_x$ for the hydrogen concentrations x=0.1, x=0.8, and x=0.9. These calculations, as mentioned above, assume that the metal sublattice is perfectly periodic while the non-metal sublattice consists of a random substitution of hydro-

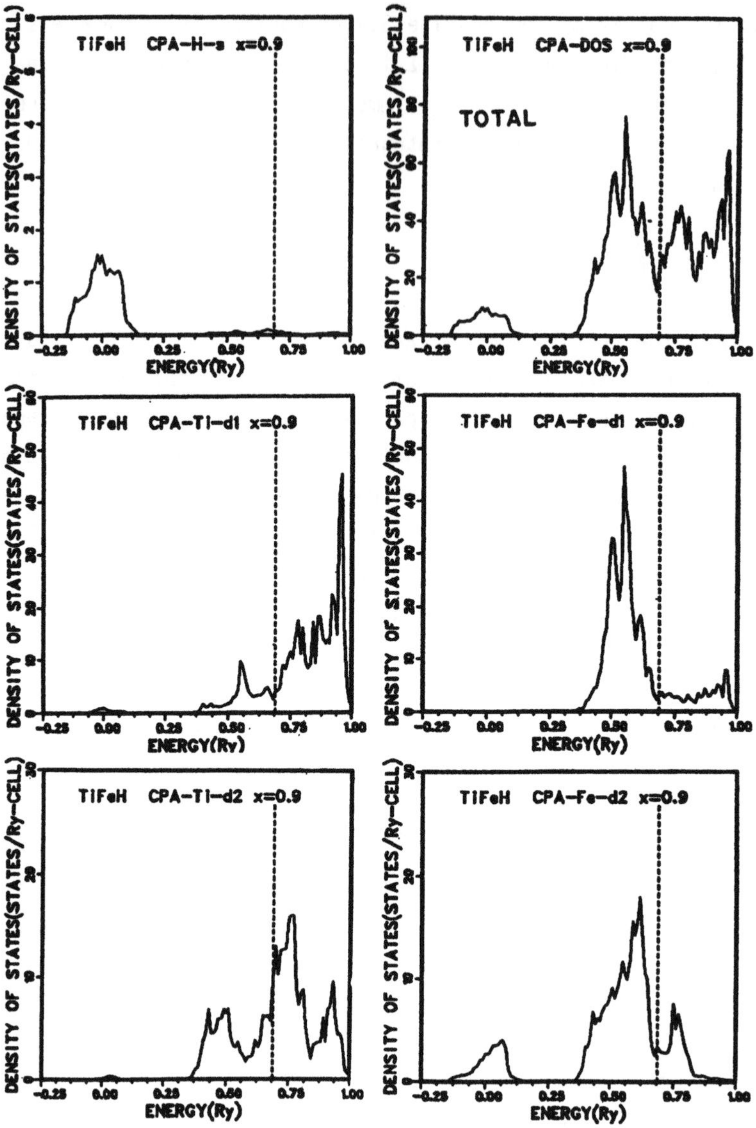

Figure 1 - Total density of states for $TiFeH_{0.9}$ shown together with the hydrogen component and the $d_1 = t_{2g}$ and $d_2 = e_g$ components of the metal sites.

gen by vacancies. The corresponding TB Hamiltonian is a 38x38 matrix which has a 36x36 block identical to that of TiFe and a 2x2 block which represents the disorder occurring between hydrogen and vacancies. The starting Hamiltonian is constructed by a SK fit to the APW band structure of TiFe and $TiFeH_{1.0}$.

In Fig 1 we show the DOS for x=0.9. We note that Ti-d states dominate the region above the Fermi level, E_F, while the Fe-d states are dominant

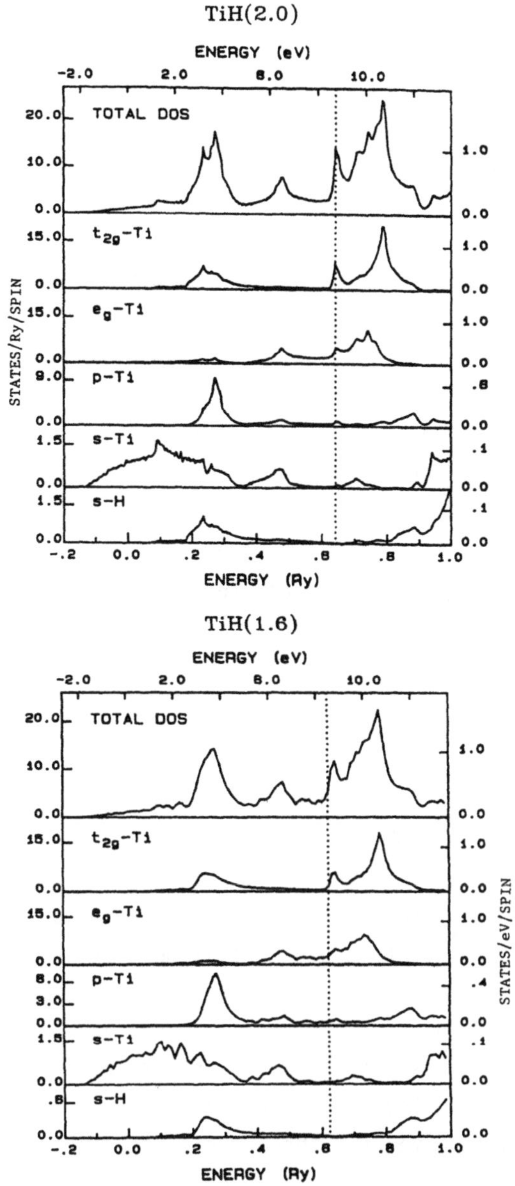

Figure 2. Densities of states for TiH_x for the hydrogen compositions x=2.0 and x=1.6.

below E_F. It is very interesting to observe that the low-lying bonding states involve mainly H and Fe-d states (of e_g symmetry as further analysis shows). This is in agreement with recent NMR studies[9] despite the expectation that the opposite might be true, given the fact that Ti forms a stable hydride while Fe does not.

Our study for $TiFeH_x$ has also shown that the DOS at E_F, $N(E_F)$, increases monotonically from TiFe to $TiFeH_{1.0}$ by about a factor of 3. This is consistent with specific heat measurements[10]. A decomposition of $N(E_F)$ into its angular momentum components reveals that all components of $N(E_F)$ increase with x except the Fe-e_g component. This is probably related to our finding that Fe-e_g states couple with H states to form the low-lying bonding states.

TRANSITION METAL DIHYDRIDES

We recently reported[11,12] APW band structure calculations of transition metal dihydrides TiH_2, VH_2, ZrH_2, NbH_2 and HfH_2 in the cubic calcium fluorite structure. These calculations are in good agreement with photoemission measurements[13] and provide the reasons for the absence of superconductivity[12] in these systems. However, TiH_2, ZrH_2, and HfH_2 show a tetragonal distortion at full stoichiometry which lowers dramatically the value of $N(E_F)$. Thus the cubic phase exists only for the substoichiometric hydrogen compositions, x, with $1.5<x<1.8$. For this reason we examined the effects of hydrogen vacancies on the DOS of the dihydrides, using again the TB-CPA theory. In this case the underlying TB Hamiltonian is an 11x11 matrix (9 metal orbitals and 2 hydrogen orbitals) determined by a SK fit to the APW results involving 39 orthogonal three-center parameters.

As an example we show in Figs. 2 and 3 the DOS of TiH_x and VH_x for x=2.0 and x=1.6. We note the following features in the DOS for x=2.0. The low-lying metal-hydrogen bonding states centered at approximately 0.3 Ry consist mainly of t_{2g}-metal, p-metal and s-H states. The states near E_F are metal d-like and contain practically no hydrogen. Looking at these DOS for x=1.6 we observe that the bonding levels are less sharp and contain considerably less hydrogen than for x=2.0. Near and above E_F the DOS show no qualitative differences between x=2.0 and x=1.6. The value of $N(E_F)$ is substantially reduced in the non-stoichiometric case as can be seen from Table I where $N(E_F)$ is given for three different concentrations. Clearly the occurrence of vacancies has a pronounced effect on the low-energy bonding states while the high-energy states are fairly insensitive to the formation of vacancies. However, the position of E_F shifts to lower energies by approximately 20 mRy from x=2.0 to x=1.6 in both TiH_x and VH_x resulting in a serious reduction of the $N(E_F)$ value. The most important difference between TiH_x and VH_x is in the value of $N(E_F)$. At x=2.0 TiH_x has a large $N(E_F)$ which is reduced by a factor of 3 at x=1.6. On the other hand the $N(E_F)$ of VH_2 has a smaller value than that of TiH_2 but is reduced at a much slower rate so that $N(E_F)$ of $VH_{1.6}$ is almost double that of $TiH_{1.6}$.

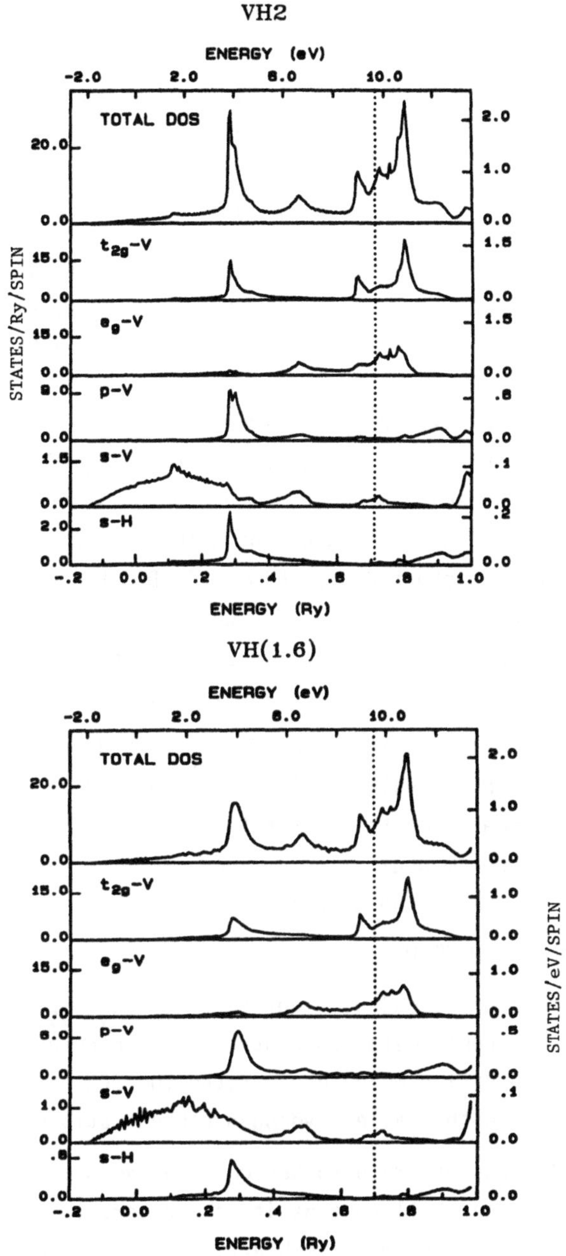

Figure 3. Densities of states for VH_x for the hydrogen compositions x=2.0 and x=1.6.

For both the Ti- and V-hydrogen systems, the CPA results show a lowering of the Fermi energy with hydrogen removal. This result, although in qualitative agreement with rigid band predictions is not even always true for substoichiometric systems (in some carbides E_F actually goes up). We find that the lowering of E_F in the Ti system from 0.643 Ry for TiH_2 to

Table I

	E_F Ry	$N(E_F)$ states/Ry/spin
$TiH_{2.0}$	0.643	14.01
$TiH_{1.8}$	0.636	11.18
$TiH_{1.6}$	0.623	4.62
$VH_{2.0}$	0.713	10.83
$VH_{1.8}$	0.710	10.74
$VH_{1.6}$	0.694	8.50

0.623 Ry for $TiH_{1.6}$ corresponds to a removal of 0.33 electrons based on a rigid band treatment of the fully stoichiometric TiH_2 DOS. This variation of E_F is in reasonable agreement with the conclusions reached by one of us[15], where to achieve agreement with the experimental specific heat data using the undistorted stoichiometric DOS the empirical relationship $\Delta E/\text{cell} = 0.72\ \Delta x$ was used. It would be interesting to see if the smaller variation with x assumed for ZrH_x in Ref. [15] would be consistent with the CPA results for this system.

The unenhanced value of $N(E_F)$ for $TiH_{1.6}$ of 0.34 states/eV/spin is in fair agreement with the value of 0.45±0.01 derived from specific heat and susceptibility measurements. Our $N(E_F)$ value for $TiH_{2.0}$ cannot be compared with the experimental value derived from the tetragonal phase which is much lower due to the distortion.

PALLADIUM-NOBLE METAL HYDRIDES

We have performed DOS calculations for $Pd_{1-y}Ag_yH_{.99}$, $Pd_{1-y}Au_yH_{.90}$ and $Pd_{1-y}Cu_yH_{.90}$ for noble metal concentrations y ranging from y=0.1 through y=0.9. In addition we have generated the DOS for various alloy concentrations with smaller hydrogen concentrations.

These calculations assume disorder on both the metal and non-metal sublattices, while each still retains its character, i.e., the two metals are randomly arranged on the metal sublattice, and vacancies occur randomly within the hydrogen (non-metal) sublattice. We proceed by generating self-consistent APW calculations for stoichiometric hydrides and then obtaining highly accurate three-center orthogonal SK fits to the energy bands. The fit is done using 10 bands, i.e., s, p, and d orbitals on the metal site and an s orbital on the hydrogen site. The CPA is then used to calculate the DOS for a given concentration, where no off-diagonal disorder is taken into account. We perform the Brillouin zone averages on a uniform mesh of 505 k points. Furthermore we set the energy scale in our CPA calculation by aligning the E_F of the stoichiometric metal hydrides.

Figure 4 shows the CPA DOS for $Pd_{0.7}Ag_{0.3}H_{0.9}$. This figure shows the total DOS of the alloy, the Pd and Ag site d-like DOS separated into t_{2g} and e_g symmetries, and the H site DOS. An important feature to note (which will be seen in the other systems as well, though to a lesser extent) is the separation of the d bands (both the t_{2g} and e_g) into well defined structures clearly attributable to Pd and Ag individually, with the Pd d bands lying nearer E_F. This split band character is not found in calculations employing the virtual crystal approximation and cannot be reproduced by any rigid-band argument; it is a true effect of alloying which is given only by the CPA theory.

The identification of the bands in this manner is made evident if we keep the H concentration fixed and vary the Ag concentration. We can clearly observe the variation in intensity of the two separate d-band features - with the deeper band being enhanced with increasing y at the expense of the higher lying Pd d-band. In addition, looking at the H-site DOS we identify low-lying bonding states (between s-H, t_{2g}-Ag and e_g-Pd) and well above E_F, the antibonding states (bonding between s-H and t_{2g}-Pd).

In analyzing effects of changes in the DOS due to changes in H concentration, x, the major effect is an increase in the H-site DOS with increasing x, in both the bonding and antibonding region. We also find an increase of the Pd-e_g DOS in the region of the bonding states along with an increase of the Pd-t_{2g} DOS in the antibonding region.

In Fig. 5 we present the DOS of $Pd_{.6}Cu_{.4}H_{.9}$. The layout is the same as in Fig. 4. The d-states are distinct in this alloy as well (although there is some small overlap) and it is clear that a rigid band approach will not suffice for $Pd_{1-y}Cu_yH_x$ systems. The low-lying bonding states are now s-H, e_g-Pd and e_g-Cu, unlike the situation of Fig. 4 which showed t_{2g}-Ag contributing to the bonding. This difference can be traced to the fact that the d-states of AgH (and indeed of pure Ag) are located much deeper than in PdH and CuH. Figure 6 shows the total DOS, the Cu and Pd site s- and p-like DOS and the H-site DOS. We observe that the s and p metal states make contributions to the metal-hydrogen bonding and antibonding, but as the DOS scale indicates these contributions are significantly smaller than those of the metal d states. At E_F the metal s-like DOS are negligible, while the metal p-like DOS are comparable to that of hydrogen. Figure 7 shows the DOS of $Pd_{.60}Au_{.40}H_{.90}$. Once again the d-bands are split less than in the Pd-Ag-H alloys. As in the copper alloys the bonding state d-like noble metal component is of e_g symmetry.

Table II. Densities of states at E_F for Pd-noble metal-hydrides

	E_f Ry	tot	Noble Metal s	Noble Metal p (states/Ry/spin)	Noble Metal t_{2g}	Noble Metal R_y	Palladium s	Palladium p (states/Ry/spin)	Palladium t_{2g}	Palladium R_y	H
$PdH_{1.0}$	.6509	3.427	-	-	-	-	.063	.422	1.823	.681	.439
$Ag_{0.1}Pd_{0.9}H_{1.0}$	.6525	3.522	.009	.046	.052	.047	.076	.363	1.834	.623	.472
$Ag_{0.3}Pd_{0.7}H_{1.0}$	.6376	3.622	.031	.131	.150	.182	.068	.272	1.450	.614	.724
$Ag_{0.5}Pd_{0.5}H_{1.0}$	.6296	3.194	.059	.206	.092	.396	.048	.172	0.918	.578	.725
$Ag_{0.7}Pd_{0.3}H_{1.0}$	.6274	2.027	.056	.199	.101	.575	.019	.074	0.152	.350	.501
$Ag_{0.9}Pd_{0.1}H_{1.0}$	.6611	0.754	.042	.145	.026	.361	.004	.011	0.022	.056	.088
AgH	.6509	0.111	.009	.015	.001	.072	-	-	-	-	.014
$PdH_{0.9}$	.6459	3.355	-	-	-	-	.072	.381	1.940	.648	.314
$Cu_{.20}Pd_{.80}H_{.9}$	.6441	3.093	.020	.105	.134	.073	.069	.321	1.498	.477	.397
$Cu_{.40}Pd_{.60}H_{.9}$	.6379	3.076	.055	.244	.228	.148	.065	.279	1.240	.366	.451
$Cu_{.60}Pd_{.40}H_{.9}$	.6326	2.622	.095	.393	.237	.188	.040	.195	.814	.208	.450
$Cu_{.80}Pd_{.20}H_{.9}$	.6319	2.016	.129	.511	.237	.183	.011	.096	.400	.076	.370
$CuH_{0.9}$	.6387	1.299	.139	.533	.265	.137	-	-	-	-	.266
$PdH_{0.9}$	.6459	3.355	-	-	-	-	.072	.381	1.940	.648	.314
$Au_{.20}Pd_{.80}H_{.90}$	.6510	3.096	.024	.108	.120	.068	.071	.291	1.513	.495	.402
$Au_{.40}Pd_{.60}H_{.90}$	.6506	3.128	.093	.254	.211	.149	.066	.237	1.253	.394	.473
$Au_{.60}Pd_{.40}H_{.90}$	.6494	2.759	.221	.340	.255	.211	.029	.138	.858	.242	.465
$Au_{.80}Pd_{.20}H_{.90}$	.6492	2.488	.482	.376	.344	.241	.050	.056	.521	.104	.104
$Au_{.99}H_{0.9}$	.6453	4.410	1.857	.320	1.577	.280	-	.002	.075	.005	.301

Ag(30)Pd(70)H(90)

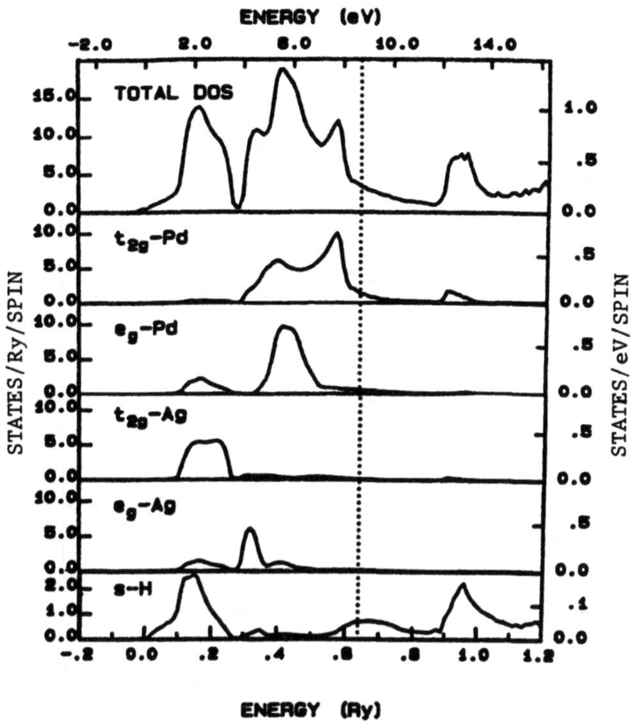

Figure 4. DOS of $Pd_{.70}Ag_{.30}H_{.90}$: total, e_g and t_{2g} for each metal, and hydrogen site contributions.

In all three materials studied, the antibonding states were found to have primarily s-H and t_{2g}-Pd character.

Table II presents the DOS evaluated at the Fermi energy for fixed hydrogen concentration and a range of *y* values for each of the palladium-noble metal-hydrogen systems considered. These results have been used[14] to calculate the electron-phonon coupling and provide an explanation of the variation (as a function of *y*) of the superconducting transition temperature T_c found in Stritzker's[16] experiments.

These experiments have shown that for Pd-noble metal-hydrogen alloys, T_c increases with the introduction of the noble metal to a maximum value, subsequently decreasing to practically 0°K at high noble metal content. A study of Table II indicates that this behavior is governed primarily through the hydrogen component of the DOS and indeed a detailed analysis[14] bears this out.

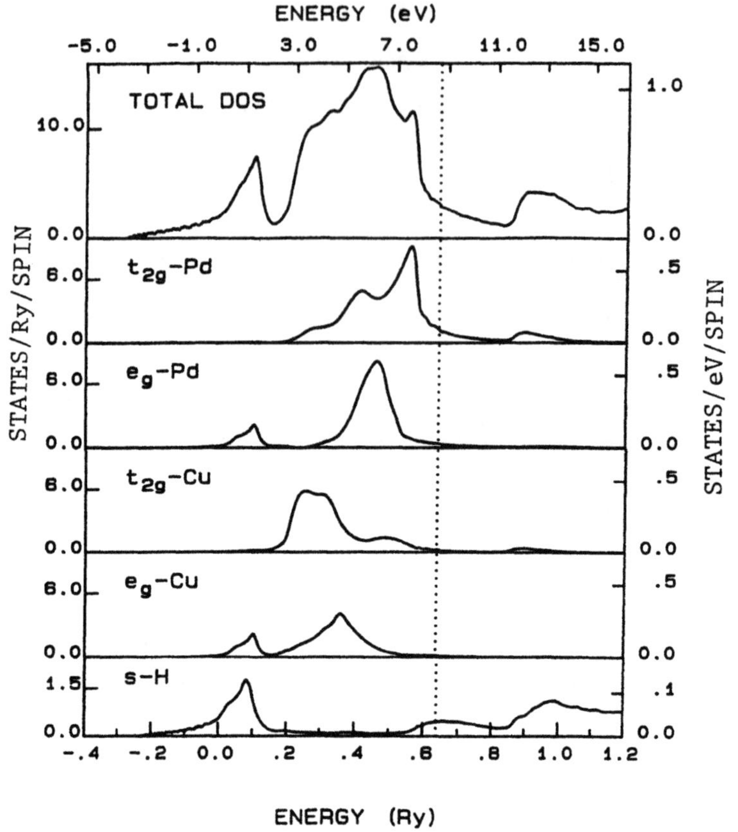

Figure 5. DOS of $Pd_{.60}Cu_{.40}H_{.90}$: total, e_g and t_{2g} for each metal, and hydrogen site contributions.

SUMMARY AND CONCLUSIONS

We demonstrate that the TB-CPA when used in conjunction with highly accurate SK fits to the self-consistent APW band structures and without the one-level approximation leads to a reliable description of the system under investigation.

In particular: a) for $TiFeH_x$ we find, contrary to expectation, but in agreement with NMR studies that Fe-d states are the main metal contributions to bonding; b) For transition metal dihydrides the variation of E_F and $N(E_F)$ with hydrogen content is consistent with specific heat and susceptibility data, and c) the results for palladium-noble metal-hydrogen show the non-rigid band character of the electronic states in these systems and have been used to explain the experimentally determined variation of T_c as a function of noble metal concentration.

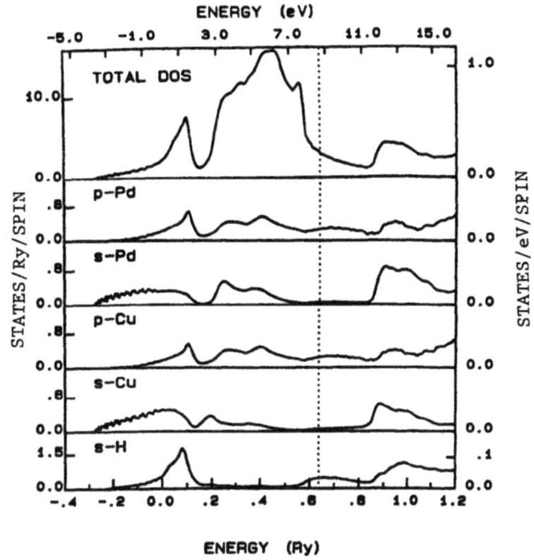

Figure 6. DOS of $Pd_{.60}Cu_{.40}H_{.90}$: total, s and p for each metal and the hydrogen site contributions.

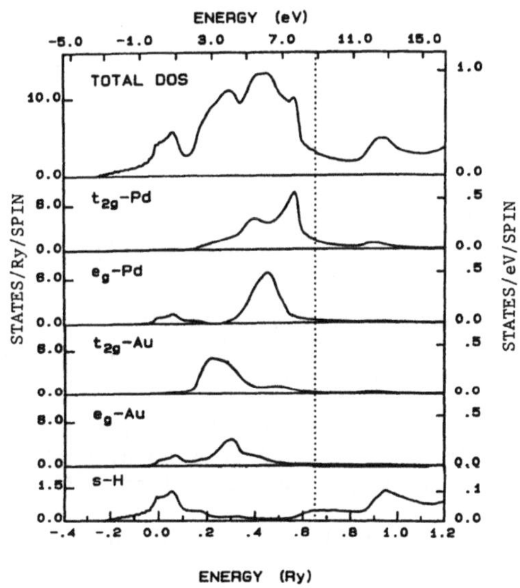

Figure 7. DOS of $Pd_{.60}Au_{.40}H_{.90}$: total, e_g and t_{2g} for total, e_g and t_{2g} for each metal, and hydrogen site contributions.

ACKNOWLEDGMENTS

We would like to thank Mrs. L. Blohm for technical assistance. The work performed at Sandia Laboratories was supported by the U.S. Department of Energy under Contract No. DE-AC04-DP00789.

REFERENCES

1. D.A. Papaconstantopoulos, "Electronic Structure of Metal Hydrides" in Metal Hydrides No. 76 of NATO ASI Series B, ed G. Bambakidis (Plenum, NY, 1981) pp 215-242.

2. P. Soven, "Coherent-Potential Model of Substitutional Disordered Alloys", Phys. Rev. 156, 809 (1967).

3. S. Kirkpatrick, B. Velicky, and H. Ehrenreich, "Paramagnetic NiCu Alloys: Electronic Density of States in the Coherent Potential Apprximation", Phys. Rev. B1, 3250 (1970).

4. G.M. Stocks, R.W. Williams, and J.S. Faulkner, "Densities of States of Cu-Ni Alloys", Phys. Rev. B4, 4390 (1971).

5. J.S. Faulkner, "Electronic States of Substoichiometric Compounds and Application to Palladium Hydride", Phys. Rev. B13, 2391 (1976).

6. G.M. Stocks, W.M. Temmerman, and B.L. Gyorffy, "Complete Solution of the Korringa-Kohn-Rostocker Coherent-Potential-Approximation Equation: Cu-Ni Alloys", Phys. Rev. Lett. 41, 339 (1978).

7. J.S. Faulkner, "The Modern Theory of Alloys", Prog. Mat. Science 27, 1 (1982).

8. D.A. Papaconstantopoulos, A.C. Switendick, "Calculations of the Electronic Properties of Substoichiometric Ti-Fe Hydride", Phys. Rev. B32, 1289 (1985).

9. R.C. Bowman, this volume.

10. R. Hempelmann, D. Ohlendorf, and E. Wicke, in "Hydrides for Energy Storage", Proc. of Inter. Sym., Geilo, Norway, eds. A.F. Andresen and A.J. Maeland (Pergamon, NY 1977) p. 407.

11. D.A. Papaconstantopoulos and A.C. Switendick, "Self-Consistent Band Structure Calculations of Titanium, Zirconium, and Hafnium Hydrides",J. Less-Common Metals, 103, 317 (1984).

12. D.A. Papaconstantopoulos, "Electron Phonon Interaction in Transition Metal Dihydrides", Proc. 17th Intern. Conf. on Low Temperature Physics Part I, p 129, North Holland 1984.

13. J.H. Weaver, D.J. Peterman, D.T. Peterson, and A. Franciosi, "Electronic Structure of Metal Hydrides, IV TiH_x, ZrH_x, HfH_x, and the fcc-fct Lattice Distortion",Phys. Rev. B23, 1692 (1981).

14. P.M. Laufer and D.A. Papaconstantopoulos, "Superconductivity in Pd-Ag-H Alloys", Physica (1985) to be published.

15. A.C. Switendick, "Electronic Structure of Group IV Hydrides and their Alloys", J. Less-Common Met 101, 191 (1984).

16. B. Stritzker, "High Superconducting Transition Temperatures in the Palladium-Noble Metal-Hydrogen System", Z. Physik 268, 261 (1974).

HYDROGEN IN DISORDERED SOLIDS: MODEL AND CALCULATIONS

R. Griessen

Natuurkundig Laboratorium, Vrije Universiteit
De Boelelaan 1081, Amsterdam
The Netherlands

I. INTRODUCTION

The metals of group IA to VA as well as the rare earths, palladium and the actinides react readily with hydrogen at room temperature[1-3] and at pressures of a few atmospheres if their surface is not contaminated by substances (e.g. oxide layers) which inhibit the hydrogen transfer from the gas phase to the bulk of the sample. Beside these ~ 40 metals there are hundreds of intermetallic compounds A_mB_n and an almost unlimited number of alloys $A_{1-y}B_y$ which react with hydrogen. For practical purposes some interest was recently also given to more complicated alloys[4-6] made of more than two types of atoms (the so-called "fruit-salad" metal hydrides). Considerable attention was also given to hydrogen in amorphous metals[7-10]. The vast majority of materials absorbing hydrogen is thus made of structurally and compositionally disordered materials.

In this paper we set up what could be called the simplest non trivial model for hydrogen absorption in a disordered solid. The paper is organized as follows. In the second section we give a simple statistical description of hydrogen in a disordered material within the mean-field approximation. The site energies as well as the short-range and long - range interaction energies which enter the model are considered in section III. The general relations derived in Sections II and III are specialized to the case of compositionally disordered face-centered-cubic alloys in section IV. Numerical results for pressure-composition isotherms of some palladium based ternary metal-hydrogen systems are given. The paper ends with conclusions about the validity of the model in section V.

II. MODEL FOR HYDROGEN IN A DISORDERED SOLID

a. Mean-field model

In a disordered solid the energy of a hydrogen (proton and its electron-screening cloud) is different for each site. Intuitively one expects that large interstitial sites coordinated by metal atoms which react easily with hydrogen will be much more attractive than small size interstitial sites coordinated by elements which do not form metal-hydrides. The energy spectrum for hydrogen in a disordered solid is thus quasicontinuous. It is entirely specified by indicating in how many sites g_i hydrogen has an energy ε_i.

When the sample is surrounded by hydrogen gas (or hydrogen fluid at high pressures) a certain number N_H of hydrogen atoms will disolve in the bulk of a given sample. At equilibrium there are N_i hydrogen occupying the g_i sites with energy ε_i. We have obviously

$$N_H = \sum_i N_i \tag{1}$$

where the summation is over all energy levels. The distribution of the N_H hydrogen atoms over the various i-sites is easily determined from the condition that the Gibbs free energy G of the entire system has a minimum when the pressure p, temperature T and total number of particles are fixed. For the minimalization of the total Gibbs free energy $G = G_{M-H} + G_{H_2}$ (where G_{M-H} and G_{H_2} are the Gibbs free energies of the metal-hydrogen system and H_2-gas (fluid), respectively) under the condition $N_H + 2N_{H_2} = N$ we use the method of Lagrange multiplicators. We require that the partial derivatives of the function

$$\tilde{G} = G_{M-H} + G_{H_2} + \alpha(\Sigma N_i + 2N_{H_2} - N) \tag{2}$$

with respect to the independent variables N_i, N_{H_2} and α vanish. This leads to

$$(\partial G_{M-H}/\partial N_j) + \alpha = 0\ , \tag{3}$$

$$(\partial G_{H_2}/\partial N_{H_2}) + 2\alpha = 0, \tag{4}$$

$$\sum_i N_i + 2N_{H_2} = 0. \tag{5}$$

From eqs. 3 and 4 follows directly that

$$\mu_H = \tfrac{1}{2}\mu_{H_2} = -\alpha \tag{6}$$

where μ_{H2} (p,T) is the chemical potential of hydrogen fluid at pressure p and temperature T.

In order to calculate $G_{M-H} \equiv U_{M-H} - TS_{M-H} + p\,V_{M-H}$ we make the following assumptions:

1^0 A given site can be occupied by at most one hydrogen at the time

2^0 Short-range H-H interaction is restricted to nearest-neighbours. The interaction energy is ε irrespective of the type of sites actually occupied by hydrogens.

3^0 The existence of long-range or, better, of <u>effective non-local H-H interactions</u> due to, for example, lattice dilation or Fermi energy shifts imply that both the site energies ε_i and the interaction energy ε depend also on the <u>total</u> number of hydrogen atoms in the sample and not only on pressure and temperature.

In the Bragg-Williams (mean-field) approximation[11],

$$U_{M-H} = \sum_i N_i\,\varepsilon_i + \frac{\varepsilon n\,N_H^2}{2M} + U_M \tag{7}$$

where n is the number of nearest-neighbour interstitial sites and

$$M = \sum_i g_i \tag{8}$$

is the total number of interstitial sites (irrespective of their type) in the sample.

For entropy we use the standard relation

$$S = k\,\ell n\,\Omega \tag{9}$$

with, in our case[12],

$$\Omega = \prod_i \frac{g_i!}{(g_i-N_i)!\,N_i!} \tag{10}$$

as each site is either empty or occupied by one hydrogen atom. By using Stirling's approximation $\ln y! \cong y \ln y - y$, and eqs. 7,9,10 we obtain from eq. 3 that

$$\mu_H = kT \,\ell n\, (x_j/(1 - x_j)) + \varepsilon_j + (E_0 + E_M + \varepsilon n)\gamma + \frac{En}{2}\gamma^2 + p\,\bar{V}_H = \tag{11}$$
$$= \tfrac{1}{2}\,\mu_{H_2} \ ,$$

in which

$$x_j = N_j/g_j \tag{12}$$

is the fraction of sites j occupied by hydrogen atoms,

$$\gamma = \Sigma\, N_j/\Sigma\, g_j = N_H/M \tag{13}$$

is the fraction of all sites occupied by hydrogen atoms,

$$E_0 = M\, \partial\varepsilon_k/\partial N_j \ ;\ E_M = M\, \partial U_M/\partial N_j;\ E = M\, \partial\varepsilon/\partial N_j \tag{14}$$

and $\bar{V}_H$ is the partial molar volume of hydrogen in the metal hydride. V_H as well as E_0, E_M and E are taken to be independent of the type of site occupied by hydrogen. This is consistent with a mean-field approach and is in fact equivalent to the assumption that V_{M-H}, U_M, ε_k and ε depend only on the total number of hydrogen atoms in the sample.

So far we have not considered the effect of hydrogen vibrations. For each type of sites one expects a different mode frequency so that[14)]

$$\mu_{vib}(\omega_j) = 3\ kT\ \ell n\ (1 - \exp\ (-\hbar\omega_j/kT)) \tag{15}$$

must be added to the chemical potential of hydrogen in the sites of type j in Eq. 11.

Solving Eq. 11 for x_j we obtain then

$$x_j = [\exp\ ((\varepsilon^0_j + f(\gamma) + p\,\bar{V}_H + \mu_{vib}(\omega_j) - \mu_H)/kT) + 1]^{-1} \tag{16}$$

in which, for simplification, the function $f(\gamma)$ represents all the concentration dependent terms, i.e.

$$\varepsilon_j + (E_0 + E_M + \varepsilon n)\gamma + \frac{En}{2}\gamma^2 = \varepsilon^0_j + f(\gamma)\ . \tag{17}$$

The energy ε^0_j is the energy of a hydrogen occupying a site of type j when no other hydrogen is present in the sample (infinite dilution limit). In the following we shall call the function $g_j(\varepsilon^0_j)$ the density-of-sites function. As the energy spectrum is a quasi-continuum (for a disordered system) we can treat ε^0_j as a continuous variable and drop the index j in eqs. 16 and 17. The thermodynamic properties of the disordered metal-hydrogen system can then be deduced from the following expression for the total fraction of occupied sites,

$$\gamma = \int_{-\infty}^{+\infty} g(\varepsilon^0)\ x\ (\varepsilon^0,p,T,\gamma,\omega(\varepsilon^0),\mu_H)\ d\varepsilon^0 \Big/ \int_{-\infty}^{+\infty} g(\varepsilon^0)\ d\varepsilon^0\ . \tag{18}$$

b. Pressure-composition isotherms

Eq. 18 is an implicit relation for γ. The pressure-composition isotherms are, however, easily calculated by means of the following simple procedure which avoids a direct solution of eq. 18 for γ. The steps are:

i) choose a value for η which is defined as

$$\eta \equiv f(\gamma) + p\,\bar{V}_H - \mu_H \;; \tag{19}$$

ii) calculate for the chosen value of η

$$\gamma = \int g(\varepsilon^0)\; x\,(\varepsilon^0,\omega(\varepsilon^0);\eta)d\varepsilon^0 \Big/ \int g(\varepsilon^0)d\varepsilon^0 \;; \tag{20}$$

iii) solve relation 19 in which $\mu_H = \frac{1}{2}\mu_{H_2}$, i.e.

$$\tfrac{1}{2}\,\mu_{H_2}(p,T) - p\,\bar{V}_H = f(\gamma) - \eta \tag{21}$$

to obtain p. This pressure corresponds to the concentration γ calculated in step ii). At pressures well below 10^3 bar the term $p\,\bar{V}_H$ can safely be neglected and

$$\mu_{H_2}(p,T) = \mu^0_{H_2}(T) + kT\,\ell n\,p\;, \tag{22}$$

with[13] $\mu^0_{H_2}(T) \cong 8.351 - 0.13025\,T + 3.4088 \cdot 10^{-5}\,(T-300)^2$. (23)

At high pressures the solution of eq. 21 requires the knowledge of μ_{H_2} (p,T) over a wide range of pressure and temperature[13].

iv) From structural information on the solid under investigation one can determine the number s of interstitial sites per atom. The concentration c of hydrogen in a binary alloy $A_{1-y}B_y$ is then

$$c = N_H/(N_A + N_B) = s\,\gamma\;. \tag{24}$$

In section IV we shall give a few examples of pressure-composition isotherms calculated by means of the procedure indicated above.

c. Enthalpy of solution

The partial molar enthalpy $\bar{H}_H$ of hydrogen in a metal is

$$\bar{H}_H = \partial(\mu_H/T)/\partial(1/T)\;. \tag{25}$$

It is advantageously obtained by taking the derivative of γ in eq. 18 with respect to 1/T at constant γ. Thus

$$0 = \frac{\partial\gamma}{\partial(1/T)}\Big|_\gamma = \int_{-\infty}^{+\infty} g(\varepsilon^0)\;\; dx/d(1/T)d\varepsilon_0 \Big/ \int_{-\infty}^{+\infty} g(\varepsilon^0)\;d\varepsilon^0\;, \tag{26}$$

in which x is expressed as a function of 1/T, μ_{vib}/T and μ_H/T so that

$$\frac{dx}{d(1/T)}\Big|_\gamma = \frac{1}{2(1+\cosh Z)}\left[\varepsilon^0 + f(\gamma) + p\,\bar{V}_H + \bar{H}_{vib} + \bar{H}_H\right], \tag{27}$$

with

$$\bar{H}_{vib} = 3\,\hbar\omega/(\exp(\hbar\omega/kT) - 1) \tag{28}$$

and

$$Z(\varepsilon^0) = (\varepsilon^0 + f(\gamma) + p\bar{V}_H + \mu_{vib} - \mu_H)/kT\;. \tag{29}$$

Finally we obtain for $\bar{H}_H$

$$\bar{H}_H = f(\gamma) + p\,\bar{V}_H + \frac{\displaystyle\int_{-\infty}^{+\infty} \frac{g(\varepsilon^0)(\varepsilon^0 + \bar{H}_{vib})d\varepsilon^0}{1+\cosh Z}}{\displaystyle\int_{-\infty}^{+\infty} \frac{g(\varepsilon^0)d\varepsilon^0}{1+\cosh Z}}\;. \tag{30}$$

The heat of solution $\Delta\bar{H}$, defined as

$$\Delta\bar{H} = \bar{H}_H - \tfrac{1}{2} H_{H_2} , \quad (31)$$

is then obtained from eq. 30 and $H_{H_2}(p,T)$. At pressures well below 10^3 bar H_{H_2} is virtually independent of pressure and[13)]

$$H_{H_2} \cong 8.506 + 0.02908 \ (T-300) \quad (kJ/molH_2) , \quad (32)$$

and the p $\bar{V}_H$ term may be neglected in eqs. 29 and 30. The heat of solution can then be calculated without having to determine the pressure in step iii) indicated above.

Equation 30 shows explicitly that there are essentially three different contributions to the partial molar enthalpy of hydrogen in a disordered system. The first term $f(\gamma)$ arises from short range H-H interactions as well as from the dependence of site energies on the total concentration of hydrogen (non-local H-H interaction). The second term is due to the influence of pressure on the lattice expansion ($\bar{V}_H > 0$) accompanying hydrogen absorption. Both terms, $f(\gamma)$ and p $\bar{V}_H$ are also present in the description of hydrogen absorption in an ordered solid. The function $f(\gamma)$ is however likely to be different in disordered and in ordered solids. In an ordered solid it may happen that the most favourable sites are not nearest-neighbour sites. Such a situation occurs for example in Pd_3Fe for which the minimum separation between two Pd_6 octahedral sites is equal to the lattice spacing (see Fig. 1a). In a disordered $Pd_{0.75}Fe_{0.25}$ alloy on the contrary, approximately 6% of the Pd_6 octahedral sites are nearest-neighbours. A direct nearest-neighbour interaction between hydrogens in these sites is thus possible (Fig. 1b).

In the third term in eq. 30, $g(\varepsilon^0)$ reduces to a set of well separated delta-functions in the case of an ordered intermetallic compound while it a quasi-continuum in a disordered system. For smeared-out density-of-sitesis function the last term in eq. 30 increases smoothly with sloping plateaus. If $f(\gamma)$ is sufficiently strongly decreasing with increasing γ these weakly sloping plateaus can be compensated so that $d\bar{H}_H/d\gamma < 0$ which is a sufficient condition for the existence of a critical point. For a wide density-of-sites function this is not possible. The question whether $T_c > 0$, i.e. whether phase separation occurs or not in a disordered metal-hydrogen system has recently been discussed by Griessen[15)] and Richards[16)].

d. Entropy of solution

The partial molar entropy of solution $\Delta\bar{S}$ of hydrogen in a metal is, by definition,

$$\Delta\bar{S} = \bar{S}_H - \tfrac{1}{2} S_{H_2} \quad (33)$$

and is equal to $\Delta\bar{H}/T$ as a consequence of the equilibrium condition $\mu_H = \tfrac{1}{2}\mu_{H_2}$.

III. ENTHALPIES OF SOLUTION AND EFFECTIVE H-H INTERACTION

With the formalism presented in section II it is possible to calculate the pressure-composition isotherms, the chemical potential μ_H, the partial molar enthalpy $\bar{H}_H$ and entropy $\bar{S}_H$ provided the density-of-sites function $g(\varepsilon^0)$, the molar volume $\bar{V}_H$, the phonon spectrum $\omega(\varepsilon^0)$, and the function $f(\gamma)$ which contains all the concentration dependence of the site energies and interaction energies are known. This large number of input parameters makes it impossible to determine $g(\varepsilon^0)$, $\bar{V}_H$, $\omega(\varepsilon^0)$, $f(\gamma)$ directly from a fit to experimental data (which in most cases are in the form of pressure-composition data or calorimetric data). There is thus a serious need for reliable approximate values of these various parameters.

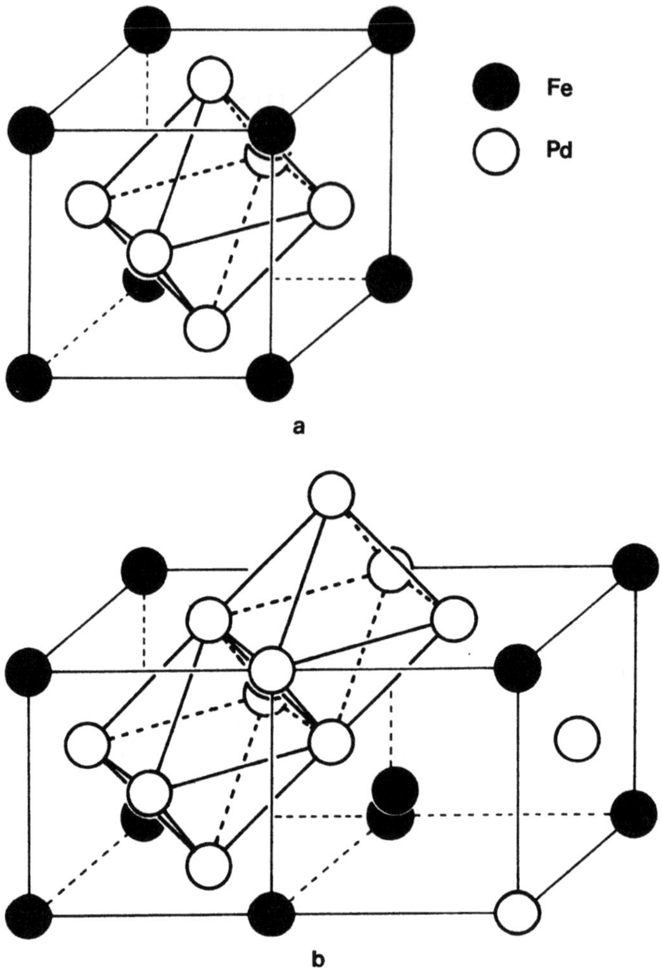

Fig. 1. Octahedral interstitial sites a) in ordered Pd_3Fe and b) in disordered $Pd_{0.75}Fe_{0.25}$.

Binary metal hydrides

In a series of articles[17-19] we have shown that the standard heats of formation of binary metal-hydrides ΔH_c are reasonably well reproduced by the relation

$$\Delta H^0 = \frac{n_s}{2} (\alpha(E_F - E_s) + \beta) \tag{34}$$

with α = 29.62 kJ/eV mol H and β = −135.0 kJ/mol H. E_F is the Fermi energy of the electrons and n_s is the number of electrons per atom in the lowest conduction band with a strong s-character with respect to the interstitial sites occupied by hydrogen atoms. E_s is the energy for which the integrated density-of-states function of the host metal is equal to $n_s/2$ electrons per atom. Except for the alkali metals (for which n_s=1) we have n_s=2 and E_s corresponds thus approximately t the centre of the lowest conduction band. One might argue at this point that in transition metals the lowest conduction band has a strong s-character at the centre of the Brillouin zone while it is essentially d-like at the boundary of the Brillouin zone as a result of s-d hybridization with one of the d-bands.

This characterization of electron state symmetry is, however, only true with respect to the host atom location. At interstitial sites d-atomic states may have a strong s-character. This point is best illustrated by considering the band structure of a fcc-metal (such as Ni, Pd, Pt, Cu, Ag, Au) along the [001] direction in reciprocal space. Within the tight-binding approximation (with nearest-neighbour interaction only) the dispersion relations for the five d-bands are simply given by

$$E_i(\vec{k}) = E_i^0 - W_i \cos(ka) \quad , \tag{35}$$

where $\vec{k}$ is the wave vector in the [001] direction, a the lattice spacing and $E_i{}^0$ and W_i are two band dependent energies (i = 1,...,5). The corresponding wave functions are of the form

$$\Psi^i_{\vec{k}}(\vec{r}) = \sum_{\vec{r}} e^{i\,\vec{k}\cdot\vec{R}}\, \phi^i_d(\vec{r}-\vec{R}) \quad , \tag{36}$$

where $\phi^i_d(\vec{r}-\vec{R})$ is an atomic d-wave-function of symmetry xy, xz, yz, x^2-y^2 or $3z^2-r^2$ centered at $\vec{R}$ in the lattice. The sum is over all host lattice positions. Now, at the centre of the Brillouin zone eq. 36 represents a simple superposition of d-wave functions. The resulting wave function has a pure d-character <u>both</u> at the host metal atom positions and at the octahedral interstitial sites (see Fig. 2a, b). At the boundary of the Brillouin zone, however, one of the bands, the band deriving from the atomic d-wave function with $3z^2-r^2$ symmetry, leads to a state with strong s-character as a result of the alternating Bloch factor (1 for $R_z = na$ and -1 for $R_z = (n+\frac{1}{2})a$ where $n \in Z$) as shown in Fig. 2c. The four other d-states keep their d-symmetry at the interstitial sites (for example see Fig. 2d). On the basis of these simple arguments it is then easily understood that the lowest conduction band in a fcc-transition metal or noble metal is almost rigidly (i.e. independently of $\vec{k}$) shifted down in energy by the attractive potential of the protons occupying octahedral interstitial sites as has been found by Gelatt, Ehrenreich and Weiss[20], Williams et al.[21] Papaconstantopoulos et al.[22] and Switendick[23,24].
Furthermore these arguments lead also directly to the conclusion that the other d-states should remain virtually unaffected by the presence of hydrogen in the metal as their wave functions have very little (or zero) amplitude at the octahedral sites.

The purpose of the brief digression given above is to justify our choice of E_s as the energy for which the integrated density-of-states is equal to 1 as indicated in Figs. 3 a to d. These figures suggest that there is a correlation between E_F-E_s and the heat of hydride formation of metals as i) E_F-E_s is small for stable hydride-forming metals such as yttrium and rather large for non-hydride-forming elements such as Ru and Ag. In this context Pd represents some sort of accident of nature: its E_F-E_s is smaller than that of Rh (although it has one electron more) because of the general trend of the d-band width to decrease rapidly on the right of the periodic table. Its E_F-E_s is also smaller than that of Ag because the 11th electron in a noble metal has to be accomodated <u>above</u> the d-band complex in the low s-p density of states region.

The correlation[17] between heat of formation and the characteristic band structure energy E_F-E_s is shown in Fig. 4. A similar correlation holds for the molar heat of solution at infinite dilution $\Delta\bar{H}_\infty$. The remarkable feature of these correlations is that they hold for transition metals as well as for noble and simple metals, rare earths and actinides.

A direct application of the semi-empirical band structure model of Griessen and Driessen (eq. 34) is to predict standard enthalpies of hydride formation of metals which absorb only negligible amounts of hydrogen under normal conditions. Driessen et al. [25] and Hemmes et al.[26] predicted on the basis of this model that formation of hydrides of Be, Cu, Ru, Ag, Cd, In,

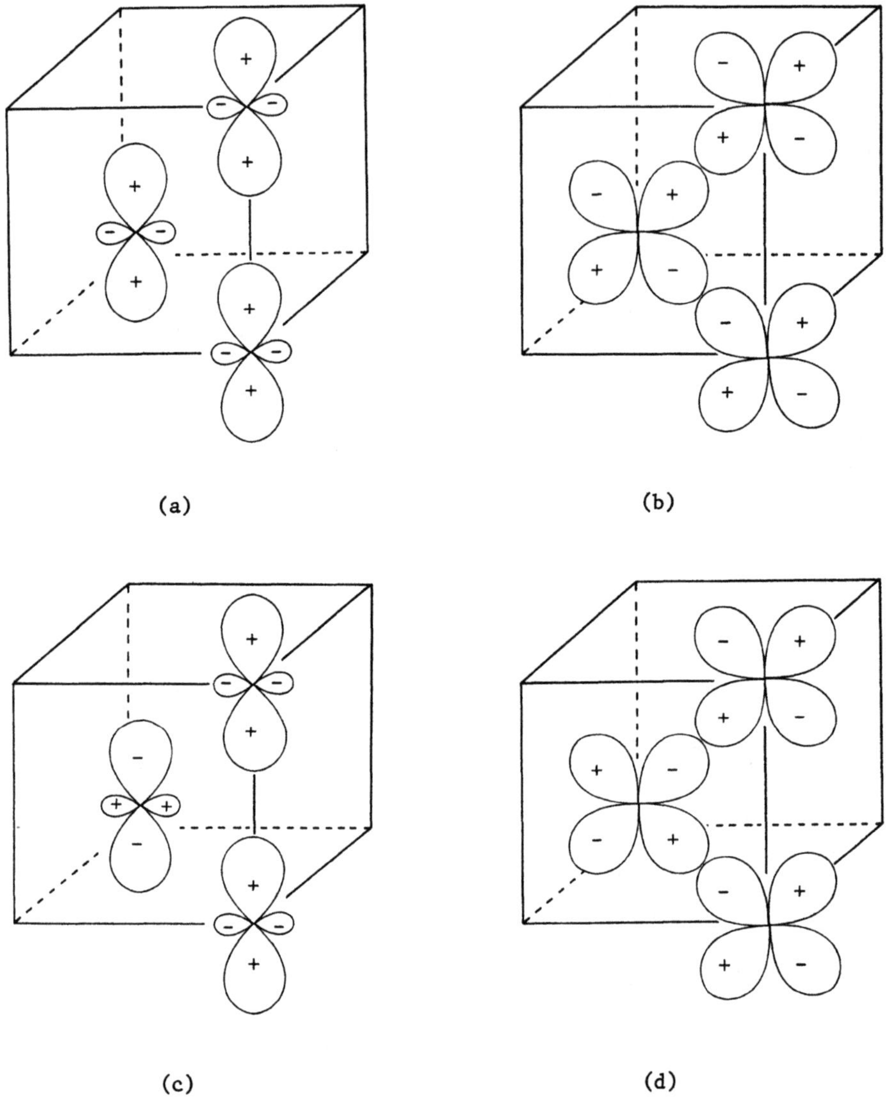

Fig. 2. Schematic representation of Bloch states with wave-vector along the [001] direction in a face-centered-cubic transition metal: a) Bloch state at $\vec{k} = 0$ deriving from the $3z^2 - r^2$ atomic d-wave function; b) as a) but for the atomic wave function yz; c) Bloch state at $\vec{k} = \vec{(0,0 2\pi/a)}$ derived from $3z^2 - r^2$; d) as c) but for yz.

W, Re, Pt and Hg should be observable under conditions ($p \gtrsim 1$ Mbar, $T \gtrsim 1000$ K) presently realizable in diamond anvil cells.

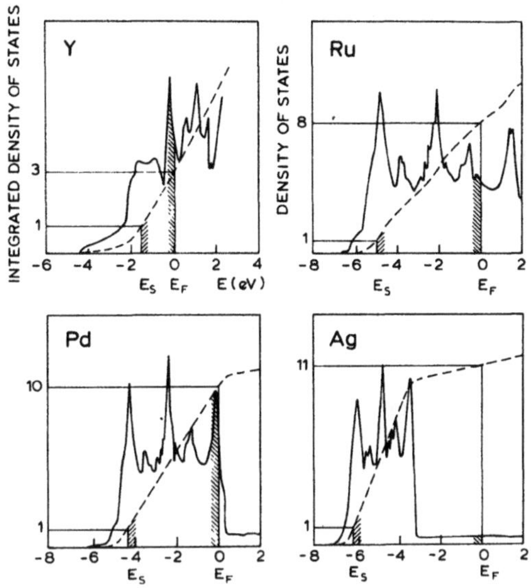

Fig. 3. Fermi energy E_F and the characteristic energy E_s for four representative 4d-metals.

We return now to the calculation of isotherms.

To actually calculate the pressure-composition isotherms of a binary metal-hydrogen system we require at least the following additional ingredients:

i) Enthalpies for all interstitial sites. The correlation indicated in Fig. 4 is only valid for the interstitial sites (I) which are easily occupied by hydrogen (e.g. octahedral sites in Pd, tetrahedral sites in Nb).
The enthalpies for hydrogen occupying less favourable sites (II) may be evaluated by means of an expression derived by Griessen and Riesterer[3),

$$\Delta\bar{H}_{II} = S_{II/I}\, \Delta\bar{H}_I + 121.2\ (S_{II/I} - 1)\ , \qquad (37)$$

with

$$S_{II/I} = \sum_{II} R^{-8.07} \Big/ \sum_{I} R^{-8.07}\ . \qquad (38)$$

The summations Σ are over the host metal neighbours coordinating an interstitial site i (i = I or II). For a fcc metal $S_{tetra/octra} = 2.131$ while for a bcc metal $S_{tetra/octra} = 0.724$.

ii) The concentration dependence of the enthalpy. According to Feenstra et al[27) the hydrogen concentration dependence of $\Delta\bar{H}$ may be calculated from the following expression

$$\frac{d\Delta\bar{H}}{dc} = -\frac{B\bar{V}_H^2}{V} + \frac{\alpha\nu}{N(E_F)}\ , \qquad (39)$$

where B is the bulk modulus, V the molar volume, and $N(E_F)$ the electronic density of states at the Fermi energy of hydride AH_c. $\bar{V}_H$ is the partial molar volume of hydrogen in AH_c. ν is the number of electrons per hydrogen atom added at E_F. Except for early transition metals $\nu=1$. By integrating eq. 39 with respect to c we obtain the variation of $\Delta\bar{H}$ with hydrogen concentration. The validity of eq. 39 can be demonstrated by using experimental values for $d\Delta\bar{H}/dc$, B, V, and $\bar{V}_H$ to calculate $N(E_F)$. The resulting density-of-states curve obtained by means of this procedure in the case of PdH_c is in remarkable agreement with the calculated density-of-states curves of Mueller et al[28] for pure Pd and Papaconstantopoulos et al.[21] for stoichiometric PdH (see Fig. 5). We consider this agreement as a convincing evidence of the existence of two non-local effective H-H interactions: one which shall be called "elastic" as it is proportional to the bulk modulus (first term in Eq. 39) and one of electronic origin (second term in Eq. 39). A similar splitting has already been proposed quite a time ago by Wagner[29] and Wicke and Brodowsky[30]. Our model, however, makes it possible to determine *quantitatively* $N(E_F)$ from basically standard experiments carried out at or above room temperature.

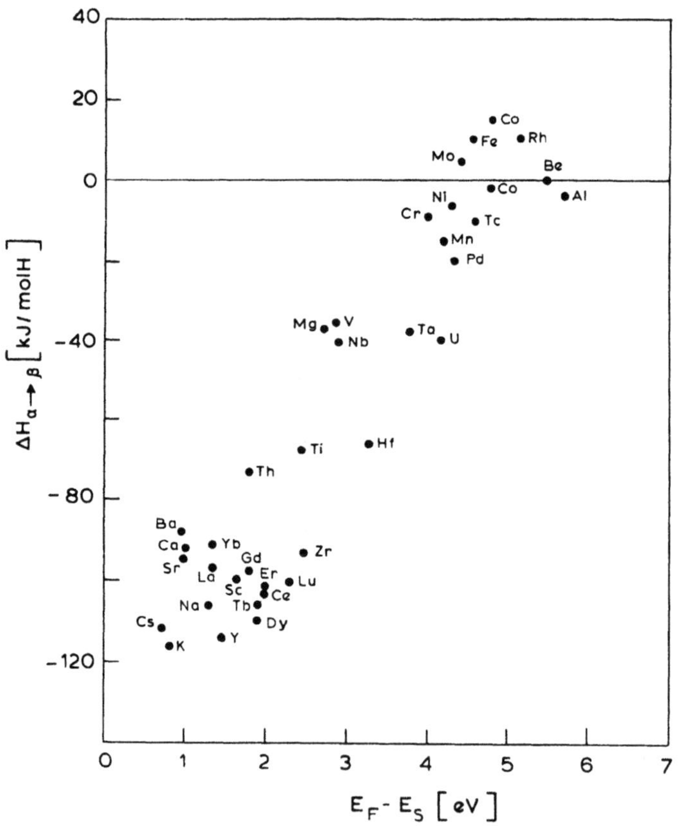

Fig. 4. Correlation between enthalpy of formation of concentrated hydrides and the characteristic band-structure energy E_F-E_S (from ref. 3).

iii) Short-range H-H interactions. Quite little is known about the origin of short range H-H interactions. As pointed out by Switendick[23,24] band structure calculations seem to indicate that there is a strong repulsion between two hydrogens separated by less than 2.1Å in a host lattice. Interesting work on short range H-H interactions has been carried out in the group of C. Hall[31-33] and K. Ross[34]. As in the present work we are not interested in the "low" temperature ($T \lesssim 250$ K) thermodynamic properties of metal hydrides, short-range H-H interactions shall be neglected.

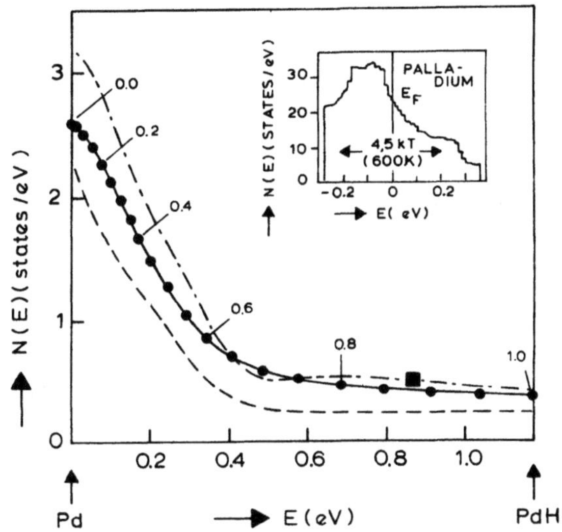

Fig. 5. Electronic density-of-states function N(E) derived from experimental values at 600 K of the bulk modulus B, partial molar volume $\bar{V}_H$ and concentration dependence of $\Delta\bar{H}$ by means of Eq. 39. The theoretical curves were obtained from refs. 22 and 28 by means of a convolution of calculated N(E) with an appropriate thermal function (600 K). For details see ref. 40.

Ternary metal hydrides

The application of the semi-empirical band structure model to ternary metal-hydrogen systems $A_{1-y} B_y H_c$ has been described in detail by Griessen et al.[3,17] and shall not be repeated here. It is important however to note that a good agreement could be obtained between experimental and model values by using the relation

$$\Delta\bar{H}(A_{1-y}B_yH_c) = \alpha\left((1-y)(E_F-E_s)_A^* + y(E_F-E_s)_B^*\right) + \beta \quad , \tag{40}$$

where $(E_F-E_s)_i^*$ corresponds to the characteristic band structure energy E_F-E_s of metal i *after* having properly scaled the density-of-states of metal A and B and *after* having allowed for a flow of electrons to insure a common Fermi energy throughout the sample (according to a procedure proposed by Cyrot and Cyrot-Lackmann[35]). Eq. 40 does not involve any additional fit parameters as is the case e.g. for the rule of reversed stability in Miedema's semi-empirical cellular model[2]. The agreement between experimental and model values is, however, substantially better than for Miedema's model[3].

The fact that Eq. 40 which contains only band structure energies reproduces experimental values adequately seems to indicate that $\Delta\bar{H}$ depends primarily on non-local characteristics of the host metal. It is, however, evident that for a correct description of site occupancies and pressure-composition isotherms, site dependent enthalpies are also required.

Hydrogen in disordered alloys

To be specific we shall in the following consider the case of compositionally disordered intermetallic alloys with a fcc-structure (see Fig. 1b). We assume that the hydrogen atoms occupy only octahedral interstitial sites. Such sites are coordinated by 6 nearest-neighbours and 8 second nearest-neighbour metal atoms. In an alloy $A_{1-y}B_y$ the various sites are specified by the number m of B-atoms on nearest neighbour positions and the number n of B-atoms on second nearest neighbour positions. For each site we calculate a canonical local enthalpy of solution $\Delta\bar{H}^0_{mn}$ by means of

$$\Delta\bar{H}^0_{mn} = \Delta\bar{H}(A_{1-y_{mn}} B_{y_{mn}} H_c) \tag{41}$$

by means of Eq. 40. The cluster effective concentration y_{mn} is given by

$$_{mn} = \frac{m + \kappa n}{6 + 8\kappa} , \tag{42}$$

where κ is a parameter which characterizes the relative influence of 1st and 2nd nearest-neighbours on the heat of solution of hydrogen in a site (m,n). As the hydrogen-metal interaction potential varies approximately as $R^{-8.07}$ (see ref. 3) where R is the separation between hydrogen and a metal atom of the host lattice one expects κ to be typically a few percent.

The embedding of a cluster (m,n) in an alloy $A_{1-y}B_y$ modifies the canonical enthalpy of the site in two respects.

a) Volume effects: The canonical enthalpy ΔH^0_{mn} corresponds to a canonical cluster of volume $\Omega^0{mn}$ which is in general different from the volume Ω_{mn} actually occupied by the cluster (m,n) in the alloy $A_{1-y}B_y$. The influence of a volume change on ΔH^0_{mn} is given by[19]

$$\frac{d\Delta\bar{H}^0_{mn}}{d \ln V} \frac{(\Omega_{mn} - \Omega^0_{mn})}{\Omega^0 mn} . \tag{43}$$

For d-band metals,

$$\frac{d\Delta\bar{H}^0_{mn}}{d \ln V} = -2.69\ \Delta\bar{H}^0_{mn} + 326 . \tag{44}$$

Very little information exists about the actual interatomic spacings in disordered alloys[36,37] so that no experimental data for Ω_{mn} in $A_{1-y}B_y$ are at hand. In the actual calculations given in the next section we assume that Ω_{mn} assumes a value intermediate between Ω^0_{mn} and the average molar volume $\bar{\Omega}$ of the alloy and introduce a parameter D such that

$$\Omega_{mn} = (1-D)\ \bar{\Omega} + D\Omega^0_{mn} . \tag{45}$$

b) Charge transfer effects:

Each cluster (m,n) has a work function Φ^*_{mn} which is in general different from the work function $\bar{\Phi}^*$ of the alloy $A_{1-y} B_y$. Differences in work functions lead to charge transfer which results in i) modifications of the *occupied* electronic band widths of each cluster and ii) rigid energy band shifts of each cluster. For a discussion of *charging* effects in alloys the reader is refered to the work of Stern [38]. As pointed out by Gelatt et al[20] shifts of the electronic bands as a result of changes in work function do not modify the enthalpy of hydride formation since *both* an electron and a proton (total charge equal to zero) experience the same potential. Changes in occupied band width, however, are directly related to $\Delta\bar{H}$ through eq. 41. Here again little is known about charge transfer in disordered alloys so that we have to introduce an additional parameter Θ to characterize the amount of charge transfer. The influence of charge transfer on $\Delta\bar{H}_{mn}$ is given by $\Theta\cdot\alpha\ (\Phi^*_{mn} - \bar{\Phi}^*)$.

After inclusion of volume and charge transfer effects we obtain then the following expression for the enthalpy of solution

$$\Delta\bar{H}_{mn} = \Delta\bar{H}^0_{mn} + \frac{d\Delta\bar{H}^0_{mn}}{d\ \ln V} \cdot \frac{\Omega_{mn}-\Omega^0_{mn}}{\Omega^0_{mn}} + \Theta\alpha(\Phi^*_{mn} - \bar{\Phi}^*) \quad . \tag{46}$$

As the density-of-sites function $g(\varepsilon^0)$ consists of δ-functions the integral in eq. 18 reduces to a sum

$$c = \Sigma \frac{g_{mn}}{\exp(Z_{mn}) + 1} \tag{47}$$

and

$$\Delta\bar{H}_H = f(c) + \frac{\Sigma\ g_{mn}(\Delta\bar{H}_{mn} + \bar{H}^{mn}_{vib})/(1 + \cosh\ Z_{mn})}{\Sigma g_{mn}/(1 + \cosh\ Z_{mn})} \quad , \tag{48}$$

with

$$Z_{mn} = (\Delta\bar{H}_{mn} + f(c) + \mu^{mn}_{vib} - \mu_H)/kT \quad , \tag{49}$$

In eq. 48 we have used the fact that in a face-centered-cubic structure there is one octahedral interstitial site per metal atom (thus s=1 in eq. 24).

In the absence of short-range order, g_{mn} for an alloy $A_{1-y}\ B_y$ is simply given by the binomial distribution

$$g_{mn} = \frac{14!}{(14-(m+n))!(m+n)!}\ (1-y)^{14-(m+n)}y^{(m+n)} \quad . \tag{50}$$

The expression for g_{mn} when short-range order (SRO) is included are complicated and will not be given here explicitly. They depend on the SRO-parameter σ which is related to the pair probabilities as follows

$$\begin{aligned} p(A-A) &= 1 - y(1-\sigma), \\ p(A-B) &= y(1-\sigma), \\ p(B-A) &= (1-y)(1-\sigma), \\ p(B-B) &= y + (1-y)\sigma . \end{aligned} \tag{51}$$

For σ=1 p(A-B) = p(B-A) = 0 and the system is completely segregated. For $y \leq 0.5$ maximum SRO is realized when $\sigma = y/(y-1)$.

Although not explicitly indicated the non-local effective H-H interaction function f(c) depends on the nominal concentration y as integration of eq. 39 gives

$$f(c) = \int_0^c \left(- \frac{B\bar{V}_H^{\,2}}{V} + \frac{\alpha\nu}{N(E_F)} \right) dc \quad , \tag{52}$$

in which B, V, $\bar{V}_H$, E_F and the electronic density-of-states function N(E) all depend both on hydrogen (c) and alloying metal (y) concentrations. For the bulk modulus[19] and the molar volume of the host metal we use weighted averages (Vegard's law) of the bulk moduli and molar volumes of the constituting metals. For N(E) the rigid -band approximation can be used when $y \ll 1$. At intermediate concentrations N(E) can be estimated by means of the Cyrot-Lackmann procedure or, better, from calculated density-of-states functions.

In section IV we show that the model presented here (together with the semi-empirical band structure model of Griessen and Driessen) leads to pressure-composition isotherms of fcc-alloys which are in agreement with existing experimental data.

IV. APPLICATIONS

The purpose of this section is twofold. First it will be shown that experimental data on hydrogen absorption in palladium-based alloys can be understood within the framework of the density-of-sites model presented here. Second, numerical results will be used to analyze the relative importance of the various ingredients entering the model.

a) $\underline{PdH_c}$. Before calculating p-c isotherms of palladium-based alloys it is necessary to have a satisfactory description of the isotherms of pure palladium. From experimental data of Blaurock[39], Feenstra et al[40], Baranowski[41] and band structure calculations for Pd and PdH (see ref. 40) we find that

$$f(c) = f_{elastic}(c) + f_{electronic}(c) \quad , \tag{53}$$

with $f_{elastic}(c) = -57c + 27c^2$ (kJ/mol H) (54)

and $f_{electronic}(c) = 15c + \sqrt{961(c-0.63)^2 + 20.4} + \sqrt{961}(c-0.63)$. (55)

Further $\Delta\bar{H}^0_\infty = -8.2$ kJ/mol H

and the Einstein temperature[42] is $\Theta_E = 700$ K.
The pressure-composition isotherms found with these parameters and eq. 18 (which reduces to the well-known Lacher-formula in the case where $g(\varepsilon^0)$ is a single δ-function) are shown in Fig. 6. Over the temperature and pressure range considered here the calculated isotherms agree fairly well with the experimental data of Wicke and Brodowsky[43].

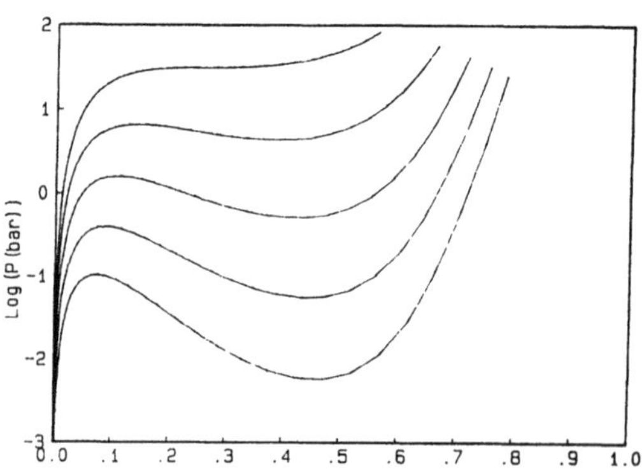

Fig. 6

Pressure-composition isotherms of PdH_c calculated with the parameters indicated above.

b) $Pd_{1-y}Ag_yH_c$.

One of the most thoroughly investigated ternary metal-hydrogen systems is Pd-Ag-H. It is experimentally known[43] that addition of Ag to Pd favours hydrogen absorption at low hydrogen concentrations and reduces the maximum solubility at higher pressures. These trends are well reproduced by the calculated isotherms shown in Figs.7 a,b for $Pd_{0.8}Ag_{0.2}$ and $Pd_{0.6}Ag_{0.4}$. The stabilisation of the hydrides at low concentrations is due to the dilation of the Pd lattice by Ag while the narrowing (and ultimately disappearance of the plateaus) is due to the gradual filling of the Pd d-band by Ag electrons.
For both alloys we used the same volume parameter $D = 0$ (see Eq.45), SRO-parameter $\sigma = 0.2$ (see Eq.51) and nearest-neighbour parameter $\kappa = 0.05$ (see Eq.42). Furthermore, as a result of the addition of 1 electron per H-atom at E_F, we have assumed that

$$f_{electronic}(c_j; Pd_{1-y}Ag_y) = f_{electronic}(c + y; Pd) ,$$

i.e. the density-of-states function N(E) is rigidly shifted with respect to E_F in order to accommodate the extra Ag-electrons.

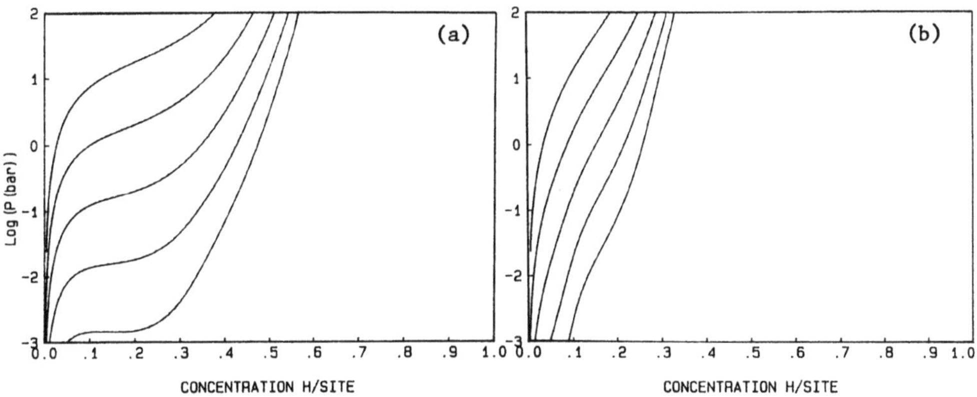

Fig. 7. Pressure-composition isotherms at 600, 480, 400, 343 and 300 K of a) $Pd_{0.8}Ag_{0.2}H_c$ and b) $Pd_{0.6}Ag_{0.4}H_c$ calculated with the multi-sites model described in this work with the parameters $D = 0$, $\sigma = 0.2$, $\kappa = 0.05$. For these alloys the effect of charge transfer is negligible ($\theta = 0$).

To identify the contributions of the various ingredients entering our model we consider the $Pd_{0.8}Ag_{0.2}$ alloy in more detail. Figs.8 a-d show the pressure-composition isotherms corresponding to values for D, σ, κ and rigid-band shift different from those used for the isotherms in Fig. 7a.

i) Local lattice spacing (see Fig. 8a)

For $D = 0$ the atoms A and B forming an alloy $A_{1-y}B_y$ occupy positions on a perfect mean-lattice with an interatomic spacing given by Vegard's law. If, for example, D is arbitrarily set equal to 0.2 the position of the atoms will depart locally from the sites of the perfect mean-lattice. A cluster consisting of A atoms only (Pd in our case) is then less expanded by Ag atoms than when $D = 0$. For the p-c-isotherms this means that the plateau pressures are increased (see eq.44).

ii) Local enthalpies (see Fig. 8b)

The behaviour of isotherms at high concentrations is influenced by the width and shape of the density-of-sites and electronic density-of states functions. The relative importance of these two factors is found by assuming, for example, that Ag does not contribute one electron to fill empty states at the top of the d-band of Pd. Then $f_{electronic}(c_j, Pd_{0.8}Ag_{0.2}) = f_{electronic}(c_j, Pd)$. The resulting isotherms are too flat around $c \cong 0.5$. At 1 bar and 300K the concentration ($c \cong 0.6$) is too large as compared to the experimental data. The fact that without rigid-band shift the critical point is lowered and the isotherms narrowed is a clear indication of the importance of site-dependent enthalpies of solution.

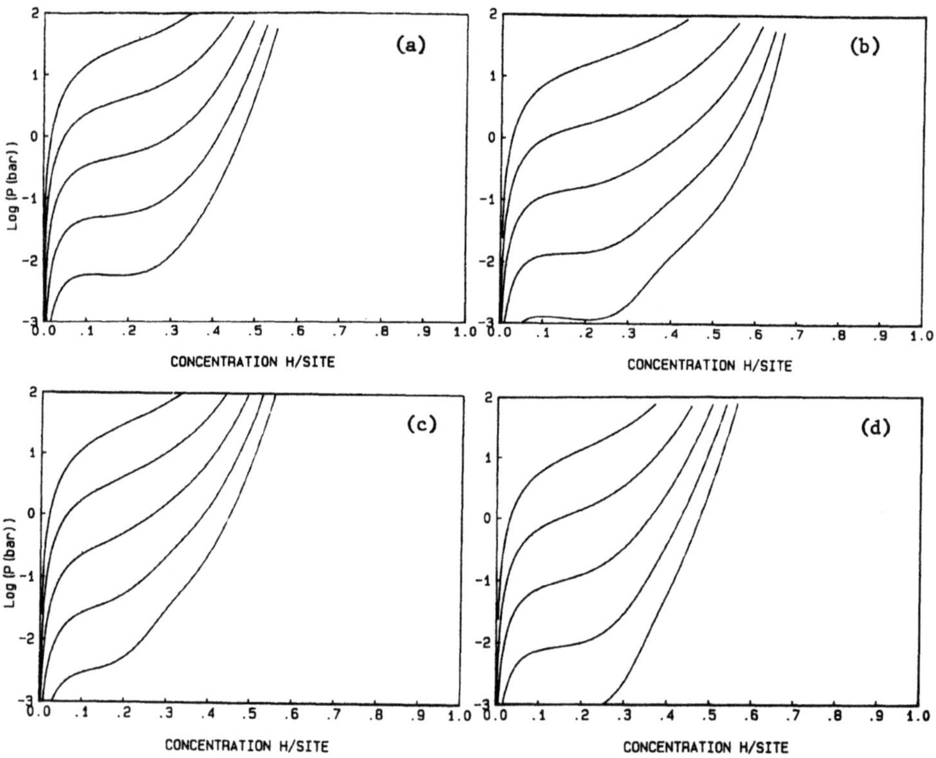

Fig. 8. Pressure-composition isotherms of $Pd_{0.8}Ag_{0.2}H_c$ at 600, 480, 400, 343 and 300K for various sets of parameters to show the importance of

a) local lattice spacing : $D = 0.2$; $\sigma = 0.2$; $\kappa = 0.05$;
b) local enthalpies : $D = 0$; $\sigma = 0.2$; $\kappa = 0.05$ but no rigid-band shift (see text) ;
c) short-range-order : $D = 0$; $\sigma = 0$; $\kappa = 0.05$;
d) 2nd nearest-neighbours : $D = 0$; $\sigma = 0.2$; $\kappa = 0$.

iii) Short-range-order (see Fig. 8c)

In the absence of SRO the pair probabilities are $P(Pd - Pd) = 0.8\,(0.84)$, $P(Pd - Ag) = 0.2\,(0.16)$, $Pd(Ag - Pd) = 0.8\,(0.64)$ and $P(Ag\text{-}Ag) = 0.2\,(0.36)$;
[The values indicated within parentheses correspond to $\sigma = 0.2$ as chosen for the calculations of the isotherms in Fig.7a]. For $\sigma = 0$ the critical point has dropped well below 300K and the lower isotherm exhibits an 'oscillation' which is not found experimentally.

iv) 2nd nearest-neighbours (see Fig. 8d)

If instead of 14 nearest-neighbours we consider only the 6 first-nearest-neighbours the p-c-isotherms resemble strongly those of Fig. 7a although the 'plateau' pressures are slightly lower. This indicates, as one expected, that the influence of 2nd nearest-neighbours is rather weak. It is however not possible to fit satisfactorily the isotherms and the value of the enthalpy of solution at infinite dilution without including the effect of 2nd nearest-neighbours.

c. $Pd_{1-y}Cu_yH_c$

The Pd-Cu system is especially interesting for a test of our model as it is found in sharp contrast to the Pd-Ag system that Cu has a retarding effect on the diffusivity of hydrogen in $Pd_{1-y}Cu_yH_c$ solutions[44]) and that Cu increases the enthalpy of solution $\Delta\bar{H}_\infty$. This increase is usually attributed to the small size of Cu atoms which lead to a decrease of the lattice spacing of $Pd_{1-y}Cu_y$ alloys with increasing y. The magnitude of the increase in $\Delta\bar{H}_\infty$ can however not be explained by invoking volume effects only[45,48])

Our calculations show that the contraction of the lattice renders interstitial sites coordinated by Pd-atoms only less favourable for hydrogen absorption than those coordinated by Pd *and* Cu-atoms. For example, for a $Pd_{0.8}Cu_{0.2}$ alloy, we find in a calculation with $D = 0.25$, $\sigma = 0.2$ and $\kappa = 0.05$ that the local enthalpy of solution for a cluster coordinated by 14 Pd-atoms is +1.02 kJ/molH while that of a cluster coordinated by 3 Pd and 3 Cu in the first shell and 8 Pd-atoms in the second shell is -7.34 kJ/molH. Cu-rich sites act therefore as trapping centers for H in agreement with the data of Kirchheim and McLellan[44]).

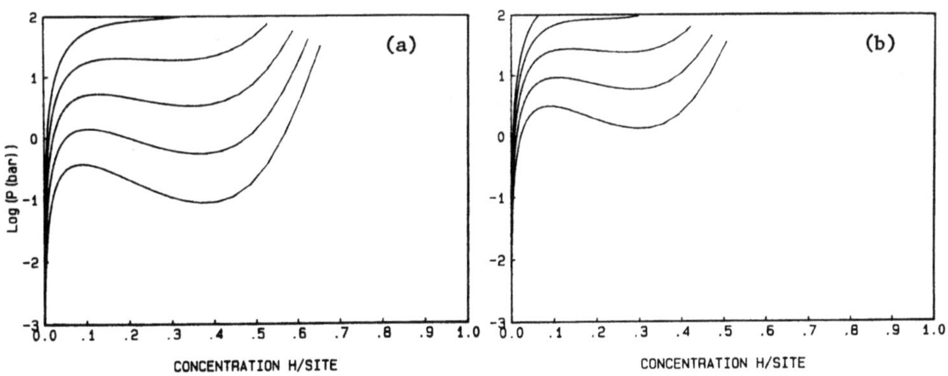

Fig. 9. Pressure-composition isotherm isotherms of $Pd_{0.8}Cu_{0.2}H_c$ at 600, 480, 400, 343 and 300K calculated with the multi-sites model with the parameters $D = 0.25$; $\sigma = 0.2$ and $\kappa = 0.05$.

The p-c-isotherms of $Pd_{0.9}Cu_{0.1}H_c$ and $Pd_{0.8}Cu_{0.2}H_c$ shown in Figs. 9a and b are in agreement with the experimental data of Burch and Buss[46]) and the heat of solution at infinite dilution agrees within a few percent with the values of Kleppa et al.[48]

V. CONCLUSIONS

In this paper we have shown that both *local* (site-dependent) enthalpies and *non-local effective H-H interactions* can be incorporated in a model for the heat of solution and pressure-composition isotherms of disordered metal-hydrogen systems. The local site-enthalpies are estimated by means of a refined version of the so-called semi-empirical band structure model for the heat of formation of metal-hydrogen systems (Griessen-Driessen-model). They depend both on the nature of the atoms coordinating a given interstitial site and on the local-lattice spacings. The variation of the heat of solution with increasing hydrogen concentration arises from i) the gradual filling of the various sites in the alloy, ii) a non-local elastic term $B\bar{V}_H^2/V$ (where B, V and $\bar{V}_H$ are the bulk modulus and the molar volume of the host metal and $\bar{V}_H$ the partial molar volume of hydrogen in the metal) and iii) a shift in the Fermi energy due to the filling of electronic states at E_F.
The model gives a good description of the pressure composition isotherms of palladium-based fcc-alloys. In this work we considered in some detail $Pd_{1-y}Ag_yH_c$ and $Pd_{1-y}Cu_yH_c$ because these systems have been thoroughly investigated in the past. Results for other palladium-based alloys shall be presented elsewhere.

To conclude, it is worthwhile to make a comment about the various parameters entering the model. The parameters α and β which relate $\Delta\bar{H}^0$ to the characteristic band structure energy (eq. 34) are the *same* for all metals. The parameter κ which is treated as fit parameter is in principle calculable once the hydrogen-metal potential is known.
As mentioned before, the strong dependence of this potential ($\sim R^{-8.07}$) implies that κ is typically a few percent, in agreement with our fitted value $\kappa = 0.05$. The parameters D and σ describe the structure of the alloy. Calculations by Froyen and Herring[36] indicate that $D \approx 0.3$ in the absence of short range-order ($\sigma = 0$). To our knowledge no calculations are at hand for the case including short-range-order. Experimental data are also scarce. It is therefore very interesting to note that pressure-composition isotherms are in principle able to provide us with information on the structure of an alloy.

ACKNOWLEDGEMENTS

Interesting discussions with Drs. A. Driessen, R. Feenstra, E. Salomons and V. Moruzzi are gratefully acknowledged.

References

1. R.B. McLellan and W.A. Oates, Acta Met. 21 (1973) 181
2. K.H.J. Buschow, P.C.P. Bouten and A.R. Miedema, Rep. Prog. Phys. 45 (1982) 927
3. R. Griessen and T. Riesterer, in Topics in Applied Physics (Springer, Heidelberg); to be published
4. F. Pourarian and W.E. Wallace, J. Solid State Chem. 55 (1984) 181
5. Y. Komazaki, M. Uchida, S. Suda, A. Suzuki, S. Ono and N. Nishimiya, J. Less-Common Metals 89 (1983) 269
6. K. Oguro, Y. Osumi, H. Suzuki, A. Kato, Y. Imamura and H. Tanaka, J. Less-Common Metals 89 (1983) 275
7. R. Kirchheim, F. Sommer and G. Schluckebier, Acta Met. 30 (1982) 1059
8. B.S. Berry and W.C. Pritchet, Phys. Rev. B24 (1981) 2299

9. See articles by R. Kirchheim, B.S. Berry, R. Bowman, A. Maeland and others (this conference)
10. F.H.M. Spit, Ph.D. thesis, University of Utrecht (The Netherlands), 1982, unpublished and F.H.M. Spit, J.W. Drijver and S. Radelaar, Scr. Metall. 14 (1980) 1071
11. K. Huang, Statistical Mechanics, John Wiley, New York (1963)
12. As a result of the impossibility for two hydrogen atoms to occupy the same site, the expression for Ω is the same as that for a gas of non-interacting fermions
13. H.K. Hemmes, A. Driessen and R. Griessen, J. Phys. C:Solid State Phys., to be published
14. The zero-point energy is already included in the site energies ε_j^o so that only the contribution of thermal phonons has to be taken into account
15. R. Griessen, Phys. Rev. B27 (1983) 7575
16. P.M. Richards, Phys. Rev. B30 (1984) 5183
17. R. Griessen and A. Driessen, Phys. Rev. B30 (1984) 4372
18. R. Griessen, A. Driessen and D.G. de Groot, J. Less-Common Met. 103 (1984) 235
19. R. Griessen and R. Feenstra, J. Phys. F:Met. Phys. 15 (1985) 1013
20. C.D. Gelatt, H. Ehrenreich and J.A. Weiss, Phys. Rev. B17 (1978) 1940
21. A.R. Williams, J. Kübler and C.D. Gelatt, Phys. Rev. B19 (1979) 6094
22. D.A. Papaconstantopoulos and A.C. Switendick, J. Less-Common Metals 88 (1982) 273 and Phys. Rev. B17 (1978) 141
23. A.C. Switendick, in Topics in Applied Physics, Vol. 28 (Springer, Heidelberg, 1978) p. 101
24. A.C. Switendick, Z. für Phys. Chem. NF 117 (1979) 89
25. A. Driessen, H. Hemmes and R. Griessen, Z. für Phys. Chem., Neue Folge (1986), to be published
26. A. Driessen, H. Hemmes and R. Griessen, Proc. of the Xth AIRAPT Int. High Pressure Conference, Amsterdam, July 1985
27. R. Feenstra, R. Griessen and D.G. de Groot, J. Phys. F:Met. Phys., to be published
28. F.M. Mueller, A.J. Freeman, J.O. Dimmock and A.M. Furdyna, Phys. Rev. B1 (1970) 4617
29. C. Wagner, Acta Metall. 19 (1971) 843
30. E. Wicke and H. Brodowsky, in Topics in Applied Physics, Vol. 29 (Springer, Heidelberg, 1978) p. 73
31. C.K. Hall, A.I. Shirley and P.S. Sahni, Phys. Rev. Letters 53 (1984) 1236; (see also W. Fenzl and J. Peisl, Phys. Rev. Letters 54 (1985) 2064)
32. M. Futran and C.K. Hall, J. Chem. Phys. 80 (1984) 383
33. A.I. Shirley, C.K. Hall and N.J. Prince, Acta Metall. 31 (1983) 985
34. R.A. Bond and D.K. Ross, J. Phys. F:Metal Phys. 12 (1982) 597
35. M. Cyrot and F. Cyrot-Lackmann, J. Phys. F:Metal Phys. 6 (1976) 2257
36. S. Froyen and C. Herring, J. Appl. Phys. 52 (1981) 7165
37. B. Lengeler, Phys. Rev. Letters 53 (1984) 74 and Solid State Comm. 55 (1985) 679
38. E.A. Stern, Phys. Rev. B5 (1972) 366 and Phys. Rev. B13 (1976) 621
39. E. Wicke and J. Blaurock, Ber. Bunsenges. Phys. Chem. 85 (1981) 1091 and J. Blaurock, Ph.D. Thesis (Münster, FRG) 1985, unpublished
40. R. Feenstra, Ph.D. Thesis, Vrije Universiteit (Amsterdam, The Netherlands) 1985, unpublished
41. B. Baranowski, in Topics in Applied Physics, Vol. 29 (Springer, Heidelberg, 1978) p. 157 (see comment in ref.27)
42. B.M. Geerken, R. Griessen, L.M. Huisman and E. Walker, Phys. Rev. B26 (1982) 1637
43. E. Wicke and H. Brodowsky, in Topics in Applied Physics, Vol. 28 (Springer, Heidelberg, 1978) p. 5 and references therein
44. R. Kirchheim and R.B. McLellan, Acta Metall. 28 (1980) 1549
45. M. Yoshihara and R.B. McLellan, Acta Metall. 31 (1983) 61

46. R. Burch and R.G. Buss, J. Chem. Soc. Far. Trans. I 71 (1975) 913 and 922
47. D. Fisher, D.M. Chisdes and T.B. Flanagan, J. Solid State Chem. 20 (1977) 149
48. O.J. Kleppa, Shamsuddin, and C. Picard, J. Chem. Phys. 71 (1979) 1656

FORMATION OF AMORPHOUS METALS BY SOLID STATE REACTIONS OF HYDROGEN WITH AN INTERMETALLIC COMPOUND

Konrad Samwer

I. Physikalisches Institut der Universität Göttingen
and SFB 126, Bunsenstraße 9, D-3400 Göttingen, FRG

ABSTRACT. A short review of a new method to produce amorphous metals in the solid state is given. The reaction of elemental hydrogen with crystalline intermetallic compounds is examined in the well studied Zr_3Rh-system. Requirements for the process and some properties of the amorphous alloy are discussed in comparison with liquid quenched metals. Similarities and differences to the solid state reaction of pure elemental metallic multilayers without hydrogen are shown.

1. INTRODUCTION

Until recently the formation of amorphous metals was confined to the rapid quenching methods either from the melt or from the vapour phase except for some very special alloys prepared from an aqueous solution. All the rapid quenching methods have in common that the removal of the heat has to be done on a timescale short compared to the time t_x necessary for nucleation and growth of the more stable crystalline counterparts. Since t_x depends on temperature, pressure, concentration and short-range-order-parameters among other, the quenching process has to be done in a duration between $10^{-2} > t_c > 10^{-7}$ s for $\Delta T \approx 500$ K in order to achieve an amorphous metal. These conditions keep usually one dimension of the alloys very small ($\lesssim 50$ µm) because the rapid solidification by an extreme heat flux can be only obtained for a nearly two dimensional alloy. A very different approach for the formation of amorphous metals is made by using the high mobility of at least one species of the alloy to obtain the same metastable state. This can be done with large equipment as in the case of ion implantation or ion mixing, with large mechanical power for instance in the ball milling process or just with thermal energy as in the case of solid state diffusion reactions. One example of the gas-solid reactions which has been recently studied to some extend[1-4] is the intermetallic compound Zr_3Rh and gaseous hydrogen.

2. WHAT IS AMORPHIZATION BY A SOLID STATE REACTION PROCESS

Although there were several hints in the literature[5,6] that the absorption of hydrogen by a metallic alloy may even lead to a phase transition into the amorphous state, no further investigations were made. This was probably due to the fact that phase transitions by hydrogen absorption into different crystalline phases even at low temperatures are a rather common effect[7]. In our experiments the amorphization of the intermetallic compound Zr_3Rh was analyzed. Fig. 1 shows the X-ray pattern of the crystalline Zr_3Rh alloy in a $Cu_3Au(L1_2)$ structure. Single phase crystalline Zr_3Rh compounds can be formed in at least two crystalline modifications by different methods. The $L1_2$ structure is observed in foils and ribbons as quenched from the melt with a quenching rate insufficient to form an amorphous alloy. Nearly single phase crystalline samples with $E9_3$ structure can be produced by annealing amorphous alloys at 360^oC-400^oC, which crystallizes the entire sample[8]. The latter has been checked by powdering and thinning the samples to avoid any surface effects and by measurements of the superconducting properties. After hydrogen absorption at 1 atmosphere H_2-pressure and 180^oC a fully amorphous $Zr_3RhH_{5.5}$ alloy has been formed

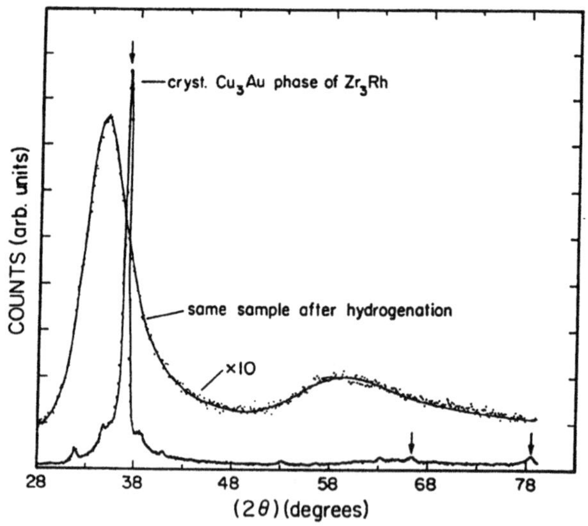

Fig. 1. X-ray diffraction pattern (Cu K_α-radiation) of a crystalline Zr_3Rh compound and the same sample in the amorphous state after solid state reactions with hydrogen.

for both crystalline modifications as shown by the typical X-ray pattern in Fig. 1 for one alloy. The reaction can be started and stopped just by lowering and raising the temperature. X-ray pattern taken in between show that the main crystalline peaks decrease in height but remain still sharp with only little change in its positions, whereas the amorphous pattern starts to appear separately. This indicates already that the amorphization process has nothing to do with a continious breaking of the crystalline grains beyond a certain limit rather than a phase transformation from one to the other state with a well defined interface. This might be already a significant difference to solid state reactions performed with ion bombardment and mechanical alloying.

To perform experiments like in our case certain requirements have to be satisfied. Again the time which is allowed to conduct the amorphization of the alloy has to be small compared with the time t_x necessary for nucleation and growth of competing crystalline phases. Since diffusion is necessary to supply the growing amorphous phase with hydrogen the reaction temperature is very critical. As can be seen in a schematic temperature-time-transition diagram (Fig. 2) the upper limit of the temperature is given by the crystallization temperature T_x which depends on time and perhaps on unknown metastable crystalline phases which might occur even earlier than the stable one. In our case reaction temperatures between 150° C < T_r < 215° C always produced amorphous alloys. The hydrogen mobility in amorphous Zr-alloys has been measured by NMR technique and is found in the order of ten magnitudes higher than the mobility of the metal atoms with activation energies $Q_H \simeq 0.4$ eV and $Q_M \approx 1.5 - 2.0$ eV

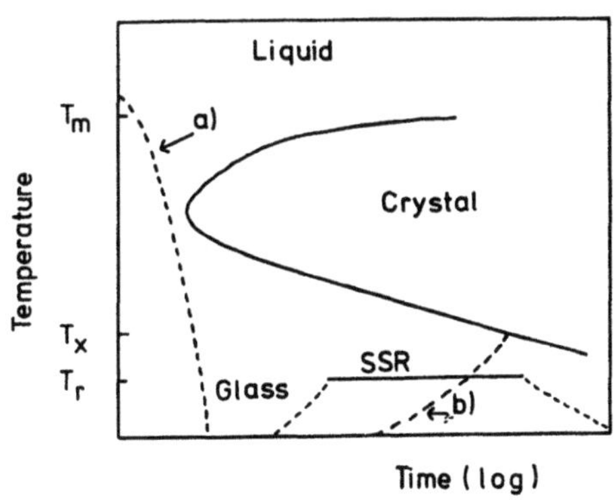

Fig. 2. Time-Temperature-Transition diagram for a) liquid quenching, b) crystallization of the amorphous alloy at T_x, SSR) amorphization by solid state reactions at T_r.

respectively[9]. This very fast diffusion at relatively low temperatures allows us to neglect any rate limiting influence of the hydrogen supply, which can be expressed by a timescale t_D very short to t_X.

The requirements for amorphization by SSR which are a low reaction temperature but an adequate diffusivity of at least one species, is completed by the absolute necessary condition of a large negative heat of mixing to obtain a thermodynamically downhill process. This can be described in a free energy diagram[2] as shown in Fig. 3. Although some parts of the diagram can be calculated using Miedema's semiempirical model[10], we still use a hypothetical diagram, which is in full agreement in those parts where the calculations are feasible. For the "pure" ZrRh-diagram the metastable crystalline Zr_3Rh-compound has a slightly lower free energy as the amorphous phase with a difference of about 10 kJ/g-atom if only the different heats of formation are taken into account. The intermetallic compound Zr_2Rh has, as shown in Fig. 3, a much lower free energy ($\Delta H_f(200^\circ\ C) = -75$kJ/g-atom) than the amorphous phase with the same composition, whereas the difference between crystalline and "amorphous" Zr is oversubcribed in Fig. 3a. In the ternary Zr-Rh-H system only a projection of a ternary to the pseudo-binary ZrRh system with different hydrogen amounts is shown in Fig. 3b. In none of our experiments we found a coherent single-phase crystalline hydride of both Zr_3Rh-structures at any temperature. Therefore the free energy parabola is drawn at a higher free energy as all others. This is corroborated by the calculation for the ternary hydride using again Miedema's approach[11]

$$\Delta H(c\text{-}Zr_3RhH_{5.5}) = \Delta H(Zr_3H_{2.75}) + \Delta H(Rh\ H_{2.75}) - \Delta H(c\text{-}Zr_3Rh)$$

for the the heat of formation of a hypothetical crystalline hydride with the

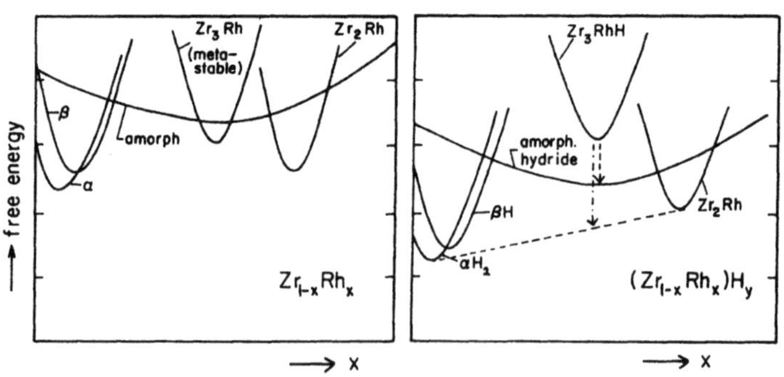

Fig. 3. Hypothetical Gibb's free energy diagram for a) "pure" $Zr_{1-x}Rh_x$ and b) $Zr_{1-x}Rh_xH_y$. Note that the quantitative differences of the free energies for the phases are arbitrarily chosen but there is qualitative agreement with prelimary calculations using Miedema's model.

same composition as the amorphous one. In the model the formation of stable ternary hydrides can be predicted following an extension of the model used to describe binary alloys and compounds mentioned above. Here hydrogen is thought to lead to a "coating" of the hydrogen attracting metal (here Zr) and a total breaking of all metal-metal "bonds" (Zr-Rh). This rule of reversed stability has certain restrictions as mentioned by Buschow et al.[12], which makes the applicability questionable here. From our simple calculations we obtain $\Delta H(c\text{-}Zr_3RhH_{5.5}) \approx +1$ kJ/g-atom, which is certainly the upper limit for the heat of formation assumming all metal-metal bonds are broken. If we take in Fig. 3b also a hydrogen free intermetallic compound of Zr_2Rh, which is a controversial matter in the literature[13], and calculate $\Delta H(ZrH_2) = -60$ kJ/g-atom for $\alpha\text{-}ZrH_2$,we obtain exactly the drawn diagram only slightly changed in the relative differences between the crystalline phases. Unfortunately the ternary phase diagram for the amorphous state is not known. This makes the calculations rather uncertain. We state here only that the reversed stability rule should give the indicated lower free energy of the amorphous hydride with respect to the $c\text{-}Zr_3RhH$ phase. To lower its free energy the system has now two different reaction paths as indicated in Fig. 3b: The usual one is the separation into two crystalline phases ZrH_2 and Zr_2Rh, which is observed for hydrogen absorption at temperatures above $215^{\circ}C$:

$$c\text{-}Zr_3Rh + H_2 \rightarrow c\text{-}Zr_2Rh + ZrH_2.$$

For lower temperatures ($150^{\circ}\,C < T < 215^{\circ}\,C$) the second possible path to lower the free energy is the amorphization of the entire sample:

$$c\text{-}Zr_3Rh + 2.5H_2 \rightarrow a\text{-}Zr_3RhH_{5.5}.$$

Here no chemical segregation occurs due to the polymorphic transformation into the amorphous state. This transformation is fast in time compared to the nucleation and growth of the two crystalline phases where diffusion of the metal atoms is necessary. The low diffusivity of the metal atoms compared to the high mobility of the hydrogen atoms connected with the attractive interaction between one metal species and the hydrogen gives the basic understanding of the SSR process.

3. THERMAL STABILITY AND COMPARISON OF SOME PROPERTIES WITH LIQUID QUENCHED METALLIC GLASSES

One major argument for the validity of any free energy diagram is the crystallization behaviour of the amorphous alloys as shown nicely by Köster and Herold[14]. To follow the different crystallization paths transmission electron microscopy experiments are well suited. For simplicity we conducted electrical resistivity measurements followed by X-ray studies.

Fig. 4 shows the crystallization of a liquid quenched amorphous Zr_3Rh alloy without hydrogen and a hydrided sample $Zr_3RhH_{3.6}$ with a slightly smaller hydrogen concentration[4]. For "pure" Zr_3Rh the crystallization temperature T_x (= 680 K) is reached when the resistance suddenly increases. X-ray diffraction pattern as well as recent DSC-measurements on the same alloys[15] verify the occurence of a metastable crystalline Zr_3Rh phase as mentioned above. The crystallographic structure of this phase seems to depend on the previous treatment[15]. A second transformation into the Zr_2Rh phase and the α-Zr phase can be seen in the resistance and the DSC-measurements. Both steps are in full accordance to our free energy diagram, where a polymorphic crystallization from the amorphous to the crystalline Zr_3Rh phase is followed by an eutectic transformation. For the hydrided alloy the crystallization path is different. Since it lowered its free energy already by the hydrogen absorption below the free energy curve of a hydrided crystalline Zr_3Rh alloy the crystallization occurs in a single step at about $T_x \approx 650$ K into crystalline ZrH_2 and a Rh-rich alloy. We note that DSC-measurements on fully hydrided samples give a $T_x \approx 609$ K and again a single step behaviour showing a huge endothermic peak around T = 660 K, which indicates a significant loss of hydrogen[9]. It is remarkable to note that the crystallization temperatures of the samples once hydrided are not very different from the "pure" amorphous Zr_3Rh alloys whereas a slight increase of the temperature above T_r (≈ 490 K) during the hydrogen absorption leads immediately to a crystalline endproduct. This documents that the necessary driving force for the segregation is reduced due to the hydrogen absorption and the phase transformation. If the hydrogen-metal bond is completed and the free energy lowered further heat treatment leads to no significant change. Therefore one can treat the problem with two different "kinds" of hydrogen atoms; one where the atoms are highly mobile (free) and "searching"

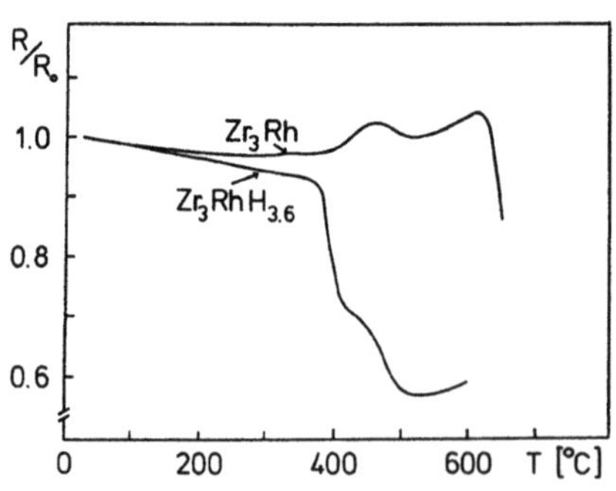

Fig. 4. Crystallization behaviour of liquid quenched amorphous Zr_3Rh and hydrided $Zr_3RhH_{3.6}$ using the temperature dependence of the normalized electrical resistance.

for a tetrahedral site build by four Zr-atoms[16] and the other, where the atoms are trapped in these sites with a distribution of activation energies for diffusion[9] and no significant desorption tendency until the entire sample crystallizes. The latter is documented by several attempts to desorp hydrogen at a temperature far below the crystallization temperature which always result in the production of crystalline ZrH_2 with a very fine grain size (< 10 nm).

In order to give more evidence that the amorphous state reached by solid state reactions has the same properties as the one reacted by rapid quenching, an amorphous liquid quenched Zr_3Rh alloy was hydrided under exactly the same conditions as the further one (180° C, 1 atom H_2). Detailed X-ray diffraction studies for all three alloys("pure" amorphous Zr_3Rh, amorphous $Zr_3RhH_{5.5}$ prepared from original amorphous Zr_3Rh (orig. amorph.) and amorphous $Zr_3RhH_{5.5}$ prepared from original crystalline Zr_3Rh (orig. cryst.)) provide the reduced radial distribution functions G(r) as described for example in Ref. 16.Fig. 5 shows that no significant differences are seen for both hydrided samples. With hydrogen absorption the first maximum of G(r), which incorporates contributions from Zr-Zr pairs and Zr-Rh pairs starts to split up in two single peaks. This is due to the increasing distances between the Zr-pairs, which become larger due to the attraction of hydrogen by the Zr atoms, whereas the Zr-Rh distances stay nearly constant. The increase of the metal-metal distances is also observed in the drop of the density. Both hydrided amorphous alloys have the same value ($\rho = 6.54$ g/cm^3) which is about 13 % less than for "pure" Zr_3Rh. This leads to a volume expansion of $\Delta V \cong 2.6 \cdot 10^{-3}$ nm^3/H, which was typically found for metallic glasses[17].

It is sometimes thought that X-ray analysis is not the ultimate proof of an amorphous state of an alloy. We therefore performed transmission

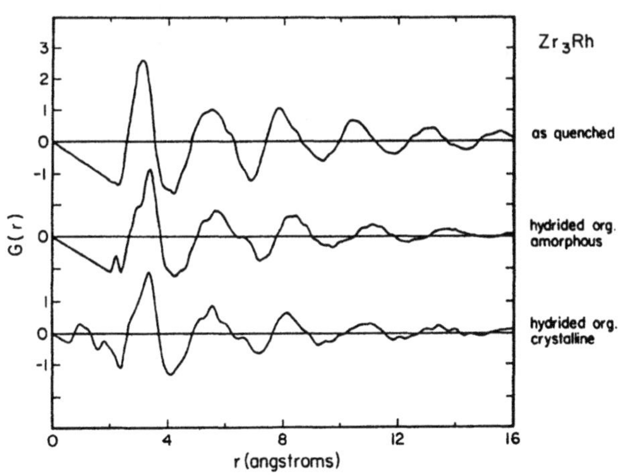

Fig. 5. Radial distribution function G(r) for amorphous Zr_3Rh and two $Zr_3RhH_{5.5}$ alloys, one produced by liquid quenching and then hydrided (orig. amorphous). The other made by SSR (orig. cryst.).

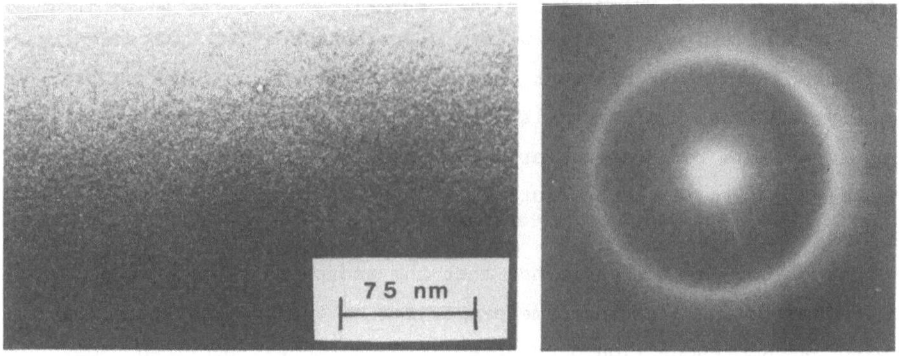

Fig. 6. TEM-picture and diffraction pattern of an originally crystalline Zr_3Rh compound in its amorphous state after SSR with hydrogen.

on electron microscopy studies on the hydrided Zr_3Rh alloys, which are difficult due to the embrittlement of the samples with hydrogen absorption. Fig. 6 shows the result of an amorphous $Zr_3RhH_{5.5}$ alloy produced by solid state reactions and its diffraction pattern. Both exhibit the characteristic pictures of an amorphous metal showing no sign of any remaining crystalline material. The "structure" seen depends on the phase contrast and is known from many other liquid quenched glasses.

The electronic properties of the two amorphous metals prepared so differently were measured by low temperature specific heat experiments in the temperature range between 1.5 K - 9 K. Fig. 7 shows the typical plot of the specific heat C divided by the temperature T versus T^2. The scattering of the data is due to the difficult handling of the brittle samples leading to a less firm contact onto the sample holder than usual. Both alloys show in this plot a linear dependence over the entire temperature range. The intercept of the y-axis gives the Sommerfeldconstant γ and the slope can be easily converted to the Debyetemperature. With hydrogen absorption the electronic density of states at the Fermi level is reduced from $D(E_f)$ = 2.43 states / [eV atom] for "pure" amorphous Zr_3Rh[19] to $D(E_f)$ = 0.58 states / [eV atom] for the orig. crystalline $Zr_3RhH_{5.5}$ alloy and $D(E_f)$ = 0.54 states / [eV atom] for the original amorphous one. This is a huge decrease by a factor of 4 but the values are still close to the one of crystalline Cu. The Debye temperature is increased by nearly a factor of 2 from $\Theta_D(Zr_3Rh)$ = 191 K to $\Theta_D(Zr_3RhH_{5.5}$ orig. cryst.) = 379 K and $\Theta_D(Zr_3RhH_{5.5}$ orig. am.) = 347 K. Althought there is a slight difference between both hydrided alloys we consider the electronic properties of the two as identical. This is confirmed by the superconducting transition temperature, which drops from $T_c(Zr_3Rh)$ = 4.4 K[19] to $T_c(Zr_3RhH_{5.5})$ = 1.03 K; = 1.21 K respectively. Both alloys become superconducting although

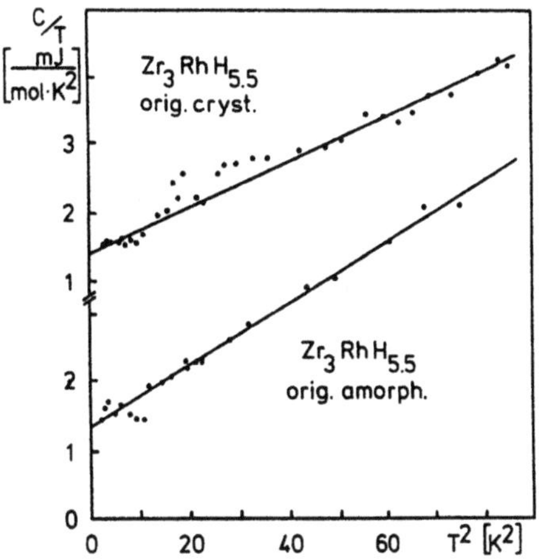

Fig. 7. Low temperature specific heat C of two amorphous $Zr_3RhH_{5.5}$ alloys prepared differently (see subscript Fig. 5).

$c = \gamma T + \beta T^3$

the density of states, which mainly determines T_c, is strongly reduced. This has been studied very recently in more details on the amorphous Zr_2PdH system, where a different electron-phonon coupling seems to be responsible for the high T_c-behaviour[20]. In summary all properties so far investigated prove that the final amorphous state is a metastable phase independent of the various ways to reach it just like it is known from the thermodynamically stable phases.

4. KINETICS OF SOLID STATE REACTIONS AND COMPARISON WITH MULTILAYER METALLIC DIFFUSION COUPLES

Finally we turn back to the mechanism of the solid state reaction in more detail. As stated earlier the process has to run on a timescale t which is small compared to t_x. Fig. 8 shows the hydrogen concentration and the electrical resistance as a function of hydrogen absorption time for $Zr_3Rh + H_2$ at 180° C. It is easily seen that already 65 % of the final hydrogen concentration is absorpted within the first 24 h. At the same time the normalized resistance taken as a very rough value for the amorphization has only changed by less than 10 %. After the first day the resistance starts to increase quite linear with time as the sample takes more and more hydrogen. After the reaction is complete the final hydrogen concentration C ≈ 1.4 H/M is reached and the resistance has increased by 75 %. From the X-ray analysis we know that the orignal crystalline Zr_3Rh phases take only very little hydrogen since the peak positions moved only very slightly and no coherent crystalline hydride has been observed. Since the amorphous phase is "growing" only slowly (not detectable in X-ray pattern after 24 h), nearly 2/3 of the hydrogen atoms have to get "stored" at the interfaces or in macroscopic voids before the formation of the amorphous phase provides suitable sites for the hydrogen atoms. Since hy-

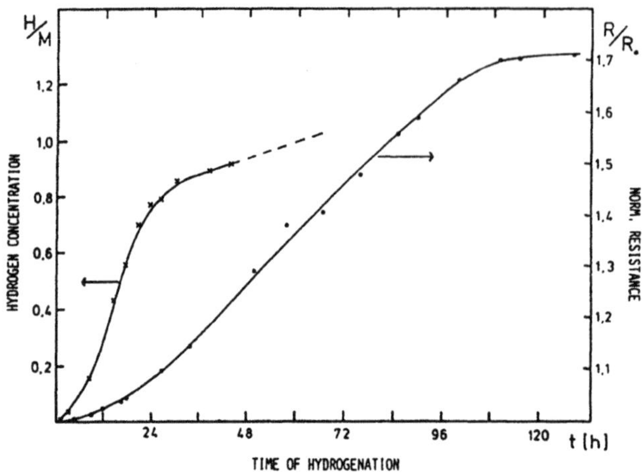

Fig. 8. Hydrogen concentration and normalized resistance versus time of hydrogen absorption for Zr_3Rh at 180° C.

drogen diffusion especially along grain boundaries is very fast the rate limiting factor for the formation of the amorphous phase must be the slow transformation from the crystalline to the amorphous state itself. In other words the moving interface between the amorphous $Zr_3RhH_{5.5}$ and the crystalline Zr_3Rh phases determines the transition rate. One might also speculate, that a certain mobility of the metal atoms is necessary for the motion of the interfaces to change the short range order of the crystalline phase with long range periodicity into the short-range-order of the amorphous metal accepting hydrogen in a tetrahedral site. A slight mobility of a metal species is reasonable because the reaction slows down very rapidly with decreasing temperature although the hydrogen atoms are still very mobile and can penetrate in Zr-based alloys even at T = 50° C easily[9]. The fascinating interplay between the different species can be nicely shown in a steady state model of Johnson et al.[21] for higher temperatures. Fig. 9 shows a schematic illustration, where hydrogen penetrates at the gas-solid boundary (B) into the crystal. At the second interface (A) the transformation into two crystalline phases is drawn (T > 215° C), where the interfacial energy σ_{12} requires a minimum grain size d_1, d_2. At higher temperatures the interdiffusion J of metal atoms allows the chemical segregation on a timescale $t_1 = d^2_{1,2}/D$ [D: interdiffusion coefficient for Zr,Rh-atoms] , whereas at low temperatures only $t_2 = L^2/D_H$ [L: typical sample dimensions, D_H: hydrogen diffusion coeff.] is dominant. Since the timescale t_x for nucleation and growth of the crystalline phases is mainly determined by t_1, amorphization by solid state reactions with a timescale larger than t_2 but smaller than t_1 can succeed. The exact law for the transformation is thought to be linear in time as one would expect for an interface limited process, but this has not been proven yet.

In the experiments of multilayer metallic diffusion couples as first

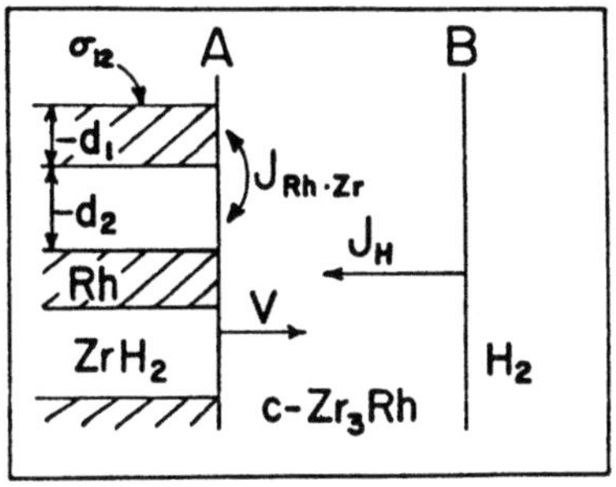

Fig. 9. (from Ref. 21) Schematic illustration of a steady state reaction front (A) ($T > 215^{o}$ C) moving at velocity v to form crystalline ZrH_2 and a crystalline Rh-alloy from crystalline Zr_3Rh with hydrogen. d_1, d_2 gives the minimm grain size required by the interface energy σ_{12} · J_{Rh-Zr}, J_H is the flux of metal and hydrogen atoms respectively. B gives the gas-solid boundary.

shown on the La-Au system[22], the basic requirements for amorphization by solid state reaction are the same. Again the reaction temperature has to be below T_x and (fast) diffusion is extremely helpful to compete on a time-scale t_D for the diffusion of one metal species with the relevant time t_x. This can only happen with the necessary driving force (ΔH_f) which comes up for diffusion and any interfacial barriers. The reaction happens at every metal-metal interface and results in a uniform, planar layer with asymmetric growth as long as both metal species are supplied[23]. Rutherford backscattering experiments show at least for longer times a $\sqrt{t}$-law for the growth of the amorphous layer as one would except for the diffusion of one species through the amorphous layer[24]. At the beginning of the reaction interfacial barriers may limit the process and change the time dependence.

In summary it was the aim of this lecture to discuss a new mechanism without rapid quenching which produces amorphous metals by solid state reactions. All parameter known so far summarize in the critical condition to be fast enough for the competing crystalline phases. The main subject was on the gas-crystal reaction were an interface limited process is expected for the reaction kinetic. This remains one on the vice versa case of the polymorphic crystallization of some metallic glasses. Pure metallic diffusion couples seem to exhibit a $\sqrt{t}$-law for the growth of the planar amorphous layers at least for longer times. This case comes close to the eutectic crystallization in the reverse subject. All amorphization processes lead into the same metastable amorphous state, which is far from being only a "frozen in" liquid. Solid state reactions are just a new way into the same minimum.

Acknowledgement

The work described has been performed in cooperation with X.L.Yeh, W.L.Johnson, J.Tebbe, R.Schulz and U.Köster, whose unselfish support and interest made the progress in this subject possible.

References

1. X.L.Yeh, K.Samwer and W.L.Johnson, Appl. Phys. Lett. 42: 242 (1983)
2. K.Samwer, X.L.Yeh and W.L.Johnson, J. of Non-Crystalline Solids 61-62: 631 (1984)
3. K.Samwer in: "Amorphous metals and non-equilibrium processing", J.v.Allmen, ed., Les Editions des Physiques, Les Ulis, (1984)
4. J.Tebbe, K.Samwer and R.Schulz in: "Rapidly Quenched Metals", S.Steeb and H.Warlimont, ed., North Holland, Amsterdam, (1985)
5. H.Östereicher, H.Clinton and H.Bittner, Mat. Res. Bull. 11: 1241 (1976)
6. A.M. van Diepen and K.H.J.Buschow, Solid State Commun. 22: 113 (1977)
7. e.g. R.M van Essen and K.H.J.Buschow, J. Less-Common Metals 64: 277 (1979)
8. A.J.Drehman and W.L.Johnson, Phys. Stat. Sol. A52: 499 (1979)
9. R.C.Bowman, e.g. this institute proceedings and references therein
10. A.K.Niessen, F.R. de Boer, R.Boom, P.F. de Chatel, W.C.M.Mattens and A.R.Miedema, Calphad 7: 51 (1983)
11. A.R.Miedema, Philips Tech. Rev. 36: 217 (1976)
12. K.H.J.Buschow, P.C.B.Bouten and A.R.Miedema, Rep. Prog. Phys. 45: 937 (1982)
13. P.P.Narang, G.L.Paul. K.N.R.Taylor and G.Wallork, in: Proc. 2nd Int. Conf. on Hydrogen in Metals, Paris (1977)
14. U.Köster and U.Herold in: "Glassy Metals I", H.-J.Güntherodt and H.Beck, ed., Springer-Verlag, Heidelberg, (1981)
15. J.S.Cantrell, J.E.Wagner and R.C.Bowman, J. Appl. Phys. 57: 545 (1985)
16. K.Samwer and W.L.Johnson, Phys. Rev. B28: 2907 (1983)
17. A.J. Maeland, e.g. this institute proceedings and references therein
18. Uwe Köster, private communications
19. P.Garoche and W.L.Johnson, Solid State Comm. 39: 41 (1981)
20. M.Kullik, G.v.Minnigerode and K.Samwer, Z. Physik B (1985), in press
21. W.L.Johnson, M.Atzmon, M.van Rossum, B.P.Dolgin and X.L.Yeh in: "Rapidly Quenched Metals", S.Steeb and H.Warlimont, ed., North Holland, (1985)
    W.L.Johnson, R.Schwarz and K.Samwer, U.S.Patent No. 8400035, (1983)
22. R.Schwarz and W.L.Johnson, Phys.Rev.Lett. 51: 415 (1983)
23. H.Schröder, K.Samwer and U.Köster, Phys. Rev. Lett. 54: 197 (1985)
24. M.van Rossum, M.A.Nicolet and W.L.Johnson, Phys. Rev. B29: 5498 (1984)

THERMAL STABILITY OF HYDRIDES OF DISORDERED AND AMORPHOUS ALLOYS

J. S. Cantrell*, R. C. Bowman, Jr.**, and G. Bambakidis***

*Chemistry Department, Miami University, Oxford, Ohio 45056
**Chemistry and Physics Laboratory, The Aerospace Corp. Los Angeles, CA 90009
***Physics Department, Wright State University, Dayton, OH 45435

INTRODUCTION

Hydrides of amorphous/glassy alloys have received considerable attention recently because of their potential applications as energy carriers, chemical storage of hydrogen, heat pumps, fuel cells, and heat engines (1,2). In all of these applications the uncharged intermetallic compound and the corresponding ternary hydride are subjected to a large number of charging and decharging cycles. A major disadvantage to using ternary hydrides which results after a relatively large number of cycles is the decomposition or disproportionation of the material so that it no longer absorbs hydrogen gas in a reversible way. It has been shown that this decomposition is a reaction by part of the ternary hydride to form the corresponding binary hydride of the more stable (stronger hydrogen-attracting component) plus free metal of the less hydrogen-attracting component (1). It is not well understood why this disproportionation reaction is significant for some systems and almost insignificant for other systems. In some cases an intermetallic compound is formed along with the more stable binary hydride instead of the free metal. Buschow, Bouten and Miedema (1) list over 100 intermetalic hydrides that have been prepared and partially characterized. There are several times this many that are known, yet few are really satisfactory for the applications listed above. This paper is limited to transition metal-transition metal type alloys where one metal (A) is the stronger hydrogen-attracting and (B) is the weaker hydrogen-attracting. Examples of A-type metals are early (IIIb, IVb, Vb such as Sc, Y, La, Ti, Zr, Hf, V, Nb, etc.) and B-type (late) (VIIIb, Ib such as Fe, Co, Ni, Cu, Rh, Pd, Ir, Pt, etc.). The alloys may be intermetallic compounds but often are all compositions that may be prepared by the melt-spinning technique and are only limited in composition by the range of glass-stability. These alloys are designated by the atom percent composition, such as $Ti_{45}Cu_{55}$ for an alloy of 45 atom percent Ti and 55 atom percent Cu. Both the crystalline (c-) and amorphous/glass (a-) alloys will be discussed in this paper. The systems that are discussed here are: a-$TiCuH_x$, c-$TiCuH_x$, c-Ti_2CuH_x, a-Zr_2PdH_x, c-Zr_2PdH_x, a-Zr_3RhH_x, where x refers to noninteger values for the hydrogen composition. In addition, the intermetallic alloys will be included in the properties discussed wherever appropriate. The experimental methods used to study thermal stability are differential scanning calorimetry (DSC), isothermal annealing, and powder X-ray diffraction (XRD). All XRD data have been taken at room temperature following quenching of the DSC or annealing studies.

Extensive studies of amorphous/glassy binary alloys have resulted in the following general conclusions regarding the formation of the amorphous phase and its recrystallization (3,4). The heat of formation of the amorphous phase is less than that of the crystalline phase which is consistent with the metastable nature of the amorphous phase. Most of the difference in the heats of formation can be interpreted as due to a reduction in the chemical short range order (CSRO) in the amorphous/glassy state. The CSRO reduction produces a less effective surrounding of an atom by dissimilar atoms and produces an energy barrier to crystallization, which is called the activation energy and is determined by a series of different heating rates used in the DSC studies. The difference in heat of formation, as shown between crystalline and amorphous materials, is much less than the activation energy for recrystallization, (E_x). E_x is diffusion-controlled and is related to the formation energy of a hole the size of the smaller of the two atomic species of the binary alloys (3). Chemical short range order also makes an important contribution in the E_x and results in an enhanced experimental E_x at concentrations where the CSRO varies rapidly near the recrystallization temperature. Because a number of the ternary hydrides discussed in this paper have both crystalline and amorphous phases that are metastable against decomposition (5,6), the ABH_x systems provide excellent examples for systematic comparisons of the behavior of crystalline and amorphous/glassy hydrides. DTA experiments (5-9) have shown that a-ABH_x crystallizes at a much lower temperature than a-AB (the unhydrided amorphous alloy). Proton NMR studies (8-11) have also indicated that the a-ABH_x is much less stable than the corresponding c-ABH_x material. This paper summarizes the reported work on the thermal stabilities of amorphous and crystalline ABH_x materials and gives quantitative comparisons of the activation energies for crystalline and amorphous/glassy metastable hydrides. Buschow has recently treated the thermodynamics of decomposition of ternary metal hydrides in terms of an activation energy for diffusion of the metal atoms (2,9). However, for the amorphous state, the enhanced hydrogen diffusion relative to the crystalline phase, observed in proton NMR studies (8,10), must also be included in treating the reduced thermal stability against thermal decomposition of the a-ABH_x relative to c-ABH_x for the same alloy hydride system.

EXPERIMENTAL

Amorphous/glassy alloys are usually prepared by two different techniques, the melt-spinning technique (2), and the anvil piston method (2). Very high purity materials (99.95% or greater purity) for both the metals and the hydrogen gas and the inert gas (Ti gettered Ar) used to continually flush the heating boat and the copper wheel, are used to avoid the effects of metal impurities and oxide surface effects as far as possible. Electrolytic charging has been used to do the hydriding in some cases (4) to provide a very easy preparation method, however a problem with homogenity of the ternary hydride may result. Both melt spinning (MS) and the anvil piston (AP) have been used in the work reported here. The hydrides were prepared by charging with gaseous hydrogen, using heat and pressure whenever necessary. Crystalline samples are prepared by employing a slower cooling rate to the melt which is often the same composition as was used to prepare the glassy material. Several different hydride compositions were studied for each binary alloy composition. The techniques of hydriding have been described elsewhere (5,6,11,12). Transition temperatures, kinetic activation energies and heats of transition were determined using DSC at temperatures between 330K and 1000K. Room temperature powder XRD measurements were made to identify the phases (including confirming the amorphous state by XRD) and determine the unit cell parameters before and after the DSC or isothermal anneal runs for most of the samples studied. The same sample was used for the XRD as was used for the DSC or isothermal anneal work. The DSC was typically done with a Perkin-Elmer model DSC-II equipped with a 3600 data station and typically eight different heating rates varying from 1.25 K/min to 160 K/min were used on each composition studied. The temperature was calibrated using the melting points of pure metals such as In and Pb, in some cases a reversible transition such as that of K_2CrO_4 was also used. Both the transition temperature and the energy scale were calibrated using these materials. In many instances it was observed that the temperature calibration was dependent upon the heating rate but that the energy scale was independent of the heating rate used. The energy scale was checked each time the heating rate was changed.

The samples studied were stored in a dry box under purified Ar gas with an oxygen concentration of less than 2 ppm before making the DSC runs. Only fresh samples were used for the DSC runs. The DSC curves of heat input vs. temperature give, in principle, the rate of enthalpy change with time as a function of temperature. The area of the peak is directly proportional to the heat evolved or absorbed by the reaction.

Powder X-ray diffraction (XRD) data were obtained on a Phillips Norelco (XRG-3000) generator and goniometer, equipped with a constant area of irradiation-type detector slit. Other XRD work was done using a Rigaku D/MAX Automated XRD. Both instruments used copper K_α radiation. DSC samples used for the XRD studies were kept hermetically sealed in gold boats to reduce the problem of oxidation until immediately before the XRD data was taken. The actual DSC samples were used for the XRD by stopping the DSC run at a temperature just above an exothermic or endothermic transition and rapidly cooling the sealed sample to room temperature (DSC cooling rate was 320 C/min). The seal was then broken and the sample was mounted on a glass slide and X-rayed at room temperature.

Indexing and Bravais lattice assignments were made by comparing the data to known reported data (13) and by trial and error and by using the Ito method (14). Unit cell parameters were determined by an iterative least-squares lattice-fit program on the indexed data. Weighting of the indexed data by intensity was often used to reduce the effect of weak data on fairly small data sets. All of the reported unit cell parameters are from converged least-squares fits of all indexed data from each XRD data set. The tolerence in two theta was usually set at 0.05 degrees for the least-squares convergence.

RESULTS AND DISCUSSION

Portions of typical DSC traces for metal alloy hydrides are shown in Fig. 1-5, which are of a-TiCu, a-$TiCuH_{1.41}$, c-$TiCuH_{0.96}$, a-Zr_2Pd, a-$Zr_2PdH_{1.91}$, a-$Zr_2PdH_{2.7}$, a-Zr_3Rh (MS-1,2,3), a-$Zr_3RhH_{3.7}$, a-$Zr_2PdH_{2.2}$, a-$Zr_2PdH_{2.9}$, a-$Zr_3RhH_{4.1}$, and a-$Zr_3RhH_{5.1}$. These three ternary alloy hydride systems illustrate many of the typical features that are used for the thermal analysis of ternary alloy hydrides that are given in this paper. Fig. 1 illustrates the lower thermal stability of the hydride (both amorphous and crystalline) compared to the glassy binary alloy and also illustrates the lower thermal stability of the amorphous hydride compared to the crystalline hydride. Typically, most alloy hydrides show an exothermic transition (a negative DSC peak), usually indicated by a T_x or by a temperature in K near the peak. In addition, the hydrided samples show a large endothermic trend or peak (positive DSC peak) at higher temperatures usually above 800K, as shown by Fig. 1,2,4,5. This effect has also been reported by many other studies, as given by (1,3,5,6,7). The $TiCuH_x$ system was also reported by Dunlap and Dini (15) and by Maeland (5,6), the Zr_3FeH_x system by Fries, et al. (16), the $ZrNi_2H_x$ system by Spit et al. (17), and the $Zr_xNi_yH_z$ system by Batalla, Altounian, and Strom-Olsen (18).

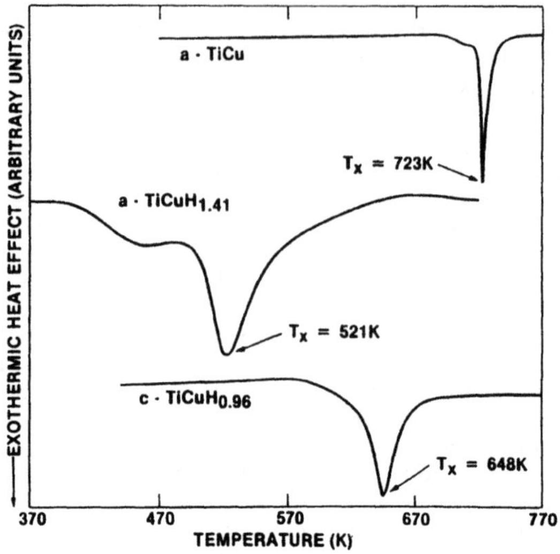

Fig. 1. Typical DSC traces at 40K/min for a-TiCu and $TiCuH_x$. T_x denotes exothermic transitions.

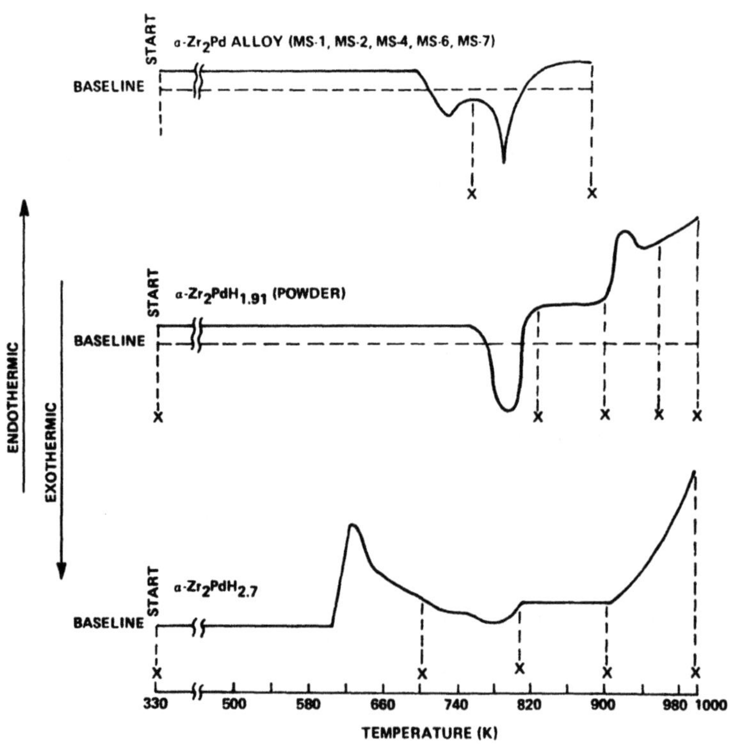

Fig. 2. DSC traces at 20K/min for $a\text{-}Zr_2Pd$, $a\text{-}Zr_2PdH_{1.91}$, and $a\text{-}Zr_2PdH_{2.7}$. X denotes DSC temp. quenched for XRD.

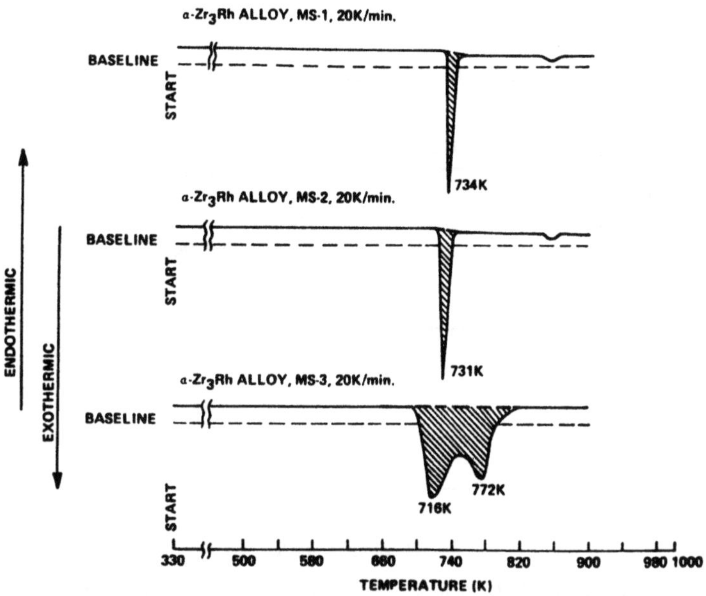

Fig. 3 DSC traces for meltspun a-Zr_3Rh. ΔH_x values are from shaded peak areas.

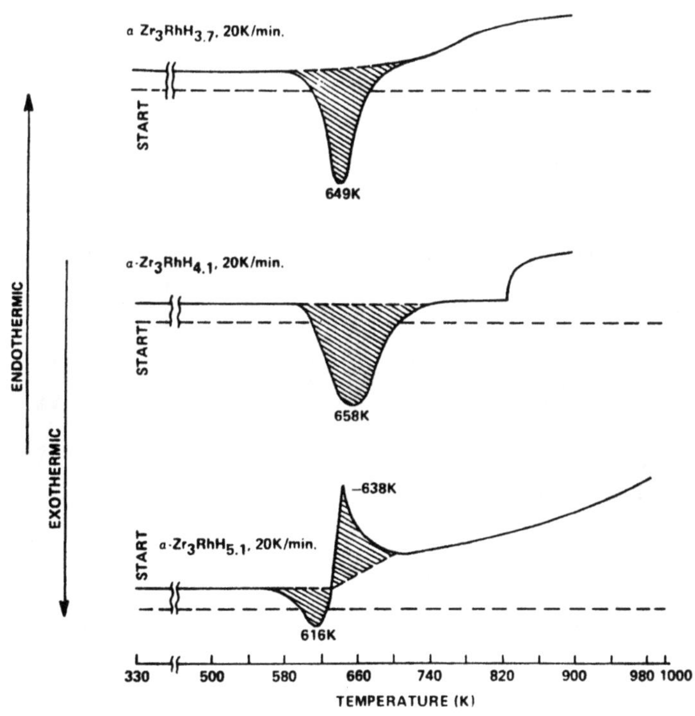

Fig. 4. DSC traces for a-$Zr_3RhH_{3.7}$, a-$Zr_3RhH_{4.1}$, and a-$Zr_3RhH_{5.1}$. ΔH_x are from shaded peak areas.

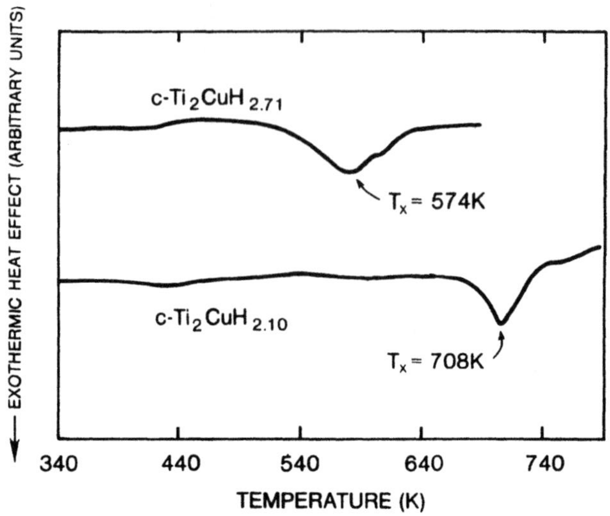

Fig. 5. Typical DSC traces at 10K/min for crystalline Ti_2CuH_x. T_x denotes exothermic transitions.

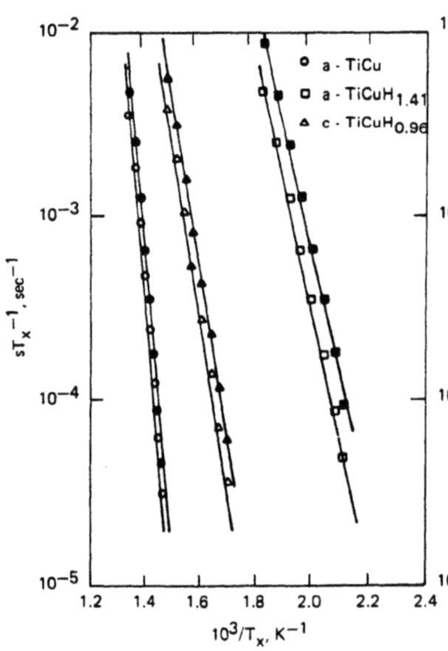

Fig. 6. E_x from $1/T_x$ vs. $\ln(s/T^2_x)$ or $\ln(s/T_x)$. • Kissinger, o Boswell.

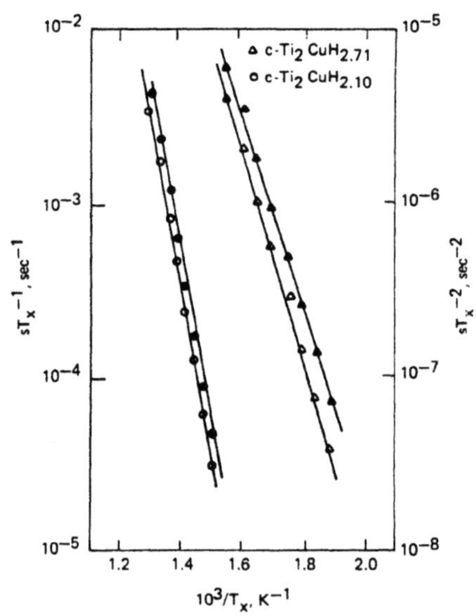

Fig. 7. E_x from $1/T_x$ vs. $\ln(s/T^2_x)$ $\ln(s/T_x)$. • Kissinger, o Boswell.

The exothermic T_x increases with the heating rate, s, and strongly depends on the hydrogen content, decreasing with increasing hydrogen content for a given heating rate. The activation energy for crystallization, E_x, was determined by both the method of Kissinger (19) and by the method of Boswell (20). The activation energy, E_x, is determined from the slope of the plot of $\ln(s/T_x)$ vs. $1/T_x$ (Boswell method) and from $\ln(s/T_x^2)$ vs. $1/T_x$ (Kissinger method). Both methods give essentially straight-line plots and very nearly the same E_x values as shown by Fig. 6 and 7.

A weaker secondary peak sometimes preceeds (lower temperature) the primary peak. This was shown for the a-TiCu and a-$TiCuH_x$ system and has been reported by Dunlap and Dini (15) and by Furlan et al. (21). Varying the heating rate did not alter the separation of these two peaks and X-ray analysis of a run stopped between the peaks and quenched to room temperature indicated that the sample was still amorphous. A similar experiment by Dunlap and Dini (15) also indicated that a quenched sample from between the two peaks was amorphous. This low-temperature peak may be the result of a change in the CSRO or a relaxation or "phase change" in the amorphous material. Some differences do appear between the DSC traces of Dunlap and Dini (15) and that reported here (21). These differences are primarily a group of 3 or 4 small exothermic peaks between 650K and 750K that are absent from that of our data. These additional peaks may be due to inhomogeneities in the hydrogen content that results from the electrolytic charging process. Our samples were prepared by charging the alloy with hydrogen gas.

When the heat of transition, ΔH_x, obtained from the area under the transition peak, was plotted vs. the heating rate, s, the samples showed a strong dependence of ΔH_x on the heating rate (see Fig. 8). Since the peak areas were calibrated for each heating rate, this dependence is not an instrumental effect. Also, this effect is not likely to be due to hydrogen evolution at the lower heating rates. Thus, this system may be undergoing a non-equilibrium relaxation involving the hydrogen (22).

REACTIONS of TiCu and Ti_yCuH_x

(1) a-TiCu $\xrightarrow{680\text{-}700K}$ γ-TiCu + δ-TiCu (approximately equal amounts)

(2) c-$TiCuH_{0.96}$ $\xrightarrow{600\text{-}710K}$ γ-TiH_x + Cu $\xrightarrow{895K}$ γ-TiH_x + $TiCu_3$ $\xrightarrow{930K}$

γ-$TiCuH_x$ + δ-TiCu + H_2

(3) a-$TiCuH_{1.41}$ $\xrightarrow{470\text{-}600K}$ γ-TiH_x + Cu $\xrightarrow{780K}$ δ-TiH_x + $TiCu_3$ $\xrightarrow{900K}$

γ-$TiCuH_x$ + δ-TiCu + H_2

(4) c-$Ti_2CuH_{2.71}$ $\xrightarrow{530\text{-}670K}$ γ-TiH_x + Cu $\xrightarrow{800K}$ γ-TiH_x + $TiCu_3$ $\xrightarrow{880K}$

γ-$TiCuH_x$ + δ-TiCu + H_2 + γ-TiH_x

Table 1 summarizes DSC results for all compositions studied (21). The values of T_x and ΔH_x given in Table 1 are for a heating rate of 20K/min. The activation energy for proton diffusion, $E_a(H)$, obtained from proton NMR studies on a-$TiCuH_{1.41}$ and c-$TiCuH_{0.96}$ is also included (10). Note from Table 1 that T_x decreases with increasing x for each type Ti_yCuH_x ternary hydride. T_x is closely correlated with E_x for both crystalline and amorphous systems. The direct variation of T_x with E_x in the recrystallization of unhydrided amorphous binary alloys is well established (3), and there is

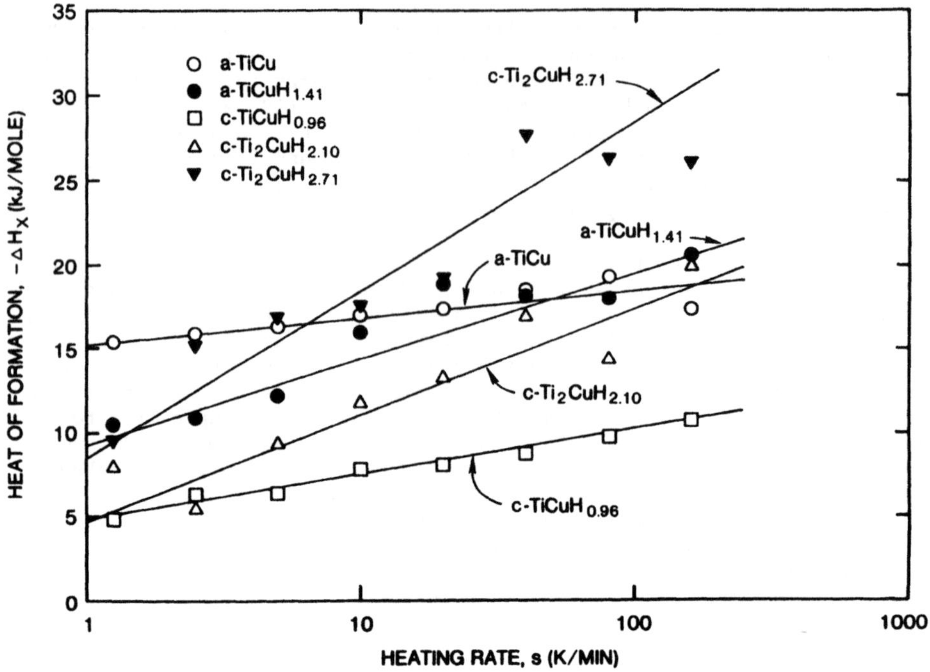

Fig. 8. Plots of ΔH_x, vs. log (s). Errors in ΔH_x are ± 10 to 20%.

TABLE 1. Thermal Data for Titanium Copper Alloys and Hydrides

Compound	T_x(K)	$-\Delta H_x$(kJ/mol)	E_x(kJ/mol)	E_a(H)(kJ/mol) (NMR)
a-TiCu	713 (700)[a]	17.4	314 (359)[a]	--
c-$TiCuH_{0.96}$	637	8.0	175	84
a-$TiCuH_{1.41}$	509 (445)	18.9	134 (78)	40
c-$Ti_2CuH_{2.10}$	721 (619)	13.2	196 (111)	--
c-$Ti_2CuH_{2.71}$	594	19.3	117	--

[a]Parentheses refer to secondary peaks

also evidence for a correlation between the stability against decomposition in the crystalline ternary hydrides RM_nH_x and the M atom diffusion activation energy (9). The low T_x and small E_x for a-$TiCuH_{1.41}$ relative to a-TiCu are most likely due to higher diffusivities in the hydride as compared to the binary alloy. The difference in the metal atom CSRO between these two systems should be small since hydrogen atoms occupy specific interstitial sites with the same local environment in crystalline and amorphous $TiCuH_x$ (10,23). On the other hand, the low T_x and small E_x for a-$TiCuH_{1.41}$ relative to c-$TiCuH_{0.96}$ are probably due to enhanced hydrogen diffusion in the amorphous material, the difference in E_x values being essentially identical to the difference in the $E_a(H)$ values (Table 1). Thus, we concluded that hydrogen diffusion was an important factor in the thermal stability of the crystalline and amorphous hydrides of TiCu, and this conclusion can most probably be extended to the hydrides of other binary alloys, at least when the two types of metal atom are comparable in size and diffusivity. Certainly, diffusion of the metal atoms is required to form the separate phases TiH_x and Cu, as identified by XRD, and this probably accounts for E_x being substantially greater than $E_a(H)$.

Fig. 2 (top trace) (24) for a-Zr_2Pd has a smaller exothermic peak preceeding the primary peak similar to that shown by the a-TiCu binary alloy. However when the DSC is stopped between the peaks and quenched, the X-ray pattern is crystalline, not amorphous, as it was for the TiCu alloys and hydrides. Furthermore, the unit cell found for the first peak is one-third that found for the second peak. The unit cell is cubic for the first peak and tetragonal for the second peak. The relationship between these two unit cells is shown by Fig. 9 (25). The two crystal systems strongly suggest that the cubic (first peak) cell becomes the tetragonal (second peak) when there is enough energy to cause the metal atoms to diffuse so that the Zr and Pd atoms are not disordered (as in the cubic unit cell) and when the metal atoms have completed the organization, the structure is the $MoSi_2$-type (see Fig. 10) as indicated by the tetragonal unit cell with Pd atoms in the body centered position and Zr atoms at one-third and two-thirds positions along the edges of the unit cell. When the lattice has completed the ordering of the Zr and Pd atoms, the (101) (Miller indices) planes of the disordered cubic unit cell become the (103) planes of the C11b ($MoSi_2$) unit cell (25). Thus, the two DSC peaks are due to a phase change in the a-ZrPd binary alloy. The smaller values of x for the alloys, a-$ZrPdH_{1.91}$, and other ternary hydrides with x less than about 2.3 do not show the cubic disordered (Zr and Pd) structure of the amorphous binary alloys. In fact these lower hydrogen concentration ternary hydrides show only one primary peak that comes almost at the same T_x as the binary alloy (unhydrided).

At higher temperatures, these ternary amorphous hydrides lose hydrogen (exothermic DSC peak) to form ZrH_x and ZrPd (as shown by XRD). The large endothermic peak at about 620-650K is identified as ZrH_x by an XRD pattern from the quenched material. The ZrPd binary alloy is only observed above 900K and it is not known whether it forms earlier with crystallites too small to diffract well or if there is a separation of Pd that then rereacts as is the case for the TiH_x and Cu that react to form $TiCu_3$ and H_2. The occurrence of T_x at lower values for the amorphous ternary hydrides compared to the amorphous binary alloys is an indication that forming the hydride has made the alloy less stable.

Fig. 3 (24) shows the three different types of DSC traces that were found for nominally the same binary alloy composition for a-Zr_3Rh. This binary alloy has a fairly complex phase behavior, MS-1 and MS-2 forms an $E9_3$ (cubic) structure at 734K, then transforms to a C16 (tetragonal, Zr_2Rh) at about 850K, but MS-3 forms a DO_e-type structure that does not transform

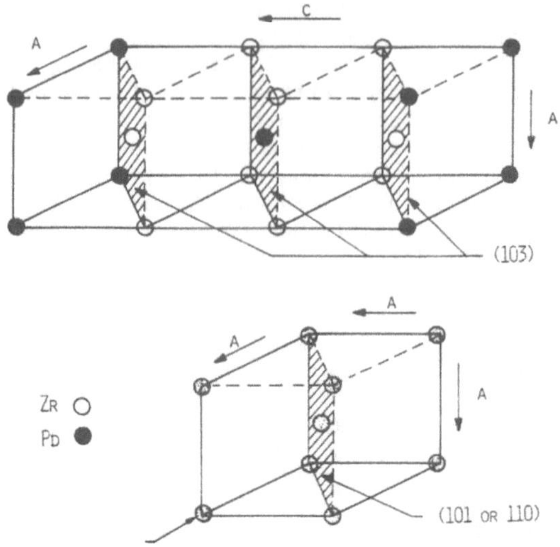

Fig. 9. Zr_2Pd unit cells with the disordered bcc and tetragonal(C11b) structures. The shaded planes emphasize the similarities of the bcc (101) and tetragonal (103) planes.

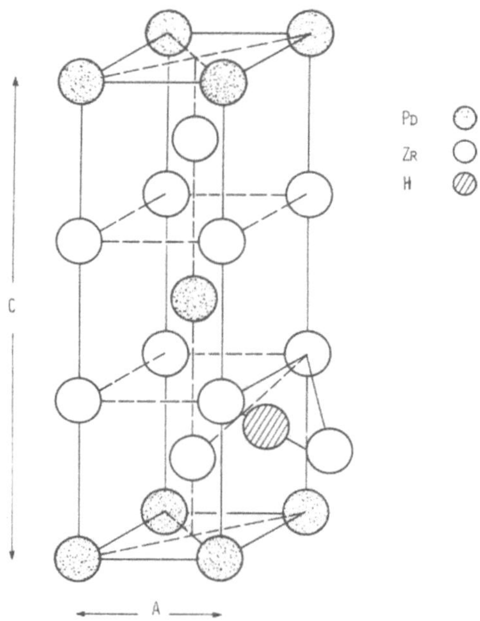

Fig. 10. The C11(b) ($MoSi_2$) form of Zr_2Pd. The striped atom (H) is shown in a tetrahedal Zr_4 site.

to C16 until above 1000K. This difference in phase behavior for essentially identical compositions is thought to be due to differences in chemical short range order (CSRO) that may have resulted from differences in melt temperatures during the melt-spin process of the glassy alloy. If the temperature of the melt was different for the three samples, this could result in less CSRO for the melt with the higher temperature, resulting in different types of nuclei and different phases when the DSC runs are made. The thermal (DSC) data for a-Zr_3Rh are summarized in Table 2.

TABLE 2. Thermal Data for Alloys of Zirconium Palladium and Zirconium Rhodium

Composition	Density	ΔH_x	T_1	E_1	T_2	E_2
	(g/cm^3)	(kJ/mol)	(K)	(kJ/mol)	(K)	(kJ/mol)
Zr_2Pd (MS-1)	7.97(5)	---	739(2)	---	804(2)	---
Zr_2Pd (MS-2)	7.85(5)	-24(2)	742(2)	290	794(2)	262
Zr_2Pd (MS-4)	7.72(9)	-23(1)	737(2)	---	771(5)	---
Zr_2Pd (MS-5)	7.84(12)	-20(2)	762(2)	---	813(2)	---
Zr_2Pd (MS-6)	7.84(7)	-24(9)	741(2)	---	793(2)	---
Zr_2Pd (MS-7)	---	-24(3)	754(2)	---	796(2)	---
Zr_3Rh (MS-1)	7.56(8)	-15(1)	734(1)	323	855(4)*	---
Zr_3Rh (MS-2)	7.53(5)	-15(1)	731(1)	356	856(5)*	---
Zr_3Rh (MS-3)	7.50(9)	-23(3)	716(2)	309	771(1)	363
Zr_3Rh (AP)	---	-14	716(1)	---	791(2)	---

*Weak, not always observed. Parentheses refer to uncertainities in last place.

The DSC of a-$Zr_3RhH_{5.1}$ is shown in Fig. 4 (24), which shows an exothermic peak at about 610K followed almost immediately by a large endothermic peak, in fact the lower values of x (approximately 4) do not show the endothermic peak. The XRD data indicate that ZrH_x is formed at the exotherm along with Rh (free metal) but the large endotherm is most likely from decomposition to form hydrogen. Since the ZrH_x is stable at these temperatures, the H_2 is probably coming from the decomposition of unreacted a-Zr_3RhH_x. No crystalline ZrRh binary alloys were found by XRD for the different DSC studies at temperatures ranging from 700K to 1000K for the a-Zr_3RhH_x materials. T_x for the ternary hydrides of a-Zr_3RhH_x all occur at lower temperatures than for the unhydrided binary alloy.

The behavior of all three systems, the TiCu alloy and ternary hydride, the Zr_2Pd alloy and ternary hydride and the Zr_3Rh alloy and hydride all exhibit similar thermal stability behavior with respect to the order of the T_x values being: amorphous ternary hydride lowest, crystalline ternary hydride second, and binary metal alloy (unhydrided amorphous) having the highest value of T_x. This is consistent with the hydride destabilizing the metal lattice and with the amorphous structure being metastable with respect to the crystal structure. The DSC exothermic activation energies, E_x, for the hydrides are lower than for the unhydrided binary alloys for all three systems (TiCu, ZrPd, Zr_3Rh) except for a-Zr2Pd(MS-2) vs. a-Zr2PdH$_{1.91}$ (see Fig. 6,7,11,12) indicating that the diffusivity of the hydrogen should be taken into

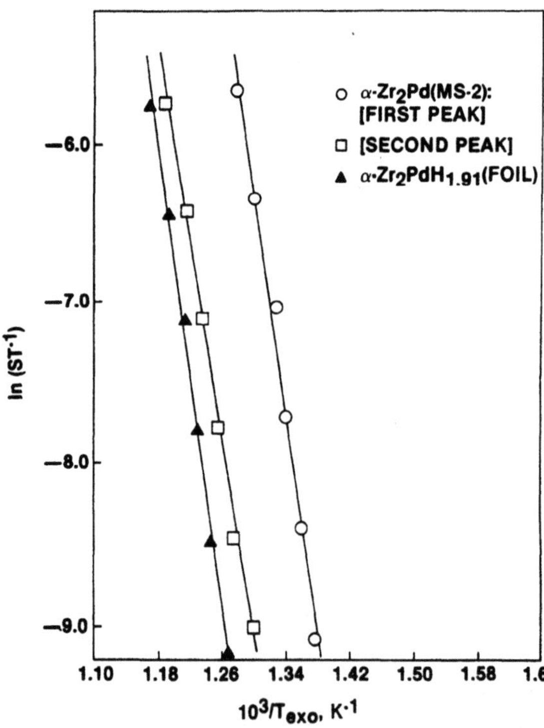

Fig. 11. E_x (Boswell) for a-Zr_2Pd (MS-2) and a-Zr2PdH$_{1.91}$.

account when considering the thermal stability of ternary alloy hydrides. Both the E_x of the TiCuH system and the E_x of the Zr_3RhH_x system are much lower than that of the binary alloy while the E_x for the ZrPd system is much closer to that of the binary alloy, also the ZrPd system forms an intermetallic, while both the TiCu and the ZrRh systems form free metal (Cu or Rh) when the ternary hydrides decompose (see Table 3). These results are in agreement with the conclusions of Buschow (9), which are: 'Most ternary hydrides RM_nH_x are metastable and tend to decompose into the stable phase consisting of the binary hydride of the strongly hydrogen-attracting component (RH_2 or RH_3) and either pure M metal or an intermetallic compound of higher M content than RM_n. This decomposition or disproportionation is described in terms of an activation energy for diffusion of metal atoms in the ternary hydrides.' In this case ZrPd was formed from Zr_2Pd which follows his conclusion that the intermetallic compound would have higher M content than RM_nH_x which was Zr_2PdH_x.

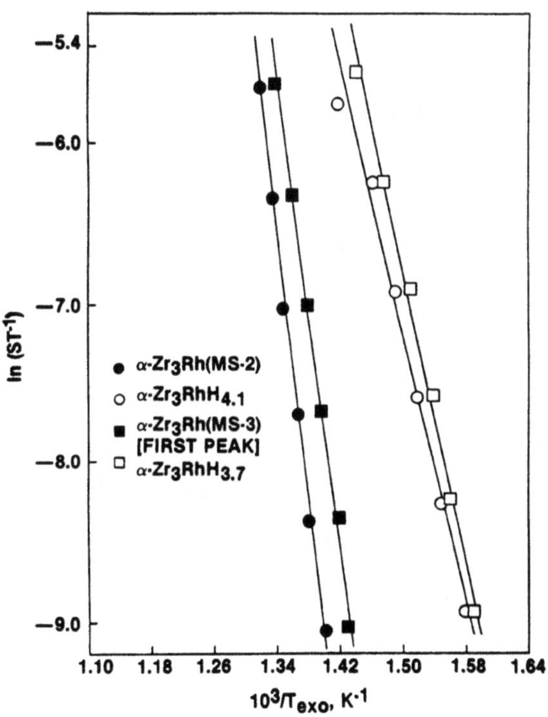

Fig. 12. E_x (Boswell) for (MS-2,3) and a-$Zr_3RhH_{4.1}$ and a-$Zr_3RhH_{3.7}$.

TABLE 3. Thermal Data for Hydrides of Zirconium Palladium and Zirconium Rhodium

Composition	x(meas.) NMR	T_x (K)	ΔH_x (kJ/mol)	E_x (kJ/mol)	T_E (K)	ΔH_E (kJ/mol)
a-$Zr_2PdH_{1.91}$	---	795(3)	-25(6)	299	910(9)	---
a-$Zr_2PdH_{1.91}$	2.2(2)	791(2)	-47(7)	---	910(9)	---
a-$Zr_2PdH_{2.0}$	2.2(2)	806(15)	---	---	950	---
a-$Zr_2PhH_{2.7}$	2.95(14)	780(12)	-10(2)	---	622(12)	150(11)
a-Zr2PdH_3	2.98(13)	785(2)	-18(1)	---	610(20)	185(16)
a-$Zr_3RhH_{3.7}$	3.7(1)	649(2)	-63(14)	193	900	---
a-$Zr_3RhH_{4.0}$	4.1(2)	658(1)	-68(8)	177	874	---
a-Zr_3RhH_5	5.0(2)	615(1)	-39(18)	---	780	---
a-Zr_3RhH_5	5.1(2)	616(2)	-35(6)	---	638	131(12)
a-Zr_3RhH_5	5.2(4)	---	---	---	569	---

REACTIONS OF Zr_2Pd and Zr_2PdH_x

(1) a-Zr_2Pd (MS-1,2,4,5,6,7) $\xrightarrow{765K}$ a-Zr_2Pd + Zr_2Pd(bcc) $\xrightarrow{800K}$ Zr_2Pd ($MoSi_2$)type

(2) a-$Zr_2PdH_{1.91}$ (powder + foil) $\xrightarrow{830-900K}$ ZrH_x + (broad peak) $\xrightarrow{920K}$ ZrH_x + ZrPd

(3) a-$Zr_2PdH_{2.7}$ $\xrightarrow{800-900K}$ ZrH_x + (broad peak) $\xrightarrow{990K}$ ZrH_x + ZrPd

REACTIONS OF Zr_3Rh and Zr_3RhH_x

(4) a-Zr_3Rh (MS-1,2) $\xrightarrow{733K}$ Zr_2Rh ($E9_3$) + αZr $\xrightarrow{840K}$ Zr_2Rh (C16) + αZr

(5) a-Zr_3Rh (MS-3) $\xrightarrow{715K}$ αZr + Rh + $Zr_2Rh(E9_3)$ $\xrightarrow{785K}$ $Zr_3Rh(DO_e)$

(6) a-Zr_3Rh(AP) $\xrightarrow{716K}$ αZr + Rh + $Zr_2Rh(E9_3)$ $\xrightarrow{790K}$ $Zr_3Rh(DO_e)$

(7) a-Zr_3RhH_x (x-3.7,4.1,5.1) $\xrightarrow{630K}$ ZrHx + (broad peak) $\xrightarrow{800K}$ ϵ-ZrH_x + δ-ZrH_x + Rh

Table 4 (21) list the XRD results for the TiCu system (both ternary hydride and the binary alloy). The primary exothermic DSC peak for a-TiCu results in approximately equal amounts of γ-TiCu and δ-TiCu at about 680-700K, which is a polymorphic phase transition according to Koster and Herold (2). A polymorphic (polymorphous) phase transition is one where the

TABLE 4. Lattice Parameters for Major Products of Heated Titanium Copper Alloys and Hydrides

Starting Composition	Maximum DSC Temp.	Major Products	Lattice	Unit Cell Parameters(Å)
a-TiCu	700K	γ-TiCu	Tetragonal	a=3.12(1) c=5.95(1)
a-$TiCuH_{1.41}$ (seal broken)	900K	δ-TiCu	Tetragonal	a=4.47(2) c=2.92(2)
	780K	$TiCu_3$	Orthorhombic	a=5.17(3) b=3.52(3) c=5.34(3)
a-$TiCuH_{1.41}$ (not broken)	900K	γ-$TiCuH_x$	Tetragonal	a=3.13(1) c=5.93(2)
c-$TiCuH_{0.96}$	930K	γ-$TiCuH_x$	Tetragonal	a=3.15(1) c=5.91(2)
	930K	δ-TiCu	Tetragonal	a=4.45(1) c=2.95(1)
c-$Ti_2CuH_{2.10}$	900K	δ-TiCu	Tetragonal	a=4.43(1) c=2.89(2)
	700K	γ-TiH_x	Tetragonal	a=4.43(1) c=4.42(10
c-$Ti_2CuH_{2.71}$	700K	γ-TiH_x	Tetragonal	a=4.44(1) c=4.42(1)
	900 K	δ- and γ-TiCu	(similar to c-$Ti_2CuH_{2.10}$ above)	

material crystallizes in only one phase which is of the same composition as the amorphous material and does not form any other phase. Most systems undergo a phase separation and crystallize into two or more phaes. Both the c-$TiCuH_{0.96}$ and the a-$TiCuH_{1.41}$ decompose to form TiH_x and Cu (free metal) at 600-700K and 500-600K respectively, showing that the crystalline material is more thermally stable than the amorphous material (requires a higher temperature to decompose). Both of these systems undergo further reaction to form $TiCu_3$ upon heating to about 800-880K and continue to react to produce γ-$TiCuH_x$ and δ-TiCu and hydrogen gas. This last reaction occurs at about 900K for both crystalline and amorphous starting material since there isn't much difference between these two materials at this temperature.

The two crystalline phases, γ-TiCu and δ-TiCu form in approximately equal amounts probably due to the inhomogenous nature of the starting material, where regions richer in Cu will form the δ-phase and regions richer in Ti will form the γ-phase. Only the γ-phase will form a hydride because the B11 structure type has layers of Ti-atoms with "body centered" Ti-atoms in every other layer, which provide tetrahedra of four Ti-atoms that can accommodate hydrogen atoms at their centers. The δ-phase does not provide Ti_4 tetrahedra. When the γ-phase forms a hydride the two theta values are decreased and the γ-phase can easily be distinguished from the δ-phase. While both the γ-and δ-phases have primitive space groups, the δ-phase is essentially a face centered cubic lattice while the γ-phase is formed from two body centered cubic cells combined to form a larger c-axis. As hydrogen enters the γ-phase the lattice expands along the a-axis and contracts along the c-axis, making the tetrahedral holes more regular. Thus the c/a ratio for the γ-phase deceases as hydrogen is added.

Both crystalline and amorphous hydrides take very similar reaction paths in hermetically sealed DSC cups for all the different heating rates studied. However, very different reaction pathways result when the system is open to flowing argon or helium as compared to the hermetically sealed gold cups used for these studies. In the latter case the hydrogen cannot escape and a reaction to form TiH_x results between 500 and 700K (the temperature being higher for crystalline than for amorphous samples but also depending on the heating rate). The reaction to form TiCu apparently proceeds through a $TiCu_3$ intermediate phase, but as Ti continues to form a compound with Cu at more elevated temperatures the hydrided form γ-$TiCuH_x$ results, with some unhydrided δ-TiCu also formed.

When DTA (differential thermal analysis) experiments had been performed open to flowing helium gas (5,6) the exothermic peaks resulting from the formation of TiH_x and Cu metal were not observed, probably due to the large endothermic peak that resulted as hydrogen was released in the decomposition of the ternary hydride.

CONCLUSIONS

Hydrogen has been found to significantly alter the thermal stabilities of metallic glasses a-Zr_2Pd and Zr_3Rh. Irrespective of the original crystallization behavior of the glassy alloys, the amorphous hydrides irreversibly disproportionate into crystalline ZrH_x and either ZrPd or Rh during the exothermic transitions. Although the hydrogen content has little influence on the exothermic DSC peak for the a-Zr_2PdH_x system, T_x was observed to systematically decrease with increasing hydrogen content in a-Zr_3RhH_x. Similar decreases in thermal stability upon hydrogenation has been recently reported for a-$TiCuH_x$ (7), a-$Zr_{1-y}Cu_yH_x$ (15), and a-$Zr_{0.76}Fe_{0.24}H_x$ (16). Enhanced diffusion in the amorphous hydrides is probably responsible for their decreased stabilities. However, the currently unique ability of a-Zr_2PdH_x to crystallize at an essentially constant temperature independent of hydrogen stoichiometry remains unexplained. The previously observed (8-10) endothermic peaks in $Zr2PdH_x$ (26,27) at high stoichiometries is attributed to hydrogen evolution from decomposition of the ternary hydride.

Five different Ti_yCuH_x compositions were studied by DSC and XRD methods. The exothermic transition temperatures, T_x, kinetic activation energies, E_x, and heats of transition, ΔH_x, were determined from the DSC data for each composition. T_x, E_x and ΔH_x were found to be dependent upon the hydrogen concentration. The Kissinger and Boswell methods of determining E_x both result in a single activation energy differing by only a few percent between the two methods. For the hydrided samples, ΔH_x depends strongly on the heating rate, indicative of a non-equilibrium relaxation process involving hydrogen. Two of the amorphous systems, a-TiCu and a-$TiCuH_{1.41}$, show a weaker secondary transition at a lower temperature, which may be due to a change in short range order. A similar secondary transition in c-$Ti_2CuH_{2.10}$ may be due to crystal structure reorganization on either the host lattice or the hydrogen sublattice.

T_x and E_x, but not ΔH_x, are closely correlated for all compositions, which is similar to what has been observed in amorphous unhydrided binary alloys and inferred in crystalline ternary hydrides. The reduced thermal stability of a-$TiCuH_{1.41}$ relative to both a-TiCu and c-$TiCuH_{0.96}$ is attributed to greater diffusivities in the amorphous hydride, with hydrogen diffusivity playing an important role.

XRD analysis of the samples that had been hermetically sealed indicates that a-$TiCuH_{1.41}$ undergoes a series of transitions as indicated in Table 3 that results, after quenching, in tetragonal γ-$TiCuH_x$ and some δ-TiCu with evolution of hydrogen when the seal ruptures. Previous DTA experiments on a-$TiCuH_x$ samples open to flowing helium gas did not detect the exothermic peak near 500K that corresponds to the formation of γ-TiH_x and Cu metal, probably due to the large endothermic peak that occurs as hydrogen is evolved from the unsealed samples. The reaction products formed at the exothermic transition are in all cases consistent with the chemical equilibrum phase diagram for Ti_yCuH_x (28,29).

ACKNOWLEDGEMENTS

The authors wish to acknowledge J. E. Wagner and T. A. Beiter for work on the zirconium palladium and zirconium rhodium systems and R. J. Furlan for work on the titanium copper system. In addition we wish to acknowledge A. J. Maeland for the titanium copper alloys and hydrides and to L. Hazelton, K. Samwer, W. L. Johnson and W. Tadlock for the zirconium palladium and zirconium rhodium alloys and hydrides.

REFERENCES

1. K. H. J. Buschow, P.C.P. Bouten and A. R. Midema, Rept. Prog. Phys. 45, 937 (1982).

2. U. Koster and U. Herold, in Glassy Metals, V. 15, H.-J. Guntherodt and H. Beck, eds. (Springer, Berlin 1981).

3. K. H. J. Buschow, J. Phys. F: Met. Phys. 14, 593 (1984) and references cited therein.

4. I. Ansara, A. Pasturel and K. H. J. Buschow, Phys. Stat. Sol. (a) 69, 447 (1982).

5. A. J. Maeland, in Hydrides for Energy Storage, A. F. Andresen and A. J. Maeland, eds. (Pergamon, Oxford 1978, p. 447.

6. A. J. Maeland, L. E. Tanner and G. G. Libowitz, J. Less-Common Mets. 74, 279 (1980).

7. R. C. Bowman, Jr., R. J. Furlan, J. S. Cantrell and A. J. Maeland, J. Appl. Phys. 56, 3362 (1984).

8. R. C. Bowman, Jr., and A. J. Maeland, Phys. Rev. B24, 2328 (1981).

9. K. H. J. Buschow, Mat. Res. Bull. 19, 935 (1984).

10. R. C. Bowman, Jr., A. J. Maeland and W.-K. Rhim., Phys. Rev. B26, 6362 (1982).

11. A. J. Maeland, in Metal Hydrides (Am. Chem. Society, Washington, D.C. 1978), p. 302.

12. A. J. Maeland, in Metal Hydrides, G. Bambakidis, ed. (Plenum, New York 1981), p. 177.

13. Joint Committee for Powder Diffraction Standards (JCPDS International Centre for Diffraction Data, Pennsylvania, 1981).

14. L. V. Azaroff and M. J. Buerger, in The Powder Method (McGraw-Hill, New York 1958), p. 106.

15. R. A. Dunlap and K. Dini, J. Phys. F: Met. Phys. 14, 2797 (1984).

16. S. M. Fries, H.-G. Wagner, S. J. Campbell, U. Gonser, N. Blaes and P. Steiner, J. Phys. F: Met. Phys. 15, 1179 (1985).

17. F. H. M. Spit, J. W. Drijver, W. C. Turkenburg and S. Radelaar, J. de Physique, 41 C8-890 (1980).

18. E. Batalla, Z. Altounian and J. O. Strom-Olsen (to be published).

19. H. E. Kissenger, Anal. Chem. 29, 1702 (1957).

20. F. G. Boswell, J. Therm. Anal. 18, 353 (1980).

21. R. J. Furlan, G. Bambakidis, J. S. Cantrell, R. C. Bowman, Jr. and A. J. Maeland (to be published).

22. K. H. J. Buschow, Acta Met. 31, 155 (1983).

23. J. J. Rush, J. M. Rowe and A. J. Maeland, J. Phys. F: Met. Phys. 10, L283 (1980).

24. J. E. Wagner, R. C. Bowman, Jr. and J. S. Cantrell (to be published).

25. J. S. Cantrell, J. E. Wagner and R. C. Bowman, Jr., J. Appl. Phys. 57, 545 (1985).

26. R. C. Bowman, Jr., J. S. Cantrell and D. E. Etter, Scripta Met. 18, 61 (1984).

27. R. C. Bowman, Jr., J. S. Cantrell, A. Attalla, D. E. Etter, B. D. Craft, J. E. Wagner and W. L. Johnson, J. Non-Cryst. Solids 61 & 62, 649 (1984).

28. T. B. Flanagan and W. A. Oates, J. Less-Common Mets. 100, 299 (1984).

29. R. Kadel and A. Weiss, Ber. Bunsenges. Phys. Chem. 82, 1290 (1978).

HYDROGEN IN Ni-Zr METALLIC GLASSES

E. Batalla, Z. Altounian, D.B. Boothroyd, R. Harris and J.O. Strom-Olsen

McGill University, Ernest Rutherford Physics Building
3600 University Street, Montreal, Quebec. H3A 2T8

In the present article we report experimental measurements on hydrogen in Ni-Zr glasses from $Ni_{33}Zr_{67}$ to $Ni_{70}Zr_{30}$. The principal aims of the research has been to see whether it was possible to identify key features of the pressure-concentration isotherms with the occupation of specific types of sites for the hydrogen in the amorphous metallic matrix. The first part of the research therefore was to establish the exact form of the reproducible pressure-concentration isotherms over the full composition range of the glassy alloys, to complement these data with X-ray diffraction thermodynamic studies of the desorption process via differential scanning calorimetry (DSC) and finally to test a hypothesis about site occupation against a computer generated model for the glassy structure.

Metallic glass ribbons of $Ni_{100-x}Zr_x$ where x = 30, 36.3, 50, 55 and 66.7 were prepared by melt-spinning; the method of sample manufacture and characterization are given elsewhere(1). Pressure-concentration isotherms were obtained using the automated volumetric apparatus whose block-diagram is illustrated in figure 1. The sample, typically 100 mg cut into 5 cm lengths, is mounted in a stainless steel holder which sits in a three-zone Lindberg furnace whose temperature may be set and controlled to ± 1K. The holder is connected to the gas handling system and pressure gauges by a capillary. The free volume inside the furnace is kept to a minimum and is about 1% of the sample volume (typically 18cc). The pressure difference between the sample and a fixed pressure reservoir was measured by a strain gauge transducer (Sensotec A-5) which could measure up to 0.1 MPa with an accuracy of ± 0.1 kPa. A change in this pressure corresponded to hydrogen absorbed by the sample and the amount

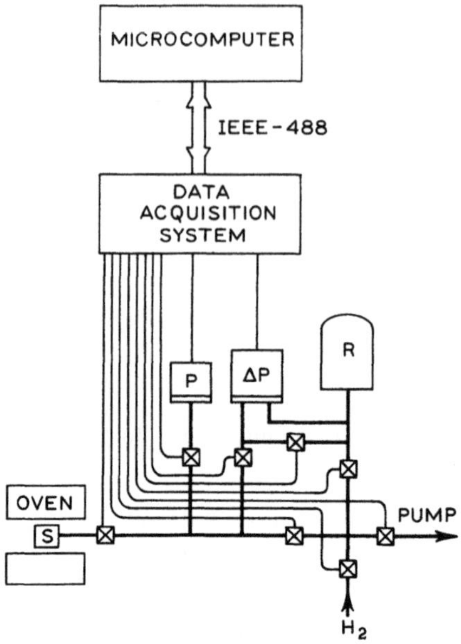

Fig. 1. Block diagram of automated gas titration system.

of absorption was calculated by the ideal gas law (the accuracy of the system did not justify using higher virial coefficients). The absolute pressure at the sample is measured by a capacitance gauge (Setra 204) whose range is 0 to 7 MPa. The entire system of charging and discharging is controlled by an Apple IIe microcomputer which records the pressures and activates the solenoid valves through a BIODATA interface (Microlink III). The temperature of the oven is set manually.

The system of charging proceeds as follows: the kinetics of adsorption in the as-made samples is poor (as has been noted by other workers), so the samples were first "activated" by exposing them to hydrogen at a pressure of 3.5 MPa for 24 hours at 450K and then at the same temperature for a similar period under vacuum (about 10^{-1} torr). After activation the pressure was increased from 0 to 0.5 MPa in about 10 steps then reduced in the same way. After each increase the computer waited until the pressure change was less than 2Pa/s before initiating the next step. The cycling was repeated until the p-c-T curves were reproducible - typically about 2-4 times. Once the reversible p-c-T was established, the sample was removed after being fully charged (the temperature first being lowered to 300K where the time constant for desorption is about 1 hour) and the hydrogen content immediately determined by heating the sample to 1000 K on one side of an oil manometer and measuring the gas evolved through the pressure change. Concentrations were found reproducible to within a few percent.

Equilibrium pressure-concentration isotherms, plotted semi-logarithmically for $Ni_{70}Zr_{30}$, $Ni_{63.7}Zr_{36.3}$, $Ni_{50}Zr_{50}$ and $Ni_{33}Zr_{67}$ are shown in figure 2 taken at various temperatures between 300K and 450K. The curves shown are for discharging; there is a small hysteresis so that the data for charging lie at slightly lower concentrations (typically about 3%). This hysteresis is probably partly intrinsic and partly instrumental - the time constants for charging at high pressures are quite long (of the order of 20 minutes per step) so that the results are adversely affected by any small absorption by the walls of the gas-handling system.

After hydrogenation, powder diffraction X-ray scans of all alloys were made. The scans showed no measurable crystallization. In all cases the mean position of the broad amorphous peak moved to lower angles, consistent with the sample dilation caused by hydrogen absorption. However the width of the peak remained unchanged which suggests that there is no significant rearrangement of Ni and Zr on the atomic scale. Hydrogen also made the samples rather more brittle but in no case did the samples lose their mechanical integrity - even after thirty or more desorption cycles. Roughly speaking, the degree of embrittlement correlates with the total quantity of hydrogen absorbed.

Several features of the data of figure 2 should be noted: the absence of any plateau in the isotherms at any of the compositions or temperatures; the relative insensitivity of the slope of the isotherms to temperature and the fact that with composition the slope of the isotherms first decreases then increases until for $Ni_{33}Zr_{67}$ they are almost vertical. It should also be remarked that during the first few cycles of charging and discharging there is a large irreversible component to the isotherm, irreversible, that is, over the pressure range 2×10^3 Pa to 6×10^5 Pa. This irreversible component increases monotonically with composition.

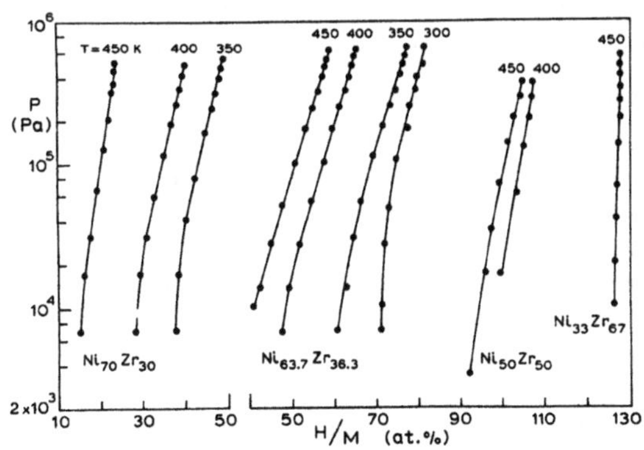

Fig. 2. P-C isotherms for amorphous Ni-Zr alloys at various temperatures.

The absence of a plateau has been reported by all previous workers[2] and seems to be a general feature of hydrogen in amorphous metals. In crystalline systems the formation of the hydride can be viewed as a co-operative phase transition of the first order in which the hydrogen condenses from a vapour like state into a liquid like state. The transition results in a plateau in the p-c isotherm. According to a simple mean-field theory due to Griessen[3], the phase transition disappears if the binding energy of the hydrogen sites is allowed to spread over a range Δ greater than half the hydrogen-hydrogen attractive energy per unit concentration - i.e. if $\Delta \geq a/2$. According to Richards[4] this criterion is too severe. However whatever the case may be, it is clear from figure 2 that either there is no transition at all or that, if there is, it occurs well below room temperature. Under these conditions the high-temperature mean-field solution of Griessen should be valid. For a single type of site of mean binding energy E_o this gives

$$\ln\left(\frac{P}{P_o}\right) = \frac{E_o - E_b - \Delta + (a + 2\Delta)c}{kT} \tag{1}$$

where E_b is one-half the binding energy of H_2.

Analysis of hydrogen in amorphous $Ni_{50}Zr_{50}$ by Suzuki et al.[5] has indicated that the hydrogen atoms most probably occupy sites tetrahedrally co-ordinated to the metal atoms, being surrounded by 4 Zr atoms at low H concentrations and by 3 Zr and 1 Ni at higher H concentrations. We suppose that the hydrogen sites in <u>all</u> amorphous Ni-Zr alloys are essentially the same - i.e. tetrahedrally co-ordinated. We believe (based on the binding energy of ZrH_2) that sites with four Zr atoms are essentially stable over our pressure range and that these sites are responsible for the irreversible hydrogen charging. It seems reasonable to start by supposing that the reversible part corresponds to filling and emptying sites with 3 Zr and 1 Ni. If we generalize Griessen's expression (1) to the case where there are n sites filled and m sites of mean energy E_o and spread Δ which are in the process of being filled, then

$$\frac{1}{2}\ln\left(\frac{P}{P_o}\right) = \frac{1}{kT}\left\{E_o - \Delta - E_b - \frac{2n\Delta}{m} + \left(a + \frac{2\Delta}{m}\right)c\right\} \tag{2}$$

If Ni and Zr are randomly distributed (i.e. if we neglect any chemical-short-range-order) then the probability of tetrahedra having 4 Zr and 3 Zr, 1 Ni atoms is given by x^4 and $4x^3(1-x)$ respectively, x being the Zr concentration. According to equation (2) the slope of the p-c isotherms is $(a + \frac{2\Delta}{m})$ so that, neglecting any variation in the hydrogen-hydrogen interaction parameter a, the slope of the isotherm should reach a minimum at the point where the probability of finding (3 Zr, 1 Ni) sites is a maximum - i.e. at x = .75. Instead, in this region the isotherms are almost vertical; the minimum slope in fact occurs between x = 0.36 and 0.50.

Furthermore when we convert the probability into the actual number of sites by multiplying by 4 (the mean number of tetrahedral holes per atom) we find a predicted hydrogen to metal atom ratio of over 2, more than twice what is observed. This last fact suggests that we should use the same criterion found in crystalline hydrides - namely that when a hydrogen atom occupies a particular site, neighbouring sites cannot be filled. In Nb-H[6], for example, all tetrahedra up to third neighbours are empty. The simplest assumption to make for amorphous metals is that only nearest neighbor tetrahedra are unoccupied (which would give a maximum hydrogen concentration of a little over one per metal atom).

If we then recalculate the number of available sites by first filling the sites with 4 Zr atoms, removing at random four tetrahedra each time, then moving into (3 Zr 1 Ni) sites in the same way, we find modified probabilities that in fact give a maximum number of 3 Zr, 1 Ni sites at about 45 at. % Zr, close to what is needed. We give no details of these calculations because they only give about half the number of sites required and also because they suffer from the more serious objection that no correlations are included between the sites removed and the sites filled. This last point means that each 4 Zr site filled can only cause the removal of 4 Zr and (3 Zr, 1 Ni) sites. for this reason we have turned to a computer generated cluster to determine more realistically the appropriate number of sites. Before examining this in detail we see the extent to which the DSC data confirms the general hypothesis of site type.

A typical DSC curve (for hydrogen desorbing from $Ni_{50}Zr_{50}$) is shown in figure 3 with arrows indicating the onset of significant events, each

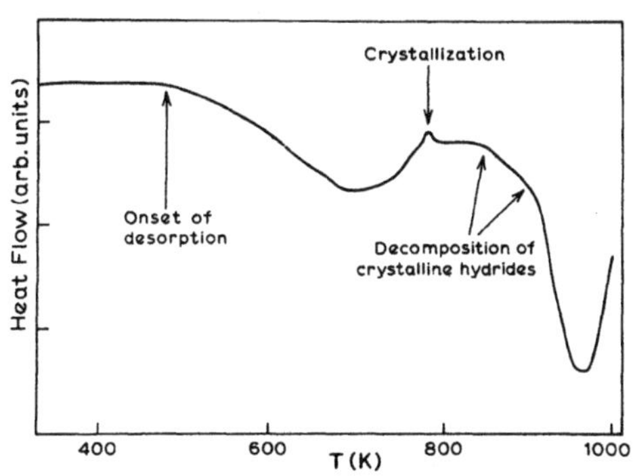

Fig. 3. DSC trace for hydrogen desorbing from amorphous $Ni_{50}Zr_{50}$.

of whose occurrence is verified independently (e.g. the onset of desorption of the amorphous hydride by the parallel process in the manometer, and the crystallization and decomposition of crystalline hydrides by X-ray analysis). Table I summarizes the DSC data. As the zirconium content is increased, the endothermic curve moves to higher temperature until by $Ni_{33}Zr_{67}$ very little hydrogen is desorbed before crystallization - which is consistent with the p-c isotherms and with the general hypothesis about occupation of 3- and 4-Zr sites. Neither the temperature of crystallization nor the corresponding enthalpy is significantly altered by the presence of hydrogen, suggesting that the hydrogen occupies similar sites, with similar mean binding energies in the glassy phase as in the crystal. However the much broader endotherm before crystallization than after indicates that there is a much greater spread of energies in the amorphous than in the crystalline phase, again as assumed in the model. A crude estimate of E_o/Δ can be obtained from the endotherms if we assume that the distribution with energy is uniform over Δ. The time constant for hydrogen release from a site of energy E is given by

$$\tau(E) = t_o \exp (E/kT) \tag{3}$$

where t_o is a characteristic time of the order of the reciprocal of the frequency of the hydrogen vibration in the site. The onset of hydrogen release occurs when $\tau(E_o - \Delta)$ reaches some critical value and stops when $\tau(E_o+\Delta)$ reaches the same value. From this we find

$$\frac{E_o}{\Delta} = \frac{T_2 + T_1}{T_2 - T_1} \tag{4}$$

T_1 being the temperature of onset and T_2 the temperature of completion. For the two compositions $Ni_{70}Zr_{30}$ and $Ni_{63.7}Zr_{36.3}$ for which all the hydrogen is released before crystallization we find $E_o/\Delta \simeq 3$.

The broad distribution of sites is also confirmed by isothermal measurements of hydrogen evolution. If the sample is maintained at some temperature T, then after time t the sites which will have released hydrogen will be approximately those for which $\tau \leqslant t$, i.e. those with energy up to $E(\tau) = kT \ln (t/t_o)$. Thus the amount of hydrogen released varies roughly as ln t assuming a flat distribution of sites with energy. In DSC what is measured is the <u>rate</u> of release which therefore should fall off as 1/t. This behavior was in fact closely followed in all alloys at temperatures below the crystallization temperature, as is illustrated in figure 4.

Table 1. Thermal data obtained from differential scanning calorimetry for hydrogenated Ni-Zr alloys.

	$Ni_{70}Zr_{30}$	$Ni_{63.7}Zr_{36.3}$	$Ni_{50}Zr_{50}$	$Ni_{45}Zr_{55}$	$Ni_{42}Zr_{58}$	$Ni_{33.3}Zr_{66.7}$
1st. endothermic peak.						
Start (K)	335	400	455	500	400	----
End (K)	645	780	800	725	730	----
Crystallization Peak (K)	840	846	778	742	731	722
		867				752
Crystallization enthalpy with H_2(KJ/mole)	3.1	4.1	2.5*	2.2*	1.9*	5.0
Without H_2(KJ/mole)	3.2	3.7	7.5	5.5	5.5	6.2
2nd endothermic peak.						
Start	----	----	840	730	800	940
End	----	----	>1000	800	900	>1000
3rd endothermic peak.						
Start	----	----	----	810	970	----
End	----	----	----	940	>1000	----

*These values are only lower bounds because the endothermic background from the hydrogen desorption interferes with a proper measurement of the exothermic enthalpy from the crystallization.

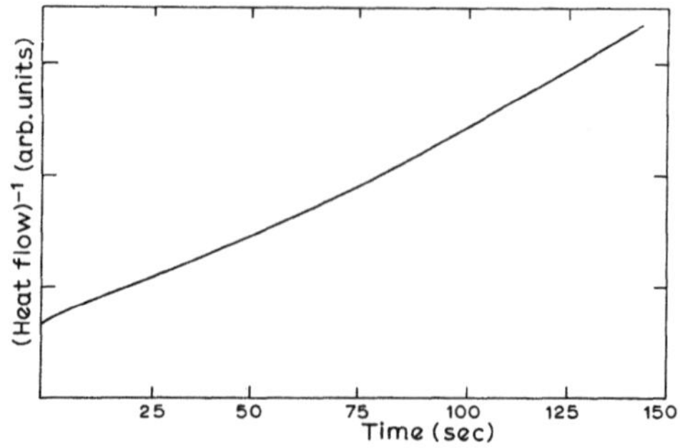

Fig. 4. Inverse of the enthalpy evolved on isothermal annealing plotted against t. The sample is $Ni_{70}Zr_{30}$ and the annealing temperature 500K.

We now examine the properties of a computer generated structural model of Ni-Zr. The model is generated according to the method of Lewis and Harris for Cu-Zr. A total of 1000 atoms in the appropriate proportion are assembled by the dense random packing of hard spheres using the Bennett algorithm[(8)], atoms being added to the cluster at the site closest to the overall center of mass. After assembly the hard sphere potential is softened to a Lennard-Jones potential and the structure allowed to relax to the local minimum of energy by means of a conjugate gradient algorithm. By adjusting the relative strength of the various interatomic potentials, a certain amount of chemical short range order is added to make the partial structure factors agree with the experimentally measured ones.

The tetrahedrally co-ordinated sites are now found in the following manner. The atoms are ordered in increasing distance from the center and for each atom all neighbours farther out are identified (this prevents double counting of the sites). The tetrahedra formed by these neighbours are examined and all those with face triangles whose angles are less than 30° or greater than 90° are rejected as having holes too small to be occupied by hydrogen. A record is kept of all holes tetrahedrally co-ordinated to either 4Zr or 3Zr and 1Ni atoms. The total distribution of such holes with distance from a reference hole for $Ni\text{-}Zr_2$ is shown in figure 5. The first peak occurs approximately at $^{a}/\sqrt{6}$ (a is the mean interatomic distance), which is appropriate for regular tetrahedra sharing a face, and the second peak at $^{a}/\sqrt{2}$, which is appropriate for tetrahedra sharing an edge. Also if the distribution of 4Zr holes and 3Zr, 1Ni holes are examined as a function of composition, then it is found that these are distributed close to x^4 and $4x^3\,(1-x)$ respectively - i.e. to the random distribution. Once the holes are identified they are now filled in the following way : first the 4Zr sites are filled, disallowing all holes

Table 2. H-site distribution of 4 Zr-type and 3 Zr, 1 Ni-type.

		$Ni_{33}Zr_{67}$	$Ni_{50}Zr_{50}$	$Ni_{60}Zr_{40}$	$Ni_{70}Zr_{30}$
Exclusion	4Zr	0.35	0.17	0.10	0.03
Distance	3Zr, 1Ni	0.35	0.28	0.23	0.14
2.1Å	Total	0.69	0.45	0.33	0.17
	4Zr	0.42	0.19	0.11	0.03
1.8Å	3Zr, 1Ni	0.72	0.41	0.36	0.19
	Total	1.14	0.60	0.46	0.23
	4Zr	0.43	0.20	0.11	0.03
1.55Å	3Zr, 1Ni	0.75	0.51	0.39	0.21
	Total	1.18	0.71	0.49	0.24
Experimental at 400K, 10 atm.		∼ 1.25	0.95	0.70	0.35

Numbers are hydrogen to total metal-atom ratio.

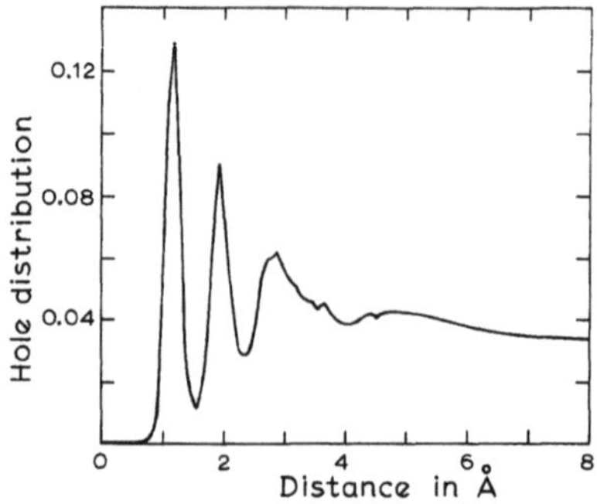

Fig. 5. Distribution of holes surrounded by 4 Zr atoms or 3 Zr and 1 Ni atoms in $NiZr_2$

within a given distance of each hole filled; then the 3 Zr, 1 Ni sites are filled in the same way. The result is shown in table 2. The first point to note is that if the exclusion distance is taken as 2.1Å (the normal figure for crystalline hydrides[8]) then the total number of 4 Zr and 3 Zr, 1 Ni sites available is far too small to account for the total hydrogen storage at <u>any</u> alloy composition. Since it is unreasonable to suppose that 2 Zr, 2 Ni are occupied at these pressures, the conclusion is that the exclusion distance must be reduced. If it is reduced to 1.55Å, (i.e. to the first minimum of the hole distribution function), then the number of holes is in close agreement with observations. However, even though the total number of sites is now reasonable, the distribution between 4 Zr and 3 Zr, 1 Ni sites is not consistent with the fact that the former are responsible for irreversible adsorption and the latter for reversible absorption, since there are far more 3 Zr 1 Ni sites present than are required from the data. The simplest way out of this dilemma is to say that, because of the spreading site energies, some of the 3 Zr, 1 Ni sites are in fact irreversibly filled or rather in equilibrium with the gas at vapor pressures lower than we achieve in our apparatus. We are currently in the process of calculating the binding energies of the various sites to test this hypothesis.

To conclude : hydrogen absorption in amorphous Ni-Zr is consistent with the hypothesis that hydrogen occupies interstitial holes tetrahedrally coordinated to either 4 Zr atoms or to 3 Zr and 1 Ni atoms provided holes closer than 1.55 Å are not simultaneously occupied. Over the pressure range 2×10^3 Pa to 6×10^5 Pa all sites surrounded by 4 Zr atoms and some sites surrounded by 3 Zr and 1 Ni atoms are irreversibly filled.

References

1. Z. Altounian, Tu Guo-Hua, and J.O. Strom-Olsen, J. Appl. Phys. 54, 3111 (1983).
2. F.M.H. Spit, J.W. Drijver, and S. Radelaar, Scripta Metall. 14, 1071 (1980); K. Aoki, A. Horata and T. Masumoto, Proceedings of the 4th International Conference on Rapidly Quenched Metals, Sendai, Japan.
3. R. Griessen, Phys. Rev. B 27, 7575 (1983).
4. P.M. Richards, Phys. Rev. B 30, 5183 (1984).
5. H. Suzuki, N. Hayashi, Y. Tomizuka, T. Fukunaga, K. Kai, N. Watanabe, J. Non-Cryst. Solids 61 & 62, 637 (1984).
6. H. Horner and H. Wagner, J. Phys. C 7, 3305 (1974).
7. L.J. Lewis and R. Harris, J. Phys. F. 13, 1359 (1983).
8. C.H. Bennett, J. Appl. Phys. 43, 2727 (1972).

MECHANICAL RELAXATION BEHAVIOR OF HYDROGENATED METALLIC GLASSES

B. S. Berry and W. C. Pritchet

IBM Thomas J. Watson Research Center
Yorktown Heights, N.Y. 10598

ABSTRACT

Two apparently quite general anelastic effects are associated with the presence of hydrogen in solution in metallic glasses. One of these is a short-range reorientation relaxation of the Snoek type; the other is the long-range Gorsky relaxation. In contrast to the situation for bcc crystalline metals, these relaxations reveal the H-interstitial not only as a dilatation center of normal strength, but also as a strong elastic dipole. Coupled with a relatively low mobility associated with migration energies of 0.25-0.5eV, the H-interstitial in metallic glasses emerges as a quasi-classical defect analogous to the heavier interstitial solutes in bcc metals. Both relaxations exhibit features that reflect the structural disorder of the glassy host. The reorientation relaxation shows considerable dispersion, the Gorsky relaxation reveals non-ideal solution behavior, and both relaxations have activation energies which characteristically decrease with increasing hydrogen concentration. The available experimental results agree well with Kirchheim's analysis for solution and diffusion in a structurally-disordered host. We point out that an apparent difficulty in explaining the strength of the reorientation relaxation may be related to a need to consider the spatial distribution of disorder, and suggest that packing constraints may lead to a near-degeneracy of neighboring interstitial sites.

1. INTRODUCTION

After more than twenty years of intensive effort, metallic glasses are of increasing importance for technological applications and a variety of specific alloys, developed mainly for their magnetic properties, are now commercially available. The awakening of a general scientific interest in hydrogen in amorphous alloys has been a relatively recent phenomenon. Whereas only a handful of papers appeared before 1980, the level of activity has since increased significantly, and has been accompanied by a widening diversity of investigative approaches. Many of these have already been introduced in surveys given by Maeland earlier in this meeting, and elsewhere.[1] Although the total effort to date is only a small fraction of that devoted to hydrogen in crystalline metals, it is already clear that the absorption of hydrogen by metallic glasses produces an interesting new type of interstitial solid solution in which structural disorder appears to exert a dominant effect on the character of the hydrogen interstitial and on its solution and diffusion behavior.

The purpose of the present paper is to describe how the use of mechanical relaxation measurements has contributed to our present understanding of hydrogen in metallic glasses. For those unfamiliar with mechanical relaxation, it may be helpful to start with a rough comparison to nuclear magnetic resonance (NMR), a topic that has already been discussed at this meeting. Proton NMR makes use of the fact that hydrogen possesses a nuclear magnetic moment. The appropriate application of static and dynamic fields is thus one way of interacting with the hydrogen atoms in a sample, and of disturbing them from an equilibrium configuration to which they will relax when the magnetic stimulus is removed. In the present case, the stimulus used to disturb the equilibrium configuration is a mechanical stress, and the coupling between stress and the hydrogen population originates from the strains produced by the insertion of hydrogen atoms into the host material. Mechanical relaxation behavior is also known as anelasticity, since it constitutes a departure from ideal elastic behavior.[2,3]

We shall show below that there are two important types of relaxation behavior associated with hydrogen in metallic glasses. One of these, known as the Gorsky relaxation, senses the hydrogen interstitial as a mobile center of dilatation and provides a convenient means of studying both long-range diffusion and non-ideality of the solution behavior. The other is a short-range reorientation relaxation that reveals the hydrogen interstitial as a strongly asymmetrical point defect or elastic dipole. This in itself is a most striking result, since it represents a major contrast to the vanishingly weak dipole strength of hydrogen in well-studied bcc crystalline metals such as niobium. Indeed, we will see that the hydrogen reorientation relaxation in metallic glasses is a disordered analog of the Snoek relaxation produced by the heavier solute interstitials carbon, nitrogen and oxygen in bcc metals. In addition to providing insight into the nature of the hydrogen interstitial, mechanical relaxation measurements also provide information on the disordered nature of the glassy host. It is generally recognized, for example, that the broadening or dispersion exhibited by the reorientation relaxation predominantly reflects a distribution of activation energies for motion of the hydrogen interstitial. In this article, we point out that an improved understanding of the reorientation relaxation may not only involve the magnitude of the structural disorder, but also the manner in which it is spatially distributed.

2. REORIENTATION RELAXATION AND THE SNOEK EFFECT

The concept of mechanical relaxation by stress-induced point-defect reorientation has been highly developed only for defects in crystalline solids[2]; nevertheless it may be explained in sufficiently general terms to include amorphous solids even though a detailed quantitative treatment is presently lacking. We consider a population of asymmetric or orientationally-distinguishable defects, each of which can change its orientation by a single atomic jump. Because of the asymmetry of the atomic displacements and corresponding strains associated with each defect, it follows that the site-occupation energies of the different orientations can be changed relative to each other by the application of a uniaxial stress, thereby throwing the defect population out of equilibrium. The subsequent repartitioning of defects amongst the available orientations to achieve a new state of equilibrium under the applied stress constitutes reorientation relaxation, and is accompanied by the time-dependent stress-strain behavior that we refer to as anelasticity.

Because of its relevance to later conclusions concerning the nature of the hydrogen reorientation relaxation in metallic glasses, it is useful to illustrate this general description with the specific example of the Snoek relaxation observed in bcc metals containing the heavier solute interstitials carbon, nitrogen or oxygen. In contrast to the situation for fcc metals, the symmetry class of the commonly-discussed interstitial sites in bcc metals is lower than the cubic crystal symmetry. This is true for the octahedral sites of Fig. 1, and for a possible alternative set of sites - the tetrahedral sites - located midway between adjacent octahedral sites. Which of these two sets happens to represent the equilibrium interstitial positions, and which corresponds to the saddle-point positions indicated in the energy diagram of Fig. 1, happens to be irrelevant to the present discussion. In either case, it may be found by inspection that the site symmetry is tetragonal, and that each set of sites therefore contains three equivalent but orientationally distinguishable subgroups, as shown in Fig. 1. When such a lower-symmetry site is occupied by an interstitial atom, the surrounding host atoms are displaced unequally and the interstitial assumes the character of an asymmetric defect or elastic dipole. A convenient representation of the net effect of these anisotropic displacements is provided by the quantity known as the λ -tensor.[2] This is derived from the tensor of macroscopic strains produced when a dilute mole fraction C of interstitials is imagined to be homogeneously dispersed amongst sites of just one orientation. These defects would cause a reference sphere of the host material to distort into an ellipsoid with principal strains ε_i (i = 1,2,3) proportional to C. The derivatives $\lambda_i \equiv \partial\varepsilon_i/\partial C$ form the principal components of the λ-tensor, and may be used to define two further quantities which we will call the dipole strength and the dilatation strength of the defect. Because of its future adaptability to amorphous hosts, the dipole strength Λ is defined here as the root-mean-square of the differences $(\lambda_i - \lambda_j)$,

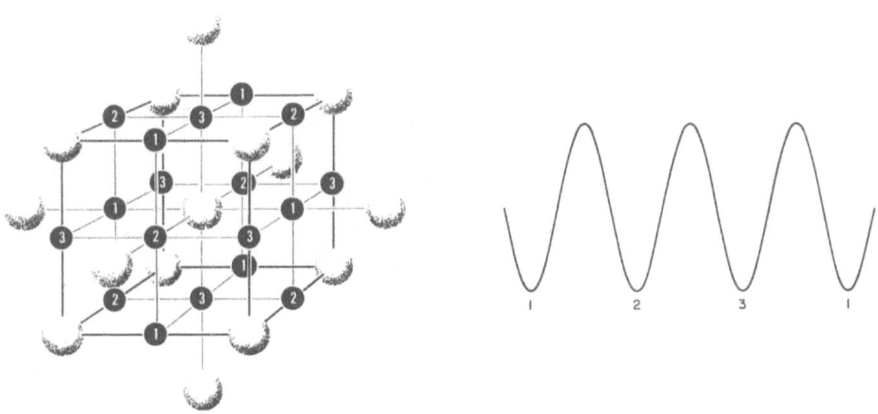

Fig. 1 Octahedral sites in the bcc lattice, labelled with a site orientation index p=1,2,3. The schematic energy diagram shown at the right emphasizes the equality of all site energies (and all saddle-point energies) in the unstressed perfect crystal. Both diagrams show that a nearest-neighbor jump automatically produces reorientation of the interstitial.

$$\Lambda \equiv \left\{\left[(\lambda_1 - \lambda_2)^2 + (\lambda_2 - \lambda_3)^2 + (\lambda_1 - \lambda_3)^2\right]/3\right\}^{1/2}, \tag{1}$$

whereas the dilatation strength is defined conventionally as the sum or trace of the principal components of the λ-tensor, and represents the volume expansion per interstitial $\Delta\vartheta$ expressed as a fraction of the atomic volume of the host, ϑ_0:

$$\Delta\vartheta/\vartheta_o = \mathrm{tr}\lambda \equiv \lambda_1 + \lambda_2 + \lambda_3. \tag{2}$$

For the case of the Snoek relaxation the tetragonal symmetry of the interstitial site leads to the simplification $\lambda_2 = \lambda_3$, and hence to

$$\Lambda = (2/3)^{1/2}\delta\lambda \tag{3}$$

and

$$\Delta\vartheta/\vartheta_o = \lambda_1 + 2\lambda_2, \tag{4}$$

where $\delta\lambda \equiv (\lambda_1 - \lambda_2)$ is frequently referred to as the shape factor of the defect.

The external manifestation of reorientation relaxation under an applied stress is the anelastic strain that accompanies a net change of orientational order. In contrast to the elastic strain, the anelastic strain develops in a time-dependent manner governed by the rate of the reorientation jump. Under a static stress, the relaxation may therefore be observed as a limited (and recoverable) creep process. Frequently, however, it is more desirable for reasons of sensitivity or convenience to observe the relaxation dynamically as a loss-peak, via internal friction measurements made as a function of temperature and/or vibration frequency.[2,3] Figure 2 shows the oxygen Snoek peak in polycrystalline thin film niobium, tested in the same vibrating-reed apparatus[4,5] used for our studies of

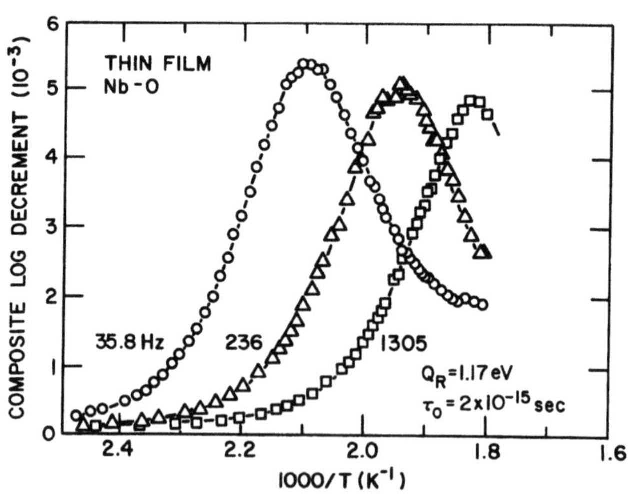

Fig. 2 Vibrating-reed internal friction measurements of the oxygen Snoek peak in 0.45μm films of polycrystalline niobium on each side of a 50μm fused silica substrate. From Ref. 4.

metallic glasses. The measure of internal friction used in this and later figures is the logarithmic decrement of free vibration decay, δ. The peaks of Fig. 2 can be fitted quite well by the simple Debye expression for a loss-peak governed by a single thermally-activated relaxation time τ_R :

$$\delta/\pi = \Delta^R \omega\tau_R/\left(1 + \omega^2\tau_R^2\right), \tag{5}$$

where

$$\tau_R = \tau_0 \exp Q_R/kT, \tag{6}$$

and k is Boltzmann's constant and T the absolute temperature. The experimental values of the prefactor τ_0 and the reorientation energy Q_R, obtained from the shift of the peak with frequency $\omega/2\pi$, are shown in Fig. 2. It should be noticed that the magnitude of τ_0 is consistent with a reorientation mechanism involving a single atomic jump. For a Debye peak, the relaxation strength Δ^R in Eq. (5) uniquely determines the peak height, since it is readily seen that $\delta/\pi = \Delta^R/2$ for the peak condition $\omega\tau_R = 1$. For the Snoek relaxation in a random polycrystalline aggregate, Δ^R is related to the defect shape-factor $\delta\lambda$ by the expression

$$\Delta^R = (4/45)\left(E\vartheta_0/kT\right)C(\delta\lambda)^2, \tag{7}$$

where E is Young's modulus and C is the mole fraction of interstitials in (dilute) solution. The influence of the $1/T$ - dependence on the relaxation strength is evident in Fig. 2. Although Eq. (7) includes the effect of orientational averaging, it is not directly applicable to an amorphous solid because no account is taken of the effect of structural disorder. Nevertheless, it is evident that the Snoek relaxation has two basic conceptual features that could apply equally well to an amorphous solid as to a bcc crystal. One feature is that individual interstitial atoms form elastic dipoles by occupancy of low-symmetry sites. The other requirement, brought out by the energy diagram of Fig. 1, is that reorientation and long-range migration occur by the same atomic jump. Based on the evidence presented below, we conclude that the hydrogen reorientation relaxation in metallic glasses meets both of these criteria, and we shall therefore describe it as being of the Snoek type.

3. THE HYDROGEN REORIENTATION RELAXATION IN METALLIC GLASSES

3.1 Introductory Experimental Results

Although internal friction studies of metallic glasses started around 1970, no well-defined peak indicative of the presence of a specific defect was observed for several years.[6] The first such peak was reported[7] in 1978 for a metal-metalloid alloy of nominal atomic composition Nb_3Ge, prepared as a glassy film by sputtering onto a liquid-nitrogen-cooled substrate. Illustrative results are shown in Fig. 3 for a detached film tested as a vibrating reed in the fundamental mode and several higher overtones. In this run, performed after a 10 min. anneal at 493K, the peak height above background $(\delta - \delta_B)$ was 0.01, or only half that observed initially prior to exposure above room temperature. Further anneals were found to eliminate the peak completely. The annealing behavior turned out to be an important clue to the identity of the defect involved, since it was soon linked to the desorption of hydrogen from the sample.[8] In

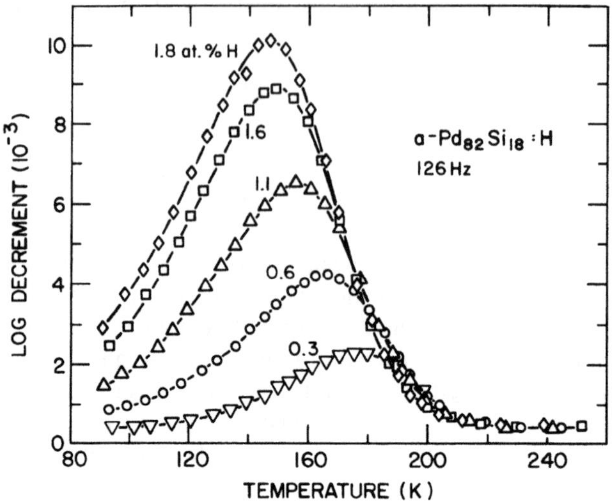

Fig. 3 The first example of a hydrogen reorientation peak in a metallic glass, discovered by vibrating-reed measurements on a detached 8μm film of Nb_3Ge accidentally contaminated with hydrogen during sputter deposition. From Ref. 8.

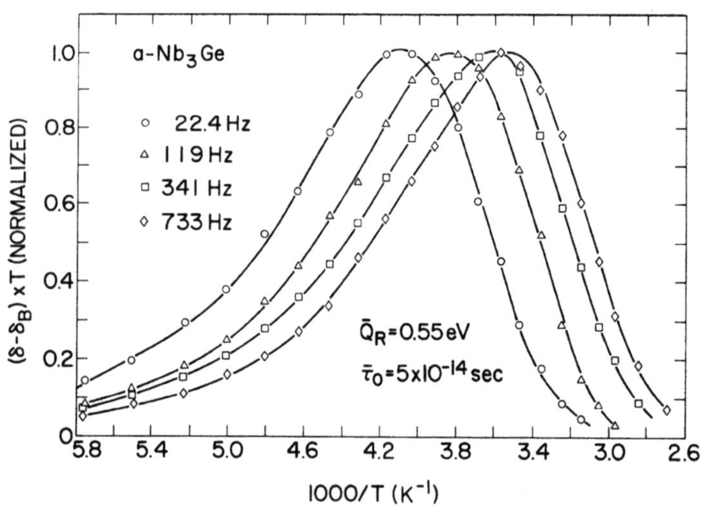

Fig. 4 Concentration dependence of the hydrogen reorientation peak in a liquid-quenched ribbon of glassy $Pd_{82}Si_{18}$.

contrast to the oxygen Snoek peaks of Fig. 2, the hydrogen-related peaks of Fig. 3 are located below room temperature and provide, from the shift with frequency, a central or most probable activation energy $\bar{Q}_R$ of 0.55eV. The peak shows asymmetry in the form of an extended low-temperature tail, and has a half-width roughly three times larger than that of a Debye peak. This broadening, coupled with the magnitude of $\bar{\tau}_0$ (Fig. 3), immediately indicates we are dealing with a short-range relaxation that is sensitive to the atomic-scale disorder of the glassy structure. Assuming the broadening to be due entirely to a distribution of activation energies, it can be estimated that this distribution, which is peaked at 0.55eV, is about 0.1eV wide on the low side and 0.06eV wide on the high side.

The discovery of the hydrogen peak in glassy Nb_3Ge was soon followed by observations of similar peaks in other metallic glasses, both of the metal-metal and metal-metalloid type.[8-14] To date, peaks have been observed in well over a dozen different alloy compositions, with activation energies $\bar{Q}_R$ falling typically in the range 0.25-0.5eV. Notable examples include such well-known binary glasses as $Pd_{80}Si_{20}$, $Fe_{80}B_{20}$, and several Ni-Zr compositions such as the $Ni_{64}Zr_{36}$ glass made prominent by the pioneering work of Spit *et al.*[15] Particularly large peaks, ($\delta \geq 0.1$) have been observed in a number of glasses capable of absorbing hydrogen to concentrations approaching 50 at.%H.

Apart from the characteristic shape and width, the hydrogen peak exhibits another important general characteristic, which is illustrated in Fig. 4 by results obtained for glassy $Pd_{82}Si_{18}$ containing various concentrations of hydrogen. As the amount of hydrogen increases, and the peak increases in height, it is typically observed that the peak shifts to *lower* temperatures. The formal interpretation of this behavior is quite straightforward; the decrease implies that the distribution of relaxation times is concentration dependent with a centroid that shifts to smaller values (faster relaxation rates) as the hydrogen concentration increases. The observation of a defect mobility which increases with increasing concentration is quite remarkable since, in crystalline solids, defect mobility is invariably retarded by the increasing importance of defect-defect interactions. However, if we consider the reorientation relaxation as a disordered Snoek relaxation, we find there is a simple explanation of this behavior, as outlined below.

3.2 Models for the Reorientation Relaxation

In considering the effects of disorder on the Snoek relaxation, it is instructive to proceed in a number of stages, as illustrated in Fig. 5. For case (a), which represents the ideal crystal, we expect (in the absence of H-H interactions) an ideal solution behavior and an activation energy Q_R which is both single-valued and concentration-independent. Case (b) represents a hypothetical solid in which only the saddle-point energies have been disordered. This introduces dispersion into Q_R, but does not cause Q_R to become concentration dependent, nor does it destroy the ideality of the solution behavior. Although this model obviously appears artificial, we should note that it is usually the one implicitly assumed when a distribution of activation energies is formally introduced to explain a broadened peak. Case (c), the opposite of (b), introduces a number of quite different features which bring us noticeably closer to reality. In general, the distribution of interstitials amongst the site-occupation levels must now be handled by Fermi-Dirac statistics. This introduces a preferential filling of low energy sites and a departure from ideal solution behavior. Moreover, since in this case there is a simple inverse correlation between the occupation energy of a particular site and the

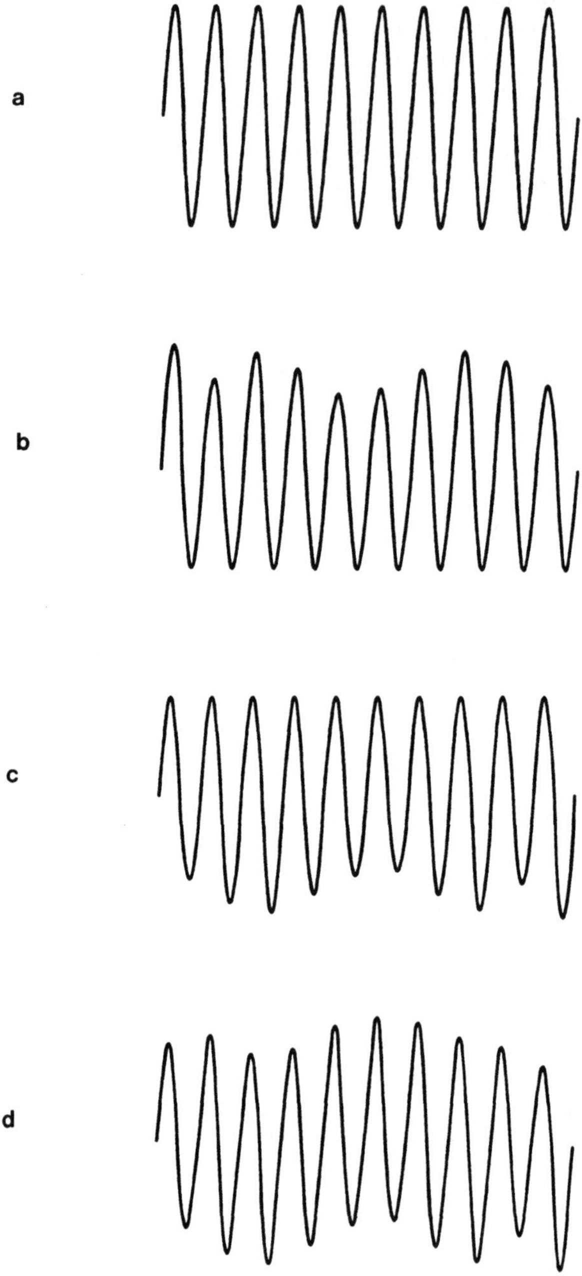

Fig. 5 Schematic energy-distance diagrams representing various types of structural disorder. (a): single-valued site-occupation and saddle-point energies (ideal crystalline solid). (b): single-valued site-occupation energy and distributed saddle-point energies. (c): distributed site-occupation energies and a single-valued saddle-point energy. (d): distributed site-occupation and saddle-point energies.

activation energy for leaving the site, this model not only introduces dispersion into Q_R but necessarily introduces a concentration dependence into $\bar{Q}_R$ of the sense observed experimentally. Although the final general case (d) could also be made to yield the same type of behavior, Kirchheim[16,17] has shown that a surprisingly good account of the solution and diffusion behavior of hydrogen in metallic glasses can be accomplished with model (c), with the assumption that the site-occupation energies follow a Gaussian distribution of the type

$$n(G) = \left(1/\sigma\pi^{1/2}\right) \exp - \left[\left(G - G^0\right)/\sigma\right]^2, \tag{8}$$

where G is a free-enthalpy of occupation of a particular site, G^0 is the most probable value of G, σ is the site distribution parameter and $n(G)dG$ is the fraction of available sites in the interval dG. Using Fermi-Dirac statistics evaluated for simplicity at absolute zero, Kirchheim has shown that the chemical potential μ of the solute species at concentration C can be written

$$\mu = G^0 \pm \sigma \text{erf}^{-1}(1 - 2C), \tag{9}$$

where the negative sign is taken for $C < 0.5$. Equation (9) is already known to give a good account of the solution behavior of hydrogen in several metallic glasses. Like the width of the internal friction peak, the distribution parameter σ shows remarkably little variation amongst glasses of different compositions and type (metal-metalloid vs. metal-metal). With the constant saddle-point assumption represented by case (c) of Fig. 5, Kirchheim has also shown that the concentration dependence of the activation energy can be written in the corresponding form

$$\bar{Q}_R = K + \sigma \text{erf}^{-1}(1 - 2C), \tag{10}$$

where the parameter K is independent of concentration. As a test of the application of Eqs. (9) and (10) to the reorientation relaxation in glassy $Pd_{82}Si_{18}$ (Fig. 4), we show

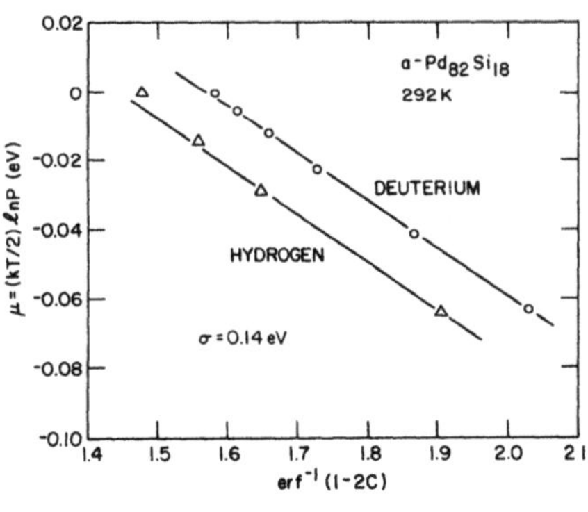

Fig. 6 Kirchheim solubility plots for hydrogen and deuterium in glassy $Pd_{82}Si_{18}$. The data were obtained from the bending-cantilever measurements of Ref. 18, combined with the value of trλ from Fig. 8.

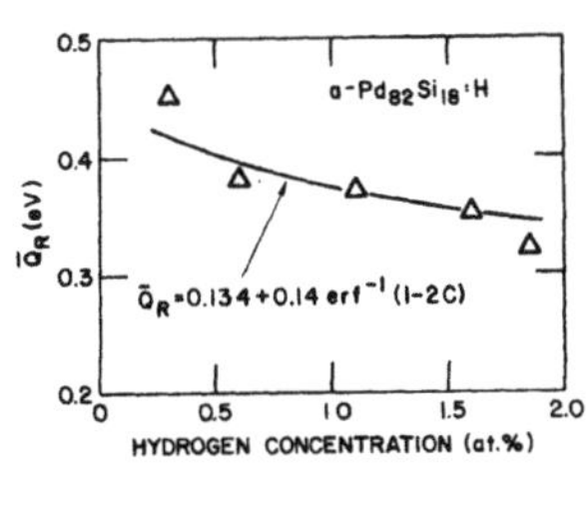

Fig. 7 Concentration dependence of $\bar{Q}_R$ for hydrogen reorientation in $a - Pd_{82}Si_{18}$. The data points are from Ref. 19. The fitted solid curve corresponds to Eq. (10) with the value of σ given by Fig. 6.

in Fig. 6 solubility measurements obtained by equilibration against known gas pressures P of hydrogen and deuterium. These data are fitted by Eq. (9) with lines of similar slope, corresponding to a value of 0.14eV for the parameter σ. This value has been used in Fig. 7 to compute the concentration dependence of $\overline{Q}_R$ predicted by Eq. (10). Based on the satisfactory agreement with previously reported results for $\overline{Q}_R$ obtained by the frequency-shift method[19], we may conclude that there is an excellent basis for explaining the reorientation relaxation in terms of the general population of hydrogen atoms present in solution. This is further supported by the strength considerations discussed below.

3.3 Strength Considerations

As may be seen from Eq. (7), one of the characteristics of the Snoek relaxation in crystalline metals is a direct proportionality between the relaxation strength Δ^R and the (dilute) interstitial concentration, C. This simple relationship is a consequence of the fact that every interstitial atom produces an identical elastic dipole. In attempting to adapt Eq. (7) to the case of an amorphous host, we must again consider the effects of structural disorder. As discussed in more detail in Section 5, it appears the major modification to Eq. (7) may result from the fact that the elementary act of reorientation now occurs between sites whose energies are not precisely equal, but are split by an

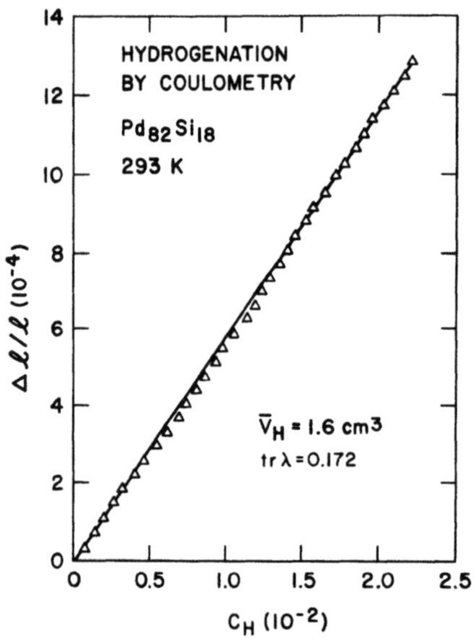

Fig. 8 Expansion strain vs. hydrogen concentration for coulometric (100% current efficient) electrocharging of $a-Pd_{82}Si_{18}$. The values of trλ and the partial molar volume $\overline{V}_H$ are calculated from the slope with the assumption of isotropic expansion.

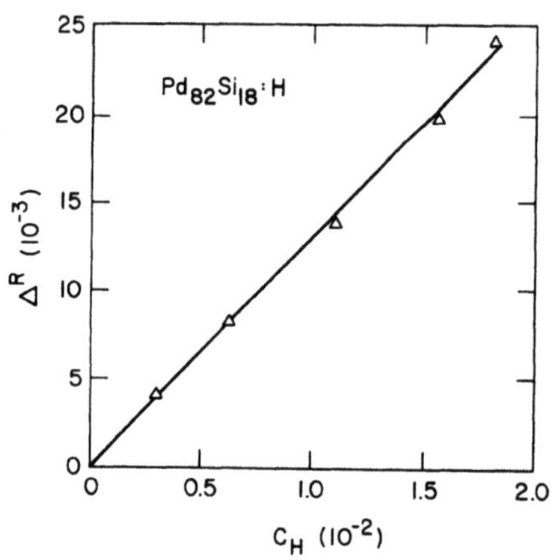

Fig. 9 Concentration dependence of the reorientation relaxation strength Δ^R for glassy $Pd_{82}Si_{18}$ containing less than 2 at.%H.

amount u. Based on the analysis of Section 5, we now write a modified version of Eq. (7) in the form

$$\Delta^R = (4/45)(E\vartheta_0/kT)f(u,T)C(\delta\lambda)^2, \tag{11}$$

where the splitting function $f(u,T)$ is ≤ 1 for $u \geq 0$.

In addition to site-splitting, another aspect of the effect of disorder relates to the expectation that the λ-tensor will become both distributed and concentration-dependent, due to the distribution in the shape and size of the interstitial sites.[20] However, presently available evidence from the concentration dependence of both Δ^R and the dilatometric expansion indicates that this effect may be relatively minor, at least at low concentrations. Figure 8 shows dilatation results obtained during slow electrolytic charging of glassy $Pd_{82}Si_{18}$ under conditions of 100% current efficiency, i.e. where the hydrogen concentration in the sample is determined by the number of coulombs deposited in the sample. The dilatation is seen to be essentially linear with concentration, corresponding to a concentration independent behavior of trλ (or, in alternative terms, of the partial molar volume $\overline{V}_H$). Interestingly, the value of 0.17 obtained for trλ from the slope of Fig. 8 is very similar to the value encountered in many crystalline hosts.[21] Indeed, the near constancy of trλ in the metallic glass may appear less surprising when it is recalled that the dilatation due to hydrogen is remarkably constant for a wide range of crystalline materials. The origin of this behavior has been discussed recently by Fukai.[22]

Returning to Fig. 8, we may argue that the concentration independence of trλ strongly implies a similar behavior for δλ, since both quantities derive from the same λ-tensor. Accordingly, we are led to conclude from Eq. (11) that the simple Snoek proportionality between Δ^R and C may be preserved for the reorientation relaxation at low hydrogen concentrations. Confirmation of this is shown by the results of Fig. 9. By reference to Eq. (11), it will be seen that the slope of this line provides a lower bound for estimation of the shape factor δλ, since $f(u,T) \leq 1$. The result so obtained, $\delta\lambda \geq 0.15$, shows that even the lower limit of δλ is remarkably similar to the value trλ=0.17, and reveals the highly asymmetrical nature of the hydrogen interstitial in metallic glasses. The finding that the dipole strength is so large implies that most of the dilatation produced by the H-interstitial occurs principally by atomic displacements in just one direction, and is consistent with increasing evidence from neutron-scattering experiments that the interstitial site for hydrogen in metallic glasses is basically of a tetrahedral character.[23,24] Just as for the case of a bcc crystal, it can easily be imagined that the major asymmetry of such a site is the tetragonality associated with the host-atom tetrahedron being geometrically irregular along just one direction. This may be illustrated by looking ahead to Fig. 18 where, for example, the dominant asymmetry of site *a* is seen to be dictated by the closer-than-regular proximity of pair 1-2 to pair 3-4.

4. THE GORSKY RELAXATION

The conclusion that the reorientation relaxation is of the Snoek type is strongly supported by observations of a second relaxation process known as the Gorsky effect. This relaxation provides an entirely different perspective on the interstitial population, since it focuses on the H-interstitial as a dilatation center rather than as an elastic dipole, and involves long-range diffusion over the thickness of the sample. From a comparison of the relaxation strengths and activation energies of these two relaxations, we are able to conclude that both of them originate from the general population of hydrogen atoms in solution.

4.1 Mechanism and Formalism

The Gorsky relaxation came into prominence through the work of Alefeld, Völkl, and their co-workers, who first recognized its potentialities for studies of hydrogen in crystalline metals.[25–27] In some cases the relaxation has been studied dynamically in flexural vibration.[28] However, it is often more conveniently observed in a quasi-static bending experiment. The mechanism of the effect involves the fact that a bending deformation produces a macroscopic stress gradient over the sample thickness and thereby induces a complementary gradient in the chemical potential of a dilatant solute species. This gradient is eliminated and equilibrium restored by a net drift of hydrogen atoms from the compressed to the expanded side of the sample, with the complementary appearance of an anelastic strain. Although the hydrogen atoms drift down a chemical potential gradient, this movement corresponds to "uphill" diffusion when simply regarded from the viewpoint of chemical diffusion in a concentration gradient, as expressed by Fick's laws. Upon removal of the applied stress, the induced concentration gradient creates a reverse chemical-potential gradient that initiates subsequent recovery, and an attendant Gorsky aftereffect. A detailed analysis for a strip of thickness *d* shows that there is only one strongly dominant Gorsky relaxation time τ_G, given by

$$\tau_G = d^2/\pi^2 D_F, \tag{12}$$

where, following Verbruggen *et al.*,[29] D_F is used to denote the chemical or Fickian diffusion coefficient of the mobile species. The existence of a single relaxation time for the Gorsky effect is a major contrast to the wide (typically 5-decade) dispersion exhibited by the reorientation peak. An expression for the relaxation strength Δ^G (defined formally as the ratio of the equilibrium anelastic and elastic strains produced by an applied stress) can be derived from classical thermodynamics. The case of an amorphous material is handled simply by considering an isotropic medium, with the result

$$\Delta^G = E\vartheta_0(\mathrm{tr}\lambda)^2/9(\partial\mu/\partial C), \tag{13}$$

where the partial derivative of the chemical potential is taken at constant stress and temperature. Since E, ϑ_0 and $\mathrm{tr}\lambda$ are all experimentally determinable quantities, measurements of Δ^G provide a means of determining $\partial\mu/\partial C$. This is a quantity of considerable interest. First, it may be used to test the predictions of Kirchheim's solution model. For, by returning to Eq. (9), we find by differentiation

$$\partial\mu/\partial C = 2\sigma\left(\partial \mathrm{erf}^{-1}y/\partial y\right), \tag{14}$$

where $y \equiv 1 - 2C$. Since σ is obtainable from solubility measurements and the derivative of the inverse error function is available from standard tables, it is possible to calculate values of $\partial\mu/\partial C$ from Eq. (14) for comparison with those found from Eq. (13). A second reason for interest in $\partial\mu/\partial C$ is that it can also be used to obtain a well-known thermodynamic factor F_T which represents a dimensionless measure of the non-ideality of the solution. The thermodynamic factor can be expressed as

$$F_T = (\partial\mu/\partial C)/(\partial\mu^i/\partial C), \tag{15}$$

where $\partial\mu^i/\partial C$ refers to an ideal solution of the same concentration. Since μ^i is given by

$$\mu^i = \mu^0 + RT \ln C, \tag{16}$$

we find immediately that $\partial\mu^i/\partial C = RT/C$, where R is the molar gas constant, and thus that

$$F_T = C(\partial\mu/\partial C)/RT. \tag{17}$$

Once F_T is known experimentally, it may be used to convert the chemical diffusion coefficient D_F to the tracer diffusion coefficient $D^\star$ by means of the relation

$$D^\star = D_F/F_T. \tag{18}$$

This procedure has already been employed in studies of crystalline Pd-H alloys by Verbruggen *et al.*[29]

4.2 Experimental Results

The first observations of the Gorsky relaxation in a hydrogenated metallic glass were reported[30] in 1981 for an electrolytically-charged sample of $Pd_{80}Si_{20}$. In contrast to the

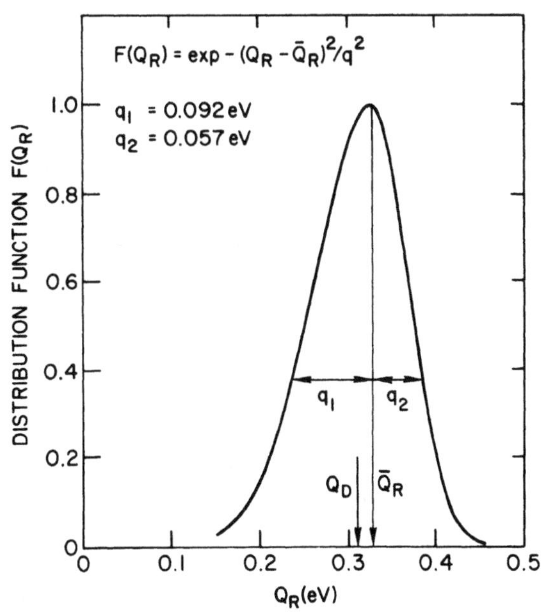

Fig. 10 Updated isoconcentration comparison of the Gorsky activation energy Q_D with an empirical distribution of activation energies $F(Q_R)$ that approximately reproduces the shape of the reorientation peak in a glassy Pd-Si alloy.

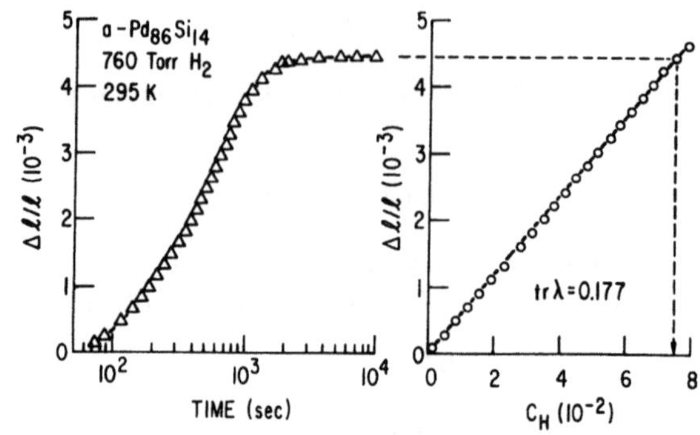

Fig. 11 Dilatometric measurements during equilibration of glassy $Pd_{86}Si_{14}$ in H_2 gas at 760 Torr pressure, and during coulometric electrolytic charging.

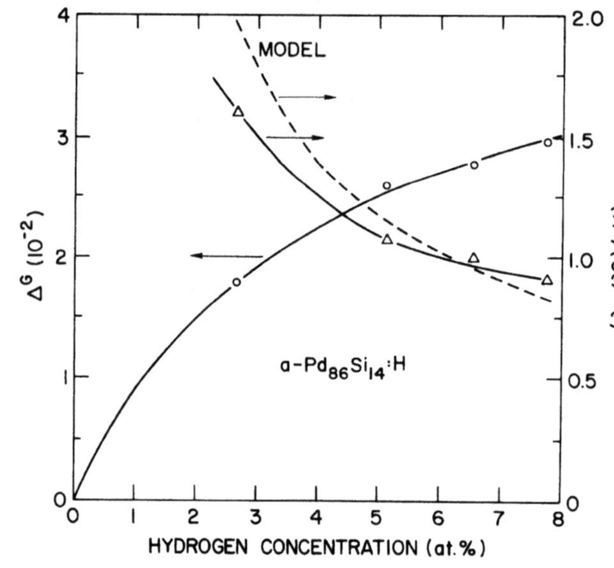

Fig. 12 Concentration dependence of the Gorsky relaxation strength in $a - Pd_{86}Si_{14}$, and a comparison of $\partial\mu/\partial C$ obtained from Eq. (13) with that calculated from Kirchheim's model via Eq. (14).

dispersion of the reorientation relaxation, the Gorsky aftereffect was indeed found to exhibit the simple exponential time dependence characteristic of a single relaxation time. The emphasis of these initial measurements was on the determination of an activation energy for long-range diffusion, Q_D, for comparison with the value of $\overline{Q}_R$ for the reorientation relaxation. An updated version of this comparison, made after the need for precise control of the hydrogen concentration was recognized, is shown in Fig. 10. The agreement between Q_D and $\overline{Q}_R$ is now closer than originally reported. More extensive measurements, on both Pd-Si and Ni-Zr alloys[19,31], have confirmed the near-equality of Q_D and $\overline{Q}_R$ and have also shown a complementary concentration dependence. This strongly supports the conclusion that the jumps involved in the reorientation relaxation also produce diffusion, as is required for a Snoek-type mechanism. We may thus predict that the reorientation and Gorsky relaxation times should show a similar isotope effect. Experimental tests of this prediction are at present quite limited and are again confined to the Pd-Si glasses. To date, the major result is that a kinetic isotope effect between hydrogen and deuterium is small and has not been clearly resolved by use of either relaxation.[19]

We turn now to some new results on the Gorsky relaxation in the lower-silicon alloy $Pd_{86}Si_{14}$. Interest in this alloy (which is close to the composition limit for quenching to

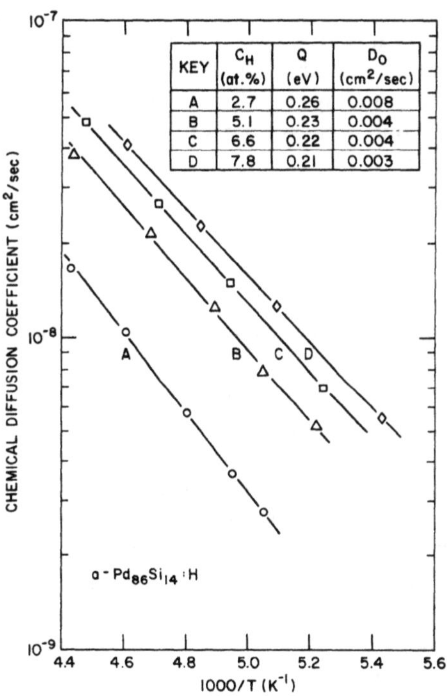

Fig. 13 Rate plots for chemical diffusion of hydrogen in $a - Pd_{86}Si_{14}$, as derived from the kinetics of the Gorsky relaxation.

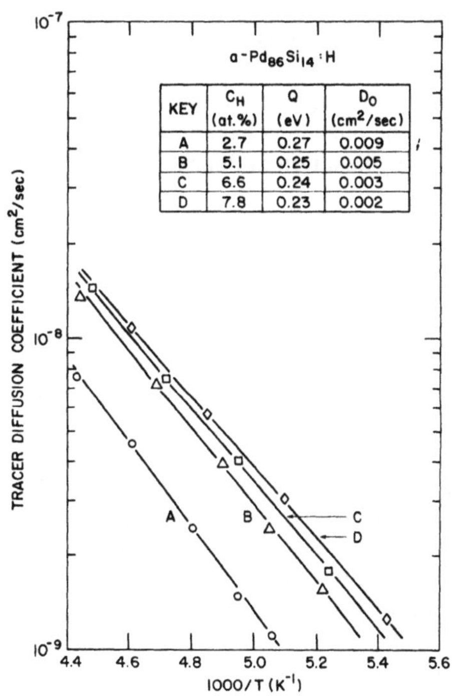

Fig. 14 Rate plots for tracer diffusion of hydrogen in $a - Pd_{86}Si_{14}$, as calculated from Fig. 13 and the thermodynamic factor obtained from the strength of the Gorsky relaxation.

the glassy state) was prompted by the substantial increase in hydrogen solubility discovered by Finocchiaro *et al.*[32] In our experiments, the hydrogen concentration obtained by equilibration against a selected gas pressure was found with the help of dilatometric measurements. As illustrated by Fig. 11, this method involves the use of a calibration curve obtained by coulometric electrolytic charging. The slope of this calibration curve also provides the value of trλ needed for the calculation of $\partial\mu/\partial C$ from measurements of the Gorsky relaxation strength Δ^G. Results for the concentration dependence of Δ^G for up to 8 at.%H are given in Fig. 12. Since trλ is a constant over this range (c.f. Fig. 11), the non-linearity of this plot is an immediate indication of non-ideal solution behavior. Figure 12 also compares the values of $\partial\mu/\partial C$ found from the Gorsky effect with those calculated from Kirchheim's solution model using Eq. (14). The agreement is rather satisfactory, and allows us to conclude that the Gorsky effect must have been produced by the general population of hydrogen atoms in solution. Use can also be made of Fig. 12 to obtain the thermodynamic factor needed to convert the chemical diffusion coefficient to the tracer diffusion coefficient. The effect of this conversion is shown by a comparison of Figs. 13 and 14. The tracer diffusion coefficient is smaller than the corresponding chemical coefficient, and shows less of a spread with concentration. Although the differences in D-values are substantial, they are seen to translate into rather minor differences in terms of the magnitude of the activation energies for chemical and for tracer diffusion.

Finally, we should mention two topics which so far are not well understood. One concerns the apparently opposite manner in which the strength of the Gorsky relaxation has been found to depend on temperature for different alloys.[31,33] The other topic concerns the puzzling characteristics reported for a second process observed to follow the Gorsky relaxation in $Pd_{80}Si_{20}$ at higher temperatures.[33]

5. DISCUSSION

In this article, we have not attempted to describe all of the ways in which anelastic measurements have proved useful for the study of hydrogen in metallic glasses. No mention has been made of the effects of annealing and eventual crystallization[13,19,34] or the strong influence of hydrogen on the magnetoelastic behavior of certain ferromagnetic glasses.[35,36] We have also refrained from describing the discovery in our laboratory of an exciting new effect which we may call hydrogen-enhanced host diffusion. Instead, we have chosen to focus exclusively on the reorientation and Gorsky relaxations as the two anelastic phenomena which together most clearly reveal the basic character of the hydrogen interstitial in metallic glasses. It is noteworthy that investigation of these relaxations is aided by a particularly fortunate combination of experimental circumstances. The relaxations are relatively strong and easy to measure; they can be observed conveniently at low temperatures where complications from concurrent structural relaxation or host-atom anelasticity can be avoided; the available sample thickness is generally ideal for quasi-static Gorsky measurements on the same sample used for dynamical measurements of the reorientation peak; and, finally, in many cases a surface oxide or surface poison provides a convenient barrier to the loss of hydrogen from the sample.

Turning to the results, we have seen that while each relaxation provides much useful information by itself, it is only by a detailed comparison that we obtain convincing evidence to support our foremost conclusion that the reorientation relaxation is of the Snoek type and the hydrogen interstitial is a strong elastic dipole. It is noteworthy that this conclusion is now supported by results from both metal-metal[31] and metal-metalloid glasses[19,30], and that a basic similarity is indicated in the nature of the interstitial site for glasses of both types.

As discussed elsewhere[19], the large dipole strength and relatively high activation energy (0.25 to 0.5eV) of the H-interstitial in metallic glasses represents a major contrast with the situation for pure bcc niobium, where a vanishingly weak dipole strength and extremely fast non-classical diffusion appear to be related to a delocalization of the H-interstitial amongst a set of tunnelling states. The quasi-classical behavior of hydrogen in metallic glasses is much more analogous to that of the heavier solute interstitials (C,N,O) in bcc metals, and leads to the conclusion that structural disorder inhibits delocalization of the H-interstitial in metallic glasses.

The remainder of this discussion will be devoted to the effects of structural disorder on the characteristics of the reorientation and Gorsky relaxations. Most of these effects are direct and obvious; however, we shall also suggest that more subtle considerations are needed to understand the strength of the reorientation relaxation. The obvious effects of disorder are the dispersion of the reorientation relaxation, the non-ideality of $\partial\mu/\partial C$ from the Gorsky relaxation, and the concentration-dependent activation energies shown by both relaxations. These aspects appear to be well accounted for by Kirchheim's model based on a set of sites with a Gaussian distribution of occupation energies, and a constant saddle-point energy. One respect in which this model is less satisfactory is the prediction of the width of the activation-energy distribution associated with the reorientation relaxation. The model predicts that this width should be given by the width of the *occupied* site distribution, whereas the width estimated experimentally is larger. Clearly, the implication of this discrepancy is that the constant saddle-point assumption is an oversimplification.

We turn now to the role of disorder on what is perhaps the most puzzling aspect of the reorientation relaxation, namely its substantial strength. As shown in Section 3.3, the calculated slope factor $\delta\lambda$ is comparable to trλ even when the limiting value of unity is assumed for the splitting factor $f(u,T)$. Smaller values of $f(u,T)$ lead to the difficulty that the calculated value of $\delta\lambda$ becomes unrealistically large. Nevertheless, as indicated by the simple 2-site models of Fig. 15, the essence of the difference between a crystal-

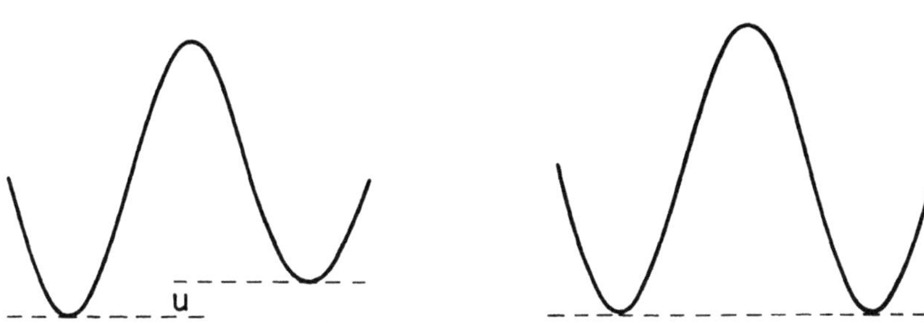

Fig. 15 Energy-distance diagrams representing basic models for point-defect reorientation in crystalline and amorphous solids. The site-splitting u is a consequence of structural disorder.

line and amorphous solid appears to reside in the disorder manifest by the site-splitting energy u . The effect of site-splitting on both the relaxation strength Δ^R and relaxation time τ_R can be calculated explicitly for the 2-site model. The relaxation time is

$$\tau_R(u,T) = \tau_0 \text{sech}(u/2kT) \exp Q_R/kT, \qquad (19)$$

and the splitting function $f(u,T)$ appearing in Eq. (11) has the form

$$f(u,T) = 4 \exp u/kT/(1 + \exp u/kT)^2. \qquad (20)$$

The calculated form of the internal friction peak obtained by insertion of Eqs. (19) and (20) into Eq. (5) is shown in Fig. 16 for a representative set of experimental parameters. These calculations reveal a pronounced suppression of the peak as u approaches 0.05eV. Since this energy is comparable to the estimated width of the occupied site distribution, we are faced with the paradox that peak suppression by site-splitting would be expected to be a major factor, thereby forcing $\delta\lambda$ to unacceptably large values. This dilemma has also been appreciated by Yoshinari *et al.*[13]

We believe the important new concept in resolving this paradox involves the *spatial distribution of disorder*, a feature that is totally ignored in site-energy distributions such as Eq. (8). Figure 17 shows two schematic energy diagrams that can be used to illustrate our argument. In Fig. 17(a), the broken curve represents the variation of site occupation energy with distance x, and can be thought of as a modulation function $U(x)$. Whereas the parameter σ in Eq. (8) involves only the amplitude distribution in $U(x)$, the actual splitting u between adjacent levels also involves the wavelength of the modulation and becomes smaller as the wavelength increases. At one extreme, therefore,

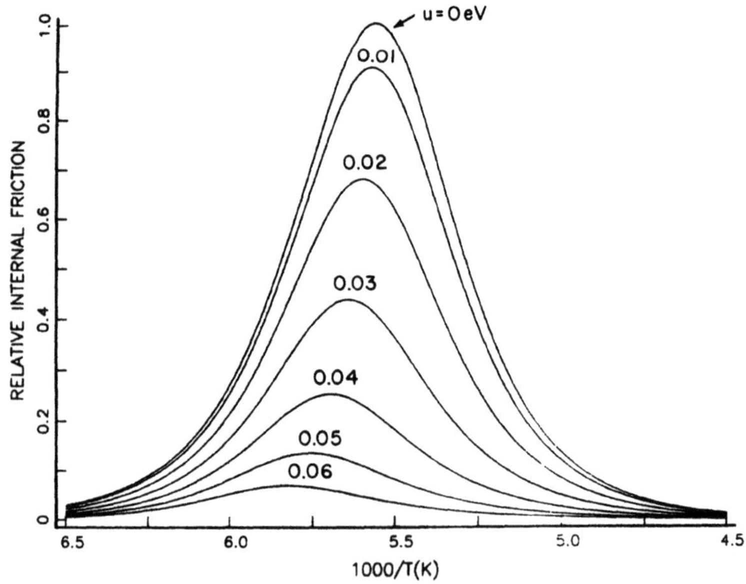

Fig. 16 The effect of the site-energy difference u on the reorientation peak corresponding to the split-level case of Fig. 15. The peaks were calculated from Eqs. (5), (19) and (20) with the following representative parameters: $Q_R = 0.4\text{eV}$, $\tau_0 = 10^{-14}$ sec , $\omega/2\pi = 100\text{Hz}$.

we might argue that the reorientation relaxation is not suppressed because the disorder occurs on a relatively long wavelength in relation to the interatomic spacing. An alternative and seemingly more attractive possibility is illustrated by Fig. 17(b), which is drawn to represent pairwise near-degeneracy of neighboring sites. In this case, the material divides into a superposition of unilevel systems [Fig. 17(a)], and suppression of the relaxation is again avoided. Although Fig. 17(b) may appear at first sight to be somewhat artificial, we believe there may be a structural basis for this model. Consider, for example, the octahedral grouping of host atoms shown in Fig. 18. In a high-symmetry form, this could be considered as part of the bcc structure, and the tetrahedral interstitial sites *a, b, c* and *d* would be energetically degenerate. In a more disordered form, this octahedral grouping is one of the simpler structural units used for the description of an amorphous solid.[20] There are a number of ways in which disorder can be introduced into Fig. 18 without removing the degeneracy of neighboring sites. For example, changes in the separation distance r_{12} between atoms 1 and 2 will change the energies of all the sites without removing the site degeneracy. Shearing the square array of atoms 3,4,5 and 6 into a parallelogram by changes in the angle ψ leads to the same result. By moving the pair 1-2 off center along a diagonal direction, e.g. 4-5, we may introduce further disorder without removing the degeneracy of sites *a* and *b*, or of *c* and *d*. It therefore appears reasonable to conjecture that the high strength of the reorientation relaxation may be caused by localized near-degeneracy of the type schematically represented by Fig. 17(b).

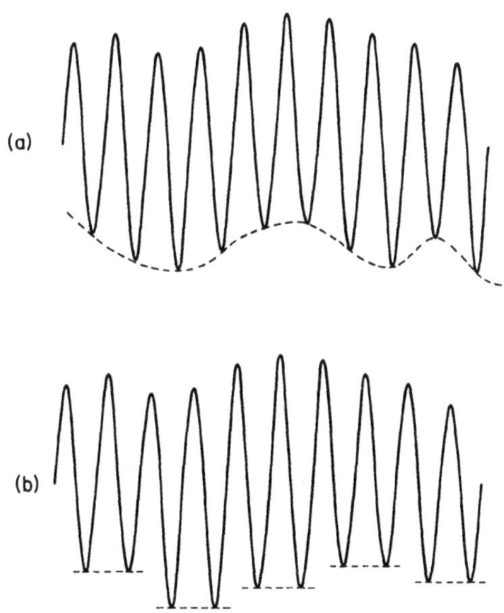

Fig. 17 Schematic energy-distance diagrams for hypothetical amorphous solids. In (a), near-degeneracy of adjacent sites is obtained only when the wavelength of the site-energy modulation (broken curve) becomes much greater than the jump distance. In (b), the sites exhibit pairwise degeneracy.

As a final comment, we may point out one further aspect of the apparent connection between structural disorder and the presence or absence of a Snoek-type relaxation. If we are correct in speculating that this difference involves the role of disorder in inhibiting delocalization of the H-interstitial, it becomes of interest to inquire how much and what type of disorder is necessary to produce localization. Specifically, since amorphous alloys contain a mixture of topological and chemical disorder, we may ask whether localization can be obtained through one of these factors alone. While we believe that the hydrogen interstitial would be a localized quasi-classical defect in pure amorphous Nb, the ability to test this prediction experimentally appears remote due to the extreme thermal instability of pure amorphous metals. On the other hand, the introduction of chemical disorder into bcc Nb is readily accomplished by alloying. It appears highly likely that the broad hydrogen peak already observed in the bcc $Nb_{50}V_{50}$ alloy[37,38] is actually of the Snoek type, and reflects a chemically-induced localization of the H-interstitial.

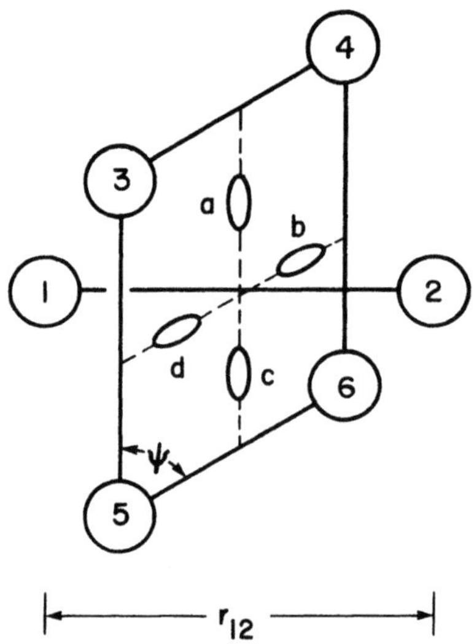

Fig. 18 An example of how structural constraints may impose near-degeneracy on the site-occupation energies of neighboring interstitial sites in an amorphous solid. The host atom tetrahedra defining the interstitial sites *a, b, c, d* all have the edge 1-2 as a common structural feature. As a consequence, near degeneracy of sites *a* and *d*, for example, only involves a near-similarity in the arrangement of pairs 3-4 and 3-5.

REFERENCES

1. A. J. Maeland, *Hydrogen in Crystalline and Non-Crystalline Metals and Alloys: Similarities and Differences*, in: "Rapidly Quenched Metals" Vol. 2, S. Steeb and H. Warlimont, eds., North-Holland, Amsterdam (1985).
2. A. S. Nowick and B. S. Berry, "Anelastic Relaxation in Crystalline Solids", Academic Press, New York (1972).
3. B. S. Berry, *Atomic Diffusion and the Breakdown of Hooke's Law*, Phys. Teach. 21(7):435 (1983).
4. B. S. Berry and W. C. Pritchet, *Vibrating Reed Internal Friction Apparatus for Films and Foils*, IBM J. Res. Dev. 19:334 (1975).
5. B. S. Berry and W. C. Pritchet, *Extended Capabilities of a Vibrating Reed Internal Friction Apparatus*, Rev. Sci. Instrum. 54:254 (1983).
6. B. S. Berry, *Elastic and Anelastic Behavior*, in: "Metallic Glasses", J. J. Gilman and H. J. Leamy, eds., Am. Soc. Metals, Metals Park, Ohio (1978).
7. B. S. Berry, W. C. Pritchet and C. C. Tsuei, *Discovery of an Internal Friction Peak in the Metallic Glass* Nb_3Ge, Phys. Rev. Lett. 41:410 (1978).
8. B. S. Berry and W. C. Pritchet, *Thermally-Activated Internal Friction Peaks in Amorphous Films of* Nb_3Ge *and* Nb_3Si , in: "Rapidly Quenched Metals III", Vol. 2, B. Cantor, ed., The Metals Society, London (1978).
9. B. S. Berry and W. C. Pritchet, *Hydrogen-Related Internal Friction Peaks in Metallic Glasses*, Scripta Metall. 15:637 (1981).
10. H.-U. Künzi, K. Agyeman and H.-J. Güntherodt, *Internal Friction Peaks in Metallic Glasses*, Solid State Comm. 32:711 (1979).
11. K. Agyeman, E. Armbruster, H.-U. Künzi, A. Das Gupta and H.-J. Güntherodt, *Hydrogen Related Internal Friction Peaks in Metallic Glasses*, J. de Physique 42:C5-535 (1981).
12. B. S. Berry and W. C. Pritchet, *Defect Studies of Thin Layers by the Vibrating-Reed Technique*, J. de Physique 42:C5-1111 (1981).
13. O. Yoshinari, M. Koiwa, A. Inoue and T. Masumoto, *Hydrogen Related Internal Friction Peaks in Amorphous and Crystallized Pd-Cu-Si Alloys*, Acta Metall. 31:2063 (1983).
14. L. E. Hazelton and W. L. Johnson, *Distribution of Activation Energies in Amorphous* Zr_2PdH_x, J. Non-Cryst. Solids 61/62:667 (1984).
15. F. H. M. Spit, J. W. Drijver and S. Radelaar, *Hydrogen Sorption in Amorphous Ni(Zr,Ti)-Alloys*, Z. Phys. Chem. (Weisbaden) 116:809 (1979).
16. R. Kirchheim, F. Sommer and G. Schluckebier, *Hydrogen in Amorphous Metals-I*, Acta Metall. 30:1059 (1982).
17. R. Kirchheim, *Solubility, Diffusivity and Trapping of Hydrogen in Dilute Alloys, Deformed and Amorphous Metals - II*, Acta Metall. 30:1069 (1982).
18. B. S. Berry and W. C. Pritchet, *Sorption of Hydrogen and Deuterium in a Glassy Pd-Si Alloy by a Bending-Cantilever Method*, in: "Rapidly Solidified Amorphous and Crystalline Metals", B. H. Kear, B. C. Giessen and M. Cohen, eds., Elsevier Science Publishing Co., New York (1982).
19. B. S. Berry and W. C. Pritchet, *Anelasticity and Diffusion of Hydrogen in Glassy and Crystalline Metals*, in: "Nontraditional Methods in Diffusion", G. E. Murch, H. K. Birnbaum and J. R. Cost, eds., The Metallurgical Society of AIME, Warrendale, PA (1984).
20. H. J. Frost, *Cavities in Dense Random Packings*, Acta Metall. 30:889 (1982).
21. H. Peisl, *Lattice Strains due to Hydrogen in Metals*, in: "Hydrogen in Metals I", G. Alefeld and J. Völkl, eds., Vol. 28 in Topics in Applied Physics, Springer-Verlag, Berlin (1978).

22. Y. Fukai, *Atomistic and Electronic Approaches to Hydrogen in Metals*, Cryst. Latt. Def. and Amorph. Mat. 11:85 (1985).
23. K. Suzuki, *Structure and Properties of Amorphous Metal Hydrides*, J. Less-Common Metals, 89:183 (1983).
24. K. Suzuki, N. Hayashi, Y. Tomizuka, T. Fukunaga, K. Kai and N. Watanabe, *Hydrogen Atom Environments in a Hydrogenated ZrNi Glass*, J. Non-Cryst. Solids 61/62:637 (1984).
25. G. Alefeld, J. Völkl and G. Schaumann, *Elastic Diffusion Relaxation*, Phys. Stat. Sol. 37:337 (1970).
26. G. Schaumann, J. Völkl and G. Alefeld, *The Diffusion Coefficients of Hydrogen and Deuterium in Vanadium, Niobium and Tantalum by Gorsky-Effect Measurements*, Phys. Stat. Sol. 42:401 (1970).
27. J. Völkl, *The Gorsky Effect*, Ber. Bunsenges. Phys. Chem. 76:797 (1972).
28. R. Cantelli, F. M. Mazzolai and M. Nuovo, *Internal Friction due to Long-Range Diffusion of Hydrogen in Niobium (Gorsky Effect)*, Phys. Stat. Sol. 34:597 (1969).
29. A. H. Verbruggen, C. W. Hagen and R. Griessen, *Gorsky Effect in Concentrated* α'*-*PdH_x, J. Phys. F:Met. Phys. 14:1431 (1984).
30. B. S. Berry and W. C. Pritchet, *Gorsky Relaxation and Hydrogen Diffusion in the Metallic Glass* $Pd_{80}Si_{20}$, Phys. Rev. B 24:2299 (1981).
31. B. S. Berry and W. C. Pritchet, *Snoek and Gorsky Relaxations in Hydrogenated Metallic Glasses*, J. de Physique (in press).
32. R. S. Finocchiaro, C. L. Tsai and B. C. Giessen, *The Young's Moduli of Glassy Pd-Si-H Alloys*, J. Non-Cryst. Solids 61/62:661 (1984).
33. A. H. Verbruggen, R. C. van den Heuvel, R. Griessen and H.-U. Künzi, *Gorsky Effect Measurements on Amorphous* $Pd_{80}Si_{20}H_x$ *between 290 and 490K*, Scripta Metall. 19:323 (1985).
34. B. S. Berry and W. C. Pritchet, *Hydrogen in a Glassy and Crystallized Pd-Si Alloy*, in: "Rapidly-Quenched Metals" Vol. 2, S. Steeb and H. Warlimont eds., North-Holland, Amsterdam (1985).
35. B. S. Berry and W. C. Pritchet, *Effect of Hydrogen on the Magnetoelastic Behavior of Amorphous Transitional Metal-Metalloid Alloys*, J. Appl. Phys. 52:1865 (1981).
36. J. P. Allemand, F. Fouquet and J. Perez, *Influence of Structural Relaxation and Hydrogen Permeation on the Magnetoelastic Effect in Iron-Based Amorphous Alloys*, in: "Rapidly-Quenched Metals" Vol. 2, S. Steeb and H. Warlimont eds., North-Holland, Amsterdam (1985).
37. C. V. Owen, O. Buck and T. E. Scott, *Indications for a Hydrogen Snoek Peak in a Nb-V Alloy*, Scripta Metall. 15:1097 (1981).
38. C. L. Snead, Jr. and J. Bethin, *Stress Induced Hydrogen Reorientation in a BCC Nb-50 at.%V Alloy*, Phys. Rev. 32B:4254 (1985).

NMR STUDIES OF THE HYDRIDES OF DISORDERED AND AMORPHOUS ALLOYS

Robert C. Bowman, Jr.

Chemistry and Physics Laboratory
The Aerospace Corporation, P. O. Box 92957
Los Angeles, CA 90009

INTRODUCTION

Nuclear magnetic resonance (NMR) spectroscopy has had extensive applications to the characterization of metal-hydrogen systems as is well documented in numerous previous review papers[1-7]. Structural information on hydrogen site occupancies can be obtained from the NMR lineshapes in the "rigid-lattice" (i.e., immobile nuclei) limit. The hyperfine interactions with conduction electrons can be monitored through Knight shifts (σ_K) and spin-lattice relaxation time (T_{1e}) contributions. The various nuclear relaxation times are usually very sensitive[1-7] to translational diffusion. Under suitable conditions, the hydrogen diffusion constants can be directly measured via various spin-echo techniques[2,3,5,6]. NMR studies have been conducted on all three hydrogen isotopes (i.e., H, D, and T) as well as many host metal nuclei (e.g., ^{45}Sc, ^{51}V, ^{89}Y, ^{93}Nb, and ^{139}La). Although most attention has been primarily focused upon the binary hydride phases (i.e., TiH_x, ZrH_x, PdH_x, etc.), the hydrides formed by crystalline alloys and intermetallics with nominal stoichiometries A_2B, AB, AB_2, and AB_5 have been the subjects of NMR measurements during the past ten years or so. This interest was mainly stimulated by the potential applications of various ternary hydrides (e.g., $TiFeH_x$, $LaNi_5H_x$, Mg_2NiH_x) as reversible hydrogen storage systems. However, very few reports of NMR experiments on amorphous hydrides formed from metallic glasses have been published to date [8-16] in spite of the recent proliferation of papers on other aspects of these materials. The relative absence of NMR results for amorphous metal-hydrogen systems is partially due to the rather low sensitivity of the technique (i.e., many spectrometers typically require at least

0.5-1.0g samples - which is often considered to be quite demanding in the metallic glass preparation field). Furthermore, few NMR experiments had been done for H/M ratios below about 0.1 in any crystalline hydride. In fact, most published NMR results[8-15] for the amorphous hydrides have been for samples with hydrogen-to-metal (H/M) atomic ratios around unity.

The intent of the present paper is to provide an overview of the information that can be gleaned from NMR measurements on transition metal hydrides. Since particular emphasis is placed upon comparison between crystalline and amorphous hydrides with similar (nominally identical) compositions, most examples are taken from the work of the author and his colleagues, which are the most extensive results currently available on these properties. It is hoped that other NMR research groups will find the present results and unresolved questions sufficiently challenging to conduct further work on this very interesting class of materials.

In order to avoid an excessively long paper the reader is referred to the excellent reviews by Cotts[1,2] for a general description of the NMR concepts and techniques as applied to metal hydrides. However, some recent advances and caveats regarding NMR experimentation on hydrides will be mentioned when appropriate. Additional background can be obtained from references 3-7. Although a few NMR results for metal nuclei in metallic glasses have been reported[17,18], no metal NMR data are apparently available for amorphous hydrides. Furthermore, the first deuteron NMR results are being presented at this conference[19] on a sample of a-$Zr_2PdD_{2.9}$. Consequently, discussions will be limited to proton NMR behavior where quadrupolar effects are completely absent since the proton has an I = 1/2 spin. Hence, the various aspects that are associated[2,5] with nuclear quadrupoles will not be considered in this review.

HYDROGEN SITE OCCUPANCIES FROM NMR LINESHAPES

The distribution of hydrogen atoms among the available interstitial sites of the metal host structure is of both fundamental and practical interest with regards to the stabilities and compositions of the various hydride phases. Although x-ray diffraction is commonly used to determine hydride structures, this technique cannot identify the hydrogen site occupancy except under extremely favorable conditions (which are quite rare). Powder neutron diffraction is the conventional method to locate deuterium atoms in metal deuteride structures. Because the incoherent neutron scattering from protons usually gives very large background signals, reliable structure determinations in the metal hydride phases are not usually possible from powder neutron diffraction. However, inelastic neutron scattering experiments have provided valuable information on hydrogen site occupancies in both crystalline and amorphous hydrides[20-23]. Unfortunately, the detailed identifications of hydrogen site occupancies from these neutron scattering measurements are not always straightforward. For example, there currently remains considerable controversy[24] with regards to the hydrogen (i.e., deuterium) locations and distributions in the extensively studied crystalline $TiFeD_x$ and $LaNi_5D_x$ phases.

When inhomogeneous line broadening contributions from magnetic impurities (e.g., ferromagnetic or superparamagnetic species that are due to surface disproportionation of ternary hydrides) and bulk sample paramagnetic demagnetization effects[25] are negligible, the low-temperature (i.e., rigid-lattice limit) proton lineshapes are exclusively determined by the nuclear dipolar interactions[1]. In transient NMR experiments the lineshape can be obtained[1] from normalized free induction decay G(t). A simple and convenient pulse sequence for G(t) measurements has been recently described[26]. Structural information is most easily deduced from the lineshape second moment (M_{2D}) defined by

$$M_{2D} = - \frac{d^2\, G(t)}{dt^2} \Big|_{t\,=\,0} \tag{1}$$

The M_{2D} parameter can be directly related to the distribution of nuclei through the well-known[1,2] Van Vleck formalism. For the specific case of a powder metal hydride with protons distributed on several (i.e.,N) inequivalent sites and no significant heteronuclear dipolar interactions with metal spins, the proton M_{2D} in units of gauss^2 can be calculated from the expression[27]

$$M_{2D} = \frac{9}{80} \left(\frac{\gamma_H h}{\pi}\right)^2 \sum_{i,j}^{N} f_i \alpha_j S_{ij} \tag{2}$$

where γ_H is the proton gyromagnetic ratio, h is Planck's constant, f_i is the fraction occupancy of protons on sites of type i, α_j is the occupancy factor on any site j and S_{ij} is the lattice summation $\sum_i r_{ij}^{-6}$ for internuclear separations r_{ij} between sites i and j. It is straightforward to generalize[1] eqn. (2) to include heteronuclear dipolar contributions. Excellent agreement between the experimental and calculated proton M_{2D}

parameters have been obtained for several binary and ternary hydrides[10,26-28], when the chosen proton site occupancies used in these calculations are consistent with the available neutron scattering results.

In order to illustrate the usefulness and limitations of hydrogen site occupancy assessments from proton M_{2D} parameters, the results for several Zr_2PdH_x samples will be discussed in some detail. This system is particularly interesting since both the tetragonal Zr_2Pd intermetallic with the $MoSi_2$-type structure and the glassy alloy (i.e., a-Zr_2Pd) readily absorb hydrogen to form crystalline and amorphous, respectively, ternary hydrides.[11,29] Several distinctive crystalline Zr_2PdH_x phases have been found[12,13,29-31], whose structures depend upon hydrogen content and preparation temperature (T_{PREP}). Since hydrogen occupies tetrahedral sites (i.e., Zr_4) in the binary ZrH_x phases and octahedral sites (i.e., Pd_6) in PdH_x, determination of site occupancy in the various ternary Zr_2PdH_x phases is needed to help understand the substantial variations in diverse properties such as the diffusion behavior and electronic structure parameters which will be described later in this paper. Table 1 provides a summary of the interstitial sites in tetragonal Zr_2Pd that could potentially contain hydrogen atoms. Neutron scattering experiments[30,32] on tetragonal c-$Zr_2PdH(D)_x$ for $x<2.0$ indicate that Zr_4(i.e., T_A sites) are the preferred sites at these compositions. However, the situation is much more complex for the crystalline phases[12-14,31] produced when $x \approx 3$ as well as for the amorphous hydrides[22,23]. The structures for several Zr_2PdH_x samples are given in Table 2 along with their preparation temperatures and the proton M_{2D} values from analyses[11,31] of their magic-echo[26] lineshapes at sufficiently low temperatures to guarantee rigid-lattice conditions.

The proton M_{2D} values in Table 2 permit semi-quantitative evaluations of the hydrogen site occupancies in the crystalline Zr_2PdH_x samples. Dipolar M_{2D} values have been calculated with eqn. (2) for a number of trial proton distributions in the tetragonal $MoSi_2$ structure. The best results are summarized in Table 3. Although the crystal structures for the HT-$Zr_2PdH_{2.9}$ and LT-Zr_2PdH_{3+} samples are not $MoSi_2$-type, the observed differences are not very large[31]. Hence, the errors in the M_{2D} calculations due to this approximation should not detrimentally influence the subsequent conclusions on site occupancy in these particular samples although they will not be quite as reliable as for those samples with the $MoSi_2$ structure. When $x< 2$, the protons preferentially occupy the T_A (i.e., Zr_4) sites. However, the experimental M_{2D} values for

Table 1. Possible Interstitial Sites for Hydrogen Occupancy In Zr_2PdH_x Phases with the $MoSi_2$-type Tetragonal (C11b) Structure

Label	Lattice Position	Nearest Neighbor Coordination	Maximum Allowed Content (x)	Comments
T_A	4d	Zr_4	2	Preferred Sites
T_B	16n	Zr_3Pd	2	Blocks Neighboring T_A,T_B,T_C sites.
T_C	8g	Zr_2Pd_2	2	Blocks neighboring T_A,T_B,T_C sites.
O_A	2b	Zr_2Pd_4	1	May be occupied at larger x
O_B	4c	Zr_4Pd_2	2	Short H-Pd separation
O_C	4e	Zr_5Pd	2	Blocks all T_B sites and short separations to T_A and T_C sites

c-$Zr_2PdH_{1.94}$ is 8% smaller than the dipolar second moment calculated for exclusive Zr_4 site occupancy. This discrepancy was removed when a small fraction of the protons was allowed to randomly occupy the Zr_4Pd_2(i.e., O_B) sites. The significant differences in the M_{2D} parameters for the two crystalline phases of Zr_2PdH_x (when x ≈ 3) can be directly related to the occupancy of only tetrahedral sites (i.e., $T_A$and T_B) in the high-temperature (HT) sample but large octahedral site occupancies (i.e., O_B) in the low-temperature (LT) samples. Changes in the proton distribu-

Table 2. Structures, Preparation Temperature (T_{PREP}), and Experimental Proton Second Moments (M_{2D}) for Zr_2PdH_x Samples at Rigid Lattice Temperatures (T).

Sample Composition	T_{PREP} (K)	Metal Structure	M_{2D} (gauss^2)	T (K)
c-$Zr_2PdH_{1.68}$	671	Tetragonal($MoSi_2$)	8.4(4)	207
c-$Zr_2PdH_{1.84}$	623	"	8.5(3)	207
c-$Zr_2PdH_{1.94}$	755	"	8.8(4)	207
HT-$Zr_2PdH_{2.90}$	746	Distorted Tetragonal	14.2(5)	288
LT-$Zr_2PdH_{3.05}$	440	Orthorhombic	10.2(3)	207
LT-$Zr_2PdH_{3.10}$	510	"	12.2(2)	207
LT-$Zr_2PdH_{3.39}$	510	"	12.3(2)	207
a - $Zr_2PdH_{1.91}$	478	Amorphous	9.9(3)	120
a - $Zr_2PdH_{2.85}$	478	"	12.2(3)	80

tions between the Zr_4 and Zr_4Pd_2 sites did not improve agreement with the experimental M_{2D} values for the LT-Zr_2PdH_x samples. It is speculated that some protons on a third site (i.e., Zr_2Pd_4) may be responsible for the larger LT-Zr_2PdH_x experimental moments. Although eqn. (2) with a three-site proton distribution can yield a dipolar second moment that reproduces the experimental result, there remains no credible way to define a unique site occupancy. This illustrates one of the major limitations of using NMR lineshapes to deduce proton occupancies in multi-site situations. Additional independent information is required to restrict

Table 3. Comparisons of Theoretical Dipolar Second Moments (M_{2D}) for Proton Lineshapes in Model Tetragonal (i.e., $MoSi_2$ - Type) Structures for Crystalline Zr_2PdH_x phases. The Ratios of the Experimental M_{2D} Values from Table 2 to the Calculated Dipolar Moments are Included.

Sample Phase	Lattice Parameters a (nm)	c (nm)	Model Proton Site Occupancies	Calculated M_{2D} (gauss^2)	Ratio
c-$Zr_2PdH_{1.68}$	0.3317	1.1642	Zr_4(0.84)	8.55	0.98
c-$Zr_2PdH_{1.84}$	0.3348	1.1571	Zr_4(0.92)	8.86	0.96
c-$Zr_2PdH_{1.94}$	0.3337	1.1582	Zr_4(0.97)	9.54	0.92
	"	"	Zr_4(0.915); Zr_4Pd_2(0.055)	(8.80)[a]	1.00
HT-$Zr_2PdH_{2.90}$	0.345	1.122	Ordered Zr_3Pd (1.00) & Zr_4 (0.90)[b]	14.00	1.01
	"	"	Ordered Zr_3Pd (0.95) & Zr_4 (1.00)[b]	14.31	0.99
LT-$Zr_2PdH_{3.05}$	0.3415	1.1442	Zr_4(1.00); Zr_4Pd_2(0.525)	9.28	1.10
LT-$Zr_2PdH_{3.10}$	"	"	Zr(1.00); Zr_4Pd_2(0.55)	9.37	1.30
LT-$Zr_2PdH_{3.39}$	"	"	Zr_4(1.00); Zr_4Pd_2(0.695)	10.01	1.23

[a] Calculated by varying Zr_4 and Zr_4Pd_2 site occupancies to yield experimental M_{2D} value.

[b] Only Zr_4 sites that are not a nearest neighbor to an occupied Zr_3Pd site can contain a proton.

the allowable combinations to relatively narrow ranges for each potential occupied site. In principle, neutron scattering could provide the necessary inputs; unfortunately, these data are not currently available for the LT-$Zr_2PdH_{\sim 3}$ samples.

Although it is formally possible to extend[33] eqn. (2) to calculate dipolar M_{2d} values for amorphous systems by converting the summation for N finite inequivalent sites to an integral in limit $N \rightarrow \infty$, it would be very difficult to properly include chemical short range order (CSRO) contributions in these calculations which are already quite complex if they are based upon a realistic distribution function[33]. Consequently, only some qualitative conclusions can be drawn from the proton M_{2D} parameters in Table 2 for a-Zr_2PdH_x. Since the experimental moments is larger for a-$Zr_2PdH_{1.91}$ than for c-$Zr_2PdH_{1.94}$, it is likely that protons only occupy tetrahedral sites in this amorphous sample. However, M_{2d} for a-$Zr_2PdH_{2.85}$ is intermediate to the values for the crystalline HT-$Zr_2PdH_{2.90}$ and LT-$Zr_2PdH_{3.05}$ samples. While long-range order should be relatively unimportant to M_{2D} due to the implicit r_{ij}^{-6} dependence, it is very difficult to separate the topological contributions from the hydrogen site occupancy distributions. However, the proton M_{2D} value for a-$Zr_2PdH_{2.85}$ does imply some proton occupancy of octahedral sites in addition to the preferred tetrahedral sites. This view is consistent with the neutron scattering results of Suzuki[22] on a-$Zr_{0.65}Pd_{0.35}D_X$ but not with the conclusion of Williams et al.[23] that hydrogen occupies only tetrahedral interstitials (e.g., mostly Zr_4 but, perhaps, some Zr_3Pd sites) in a-$Zr_2PdH_{\sim 3}$. Coordinated NMR and neutron experiments on well-characterized crystalline and amorphous $Zr_2PdH_{\sim 3}$ samples are needed to definitively establish the proton site occupancies in these phases.

DIFFUSION BEHAVIOR FROM PROTON RELAXATION TIMES

The nuclear relaxation times T_1 (spin-lattice), $T_{1\rho}$ (rotating frame), and T_2 (spin-spin) generally provide the most versatile means of assessing diffusion behavior from NMR studies[1-7]. For the protons in metal hydride phases, these relaxation times are usually dominated over wide temperature ranges by the temperature dependent modulation of homonuclear and/or heteronuclear dipolar interactions induced from various hydrogen jump processes among the interstitial sites. Both localized motion (e.g., protons jump among a cluster of equivalent or inequivalent interstitial sites in specific regions of the structure) and the long range hopping motion that translates individual protons throughout the material contributed to dipolar relaxation. However, the nuclear relaxation times are effected by other processes in addition to diffusion. For example, there can be three likely contributions to the measured spin-lattice relaxation rate for protons

$$T_1^{-1} = T_{1d}^{-1} + T_{1e}^{-1} + T_{1p}^{-1} \tag{3}$$

where T_{1d} is the diffusion (i.e., dipolar) term, T_{1e} is the conduction electron hyperfine term, and T_{1p} represents paramagnetic relaxation from local magnetic moments. Until rather recently it was usually assumed that the T_{1p} contribution to eqn. (3) was negligible in most samples of nominally good purity unless the material had a strongly paramagnetic character (e.g., UH_3, CeH_x, etc.). However, the work of Barnes, et al.[34-36] has clearly shown that even quite small (e.g., 1-2 ppm Gd or Ce impurities in some cases) quantities of paramagnetic impurities will significantly alter the magnitude and temperature dependence of the T_1 relaxation times. If the paramagnetic contribution is not properly recognized, misinterpretation of the T_1 data can easily occur[34]. Although the $T_{1\rho}$ and T_2 relaxation times have similar contributions as in eqn. (3), the dipolar term is generally dominant in most commonly encountered situations. Consequently, the paramagnetic contributions usually are not as manifest for these parameters[34-36].

The dipolar relaxation times are indirectly[2] related to diffusion coefficients (D) through the correlation -time functions $G^{(q)}(t)$. For the case of only proton-proton dipolar interactions, the general expressions[1-3] are

$$(T_{1d})^{-1} = (3/2)\ C_H\ [J^{(1)}(\omega_0) + J^{(2)}(2\omega_0)] \tag{4}$$

$$(T_{1\rho})^{-1} = (3/8)\ C_H\ [J^{(0)}(2\omega_1) + 10\ J^{(1)}(\omega_o) + J^{(2)}(2\omega_0)] \tag{5}$$

$$(T_2)^{-1} = (3/8)\ C_H\ [J^{(0)}(0) + 10\ J^{(1)}(\omega_0) + J^{(2)}(2\omega_0)] \tag{6}$$

where $C_H = \gamma_H^4 h^2\ I(I + 1)/\ 4\pi^2$; ω_0 is the Larmor frequency in the laboratory magnetic field H_0, $\omega_1 = \gamma_H H_1$ for the applied rf field H_1; and $J^{(p)}(\omega)$ are Fourier transforms of the time-dependent correlation functions

$$J^{(p)}(\omega) = \int_{-\infty}^{\infty} G^{(q)}(t) \exp(-i\omega t)\, dt. \tag{7}$$

Although these expressions are formally correct, the analysis of the experimental relaxation times require tractable, yet physically realistic, models to represent the underlying microscopic motions of the nuclei. The most common and simplest expression for G(t) is the exponential correlation time approximation that is widely known[1-7] as the BPP model. This model gives

$$J^{(p)}(\omega) = \frac{2\,\tau_c\, G^{(q)}(0)}{1 + \omega^2\,\tau_c^2} \tag{8}$$

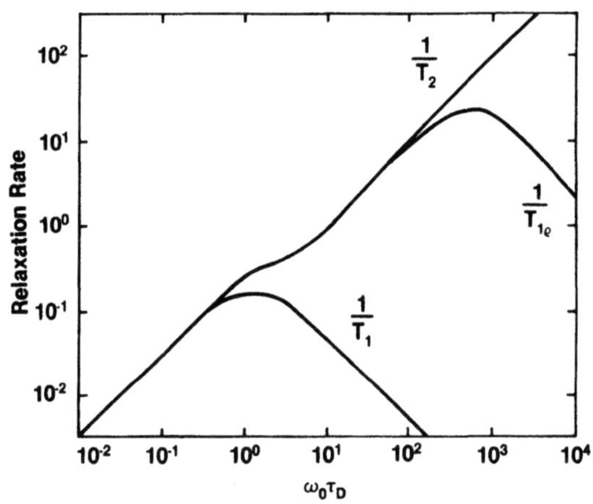

FIG. 1 BPP model representation of relaxation rates T_1, $T_{1\rho}$, and T_2 as functions of $\omega_o\tau_D$ when $\omega_o = 500\omega_1$, (after Cotts[2]).

for each of the individual terms in eqns.(4-6). Here, τ_c is the diffusion correlation time and is one-half the mean time between jumps (i.e., τ_D) when only proton-proton interactions are involved[2-5]. Fig. 1 shows typical dependences[2] of the reciprocals of the T_{1d}, $T_{1\rho}$, and T_2 relaxation times upon τ_D in terms of the BPP model where τ_D follows the Arrhenius relation

$$\tau_D(T)^{-1} = 2A \exp(-E_a/k_BT) \tag{9}$$

with an activation energy E_a and pre-exponential factor A. Although the BPP model often yields τ_c values with absolute errors greater than a factor of two, the E_a values have been usually reliable to within 10% or better when direct comparisons are possible[2,3,5]. More accurate lattice specific calculation procedures for $G^{(q)}(t)$ have been developed[2,3,5,37], but their numerical solutions are restricted to a few high symmetry structures. Consequently, they are not currently applicable to those hydrides with complex or amorphous crystal structures. Furthermore, these lattice specific models exhibit[37] nearly the same temperature and frequency behavior as the simpler BPP model and cannot explain the anomalous dependences that are being experimentally found in an increasing number of metal-hydrogen systems that includes the amorphous hydrides[8,10,11,14-16]. Although there have been some attempts[16,38,39] to analyze the non-Arrhenius behavior of proton relaxation times through distribution functions of individual BPP models, the physical validity of these approaches has not be firmly established. Consequently, the present discussion will emphasize a conventional BPP-type analysis of the proton relaxation times for the amorphous hydrides and their crystalline counterparts. However, these assessments are subject to subsequent modification as additional experimental and theoretical results become available.

A variety of techniques have been used to monitor hydrogen diffusion in metallic glasses[40,41] when the hydrogen contents are small (i.e., $H/M < 0.1$). However, very few studies [42], other than the NMR measurements, are presently available for the higher hydrogen contents. The emphasis of most[8-15] of the NMR studies has been to compare crystalline and amorphous hydrides with similar (if not essentially identical) compositions. While hydrogen trapping at defects or impurities can significantly influence hydrogen diffusion at the low concentrations[40,41], these effects should be minimal in the samples that have been studied to date by NMR techniques.

The first proton NMR experiments of diffusion in an amorphous hydride were done[8,10] on $TiCuH_x$ where various relaxation times were measured. Fig. 2 compares the temperature behavior of τ_c^{-1} for c-$TiCuH_{0.94}$ and a-$TiCuH_{1.4}$ that were obtained[10] by BPP analyses of the proton τ_{1p} data. These results are also compared with similar parameters[43] for cubic $TiH_{1.90}$. Significant differences in hydrogen diffusion behavior are immediately apparent in Fig. 2. A structural restriction accounts[7,10] for the smaller τ_c^{-1} values and larger activation energies E_a when c-$TiCuH_{0.94}$ is compared to $TiH_{1.90}$. Presumably, the structural disorder in $a\text{-}TiCuH_{1.4}$ eliminates these constraints on the proton jump processes to produce the enhanced mobilities and decreased E_a values. When protons occupy only Ti_4 sites (i.e., $c\text{-}TiCuH_{0.94}$ and $TiH_{1.90}$), the τ_c^{-1} data closely follow the Arrhenius relations over the entire observable temperature range. However, very distinct breaks occur[8,10] for all the diffusion sensitive proton relaxation times in $a\text{-}TiCuH_x$. Since these breaks do not appear to be the gradual deviations expected[44] for a continual distribution of jump activation energies, a relatively few jump paths probably contribute to hydrogen diffusion. It is likely that short-range order in the local structure of the amorphous hydride is responsible for this behavior.

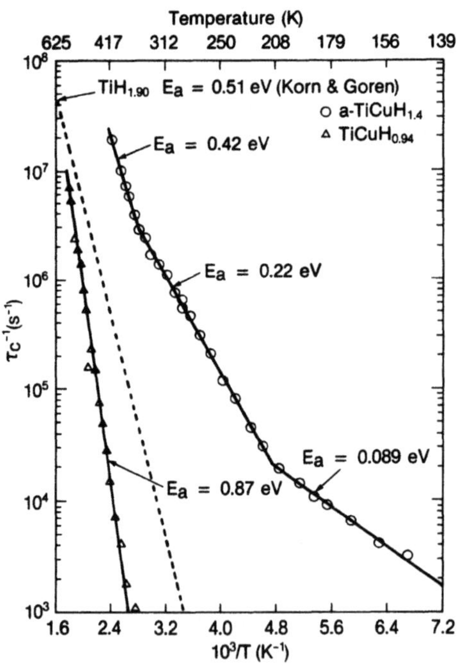

FIG. 2 Comparison of diffusion jump rates τ_c^{-1} for crystalline and amorphous $TiCuH_x$ from BPP model analyses of proton $T_{1\rho}$ data[10]. The dashed line represents fit to similar results for crystalline $TiH_{1.90}$ by Korn and Goren[43].

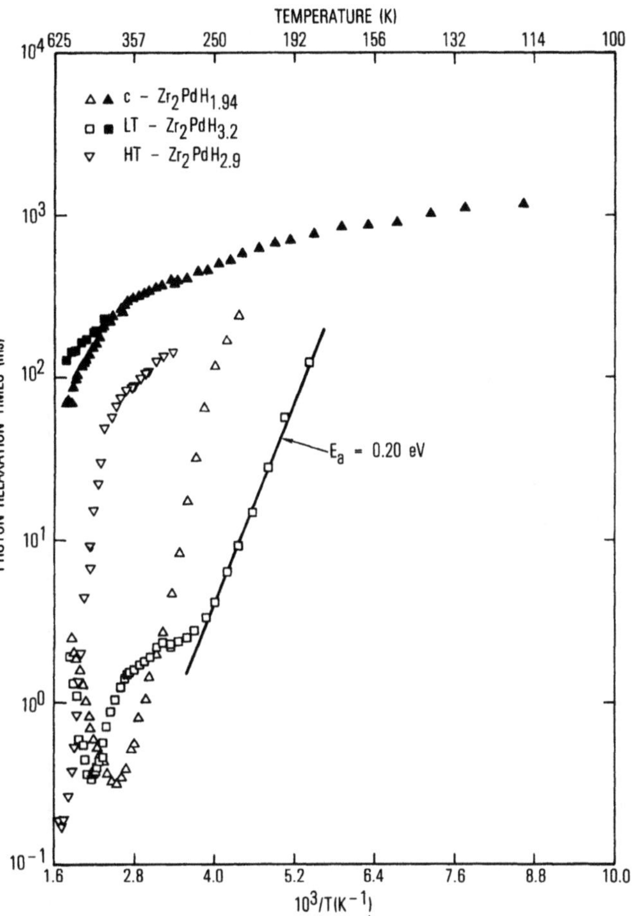

FIG. 3 Proton relaxation times T_1 (filled symbols) and $T_{1\rho}$ (open symbols) for three crystalline Zr_2PdH_x samples.

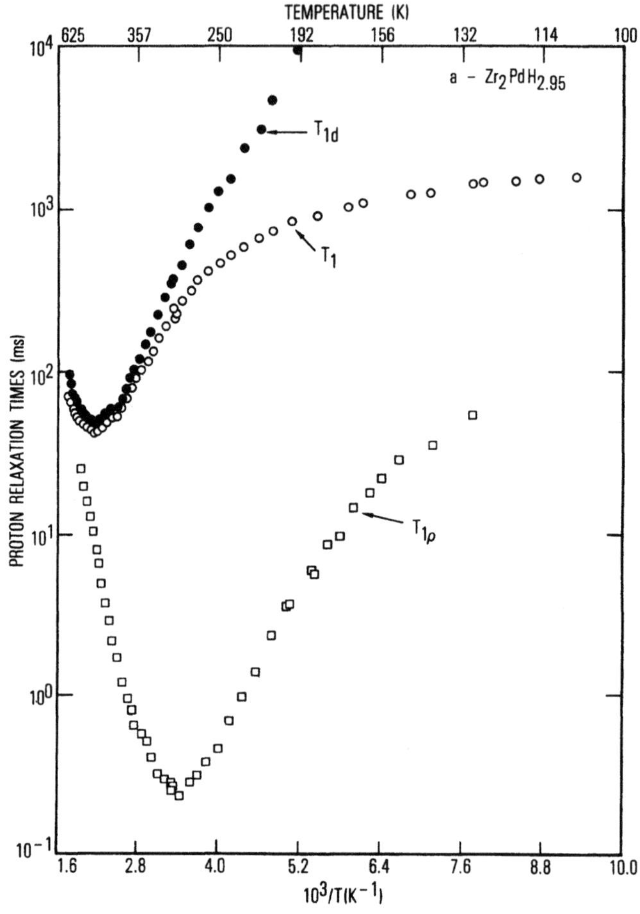

FIG. 4 Proton relaxation times for glassy a-$Zr_2PdH_{2.95}$.

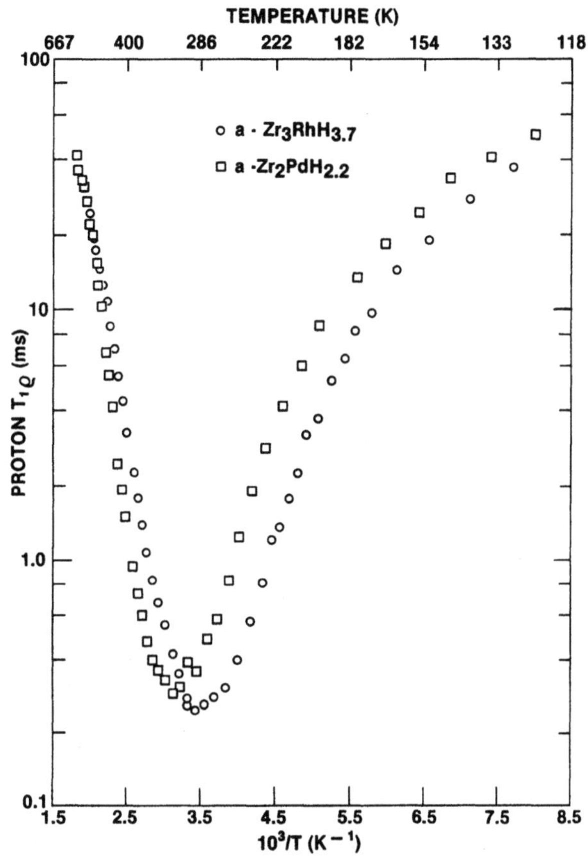

FIG. 5 Proton $T_{1\rho}$ relaxation times for amorphous a-$Zr_2PdH_{2.2}$ and a-$Zr_3RhH_{3.7}$.

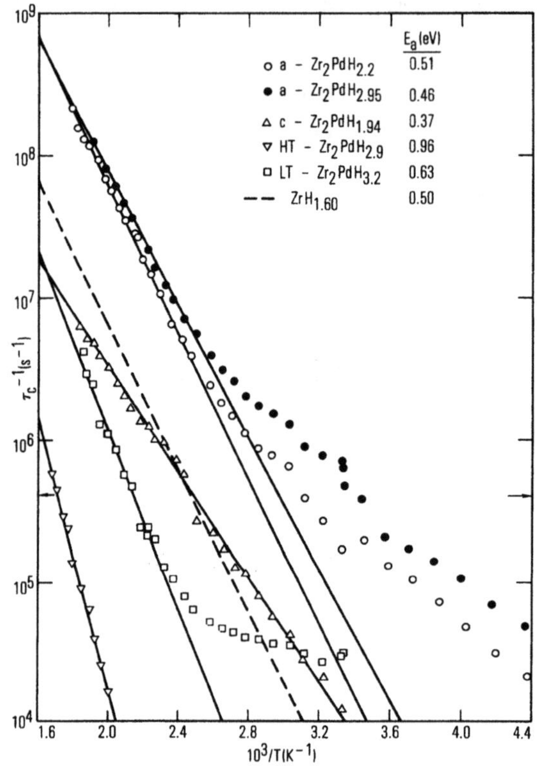

FIG. 6 Temperature dependences of τ_c^{-1} parameters from BPP analyses of proton $T_{1\rho}$ data for crystalline and amorphous Zr_2PdH_x as well as similar curve (dashed line) for cubic $ZrH_{1.60}$ sample[46].

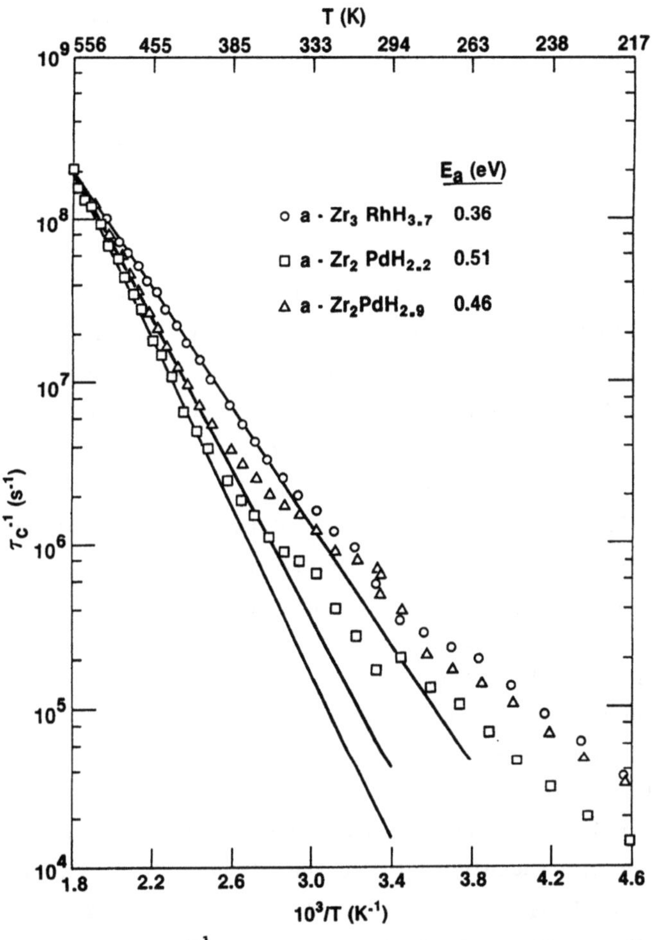

FIG. 7 Comparison of τ_c^{-1} parameters from BPP analyses of proton $T_{1\rho}$ data for the three Zr-based amorphous hydrides.

The several phases and hydrogen site occupancies in Zr_2PdH_x, as described in the previous section, allow for most interesting comparisons of diffusion behavior among the crystalline and amorphous ternary phases[11,13-15,45] as well as with the crystalline binary hydrides ZrH_x(Ref. 46) and PdH_x (Refs. 47 and 48). The temperature dependences of the proton T_{1d} and $T_{1\rho}$ relaxation times[13,45] for three crystalline Zr_2PdH_x samples, whose structural properties are given in Table 2, are presented in Fig. 3. Similar results[14,15] are given in Fig. 4 for glassy a-$Zr_2PdH_{2.95}$ while Fig. 5 compares the $T_{1\rho}$ data[15,45] for a-$Zr_2PdH_{2.2}$ and a-$Zr_3RhH_{3.7}$. The temperature dependences of the τ_c^{-1} parameters that had been obtained with BPP model analyses[13,45] of the $T_{1\rho}$ data are presented in Figs. 6 and 7. Although the details of the hydrogen diffusion mechanisms are not completely understood for the various Zr_2PdH_x phases, many interesting features have been identified. In the crystalline c-Zr_2PdH_x samples with $x < 2$, the τ_c^{-1} parameters are seen[13] to obey Arrhenius relationships that yield E_a values very similar to the ZrH_x phases[46]. These results are consistent with the predominantly Zr_4 site occupancies in c-Zr_2PdH_x samples when $x < 2$ and further imply similar jump processes[7] in both structures. However, two distinctive diffusion mechanisms are apparently occuring in the LT-$Zr_2PdH_{3.2}$ sample. The low-temperature (i.e., when $T < 250K$) $T_{1\rho}$ data in Fig. 3. yield an E_a of 0.20 eV for LT-$Zr_2PdH_{3.2}$ that is nearly identical to the diffusion activation energy (i.e., 0.22eV) for proton diffusion[47,48] among the octahedral sites in PdH_x. However, a BPP analysis of the proton $T_{1\rho}$ data above 460K gives a much larger E_a value (i.e., 0.63eV) for the same LT-$Zr_2PdH_{3.2}$ sample as shown in Fig. 6. Furthermore, the unusual behavior of the LT-$Zr_2PdH_{3.2}$ proton $T_{1\rho}$ relaxation times, which are nearly independent of temperature between these two regions, is noted to be remarkably similar to recently published results[49,50] for diffusion processes between two inequivalent lattice sites. Since the proton M_{2D} analysis for LT-$Zr_2PdH_{3.2}$ has indicated predominantly Zr_4 and Zr_4Pd_2 interstitial site occupancies, it is proposed that the smaller E_a value reflects motion of those protons distributed among the Zr_4Pd_2 sites while the larger activation energy includes diffusion through the Zr_4 sites as well. The very large E_a value obtained for HT-$Zr_2PdH_{2.9}$ is probably due to the ordering of protons on the Zr_4 and Zr_3Pd sites. Hence, diffusion requires an additional defect formation energy contribution similar to the one recently suggested[46] for the high E_a value in $ZrH_{1.997}$. The elimination of low barrier jump paths for the protons is also possible[13] in the distorted HT-$Zr_2PdH_{2.9}$ structure.

From comparisons of the τ_c^{-1} parameters in Figs. 6 and 7, significantly enhanced proton mobilities are indicated for the glassy hydrides a-Zr_2PdH_x and a-$Zr_3RhH_{3.7}$ when compared with the crystalline ZrH_x and Zr_2PdH_x phases. The parameters A and E_a for these samples as well as the TiH_x and $TiCuH_x$ samples are summarized in Table 4. It is noteworthy that the proton diffusion parameters are very similar for the three glassy systems, which would not have been predicted beforehand since other properties such as thermal stabilities[51] show substantial differences.

Table 4. Proton Diffusion Parameters from BPP Model Analyses of $T_{1\rho}$ Relaxation Times (High Temperature Limits for Samples with Non-Arrhenius Behavior).

Hydride	Structure	A (s^{-1})	E_a (eV)	Temperature Range (K)	Data Source (Ref.)
$TiH_{1.90}$	Cubic(fcc)	7.9×10^{11}	0.51(1)	270-660	44
c-$TiCuH_{0.94}$	Tetragonal	4.4×10^{14}	0.87(3)	360-560	10
a-$TiCuH_{1.4}$	Amorphous	2.8×10^{12}	0.42(3)	355-415	10
$ZrH_{1.60}$	Cubic(fcc)	2.1×10^{11}	0.50(1)	300-564	46
$ZrH_{1.90}$	Tetragonal	3.7×10^{10}	0.45(1)	368-554	46
c-$Zr_2PdH_{1.94}$	Tetragonal	1.7×10^{10}	0.37(1)	260-540	13
HT-$Zr_2PdH_{2.90}$	Tetragonal(?)	6.9×10^{13}	0.96(1)	440-595	13
LT-$Zr_2PdH_{3.2}$	Orthorhombic	2.4×10^{12}	0.63(3)	470-544	45
a-$Zr_2PdH_{2.2}$	Amorphous	8.8×10^{12}	0.51(2)	425-558	15,45
a-$Zr_2PdH_{2.9}$	Amorphous	3.1×10^{12}	0.46(2)	401-525	13,45
a-$Zr_3RhH_{3.7}$	Amorphous	3.6×10^{11}	0.36(1)	442-566	15,45

Perhaps, a more-or-less common diffusion mechanism is operative for these amorphous hydrides that is rather insensitive to the metal constituents or hydrogen stoichiometry. However, a decrease in E_a with increasing hydrogen content is seen for glassy a-Zr_2PdH_x samples which is opposite to the trend observed for the crystalline phases. Changes in the proton site occupancies[22] with stoichiometry may be responsible where more shallow potential wells (i.e., sites with lower diffusion barriers) are filled as the hydrogen content is increased. Unfortunately, a firm conclusion on this matter cannot be made from the currently available NMR relaxation time data. Significant positive deviations from the Arrhenius expression can also be seen in Figs. 6 and 7 for the amorphous hydrides at lower temperatues. In contrast all of the crystalline ZrH_x and Zr_2PdH_x samples, except the previously discussed LT-$Zr_2PdH_{3.2}$ sample, follow the Arrhenius relation for the entire common temperature range. Reduced activation energies have been obtained[13,14] over limited portions of the lower temperatures. As in a-$TiCuH_x$, distinct changes in diffusion properties rather than a continuous distribution of activation energies are believed to be responsible for the non-Arrhenius behavior in the amorphous Zr - based hydrides. It is likely that the smaller E_a values correspond to some types of localized motion. However, more complete information on proton site occupancies is needed to give a clearer picture of the diffusion processes. Furthermore, a better analytical approach than the simple BPP model currently being used is probably required to properly address multi-step jump processes among inequivalent sites.

ELECTRONIC STRUCTURE FROM PROTON NMR STUDIES

The effects of hydrogen on the electronic and magnetic properties of crystalline and amorphous metals have received considerable attention[7,12,22,52-56]. The traditional measurements of magnetic susceptibility, specific heat, and superconductivity parameters can provide information on the total density of electronic states at the Fermi level $N(E_F)$ while Mossbauer and NMR spectroscopies are related to the portion of $N(E_F)$ projected on the resonant nucleus. Within the past few years, photoelectron and x-ray spectroscopy methods have become extremely valuable in the determination of electronic structures of metallic glasses[57] and metal hydrides[52] since they permit more-or-less direct comparisons with the valence band state densities calculated by first-principles theory[57,58]. The formation of the metal-hydrogen chemical bond generally produces a new band of states in the valence region below the original

Fermi level and often shifts the Fermi level upwards. Hence, substantial changes in $N(E_F)$ upon hydrogenation have been theoretically predicted[58] and experimentally observed in numerous metal hydrides[2,7,22,52-56]. Proton NMR provide unique insights on the electronic properties at the the hydrogen site which can be calculated, in principle, by band theory methods[57,58] but are usually inaccessible to most other experimental techniques. Although qualitative evaluations of local electron densities from proton NMR experiments are feasible, a quantitative analysis is not currently possible since the required hyperfine parameters are not sufficiently reliable. Nevertheless, useful correlations between proton NMR data and electronic structure properties have been made[2,7]. This latter approach will be briefly illustrated with examples from the $Zr_{1-y}Pd_yH_x$ system. Similar results on other binary and ternary hydrides are referenced in the reviews by Cotts [2] and Bowman[7].

Within the framework of the usual partition model widely used[59] to analyze hyperfine interactions in transition metal systems, the proton Knight shift (σ_K) and conduction electron to spin-lattice relaxation (T_{1e}) are assumed to consist of two major contributions as

$$\sigma_K = 2\mu_B \left[H_{hf}(s)N_s(E_F) + H_{hf}(d)N_d(E_F)\right] \tag{10}$$

$$(T_{1e}T)^{-1} = 2\,h\,\gamma^2 k_B\left\{\left[H_{hf}(s)N_s(E_F)\right]^2 + q\,\left[H_{hf}(d)N_d(E_F)\right]^2\right\} \tag{11}$$

The new quantities in these expressions include the Bohr magnetron μ_B and the d-orbital degeneracy factor q, which is directly proportional[59] to the fractional t_{2g} character at the Fermi surface and $0.2 < q < 0.5$ for cubic symmetry. $N_s(E_F)$ and $N_d(E_F)$ are the s-band and d-band density of electron states at the Fermi level, respectively. The two hyperfine fields at the resonance nuclei are the positive $H_{hf}(s)$, which is caused by the direct Fermi contact interaction with the unpaired s electrons at E_F, and the negative $H_{hf}(d)$ due to the polarization of spin-paired proton s orbitals in valence band by the unpaired Fermi level d electrons associated with the transition metals. Although the contact term is much larger for a single unpaired electron[59], the large $N_d(E_F)$ values in most transition metal hydrides[52,58] usually insures large (even dominant) polarization contributions to the T_{1e} and σ_K parameters. Because of the sign difference between $H_{hf}(s)$ and $H_{hf}(d)$, significant cancellations can occur in the proton Knight shifts; however, the two contributions are always additive for the T_{1e} relaxation rates. Qualitative assessments of the electronic structure at the hydrogen sites are possible with a generalization of the Korringa[60] expression in the form

$$q_E = C_K/(\sigma_K^2 T_{1e}T) \tag{12}$$

where $C_K = h\,\gamma_e^2/(8\pi^2 k_B \gamma_H^2)$. When σ_K is positive and $q_E \approx 1.0$, the s-band Fermi contact term is the dominant proton hyperfine interaction. When $q_E > 1$ for positive shifts, substantial polarization contributions will also be present. In contrast, a negative Knight shift with q_E =q is strong evidence for only the d-band polarization interaction which clearly indicates $N_d(E_F) \gg N_s(E_F)$ since $H_{hf}(s) > H_{hf}(d)$.

The usefulness of eqns. (10-12) to qualitatively identify the character of the Fermi level electronic states in metal hydrides will be briefly illustrated with the results compiled in Table 5 for ZrH_x, PdH_x,

Table 5. Comparison of Proton Knight Shifts (σ_K), $(T_{1e}T)^{-1/2}$ Parameters, and Korringa parameters (q_E) for ZrH_x, PdH_x, and Zr_2PdH_x samples.

Sample	σ_K (ppm)	$(T_{1e}T)^{-1/2}$ $(sK)^{-1/2}$	q_E	Data Sources (Ref.)
$ZrH_{1.60}$	-17	0.042	1.6	61
$ZrH_{1.80}$	-51	0.064	0.42	61
$ZrH_{2.00}$	-40	0.047	0.37	61
$PdH_{0.65}$	+20	0.094	5.6	12
$PdH_{0.76}$	+29	0.120	4.4	12
c-$Zr_2PdH_{1.94}$	-29	0.088	2.7	12
LT-$Zr_2PdH_{3.2}$	-22	0.075	3.3	12
HT-$Zr_2PdH_{2.9}$	-55	0.077	0.50	12
a-$Zr_2PdH_{1.89}$	-17	0.072	4.8	12
a-$Zr_2PdH_{2.88}$	-12	0.070	9	12

and Zr_2PdH_x. The negative σ_K values measured[61] in ZrH_x are entirely consistent with the small $N_s(E_F)$ predicted from the band structure calculations[62,63] for cubic and tetragonal $ZrH_{2.0}$. The $(T_{1e}T)^{-1/2}$ maximum for $ZrH_{1.80}$ has been related[61] to the Fermi level moving across a peak in the density of states over the hydrogen concentration range $1.6 < x < 2.0$. Although the proton q_E values for $ZrH_{1.80}$ and $ZrH_{2.0}$ indicate only d-electron polarization hyperfine interactions at the hydrogen sites, a finite s-band contact contribution is apparently present in $ZrH_{1.60}$. The positive σ_K for the PdH_x samples indicate that their proton hyperfine interactions are dominated by unpaired Fermi level s-electrons. This assertion agrees with the band theory calculation[64] which place E_F just above the Pd d-bands in a region composed mainly of s-electron states. However, the rather large q_E values for the PdH_x samples also imply the presence of the d-band polarization hyperfine interaction whose relative contribution decreases with increasing hydrogen content. Negative proton Knight shifts were obtained for all the Zr_2PdH_x samples given in Table 5. However, large q_E values were deduced for each material except HT-$Zr_2PdH_{2.9}$ whose q_E is within the range expected for just a polarization hyperfine interaction.

Since no electronic band structure calculations are currently available for the Zr_2PdH_x phases, detailed quantitative identification of the causes for relatively large $N_s(E_F)$ values at the proton sites is not currently possible. However, differences in the interstitial site occupancy will certainly have an important role. This is clearly evident when the parameters in Table 5 for LT-$Zr_2PdH_{3.2}$ with Zr_4 and Zr_4Pd_2 proton site occupancies are compared to the NMR values for HT-$Zr_2PdH_{2.9}$ where only tetrahedral $Zr_{4-y}Pd_y$ sites are believed to be occupied. However, significantly larger q_E values are also found for the a-Zr_2PdH_x, which are primarily caused by their small negative σ_K values. This behavior cannot be readily related to only differences in proton site occupancies since the proton M_{2d} parameters are very similar for the a-Zr_2PdH_x and c-Zr_2PdH_x samples. Although previous comparisons[57,65] of hydrogen-free amorphous and crystalline $Zr_{1-y}Pd_y$ alloys have indicated minor differences for the valence band N(E) curves, recent magnetic susceptibility measurements[53] of Zr_2Pd alloys do suggest different $N(E_F)$ values for the crystalline and glassy phases. However, there is strong theoretical and experimental evidence[57,65] that the Pd d-band states lie several eV below E_F in both types of $Zr_{1-y}Pd_y$ alloys while the Fermi level falls in the states due primarily from Zr d-bands. This view is further supported by the recent magnetic resonance and susceptibility studies of Eifert, et al[17,66] whose results also indicated that $N_d(E_F)$ in a-Zr_2Pd is almost 30% larger than $N_d(E_F)$ in crystalline Zr metal. However, the larger proton $(T_{1e}T)^{-1/2}$ parameters for the Zr_2PdH_x phases relative to the ZrH_x samples cannot be entirely due to larger $N_d(E_F)$ in the Zr_2PdH_x phases since the Knight shifts are much more negative in ZrH_x. Since $H_{hf}(s) > H_{hf}(d)$, only slightly larger $N_s(E_F)$ values in the Zr_2PdH_x phases are needed to explain their larger $(T_{1e}T)^{-1/2}$ and q_E parameters. The Fermi level s-electron states in the Zr_2PdH_x phases probably arise from the Pd atoms although the Pd d-bands lie well below E_F. If the Pd s-band hyperfine contributions are greater for protons in Zr_4Pd_2 sites rather than in the tetrahedral sites, the larger q_E and less negative σ_K value in LT-$Zr_2PdH_{3.2}$ compared to HT-$Zr_2PdH_{2.9}$ can be understood. However, this explanation is probably not totally valid for the similiar features in the a-Zr_2PdH_x samples where protons preferentially occupy[22,23] tetrahedral sites. Several independent studies[22,53,56] show that $N_d(E_F)$ apparently decreases in a-Zr_2PdH_x as the hydrogen content increases. Thus, the less negative σ_K and larger q_E values in Table 5 for the two a-Zr_2PdH_x samples are probably associated with reduced $N_d(E_F)$ in these glassy hydrides. Since a very similar $N_d(E_F)$ reduction has been inferred[10] for a-$TiCuH_x$, this may be a characteristic feature of amorphous transition metal hydrides and should be explored in other glassy systems.

CONCLUSIONS

This paper has presented a review of the proton NMR results that directly compare the properties of crystalline and amorphous metal hydrides with similar compositions. The types of information on structure, diffusion behavior, and electronic properties that can be extracted from the experimental proton NMR parameters have been described. While these studies have shown some rather remarkable differences between the crystalline and amorphous hydride phases, most of the current evaluations have been of a qualitative nature. Additional, more detailed, experiments and supporting quantitative analyses are needed to explain the specific sources of these differences. The existing NMR data should be extended to other suitable metal-hydrogen systems while as many complimentary techniques as feasible should also be brought to bear on these issues.

REFERENCES

1. R.M. Cotts, Ber. Bunsenges. Phys. Chem. 76: 760 (1972).
2. R.M. Cotts, in "Hydrogen in Metals I - Basic Properties", G. Alefeld and J. Volkl, eds., Springer-Verlag, Berlin (1978) p. 227.
3. R.C. Bowman, Jr., in "Metal Hydrides", G. Bambakides, ed., Plenum, New York (1981), p.109.
4. R.G. Barnes, in "Nuclear and Electron Spectroscopies Applied to Material Sciences", E.N. Kaufmann and G. Shenoy, eds., Elsevier, Amsterdam (1981), p. 19.
5. E.F.W. Seymour, J. Less-Common Met. 88: 323 (1982).
6. R.M. Cotts, in "Proc. Int. Symp. on Electronic Structure and Properties of Hydrogen in Metals", P. Jena and C.B. Satterthwaite, eds., Plenum, New York (1983) p. 451.
7. R.C. Bowman, Jr., Hyperfine Inter. (In Press).
8. R.C. Bowman, Jr. and A.J. Maeland, Phys. Rev. B 24: 2328 (1981).
9. P. Panissod and T. Mizoguchi, in "Proc. Rapidly Quenched Metals IV", T. Masumoto and K. Suzuki, eds., Japan Inst. Metals, Sendai (1982) p. 1621.
10. R.C. Bowman, Jr., A.J. Maeland, and W.-K. Rhim, Phys. Rev. B 26: 6362 (1982).
11. R.C. Bowman, Jr., M.J. Rosker, and W.L. Johnson, J. Non-Cryst. Solids 53: 105 (1982).
12. R.C. Bowman, Jr., W.L. Johnson, A.J. Maeland, and W.-K. Rhim, Phys. Lett. A 94:181 (1983).
13. R.C. Bowman, Jr., A. Attalla, A.J. Maeland, and W.L. Johnson, Solid State Commun. 47:779 (1983).
14. R.C. Bowman, Jr., J.S. Cantrell, A. Attalla, D.E. Etter, B.D. Craft, J.E. Wagner, and W.L. Johnson, J. Non-Cryst. Solids 61 & 62: 649 (1984).
15. R.C. Bowman, Jr., J.S. Cantrell, E.L. Venturini, R. Schulz, J.E. Wagner, A. Attalla, and B.D. Craft, in "Rapidly Quenched Metals", S. Steeb and H. Warlimont, eds., Elsevier, Amsterdam (1985) p. 1541.
16. K. Dolde, R. Messer, and U. Stolz, in "Rapidly Quenched Metals", S. Steeb and H. Warlimont, eds. Elsevier, Amsterdam (1985) p. 1553.
17. H.J. Eifert, B. Elschner, and K.H.J. Buschow, Phys. Rev. B 25: 7441 (1982).
18. H.R. Khan and K. Lueders, Phys. Stat. Sol. (b) 108:9 (1981).
19. V.P. Bork, P.A. Fedders, R.E. Norberg, R.C. Bowman, Jr. and E.L. Venturini, Proceedings of this conference.
20. D.K. Ross, P.F. Martin, W.A. Oates, and R. Khoda-Bakhsh, Z. Physk. Chem. N.F. 114:221 (1979).
21. J.J. Rush, J.M. Rowe, and A.J. Maeland, J. Phys. F: Met. Phys. 10: L283 (1980).
22. K. Suzuki, J. Less-Common Met 89: 183 (1983).
23. A. Williams, J. Eckert, X.L. Yeh, M. Atzmon, and K. Samwer, J. Non-Cryst. Solids 61 & 62: 643 (1984).
24. K. Yvon, J. Less-Common Met. 103: 53 (1984).
25. L.E. Drain, Proc. Phys. Soc. 80: 1380 (1962).
26. R.C. Bowman, Jr. and W.-K. Rhim, J. Magn. Reson. 49: 93 (1982).
27. D.L. Anderson, R.G. Barnes, D.T. Peterson, and D.R. Torgeson, Phys. Rev. B 21: 2625 (1980).
28. R.C. Bowman, Jr., E.L. Venturini, and W.-K. Rhim, Phys. Rev. B 26: 2652 (1982).
29. A.J. Maeland and G.G. Libowitz, J. Less-Common Met. 74: 295 (1980).
30. A.J. Maeland, J. Less-Common Met. 89: 173 (1983).
31. R.C. Bowman, Jr., J.S. Cantrell, and A.J. Maeland, to be published.
32. A.J. Maeland, J.J. Rush, and A. Santoro, to be published.
33. I. Bakonyi, L. Takacs, and K. Tompa, Phys. Stat. sol. (b) 103: 489 (1981).
34. M. Belhoul, G.A. Styles, E.F.W. Seymour, T.-T. Phua, R.G. Barnes,

D.R. Torgeson, and D.T. Peterson, J. Phys. F: Met. Phys. , 12: 2455 (1982).
35. T.-T. Phua, B.J. Beaudry, D.T. Peterson, D.R. Torgeson, R.G. Barnes, M. Belhoul, G.A. Styles, and E.F.W. Seymour, Phys. Rev. B 28: 6227 (1983).
36. M. Belhoul, G.A. Styles, E.F.W. Seymour, T.-T. Phua, R.G. Barnes, D.R. Torgeson, R.J. Schoenberger, and D.T. Peterson, J. Phys F: Met. Phys. 15: 1045 (1985).
37. C.A. Sholl, J. Phys. C: Solid State Phys. 14: 447 (1981).
38. J. Shinar, D. Davidov, and D. Shaltiel, Phys. Rev. B 30: 6331 (1984).
39. J. Shinar, J. Less-Common Met. 104: 87 (1984).
40. B.S. Berry and W.C. Pritchet, in "Nontraditional Methods in Diffusion", G.E. Murch, H.K. Birnbaum, and J.R. Cost, eds., Metallurgical Society of AIME (1984) p.83.
41. G.G. Libowitz and A.J. Maeland, J. Less-Common Met. 101:131 (1984).
42. L.E. Hazelton and W.L. Johnson, J. Non-Cryst. Solids 61 & 62: 667 (1984); and Y.S. Lee and D.A. Stevenson, J. Non-Cryst., Solids 72:249(1985).
43. C. Korn and S.D. Goren, Phys. Rev. B 22:2727 (1980).
44. P.M. Richards, Phys. Rev. B27:2059 (1983).
45. R.C. Bowman, Jr. and B.D. Craft (to be published).
46. R.C. Bowman, Jr. and B.D. Craft, J. Phys. C.: Solid State Phys. 17:L477 (1984).
47. G.K. Schoep, N.J. Poulis, and R.R. Arons, Physica 75: 297 (1974).
48. R.R. Arons, H.G. Bohn, and H. Lutgemier, Solid State Commun. 14: 1203 (1974).
49. T.Y. Hwang, R.T. Schoenberger, D.R. Torgeson, and R.G. Barnes, Phys. Rev. B 27:27(1983).
50. G.A. Jaroszkiewicz and J.H. Strange, J. Phys. C:Solid State Phys. 18:2331(1985).
51. J.S. Cantrell, R.C. Bowman, Jr., and G. Bambakidis, Proceedings of this Conference.
52. L. Schlapbach, J. Osterwalder, and T. Riesterer, J. Less-Common Met. 103: 295 (1984).
53. E.L. Venturini, R.C. Bowman, Jr., and J.S. Cantrell, J. Appl. Phys. 57:3542 (1985).
54. J. Tebbe, K. Samwer, and R. Schulz, in "Rapidly Quenched Metals", S. Steeb and H. Warliment, eds. Elsevier, Amsterdam (1985) p. 1581.
55. S.M. Fries, H.-G. Wagner, S.J. Campbell, U. Gonser, N. Blaes, and P. Steiner, J. Phys. F: Met. Phys. 15:1179(1985).
56. M. Kullik, G. V. Minnigerode, and K. Samwer, Z. Phys. B - Condensed Matt. 60:357 (1985).
57. V.L. Moruzzi, P. Oelhafen, A.R. Williams, R. Lapka, H.-J. Guntherodt, and J. Kubler, Phys. Rev. B 27: 2049 (1983).
58. M. Gupta, J. Less Common Met. 101:35(1984).
59. G.C. Carter, L.H. Bennett, and D.J. Kahan, "Metallic Shifts in NMR:, Pergamon, Oxford, 1977.
60. J. Korringa, Physica 16:601(1950).
61. R.C. Bowman, Jr., E.L. Venturini, B.D. Craft, A. Attalla, and D.B. Sullenger, Phys. Rev. B 27: 1474 (1983).
62. M. Gupta and J.P. Burger, Phys. Rev. B 26: 2652 (1982).
63. A.C. Switendick, J. Less-Common Met. 101: 191 (1984) and A.C. Switendick, J. Less-Common Met. 103:309 (1984).
64. D.A. Papaconstanopoulos, B.M. Klein, E.N. Economu, and L.L. Boyer, Phys. Rev. B. 17:141 (1978) and D.A. Papaconstanopoulos, B.M. Klein, J.S. Faulkner, and L.L. Boyer, Phys Rev. B 18: 2784 (1978).
65. R.H. Fairlie, W.M. Temmerman and B.L. Gyorffy, J. Phys. F: Met. Phys. 12: 1641 (1982).
66. H.-J. Eifert, B. Elschner, and K.H.J. Buschow, Phys. Rev. B 29:2905 (1984).

DEUTERON MAGNETIC RESONANCE IN $a\text{-}Zr_2PdD_{2.9}$*

V. P. Bork, P. A. Fedders, and R. E. Norberg
Washington University, St. Louis, Missouri 63130

R. C. Bowman, Jr.
Aerospace Corporation, Los Angeles, California 90009

E. L. Venturini
Sandia National Labs, Albuquerque, New Mexico 87185

INTRODUCTION

Deuteron Magnetic Resonance (DMR) measurements at 30.7 MHz have been performed on an $a\text{-}Zr_2PdD_{2.9}$ sample between 19 and 460 K. A Bruker CXP 200 spectrometer and a cryogenic probe were used to examine Fourier transform line shapes and relaxation rates. Spin-lattice relaxation rates ($\Gamma_1 = T_1^{-1}$) were obtained by the saturation-recovery method. Transverse relaxation rates ($\Gamma_2 = T_2^{-1}$) were measured from line shapes which were Gaussian in the rigid lattice and Lorentzian when well narrowed. A quadrupole echo sequence was used to measure Γ_1 and Γ_2 for temperatures less than 345 K. In the rotating reference frame spin-lock rates $\Gamma_{1\rho}(SL)$ were measured with H_1 locking fields of 10 and 40 G. Below 170 K, decay of quadrupolar order was observed with Jeener-Broekaert[1] pulse sequences.

The metallic glass $a\text{-}Zr_2Pd$ (batch identification RCB-5) had been produced by a previously-described[2] helium atmosphere melt-spinning method from a stoichiometric crystalline alloy. The amorphous deuteride was prepared by slowly reacting 2 to 3 cm strips of the glassy ribbon with approximately one atmosphere of D_2 gas over a total period of six days. The maximum reaction temperature of 450 K was held for two days and was followed by slow cooling to room temperature. A final composition of $a\text{-}Zr_2PdD_{2.7}$ was calculated from the total pressure drop, but similar preparation of a nominal $a\text{-}Zr_2PdH_{2.7}$ sample showed[3] $H_{2.9}$ upon NMR spin count. The magnetic susceptibility of a small part of the ribbon pieces was measured between 7 and 300 K with an S.H.E. SQUID magnetometer, as had been done on the original glassy alloy.[4] The NMR sample was prepared by grinding the remaining ribbons into a powder under a purified argon atmosphere. The powder was immediately loaded into a 7 mm o.d. glass tube, evacuated, and flame sealed. The NMR sample consisted of 210 mg of amorphous deuterated alloy.

---

*Supported in part by NSF Grant 85-03083 and, at Sandia, by USDOE under DE-AC04-76-DP00789.

Figure 1 summarizes the temperature dependences of the DMR relaxation rate data. The deuteron rates are plotted as solid symbols and, for comparison, open symbols indicate corresponding rates reported[5,6] for protons in a-$Zr_2PdH_{2.9}$. At the top of the figure, Γ_2 in each case shows motional narrowing from the rigid lattice. Next a set of rotating frame spin-lock

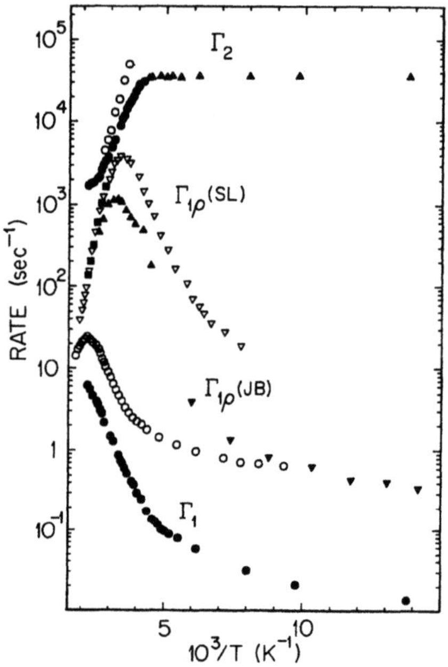

Figure 1. Nuclear relaxation rates for protons (open symbols) and deuterons (solid symbols). Spin lock rates are measured for H_1 fields of 7.3 G (∇), 10 G (■), and 40 G (▲).

data show $\Gamma_{1\rho}$(SL) for protons at 7.3 G H_1 and for deuterons at 10 and 40 G (solid squares and triangles, respectively). Solid inverted triangles indicate DMR Jeener-Broekaert $\Gamma_{1\rho}$(JB) results below 170 K. At the bottom of the figure laboratory frame rates Γ_1 are shown for both samples. Clearly there is an overall similarity between the H and D rates. There are, however, significant differences.

Figure 2 presents, on an expanded scale, the Γ_2 data of Fig. 1. The proton dipolar $\Gamma_2(H)$ narrows exponentially from its (primarily H-H) 1.1×10^5 sec^{-1} rigid lattice value (indicated by a horizontal line). The present DMR results show a Gaussian rigid lattice value (triangles) of $\Gamma_2(D) = 3.6 \times 10^4$ sec^{-1}, much too large to be of dipolar origin. The principal static broadening for deuterons is quadrupolar and arises primarily from interaction with adjacent Zr and Pd. The DMR line narrows near 240 K and quickly becomes Lorentzian (solid circles). The $\Gamma_2(D)$ data narrow with an activation energy of 0.16 eV, significantly smaller than the corresponding 0.25 eV for the proton narrowing. The difference may arise from longer range correlations for the deuteron quadrupolar interaction.

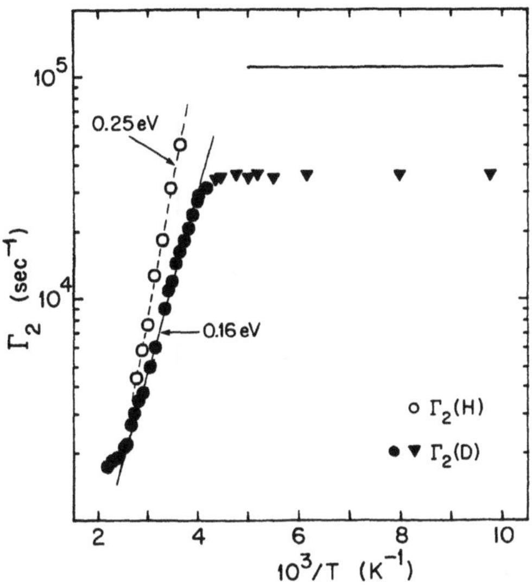

Figure 2. Line narrowing begins near 230 K for both H (circles) and D (solid circles).

The Γ_1 data at the bottom of Fig. 1 level off towards slower variations at lower temperatures, as the nuclear spin-lattice relaxations become dominated by contact with conduction electrons. To a reasonable approximation this limiting nuclear relaxation rate Γ_{1e} is proportional to T. Figure 3 presents a log-log plot of the temperature dependence of the $\Gamma_1(H)$ and $\Gamma_1(D)$ data of Fig. 1. The DMR rates extend to low enough temperature to provide a confident determination of the product [$T_{1e}(D)$ $T = 5340$ sK]. The proton $\Gamma_1(H)$ indicate, with somewhat less certainty, a product [$T_{1e}(H)$ $T = 176$ sK]. If the two samples were identical except for isotopic substitution then to first approximation one would expect the ratio $\Gamma_{1e}(H)/\Gamma_{1e}(D) = \gamma_H^2/\gamma_D^2 = 42.44$. The observed ratio is $5340/176 = 30.3$. Evidently the two samples are only imperfectly comparable insofar as conduction electron relaxation of nuclei is concerned.

As is usual in Pd systems, the inclusion of H or D in a-Zr_2Pd produces significant decreases in the bulk magnetic susceptibility $\chi(T)$. Figure 4 shows results for four ribbon samples. There are $\Delta\chi$ variation $\sim 10^{-7}$ cm^3/g for the a-Zr_2Pd alloys.[4] Inclusion of $x \simeq 2.9$ H and D produces 50% reduction in χ. Since $\chi(T)$ for a-$Zr_2PdD_{2.9}$ is larger than for a-$Zr_2PdH_{2.9}$, there may be an isotope effect. However differences in stoichiometry also may be responsible. Wiley and Fradin have reported[7] an isotope effect in electron-induced nuclear relaxation in stoichiometric PdH and PdD. However it corresponds to only a 4% relative increase in $\Gamma_{1e}(D)$. The $\chi(T)$ increase below 20 K probably arises from Fe impurities.

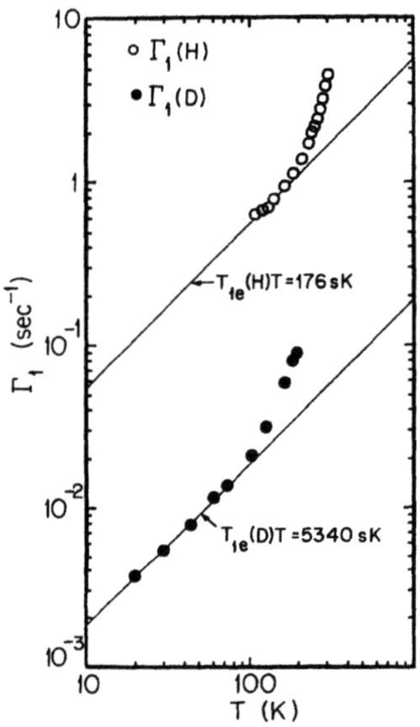

Figure 3. A log-log plot of Γ_1 shows the H and D relaxation by conduction electrons at low temperatures.

After subtraction of the observed nuclear relaxation by conduction electrons, the residual rate $\Gamma_1' = \Gamma_1 - \Gamma_{1e}$ reflects relaxation associated with atomic motions. Figure 5 presents these residual rates $\Gamma_1'(D)$ and $\Gamma_1'(H)$ for the data of Figs. 1 and 3. Between 250 and 400 K the Γ_1' rates for both D and H show the same average activation energy 0.15 eV. Below 160 K the $\Gamma_1'(D)$ data reflect a deuteron hopping with a lower activation energy near 0.058 eV. The proton $\Gamma_1'(H)$ data do not extend into this region, because of $\Gamma_{1e}(H)$ limitations (Fig. 3).

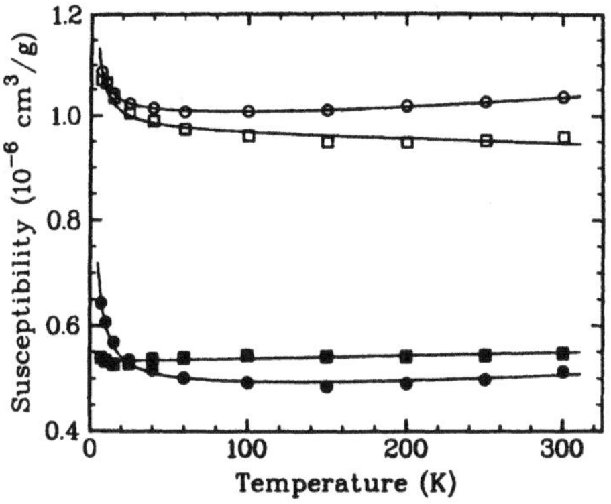

Figure 4. Bulk magnetic susceptibilities for four alloy samples. o a-Zr_2Pd (WLJ-3), □ a-Zr_2Pd (RCB-5), ● a-$Zr_2PdH_{2.9}$ from (WLJ-3), ■ a-$Zr_2PdD_{2.9}$ from (RCB-5).

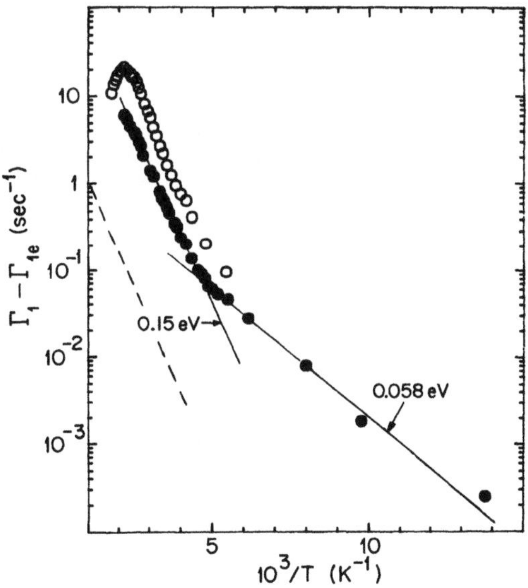

Figure 5. H and D laboratory frame spin-lattice relaxation, after subtraction of electron-induced relaxation. The dashed line indicates D dipolar relaxation rates calculated from the H data.

The dashed line in Fig. 5 indicates the dipolar rate $\Gamma'_{1d}(D)$ calculated for deuterons on the basis of the $\Gamma'_1(H)$ data. The proton relaxation is dominated by H-H interactions, but deuteron dipolar interactions with ^{91}Zr and ^{105}Pd are included with D-D relaxation in the computed $\Gamma'_{1d}(D)$ rate. Assuming that the degree of neighboring site occupancy exclusion is the same for protons and deuterons, the calculated ratio of dipolar rates is $\Gamma'_{1d}(H)/\Gamma'_{1d}(D) = 230$. The calculated dipolar $\Gamma'_{1d}(D)$ provides a negligible contribution to $\Gamma'_1(D)$, which is overwhelmingly quadrupolar in origin. Its interaction strength reflects deuteron site occupancy and quadrupolar interaction with nearby Zr and Pd.

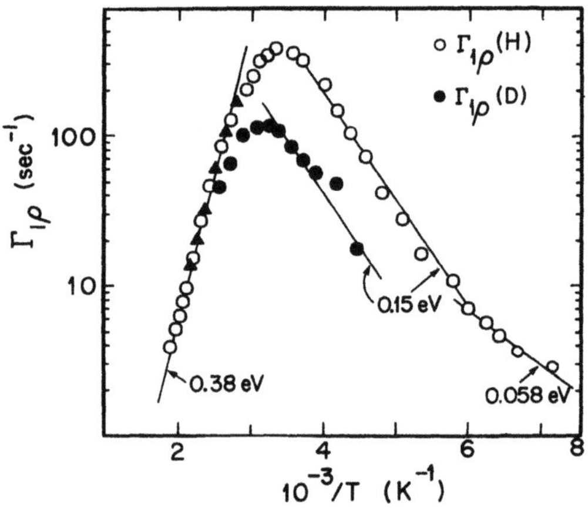

Figure 6. Spin-lock $\Gamma_{1\rho}$ rates for H (7.3 G H_1, open circles) and D (40 G H_1, solid circles; 10 G H_1, solid triangles)

Figure 6 compares rotating frame spin-lock relaxation rates for deuterons and protons. The lines drawn through the data between $10^3/T = 3.5$ and 5.5 in the two samples are parallel to each other and to the line drawn in Fig. 5. Here an activation energy of 0.15 eV appears to fit the Γ'_1 and $\Gamma_{1\rho}$ data for both D and H. At temperatures below 167 K ($10^3/T = 6$) the proton $\Gamma_{1\rho}(SL)$ data are fitted by an 0.058 eV line parallel to that drawn through the low temperature $\Gamma'_1(D)$ data in Fig. 5. Deuteron 40 G H_1 spin-lock measurements (Fig. 6) were restricted to the vicinity of the maximum because of transmitter power limitations. Smaller H_1 fields were unable to maintain a spin temperature over the rigid lattice line and non-exponential rotating frame decay envelopes were observed. In the weak collision regime, single exponential relaxation are observed for 10 and 40 G H_1. $\Gamma_{1\rho}(D)$ and $\Gamma_{1\rho}(H)$ data show similar activation energies $\simeq 0.38$ eV in this warm region.

The solid circles in Fig. 7 show the Jeener-Broekaert deuteron rates $\Gamma_{1\rho}$(JB) (Fig. 1) over the full data range down to 19 K. The smooth curve is drawn as a guide to the eye. At the bottom of Fig. 7 the open circles and straight line show again, for comparison, the laboratory frame Γ_1(D) and limiting low temperature Γ_{1e}(D) fit from Fig. 3. The $\Gamma_{1\rho}$(JB) rate cannot reflect dynamic relaxation because the low temperature $\Gamma_{1\rho}$(JB) rate is more than 25 times Γ_1. $\Gamma_{1\rho}$(JB) rather represents an inhomogeneous

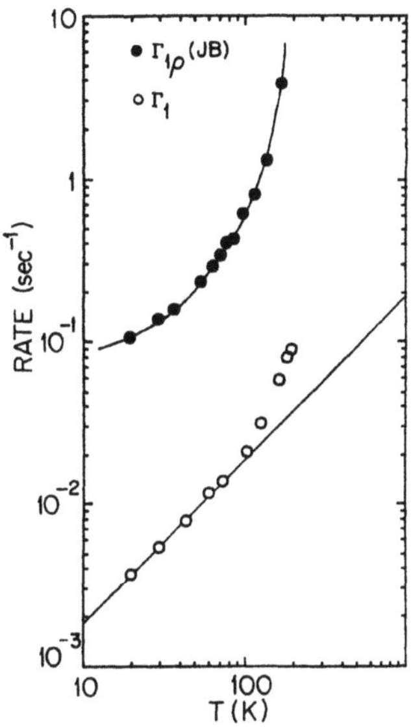

Figure 7. Jeener-Broekaert $\Gamma_{1\rho}$(D) rates (solid circles) down to 19 K. Laboratory frame rates Γ_1(D) from Fig. 3 also are shown (open circles).

broadening associated with a distribution of quadrupolar frequencies $f(\omega_q)$. The observed $\Gamma_{1\rho}$(JB) rates were single exponentials and probably reflect long range electric field gradients from point-like sources. The magnitude and temperature dependence of $\Gamma_{1\rho}$(JB) both decrease as the temperature is decreased. The decreasing rate reflects changes at intermediate range, significantly larger than an atomic scale. It may arise[8] from local concentration gradients and/or strain fields.

Dynamic hole-burning and selective relaxation measurements were made on the DMR line at seven temperatures between 19 and 160 K. Figure 8 presents some typical examples of hole-burning spectra at 18, 82, and 160 K. The experimental sequence consisted of an initial multiple pulse saturating train followed by a relaxation interval close to T_1. Then, in the cases

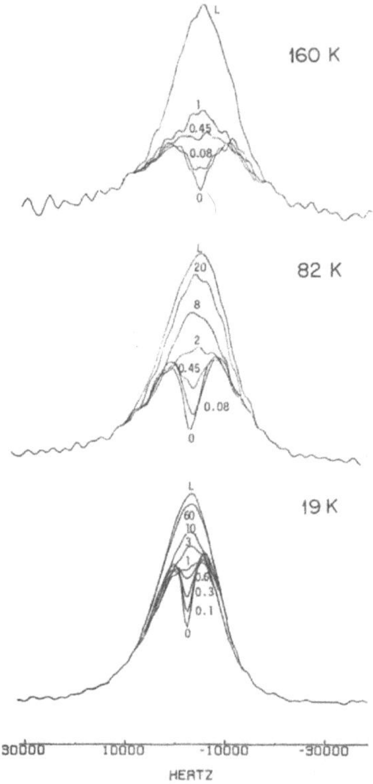

Figure 8. DMR line shapes at various waiting times after hole-burning at 19, 82, and 160 K.

shown in Fig. 8, a 0.3 sec 0.25 G H_1 hole-burning field was applied on-resonance. After a variable delay time T_a, a two-pulse quadrupole echo interrogation sequence was applied. The resulting FTQE line shapes are plotted, as functions of delay times T_a, in Fig. 8. The upper line shape (marked L) at each temperature was obtained at a time T_1 after saturation, without hole-burning.

The HWHM width of the hole is nearly independent of temperature below 100 K, ranging from 0.8 to 1.0 sec^{-1} between 19 and 100 K. At higher temperatures the hole broadens more rapidly as deuteron hopping begins to contribute and brings a distribution of isochromats into play. The extra hole width at 160 K corresponds to a calculated D hopping frequency of $2.6 \times 10^3\ sec^{-1}$.

Figure 9 shows a typical hole-burning magnetization recovery sequence (that for 82 K , from Fig. 8). The difference from equilibrium magnetization $[M_o - M(T_a)]$ is plotted as a function of T_a. The inhomogeneously-saturated hole fills in first, at a rate of 1.3 sec^{-1}, much less than Γ_2 and much larger than Γ_1. This is followed by a homogeneous relaxation of the entire resonance line. The sloping line in Fig. 9 is drawn to correspond to T_1 = 65 sec, from the Γ_1 data of Fig. 1.

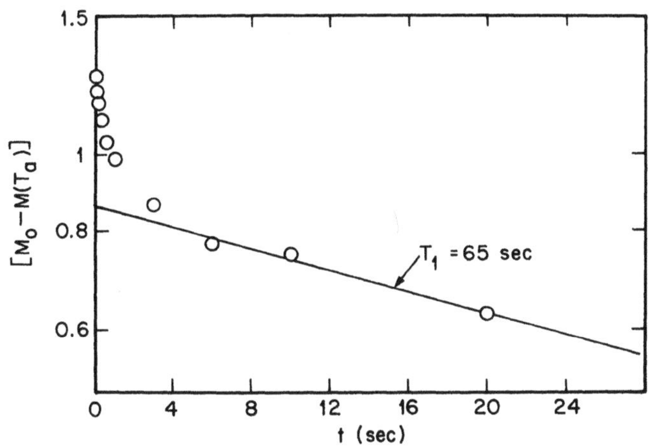

Figure 9. Magnetization recovery after hole-burning at 82 K.

CONCLUSIONS

DMR measurements in a-$Zr_2PdD_{2.9}$ confirm the existence of different activation energies in different temperature regimes, first postulated from a-Zr_2PdH_x results. The existence of definite breaks in the temperature dependences requires a more detailed interpretation than can be provided by an appeal to a continuous distribution of motional correlation functions. Jeener-Broekaert relaxation rates indicate the presence of slow rearrangements at temperatures down to 19 K. Hole-burning experiments provide independent determinations of deuteron hopping frequencies. In general the deuteron relaxation rates reflect modulation of metal-deuteron quadrupole couplings, and may yield site occupancy information upon subsequent analysis.

REFERENCES

1. J. Jeener and P. Broekaert, Phys. Rev. 137, 232 (1967).
2. J. S. Cantrell, J. E. Wagner, and R. C. Bowman, Jr., J. Appl. Phys. 57, 545 (1985).
3. R. C. Bowman, Jr., J. S. Cantrell, E. L. Venturini, R. Schulz, J. E. Wagner, A. Attala, and B. D. Craft, in *Rapidly Quenched Metals*, edited by S. Steeb and H. Warlimont, (Elsevier, Amsterdam, 1985), p. 1541.
4. E. L. Venturini, R. C. Bowman, Jr., and J. S. Cantrell, J. Appl. Phys. 57, 3542 (1985).
5. R. C. Bowman, Jr., J. S. Cantrell, A. Attalla, D. E. Etter, B. D. Craft, J. E. Wagner, and W. L. Johnson, J. Non-Cryst. Sol. 61, 649 (1984).
6. R. C. Bowman, Jr., M. J. Rosker, and W. L. Johnson, J. Non-Cryst. Sol. 53, 105 (1982).
7. C. L. Wiley and F. Y. Fradin, Phys. Rev. B 17, 3462 (1978).
8. J. Shinar, D. Davidov, and D. Shaltiel, Phys. Rev. B 30, 6331 (1984).

HYDROGEN DIFFUSION IN AMORPHOUS $Pd_{80}Si_{20}H_3$ -

A QUASIELASTIC NEUTRON SCATTERING STUDY

R. Hempelmann, G. Driesen and D. Richter

Institut fur Festkörperforschung, Kernforschungsanlage Jülich, 517 Jülich, W - Germany

ABSTRACT

Hydrogen diffusion in amorphous $Pd_{80}Si_{20}H_3$ apparently does not involve a continuous distribution of hydrogen jump rates, but consists of two well-separated classes of jump rates. A two state trapping model allows a semi-quantitative description of our quasielastic neutron scattering data.

INTRODUCTION

For the investigation of the spatial and temporal development of the microscopic hydrogen diffusion mechanism in metals, quasielastic neutron scattering (QNS) is a unique tool. It has been applied first to simple systems like α-PdH_x[1] or α-NbH_x[2] where the H sublattices consist of isoenergetical interstitial sites ("one-level-systems") . A more complex but still ordered H sublattice is found in YH_x[3] with the hydrogen atoms diffusing over octahedral and tetrahedral interstices ("two-level-systems") . In Nb doped with N and H, $\underline{Nb}N_xH_y$, the interstitial sites close to the N impurities are energetically deepened and act as traps for the diffusing hydrogen atoms; thus hydrogen diffusion in $\underline{Nb}N_xH_y$ represents to a first approximation a two-state process[4] : i) in the "free" state the H atoms propagate through the undisturbed regions of the lattice (spending most of their time self-trapped at normal bcc tetrahedral sites), until they are trapped by an impurity; ii) in the "trapped" state they remain in the trap until they are released by thermal activation. In intermetallic hydrides like $Ti_{1.2}Mn_{1.8}H_3$ - where the hydrogen atoms are distributed over several energetically different interstices - the hydrogen diffusion is also characterized by trapping effects, but in addition, due to the high hydrogen concentration, blocking effects become important. A satisfactory description of this behaviour is possible in the framework of a three-state - model[5).

For the hydrogen diffusion process in amorphous metals one intuitively expects a continuous distribution of jump rates. This idea has quantitatively been formulated by Kirchheim[6) (Gaussian model). The surprising result of the QNS study of amorphous $Pd_{80}Si_{20}H_3$ described here, however, is the finding that the hydrogen diffusion process apparently consists of two well separated classes of jump rates (i.e. a bimodal distribution, not a continuous one). In the following we give an intermediate report about this work.

EXPERIMENTAL

The amorphous $Pd_{80}Si_{20}$ sample was prepared by the melt spinning technique. Examination by neutron and X-ray diffraction showed no detectable crystalline fraction. The recrystallisation temperature, as checked by differential thermoanalysis, is 640 K. The sample was hydrogenated by exposing it to 47.7 bar H_2 gas at 373 K within the QNS aluminium sample holder, which was sealed after thermal equilibrium was reached. The resulting composition was $Pd_{80}Si_{20}H_{2.6}$ at 373 K and $Pd_{80}Si_{20}H_{2.9}$ at 332 K, according to pressure composition isotherms by Finocchiaro[7).

The QNS experiments were performed at the high flux reactor of the Institute Laue Langèvin in Grenoble, France. In order to cover a wide dynamical range we employed two spectrometers with different energy resolution. The back-scattering spectrometer IN10 lets us study jump rates between 10^8 and $5 \cdot 10^9$ s^{-1} whereas the time-of-flight (TOF) spectrometer IN5 - operated at 7.84 Å incident neutron wavelength - measures jump rates faster than 10^{10} s^{-1}.

IMMEDIATE RESULTS

Fig. 1 displays typical QNS spectra recorded at IN5 and IN10. The scattering contribution of the H_2 gas and the background scattering of the empty sample holder have been determined separately and already subtracted. Without quantitative evaluation, examination of the spectra indicates two distinctly different classes of jump rates : slow jump rates which cause the broadening of the IN10 spectrum and which appear as an elastic line in the IN5 spectrum, and fast jump rates which give rise to the quasielastic contribution in the IN5 spectrum. These jump rates differ by about a factor 100 .

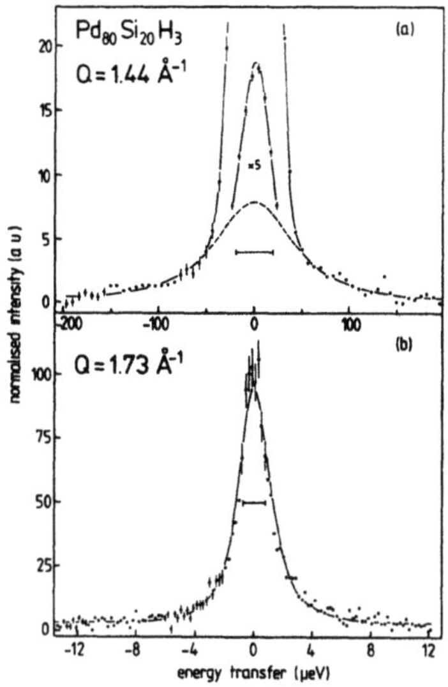

Fig. 1.: QNS spectra of H in amorphous $Pd_{80}Si_{20}H_3$ at 273 K
a) TOF result (IN5), solid line: fit with a superposition of an elastic and a quasielastic scattering contribution (dashed line)
b) backscattering result (IN10). The bars indicate the resolution

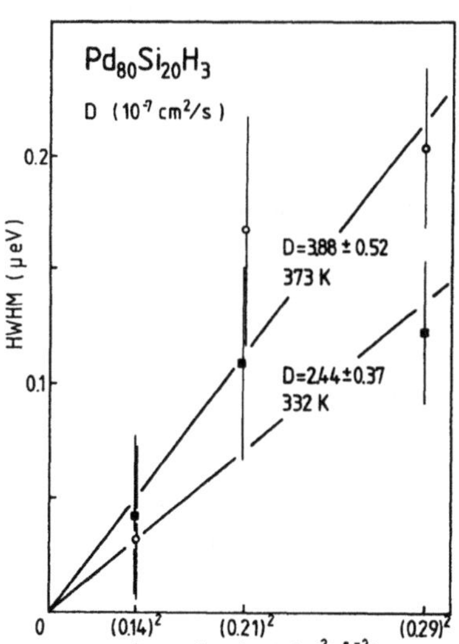

Fig. 2 : Linewidths at small Q, plotted vs Q^2

MODEL-INDEPENDENT DATA EVALUATION

Independent of the host system and of the diffusion mechanism, at sufficiently small Q the diffusional scattering function is a single Lorentzian with the linewidth (HWHM)

$$\Lambda = \hbar D Q^2 \tag{I}$$

where D is the macroscopic tracer diffusion coefficient (which at the low H concentration of our experiment is very close to the chemical diffusion coefficient). This is the so-called Q^2 law, the corresponding plot is shown in Fig. 2 . From the slope we obtain the diffusion coefficient $2.4\cdot10^{-7}$ cm^2/s at 332 K , which compares well with the Gorsky effect results $4.8\cdot10^{-7}$ cm^2/s [8] and $7\cdot10^{-8}$ cm^2/s [9] and the electrochemical result $4.1\cdot10^{-8}$ cm^2/s [10]. From the observation of a reasonable diffusion coefficient and thus from the fact that the scattering function at small Q consists of only one component, we conclude that all hydrogen atoms participate in the diffusion process.

At large Q we can describe our QNS data by a superposition of a broad and a narrow Lorentzian; the latter appears as an "elastic" scattering contribution in the TOF spectra because of the larger resolution widths. The dependences of the resolution corrected linewidths on the momentum transfer Q are presented in the lower part of Fig. 3 , while in the upper part we show the relative weight of the narrow component in the TOF spectra. Astonishingly we observe only a very weak temperature dependence particularly for the broad linewidth. This suggests that instead of two single Lorentzians (jump rates) the spectra consist of two distributions of Lorentzians (jump rates). If one describes a distribution of Lorentzians by a single Lorentzian, the resulting linewidth represents only a fraction of the distribution, and this fraction changes with Q and T, thus obscuring the T dependence.

The vanishing weight of the broad component at small Q (compare validity of Q^2 law as shown above) indicates that the fast hydrogen jump process in amorphous $Pd_{80}Si_{20}$ occurs across spatially confined regions. It is interesting to note, that the weight of the broad component at large Q increases with T. Apparently the fast moving state is reached by thermal activation. Thus the weight function in Fig. 3 may be interpreted as a superposition of the elastic incoherent structure factor (EISF) for the fast motion, which eventually decays to zero at large Q, and a Q independent fraction from those protons which do not participate in the fast jumps. Thereby the EISF may be visualized as the Fourier transform of the spatial region visited by the proton in the fast jumping mode. Since at $Q = 0.75\ Å^{-1}$ most of its decay has already taken place, the spatial extent of the fast motion has to be considerable. For example if we compare our result with the theoretical EISF for diffusion in the bulk of a sphere, its diameter would be of the order of 10 Å .

A possible explanation of the fast jump processes could be a rapid back-and-forth hopping in extended traps. Such trap regions could occur either as random configurations of adjacent sites with a low saddle point energy between them, or due to structural pecularities. In the latter case we could, e.g., consider the existence of local order in the glass similar to that in crystalline Pd_3Si . This intermetallic phase has distorted Pd_6 octahedra, as shown in Fig. 4 . In view of their distortion these sites could be con-

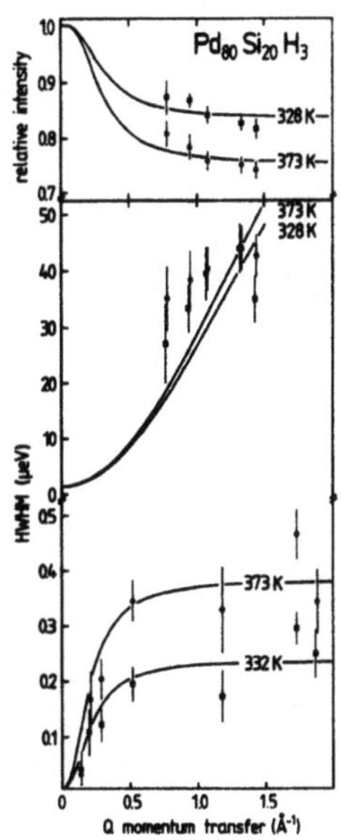

Fig. 3 :
upper part: weight of the elastic component in the IN5 spectrum as a function of temperature and momentum transfer ;

lower part: T and Q dependence of the quasielastic linewidth (HWHM) observed at IN5 and IN10; the solid lines represent a fit of the two-state-model to these data.

sidered as two square-pyramidal interstices sharing the basal plane. A rapid back-and-forth jumping of H atoms between these two sites would then be plausible. This does not appear likely, however, since the spatial extension suggested by our results for the fast localized motion is too large for such a two-site process, and secondly the temperature dependence of the weight function (upper part of Fig. 3) is not consistent with this hypothesis (the fast motional state is reached by thermal activation from sites of lower energy).

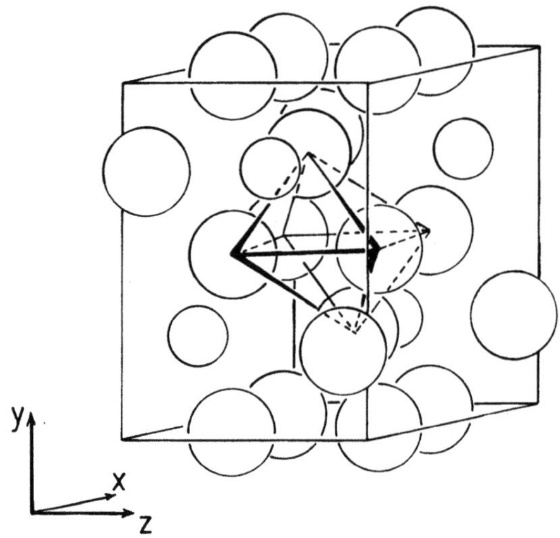

Fig. 4 : Structure of Pd_3Si
(large circles Pd, small circles Si)

Summarizing this qualitative discussion, the following physical picture emerges: Hydrogen diffusion in amorphous $Pd_{80}Si_{20}$ comprises two distinctly different groups of jump rates separated by two orders of magnitude. The presence of "localized" short-range diffusion is not consistent with the Gaussian model mentioned in the introduction. The fast jumping is spatially limited to regions of about 10 Å and is reached by thermal activation from sites of lower energy which we henceforth call traps. We interprete the slow jump processes as jumps out of these traps and the fast jump processes as jumps in between these traps.

TWO - STATE - MODEL

The motional behaviour outlined above is analogous to that observed for hydrogen diffusion in $\underline{NbN_xH_y}$ [4)], in Nb itself at high temperatures[2)] and in $Ti_{1.2}Mn_{1.8}H_3$ [5)]. Hence we attempt to use a relatively simple two-state-model for the microscopic description of the hydrogen diffusion process in amorphous $Pd_{80}Si_{20}$: In the free state a hydrogen atom propagates over a limited number of energetically higher sites with the transport jump rate τ^{-1} ; after a time τ_1 , i.e. with a trapping rate τ_1^{-1} , it is captured in a trap site and leaves this site with the (comparatively) slow escape rate τ_o^{-1} . The model contains two geometrical lengths, the mean individual jump length d and the mean free path in the free state s, which are related by $s^2/\tau_1 = d^2/\tau$. The resulting scattering function[4)]

$$S_{inc}(Q,\omega) = R_1 \frac{\Lambda_1/\pi}{\Lambda_1^2 + (\hbar\omega)^2} + (1-R_1) \frac{\Lambda_2/\pi}{\Lambda_2^2 + (\hbar\omega)^2} \quad (2)$$

consists of two Lorentzian peaks with closely related widths and weights :

$$\Lambda_{1,2} = \frac{\hbar}{2} (\tau_o^{-1} + \tau_1^{-1} + f(Q)) \pm \sqrt{(\tau_o^{-1} + \tau_1^{-1} + f(Q))^2 - 4f(Q)\,\tau_o^{-1}} \quad (3)$$

$$R_1 = \frac{1}{2} + \frac{1}{2} \frac{f(Q)\frac{\tau_1-\tau_o}{\tau_1+\tau_o} - \tau_o^{-1} - \tau_1^{-1}}{\sqrt{(\tau_o^{-1} + \tau_1^{-1} + f(Q))^2 - 4f(Q)\,\tau_o^{-1}}} \quad (4)$$

In view of the topological disorder, the diffusion in the free state is considered to be liquid-like; hence according to Chudley and Elliott[II)]

$$f(Q) = \frac{1}{\tau} \left(1 - \frac{\sin Q d}{Q d} \right) \quad (5)$$

The result of a fit with this model to the observed linewidths and weights is represented by the solid lines in Fig. 3 . It is obvious from the fit that the model allows a semiquantitative description of the hydrogen diffusion process; the resulting microscopic parameters are listed in Table I .

The effective long range diffusion coefficient D_{eff} is reduced relative to the faster diffusion coefficient in between traps $D = d^2/6\tau$ by the factor

Table I : Microscopic parameters for H diffusion in amorphous $Pd_{80}Si_{20}H_3$

	332 K	373 K
transport rate τ^{-1}	$(8.8 \pm 0.8)\ 10^{10}\ s^{-1}$	$(9.5 \pm 0.7)\ 10^{10}\ s^{-1}$
trapping rate τ_1^{-1}	$(1.8 \pm 0.1)\ 10^{9}\ s^{-1}$	$(1.7 \pm 0.1)\ 10^{9}\ s^{-1}$
escape rate τ_o^{-1}	$(3.6 \pm 0.3)\ 10^{8}\ s^{-1}$	$(5.9 \pm 0.5)\ 10^{8}\ s^{-1}$
eff. diff. coeff. D_{eff}	$(7.5 \pm 2.4)\ 10^{-7}\ cm^2/s$	$(12.3 \pm 3.8)\ 10^{-7}\ cm^2/s$
individual jump length d	(1.75 ± 0.25) A	
mean free path s	(11.5 ± 1.3) A	

$\tau_1 / (\tau_1 + \tau_0)$ where τ_1 is the time for fast diffusion within the time unit $(\tau_1 + \tau_0)$. D_{eff} agrees with our estimates from the Q^2 law (Fig. 2) satisfactorily. In addition our evaluation is self-consistent with respect to the mean free path s .

DISCUSSION

QNS data at small momentum transfer are directly related to the macroscopic diffusion coefficient. Comparing our effective diffusion coefficient with literature data, we get reasonable agreement with Gorsky effect results[8] which extrapolated to 332 K and 373 K give $D = 4.8 \cdot 10^{-7}$ and $12.6 \cdot 10^{-7} cm^2/s$; whereas the electrochemical diffusion coefficient at 332 K , $D = 4.1 \cdot 10^{-8}\ cm^2/s$[10] , seems to be a little low.

QNS data at large momentum transfer give insight into the diffusion mechanism (single jump events). Here we observe striking similarities to the H diffusion process in crystalline intermetallic compounds[5] , which also involves trapping and escape processes. Recent NMR papers on H diffusion in amorphous TiCu[12] and amorphous Zr_2Pd[13] report kinks in the Arrhenius representations of the correlation rates τ_c^{-1} . The temperature dependence of these NMR τ_c^{-1} data is very similar to the corresponding curves obtained from hydrogen diffusion in intermetallic hydrides like $LaNi_5H_6$[14] . In particular for $LaNi_5H_6$ several QNS studies have demonstrated that the H diffusion process comprises well separated classes of H jump rates[15,16]. Thus the NMR data may also be interpreted as a signature of distinctly different groups of H jump rates.

It should be pointed out, however, that there are at least two shortcomings to our data evaluation procedure: i) The two types of H sites involved in the diffusion process must unfortunately remain unspecified until we have the necessary structural information. A corresponding neutron spectroscopic study of the local environment of interstitial hydrogen in these materials is presently in progress. ii) Due to the disorder in the host system we cannot, of course, expect two discrete jump rates in amorphous $Pd_{80}Si_{20}H_3$ but rather two well separated distributions of jump rates or linewidths. That our two-state-model takes into account only two distinct jump rates, is certainly a crucial simplification and most probably obscures the T dependence of the jump rates.

CONCLUSION

We have performed a QNS study over a wide dynamical range on the H diffusion in amorphous $Pd_{80}Si_{20}$. Our most prominent result is the discovery of two distinctly separated regimes of jump rates instead of the expected continous distribution. From our model-independent data evaluation we get reasonable H diffusion coefficients and we derived that the fast H motion is spatially restricted to regions of about 10 Å diameter. From the temperature dependence of the weight of the broad component we conclude that the fast localized motion is due to thermal activation from a trapped state and not to back-and-forth hopping in extended traps. A satisfactory description of the QNS data is possible in the framework of a two-state-model for amorphous $Pd_{80}Si_{20}$ in which fast local H diffusion is interrupted frequently by trapping events.

ACKNOWLEDGEMENT

We thank Dr. R. Kirchheim for the preparation of the amorphous metal and for many stimulating discussions. We are also grateful to Drs. I. Anderson and A. Magerl for their help during the neutron scattering experiments in Grenoble and to Dr. J. J. Rush for a critical review of the manuscript.

REFERENCES

1) J. M. Rowe, J. J. Rush, L. A. de Graaf and G. A. Ferguson, Phys. Rev. Lett. 29 , 1250 (1972)

2) V. Lottner, J. W. Haus, A. Heim and K. W. Kehr, J. Phys. Chem. Solids 40 , 557 (1979)

3) I. S. Anderson, A. Heidemann, J. E. Bonnet, D. K. Ross, S. K. P. Wilson and M. W. McKergow, J. Less-Common Met. 101 , 405 (1984)

4) D. Richter and T. Springer, Phys. Rev. B 18 , 126 (1978)

5) R. Hempelmann, D. Richter and A. Heidemann, J. Less-Common Met. 88 , 343 (1982)

6) R. Kirchheim, F. Sommer and G. Schluckebier, Acta metall. 30 , 1059 (1982)

7) R. S. Finocchiaro et al. , Proc. Rapidly Solidified Amorphous and Crystalline Alloys, ed. by B. H. Kehr, B. C. Griessen and M. Coken, Boston 1983 , p. 243

8) B. S. Berry and W. C. Pritchet, Scripta Met. 15 , 637 (1981) and Phys. Rev. B 24 , 2299 (1981)

9) A. H. Verbruggen, R. C. van den Heuvel, R. Griessen and H. U. Künzi, Scripta met. 19 , 323 (1985)

10) U. Stolz, R. Kirchheim, J. E. Sadoc and M. Laridjani, J. Less-Common Met. 103 , 81 (1984)

11) C. T. Chudley and R. J. Elliott, Proc. Phys. Soc. London 77, 353 (1961)

12) R. C. Bowman, A. J. Maeland and W.-K. Rhim, Phys. Rev. B 26,6362 (1982)

13) R. C. Bowman, J. S. Cantrell, A. Attala, D. E. Etter, B. D. Craft, J. E. Wagner and W. L. Johnson, J. Non-Crystalline Solids 61&62 , 649 (1984)

14) F. E. Spada, H. Oesterreicher, R. C. Bowman and M. P. Guse , Phys. Rev. B 30 , 4909 (1984)

15) D. Richter, R. Hempelmann and L. A. Vinhas, J. Less-Common Met. 88 , 353 (1982)

16) C. Lartigue, A. Percheron-Guegan, J. C. Achard, M. Bee and A. J. Dianoux, J. Less-Common Met. 101, 391 (1984)

NEUTRON VIBRATIONAL SPECTROSCOPY OF DISORDERED METAL-HYDROGEN SYSTEMS

R. Hemplemann[1] and J. J. Rush[2]

[1]Institut für Festkörperforschung, Kernförschungsanlage Jülich, 517 Jülich, W. Germany

[2]Institute for Materials Science and Engineering
National Bureau of Standards
Gaithersburg, MD 20899

ABSTRACT

A review is presented of some recent applications of neutron vibrational spectroscopy to the study of disordered metal-hydrogen systems. The examples discussed cover a range of systems from "simple" dilute solutions in bcc or fcc metals to amorphous alloy hydrides. It is shown that neutron inelastic scattering studies of the vibrational density of states provide a powerful and sensitive probe of the local potentials and bonding sites of hydrogen in metals and often reveal critical information on the novel microscopic physical properties and behavior of disordered metals-hydrogen systems, including those influenced by interstitial or substitutional defects.

1. INTRODUCTION

The term "disorder" in metal-hydrogen systems covers a wide range of materials and phenomena. This paper reviews the use of inelastic neutron scattering in the study of hydrogen in metals in a variety of systems discussed according to the complexity of their disorder. We shall start with the simplest form of disorder, the solid solution of hydrogen in simple metals, and shall proceed to studies of almost totally disordered systems, namely hydrogen in amorphous metals. For each system, emphasis will be on a different physical phenomenon to the understanding of which neutron spectroscopic investigations have contributed. So the section on the disordered solid solution of H in "simple" metals (Section 3) deals mainly with the local hydrogen bonding and potential. Additionally, we discuss the isolated mass defect which occurs when a deuterated sample is doped with a small amount of light hydrogen.

In Section 4, disorder is also introduced into the host lattice by means of doping with impurities. For the example of $\underline{Nb}N_xH_y$ we discuss the spectroscopic investigation of the nature of the trapped site, and in the case of $\underline{Nb}Cr_xH_y$, the repopulation with temperature of H among trapped and untrapped states will be described, along with information gained about the energetic structure of the H sublattice. Section 5 concerns H in intermetallic compounds and gives an example of an H site determination by quantitative arguments. Finally systems with the highest degree of disorder are covered in Section 6, where neutron spectroscopic studies of hydrogen in amorphous metals are discussed; here the interstitial H atoms are used as probes of the local environment of the interstitial sites. Finally, we try to sum up the merits of neutron vibrational spectroscopy applied to disordered metal hydrogen systems and risk a brief look into the future.

In order to provide some perspective on the neutron scattering method, we begin our review with a short section on vibrational spectroscopy by incoherent neutron scattering.

2. NEUTRON SPECTROSCOPY

Due to their masses H atoms on interstitial sites in metals perform high frequency optical vibrations which are superposed on the much slower and thus energetically well separated acoustic vibrations of the host lattice. If we disregard direct hydrogen-hydrogen interactions, which is a correct assumption for hydrogen in dilute α phases and a reasonable approximation for many metal hydrides, then the interstitially dissolved hydrogen atoms can be considered as single independent three-dimensional Einstein oscillators, each with three vibrational degrees of freedom. In disordered metal-hydrogen systems the hydrogen atoms are generally distributed over one or more types of site and independently contribute to the neutron scattering intensity. For each H site the incoherent inelastic neutron scattering (IINS) intensity is proportional to the double differential cross section[1]

$$\frac{\partial^2 \sigma_{sin}}{\partial\omega\partial\Omega} = \frac{\sigma^{tot}}{4\pi} \frac{k_f}{k_i} N_H S_{inc}(\bar{Q},\omega) \qquad (1)$$

where σ^{tot} is the total neutron scattering cross section of hydrogen, k_i and k_f the initial and final neutron wave vectors, N_H the number of H atoms on the respective site, and $\bar{Q} = (\bar{k} - \bar{k}_f)$ and ω are the wavevector and frequency of the various normal modes.

If we assume a harmonic form of the hydrogen potential, the scattering function is given by[1]

$$S(\underset{\sim}{Q},\omega) = e^{-2W(\underset{\sim}{Q})} e^{\hbar\omega/2k_BT} \sum_{\substack{n\\m\\\ell}=-\infty}^{\infty} I_n(y_1)\, I_m(y_2) I_\ell(y_3) \cdot$$

$$\delta(\hbar\omega+(n\hbar\omega_1+m\hbar\omega_2+\ell\hbar\omega_3)) ,$$

$$\text{with} \quad y_i = \frac{\hbar Q^2}{2m_H\omega_i} \operatorname{csch} \frac{\hbar\omega_i}{2k_BT} . \qquad (2)$$

In this equation the first term is the Debye Waller Factor[2], the second one is the detailed balance term which interrelates the intensity of energy gain and energy loss processes, and n, m and ℓ are the quantum numbers for the three normal modes of the oscillator.

For the vibrational spectrum, we therefore expect three fundamental peaks for each type of H site; depending on the point symmetry of the respective interstitial sites, two or even all three peaks can be degenerate. The lineshape of vibrational peaks can be Gaussian or Lorentzian-like, depending on whether static disorder or lifetime affects the linewidth.

In many cases, Eq. 2 can be considerably simplified: for small argument, y_i, viz. for $k_BT \ll \hbar\omega_i$, the modified Bessel function can be expanded as

$$I_n(y_i) = (1/2\ y_i)^n/n!\ , \tag{3}$$

which - when we restrict ourselves to fundamental vibrations and neutron energy loss processes - results in

$$S(\underline{Q},\omega) = e^{-2W(\underline{Q})} \sum_{i=1}^{3} \frac{\hbar^2 Q^2}{2m_H h\omega} [n_B(\omega) + 1]\ \delta(\hbar\omega - \hbar\omega_i)\ , \tag{4}$$

where the first term in the sum is called the phonon form factor and n_B is the Bose factor, which for $k_BT \ll \hbar\omega$ vanishes. For neutron energy gain processes, the "+1" in the square brackets is missing, and measurements with $k_BT \ll \hbar\omega$ suffer then severely from intensity problems.

If the oscillator is only weakly anisotropic, Eq. 4 is transformed into

$$S(Q,\omega) = e^{-2W(Q)} \sum_{i=1}^{3} \frac{\hbar^2 Q^2}{6m_H h\omega}\ \delta(\hbar\omega - \hbar\omega_i)\ , \tag{5}$$

with

$$2W(Q) = \frac{\hbar^2 Q^2}{6m} \sum_{i=1}^{3} \frac{\coth(\hbar\omega_i/2k_BT)}{h\omega_i}\ .$$

The main instrument for this type of neutron scattering experiment at steady state neutron sources (research reactors) is the beryllium-filter-spectrometer, where incident neutron energies >25 meV are scanned and the energy of the scattered neutrons is analyzed at a constant value of ≈3 meV by means of a cold beryllium filter. For this type of spectrometer, the observed scattering intensity is to good approximation directly proportional to the H content, so each type of H site in the sample gives rise to three peaks (sometimes degenerate) with equal intensity. Because of this simplicity and because of the analogy to IR and Raman spectroscopy, the incoherent inelastic neutron scattering (IINS) technique is often called neutron spectroscopy, either focussed on vibrational excitations, as in the present system, or more generally on vibrational or rotational modes for condensed molecular systems.

The energetic range accessible to neutron spectroscopy has recently been considerably expanded, when the spallation neutron sources at Argonne and Los Alamos (USA), Tsukuba (Japan) and Didcot (UK) went into operation.[3] They allow energy transfers larger than 1 eV, and this might be useful in the future.

3. DISORDERED SOLID SOLUTION OF H IN "SIMPLE" METALS

3.1 Hydrogen Potential

Neutron spectroscopy on hydrogen in metals directly measures the frequencies of the hydrogen vibrations which are related to the hydrogen potential. Considerable progress in achieving a quantitative understanding of metal-hydrogen potentials has been achieved recently for ordered binary metal hydrides[4-7] particularly by comparing the fundamental vibrations of hydrogen isotopes with their higher harmonics which have been measured in some cases up to fifth order at neutron spallation sources.[8]

In general less quantitative insight has been achieved thus far in dilute metal-hydrogen systems. Because of its simplicity (cubic symmetry, regular octahedral site occupancy) and its importance as a prototype metal-hydrogen system we will first consider recent results on α-PdH_x[9]. Let us consider a hydrogen atom on the cubic octahedral site, as in fcc palladium. If we allow for small deviations from harmonicity and regard only the lowest order non-vanishing anharmonic terms, the hydrogen potential near the equilibrium position can be described phenomenologically as

$$V(x,y,z) = c_2(x^2+y^2+z^2) + c_4(x^4+y^4+z^4) + c_{22}(x^2y^2+y^2z^2+z^2x^2). \quad (6)$$

In this expression, the second term only influences the magnitude of the energies, whereas the third term also brings about a coupling between the vibrations in the three normal directions. If the corrections to the harmonic shape of the potential are sufficiently small, the energy eigenvalues E_{nm} for a particle of mass m oscillating in this potential are obtained in first order perturbation theory as

$$E_{nm\ell} = \hbar\omega_o(n+m+\ell+3/2) + \beta(n^2+n+m^2+m+\ell^2+\ell+3/2)$$
$$+\gamma\{(2n+1)(2m+1)+(2m+1)(2\ell+1)+(2\ell+1)(2n+1)\}, \quad (7)$$

where

$$\omega_o = \sqrt{c_2}/m \ , \ \beta = \frac{3\hbar^2 c_4}{4mc_2} \ , \ \gamma = \frac{\hbar^2 c_{22}}{8\ c_2 m}$$

and where n, m and ℓ are the quantum numbers for the vibrations in the x, y, and z directions, respectively. Hence the first excitation energies $\varepsilon_{nml} = E_{nml} - E_{ooo}$ are:

$$\varepsilon_{100} = \varepsilon_{010} = \varepsilon_{001} = \hbar\omega_o + 2\beta + 4\gamma \text{ (fundamental vibration)},$$
$$\varepsilon_{200} = \varepsilon_{020} = \varepsilon_{002} = 2\hbar\omega_o + 6\beta + 8\gamma \text{ (first overtone)},$$
$$\varepsilon_{110} = \varepsilon_{011} = \varepsilon_{101} = 2\hbar\omega_o + 4\beta + 12\gamma \text{ (combination mode)}. \quad (8)$$

Experimentally, vibrational modes of H in Pd $H_{0.005}$ were observed at 69 meV, 138 meV and 156 meV, as shown in Fig. 1. (These measurements and those on other dilute H in metal systems discussed below, were recently made possible by the use of a crystal spectrometer using a highly sensitive beryllium detector analyzer.) By a comparison of these results with Eq. 8, the parameters $\hbar\omega_o$, β and γ could be determined for α - PdH_x and resulted in $\hbar\omega_o = 50$ meV, $\beta = 9.5$ meV and $\gamma = 0$. The

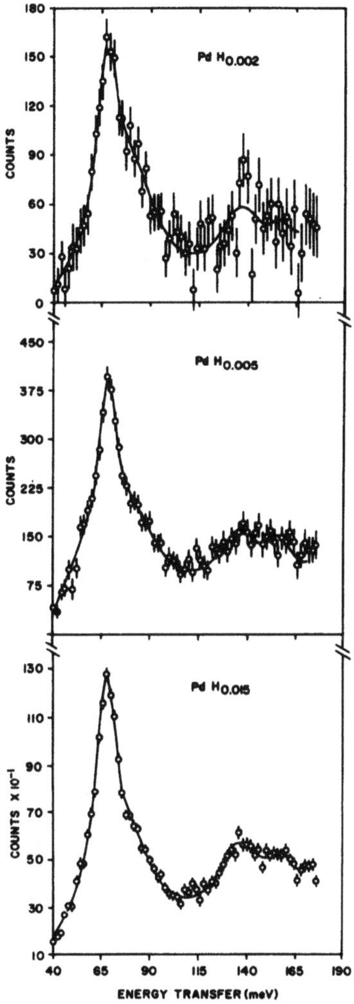

Fig. 1. Vibrational spectra of H in α PdH_x

vanishing γ value indicates a negligible coupling of the H vibrations in different directions. A positive β value and thus a positive c_4 in Eq. 6 means that in the energy range of the neutron scattering experiment, the hydrogen potential is stiffer than a harmonic potential. The harmonic frequency ω_o and the anharmonicity parameter β, both corrected for the isotope mass ratio, also describe satisfactorily the fundamental frequency of α - PdD_x. Since one anharmonicity parameter determines the frequency of the overtone and the isotope effect of the fundamentals, the hydrogen potential is equal for both isotopes: there is no evidence for different electronic forces.

These results have important consequences for the understanding of the superconductivity of the PdH(D) system. It has been shown[10] that the existence of relatively high T_c's in this system (T_c = 9K for $PdH_{1.0}$ and T_c = 11K for $PdD_{1.0}$) is the result of a strong coupling between electrons and optical phonons, the first time that a strong electron phonon interaction has been shown to involve optical modes. The neutron results in Fig. 1 have demonstrated clearly that the origin of the anomalous isotope effect is not due to changes in the electronic band structure[11] which produces a different potential for PdH and PdD. Instead, the neutron spectroscopic results confirm the view of Ganguly[13-14] that the anharmonicity of the optic hydrogen vibrations brings about the anomalous T_c.

Since the energy levels of the anharmonic oscillator H in Pd differ from those of an harmonic oscillator, the oscillatory partition function also differs from the harmonic value. Because of the missing coupling term it can be evaluated as the third power of a one dimensional partition function:

$$Z = [\sum_n \exp(-\varepsilon_n/k_B T)]^3 , \qquad (9)$$

with

$$\varepsilon_n = \hbar\omega_o(n+1/2) + \beta(n^2+n+1/2) . \qquad (10)$$

For the oscillatory part of the entropy(at 400K, e.g.) this results in 1.25 k_B instead of the harmonic value 1.37 k_B. The total entropy of solution then diminishes by about 5% (this effect increases with decreasing temperature). Thus it is evident that for a quantitative microscopic understanding of thermodynamic properties of metal hydrogen systems the anharmonicity of the hydrogen potential has to be taken into account. Along these lines, recent thermodynamic (p-c-T) measurements by Lässer[15] showed that the agreement between theory (statistical thermodynamics) and experiment is improved when taking into account the anharmonicity of the potential.

Concerning a more general theoretical understanding of the H potential in metals, we emphasize again that the power expansion mentioned above is just a phenomenological description. In a much more sophisticated approach by Fukai and Sugimoto[16,17,40] the H potential is taken as a superposition of Born-Meyer potentials centered at the metal atom position; then the Schrödinger Equation is solved for the H atoms vibrating in this potential. A single Born-Meyer potential turned out to be inadequate, so two Born-Meyer potentials were taken at each metal atom. However, this approach is also purely phenomenological (the parameters of the Born-Meyer potentials are fitted to the experimental vibrational frequencies), and in fact the experimentally observed overtone frequencies are not reproduced. A considerable theoretical advance has recently been achieved by Ho et al,[18] who performed first-principles total energy calculations for ordered β-NbH with use of the pseudopotential approach within the local density-functional formalism. The total energy was successively calculated for stepwise displacements of the hydrogen sublattice relative to the metal sublattice ("frozen phonon"), and thus the hydrogen potentials in the stiff and the soft directions of the tetragonal tetrahedral site were obtained. The H frequencies (fundamentals and first overtones) in these two (one-dimensional) potentials are in excellent agreement with experimental results for ordered refractory-metal hydrides. However, as described below they do not provide a satisfactory description of the unusual vibration spectra observed for H bound as interstitial defects in bcc metals.

3.2 Hydrogen as Local Mass Defect

Light-atom mass defects substituted in an otherwise perfect crystal destroy the translational symmetry of the system and at the same time strongly interact with the host medium. This interaction produces local-vibrational modes which can be shifted significantly from the lattice vibrations of the host. The frequency ω_ℓ of the localized mode is given by[19,20]

$$\frac{1}{\omega_\ell^2} = \frac{M_h - M_\ell}{M_\ell} \int \frac{g(\omega)d\omega}{\omega_\ell^2 - \omega^2} \qquad (11)$$

where $g(\omega)$ is the normalized density of states of the host lattice and $M_{\ell(h)}$ is the mass of the defect (host) atom. Usually, the insertion of a mass-defect atom requires the insertion of chemically and thus electronically different atoms in the crystal which introduces a local electronic perturbation along with the dynamical perturbation due to dissimilar masses. However, if hydrogen is introduced into a deuterium

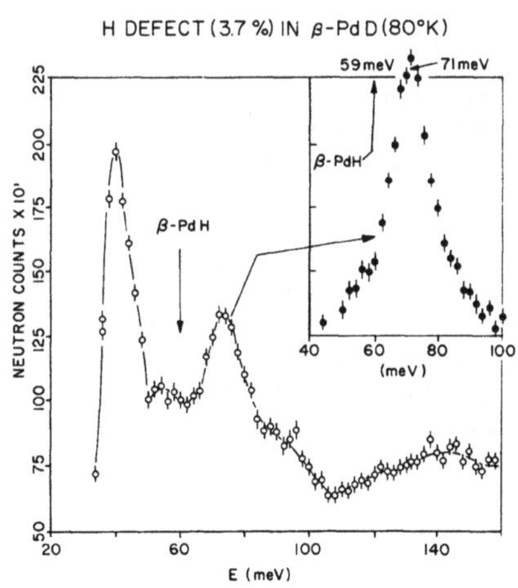

Fig. 2. Neutron energy loss spectrum for β-palladium deuteride containing 3.7 at. %H. The spectrum measured for the deuteride before introducing the H defects is also shown as a dashed line. The position of the dominant density-of-states peaks for β-$PdH_{0.6}$ is indicated by the arrow.

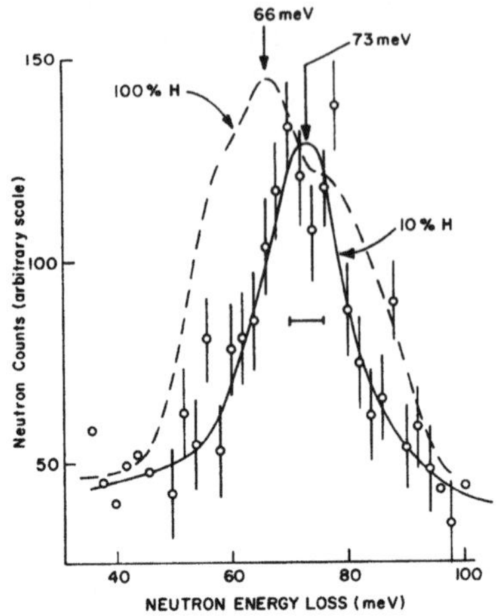

Fig. 3. Neutron energy loss spectra at 80K for a saturation coverage of H on Pt black (dashed line) and of 90%D and 10%H (experimental points and solid line).

sublattice, a huge relative mass difference is achieved without electronic distortion. The palladium-hydrogen system again offers an excellent prototype to study such a pure mass defect. Fig. 2 presents recent neutron spectroscopic results[21] for β-$PdD_{0.6}$ and for the same deuteride sample doped with 3.7 at. % H, recorded at 80 K. While the deuteride density of vibrational states is unaffected by the hydrogen defects, a strong additional peak is observed at (71±1) meV in the spectrum of the H doped samples. (See insert in Fig. 2). This peak is far displaced from the density-of-states peak of β-$PdH_{0.6}$, and is even higher in energy than that observed for dilute interstitial hydrogen in α-phase palladium (Fig. 1). Using the measured deuteride density of states as shown in Fig. 2, Eq. 11 predicts a local-mode energy of 70.5 meV. The excellent agreement with the experimental value demonstrates that the simple mass-defect theory is sufficient to describe the observed effect.

A recent application of this isotope dilution neutron spectroscopy[22] concerns H atoms chemisorbed on the surface of Pt. Fig. 3 shows the vibrational spectra of 100% H and of 90% D + 10% H on Pt. Due to a high incoherent neutron scattering cross section, H atoms on substrates with large surface areas can be investigated by means of neutron vibrational spectroscopy[23], whereas D atoms are much more difficult to probe. So for the H isotope mixture coverage in Fig. 3 only the H defect mode is visible and, in order to apply Eq. 11, the pure 100% H spectrum was scaled down by $\sqrt{2}$ to derive the deuteride density of states, assuming harmonic forces.

Using this scaled spectrum, Eq. 11 predicts an H oscillator energy of 72.8 meV, in excellent agreement with the peak observed at (73±1) meV for 10% H. The observed "defect" peak is broadened beyond instrumental resolution, since defects at this concentration are far from isolated. The comparison demonstrates that the complex spectrum for the H overlayer on Pt is primarily due to dispersion of a vibrational mode associated with a single type of site on the Pt surface (which in turn reflects strong dynamic interaction of H atoms), and not to the existence of different surface H sites or surface impurities.

3.3 Hydrogen in the α-Phase of bcc Metals

Recently measurements have also been made for the first time of local modes of hydrogen in bcc refractory metals at concentration ≲1%. These spectra, which are shown in Fig. 4, are quite different in character from the results for α-PdH_x and reveal several highly unusual features. First, the widths of the observed vibration peaks are in every case considerably wider (> 10 meV) than the instrumental resolution, in contrast to the very narrow linewidths anticipated for the vibrations of dilute interstitial defects. The spectral peaks (two modes are expected for H in a bcc tetrahedral site, one at low-energy and a doubly degenerate mode at higher energy) are most broadened for $VH_{0.012}$, with somewhat smaller but still substantial widths observed for H in Nb and Ta. The "peak" energies in meV observed at 295 K for the three α-phase samples are: 106 and ~170 ($VH_{0.012}$), 107 and 163 ($NbH_{0.03}$) and 114 and 154 meV ($TaH_{.037}$). A second unusual feature of these results is that the high energy mode peak for TaH_x (Figure 4) increases in energy from 154 to 163 meV as x is changed from 0.03 to 0.18, while the lower energy mode remains unchanged. A third striking aspect, not shown in the figure, is the anomalous shift of the lower excitation peak in $NbH_{0.005}$ to higher energy as the temperature is lowered toward the α-δ phase boundary at ~200°K.

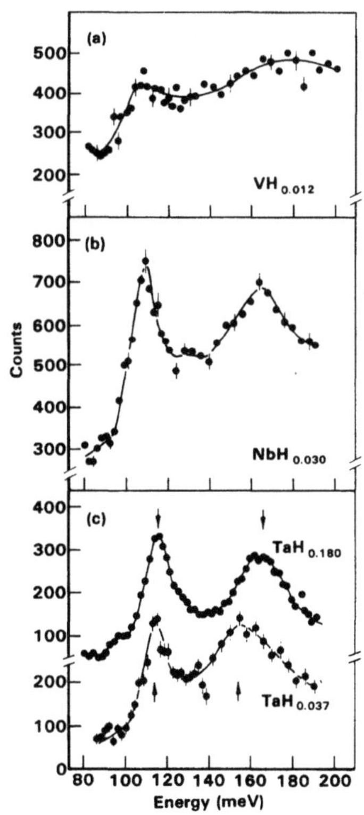

Fig. 4. Spectra of H local modes for $VH_{0.012}$, $NbH_{0.03}$ and TaH_x at 295°K.

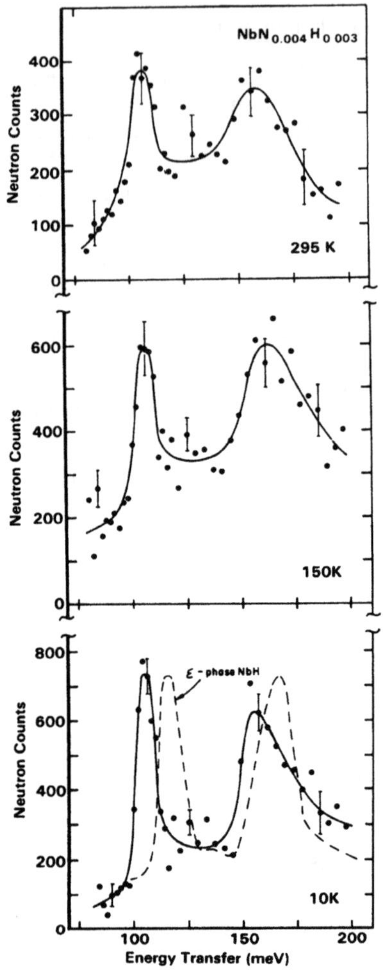

Fig. 5. Spectra from $NbN_{0.004}\ H_{0.003}$ at 3 different temperatures; dashed curve represents ε phase results.

After careful analysis of these results and comparison with theoretical predictions[5], it was concluded that the broad vibrational linewidths are associated with a delocalization of the excited vibrational states of the interstitial defects. In fact an estimate of the ground state tunnel splittings for self trapped protons in refractory metals suggests excited state "splittings" ~10 meV, which are in qualitative agreement with the observations. The anomalous shift of the low-energy density-of-states band for $NbH_{.005}$ toward the position of the ordered (δ) phase peak suggests the existence of fluctuating short-range order similar to the ordering in the low temperature phase.

4. HYDROGEN IN METALS DOPED WITH IMPURITIES

4.1 Hydrogen Determination by Qualitative Arguments

Interstitial impurities such as oxygen or nitrogen in niobium act as trapping centers for dissolved hydrogen and suppress hydride precipitation at low temperatures.[26] Figure 5 shows the vibration spectra of hydrogen atoms trapped by nitrogen impurities at various temperatures.[27] The most striking feature of these results is the invariance of the peak position with temperature. The frequencies at 295K, where less than half of the hydrogen is trapped[28] show very little change down to 10K, where all protons are in the trapped state. So the local vibrations of hydrogen atoms trapped at nitrogen (or oxygen[27]) impurities in niobium closely resemble the frequencies of the dilute α-NbH_x system and provide clear evidence that the sites occupied by trapped protons are strongly related to tetrahedral positions in a relatively undisturbed niobium environment. In particular, there is no evidence of any direct electronic force between the impurity and the hydrogen atom which would severly affect the tetragonal symmetry of the force field seen by the hydrogen atom. The conclusion of tetrahedral site occupation and the lack of evidence for triangular site occupation disagrees with the earlier suggestions[29] that hydrogen atoms are distributed among four tetrahedral and four triangular sites in the (100) plane as proposed earlier by Birnbaum and Flynn[30] for hydrogen in pure niobium. The experimental results are also inconsistent with the proposal of Fukai and Sugimoto[16] that the hydrogen atoms are in the 4T configuration, i.e., each hydrogen atom is delocalized over the four tetrahedral sites on a cube face and forms a common self-trapping distortion leading effectively to the symmetry of an O site. Finally, the location of the hydrogen atom between the T and the O sites as proposed on the basis of channeling experiments on TaN_xD_y[31] can be excluded.

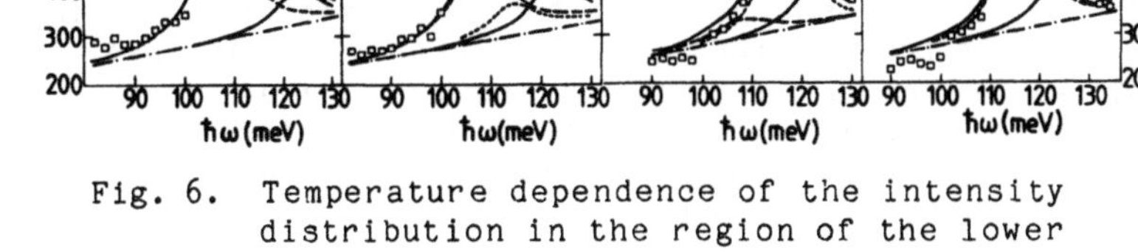

Fig. 6. Temperature dependence of the intensity distribution in the region of the lower fundamental vibration in $NbCr_{0.009}H_{0.009}$. The upper line represents the result of a fit with the combined trapping and precipitation model (see text). The broken line represents the α-phase contribution, the dashed line the intensity from protons in the hydride phase, while the lower solid line displays the intensity originating from protons in the trap. Finally, the dot-dashed line shows the background level.

4.2 Hydrogen Occupancy

An instructive example of the capability of neutron spectroscopy to detect quantitatively the occupancy of a certain interstitial site is the work by Richter et al.[32] on hydrogen trapping in $Nb_{0.99}\,Cr_{0.01}\,H_{0.009}$. Figure 6 presents the results of this experiment and shows the gradual intensity transfer from a high-temperature peak around 106 meV to a higher energy peak at lower temperatures. A second peak develops near 123 meV and grows with decreasing temperature. At 210K separate peaks are not distinguishable. Below 210K a new center of vibrational intensity develops around 117 meV which takes up all intensity at low T. Obviously, in this system both trapping and hydride precipitation take place: the binding energy ΔE at the trap is apparently smaller than the formation enthalpy ΔE of the precipitate. In order to describe the data quantitatively, the temperature dependence of the boundary H solubility in the α phase was taken to be

$$C_\alpha^{max} = C_o \exp(-\Delta H/k_B T) \quad , \qquad (12)$$

with C_o = 4.5 [H]/[Nb] and ΔH = 120 meV was taken from Pfeiffer and Wipf[26] as well as their result that each impurity is associated with one trapping site. In thermal equilibrium the chemical potentials of trapped and dissolved protons must be equal,

$$KT\ell n \frac{C_\alpha}{6} = -\Delta E + kT\ell n \frac{C_x}{C_i - C_x} \quad . \qquad (13)$$

C_α and C_t are the H concentrations in the α phase and in the trapped state, respectively, and C_i is the impurity concentration. For the α phase it was taken into account that in Nb the number of tetrahedral sites is 6 times higher than the number of Nb atoms and that $C_\alpha \ll 6$. From Eq. 13, the temperature dependence of C_t can be determined if one replaces C_α by C_α^{max} (Eq. 12) at those (low) temperatures where the C_α calculated from Eq. 13 exceeds the H solubility limit in the α phase. The result of a fit with this model is indicated by the various lines in Fig. 6. It is evident that the combined trapping and precipitation model allows a good description of the experimental data. In particular it accounts for the gradual growth of the trapping peak below 295K as well as for the sudden intensity increase of the central γ phase peak below 210K. For the trapping energy, a value of ΔE = 105 ± 10 meV is derived, which is slightly smaller than the formation enthalpy of the hydride precipitate mentioned above.

In summary, these examples of hydrogen trapping studies demonstrate that neutron spectroscopy can provide direct information about the type of H sites involved, i.e. about the geometric structure of the H sublattice, and about the energy differences between the different sites, i.e. about the energetic structure of the H sublattice.

5. HYDROGEN SITES IN INTERMETALLIC COMPOUNDS

A considerable number of intermetallic hydrides have been investigated by means of neutron spectroscopy, e.g. $LaNi_5H_x$, $TiFeH_x$, TiCuH, $ZrCoH_x$, $ZrNiH_x$, $Ti_{1.2}Mn_{1.8}H_x$, $ZrBe_2H$, Zr_2NiH_x and others.[2] The interpretation of the data, however, is mostly qualitative. In the following we present recent results on $LaNi_5H_x$, for which a more quantitative evaluation of neutron spectroscopic data has been made.[33]

The localized mode spectrum of β-$LaNi_5H_6$ is rather broad, with scattering intensity ranging from 60 to 180 meV; separate peaks are not visible. Similar large widths of the localized modes are observed in practically all intermetallic hydrides and may be partially due to mutual H-H interactions (dispersion). In addition, however, activation-induced defects make a substantial contribution to the linewidth. If however, an α phase sample is prepared without ever being transformed to any hydride phase, then well defined vibrational peaks are oserved, e.g., in $LaNi_5H_{.15}$[33] or in $ZrNiH_{0.6}$.[34]

For the assignment of the different vibrational peaks to certain hydrogen sites a simple lattice dynamical model was used. The frequencies of localized hydrogen vibrations in metals are obtained by solving the eigenvalue problem

$$\underline{D}\ \underline{\hat{u}} = \omega^2 \underline{\hat{u}} \tag{14}$$

where the eigenvectors $\underline{\hat{u}}$ indicate the directions of the vibrational displacements. To a good approximation we can take the metal atoms as completely immobile; then $\underline{D}$ is a 3 x 3 dynamic matrix. The elements D_{ij} of this matrix represent the force acting on the hydrogen atom in the direction i if it is displaced in the direction j. The force constants are generally expressed by appropriate second derivatives of the potential. The restoring force $\underline{F}$ can be modelled approximately by longitudinal springs $\underline{S}^n = f^n \underline{\hat{S}}^n$ to the Z metal atoms of the first coordination shell (e.g., octahedron);

$$\underline{F} = \sum_{n=1}^{Z} f^n (\hat{u} \cdot \hat{s}^n) \quad \hat{s}^n \ . \tag{15}$$

The vectors $\hat{s}^n$ indicate the directions of the Z springs; their magnitude is given by the distance-dependent force constants

$$f^n = f^n(r^n) = \left(\frac{c^M}{r^n} \right)^2 , \tag{16}$$

where r^n is the distance between the hydrogen atom and the metal atom n and the values c^M characterize the hydrogen-metal interaction; the latter are adjustable parameters. The $1/r^2$ dependence of f was chosen in order to reproduce the 1/r dependence of the hydrogen frequencies. This dependence has been observed experimentally in a large variety of binary metal hydrides with the CaF_2 structure[35,36] and has also been proposed on the basis of theoretical considerations as outlined at the end of section 3.1.[36] Simple algebra transforms Eq. 15 into the form of

Eq. 14 with

$$D_{ij} = \sum_{N=1}^{z} \hat{s}_i^n \hat{s}_j^n f^n \qquad i,j = 1,2,3 \ . \qquad (17)$$

For α-$LaNi_5H_{0.15}$ the atomic coordinates and the metal-hydrogen distances for all hypothetical sites were obtained from the well-known structure of the host lattice (space group P6/mmm). Using two adjustable parameters, C^{La} and C^{Ni}, a frequency distribution can be calculated and compared with the experimental spectrum (Fig. 7). The result of this comparison suggests that hydrogen in α-$LaNi_5H_{0.15}$ occupies both octahedral 3f and the tetrahedral 6m sites, in contrast to the results of previous neutron diffraction work[37], in which only the octahedral site was detected. Thus it can be seen that at low H concentrations neutron spectroscopy can often be superior to neutron diffraction in the determination of H sites.

6. HYDROGEN AS A LOCAL PROBE IN AMORPHOUS METALS

The capability of neutron diffraction to determine hydrogen (deuterium) positions in metals is lost when the system of H sites to be investigated does not exhibit translational symmetry. Therefore structural investigations on hydrogenated amorphous metals by means of neutron scattering can only be interpreted in terms of pair correlation functions. As an example, we present for glassy Ti_2NiD_x[40] in Fig. 9 the Bathia-Thornton type neutron total structure factor S(Q) defined as

$$\left(\frac{d\sigma_{coh}}{d\Omega}\right)_{total} = N \langle b^2 \rangle \, S(Q) \qquad (18)$$

[where $(d\sigma_{coh}/d\Omega)$ is the angular differential cross section for coherent neutron total scattering and $\langle b^2 \rangle$ is the average of the coherent neutron scattering lengths of the constituent nuclei in the glass], together with the reduced atomic distribution function $G^1(r)$ obtained as the

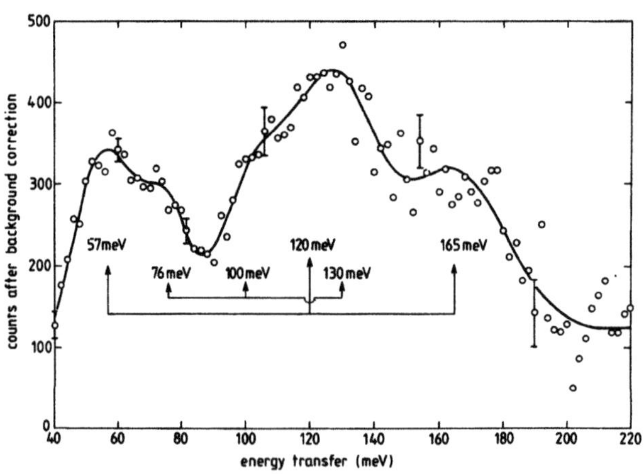

Fig. 7. Localized hydrogen vibration in virgin bulk α-$LaNi_5H_{0.15}$.

Fourier transform of S(Q) truncated at $Q_{max} = 25\ Å^{-1}$, according to

$$G^1(r) = 2/\pi \int_0^{Q\,max} Q\{S(Q)-1\}\sin(Qr)dQ \quad . \tag{19}$$

Ti_2Ni as a host metallic glass is a favorable system for this kind of investigation because the scattering lengths $b_{Ti} = -0.34\cdot10^{-12}$cm and $b_{Ni} = 1.03\cdot10^{-12}$cm have opposite sign and bring about a very small value of $\langle b\rangle = 0.11$; thus $NiTi_2$ is nearly a so-called "neutron zero alloy", i.e., the Laue background scattering is very low and atom-atom correlations are more easily detectable. The precision of the Fourier transformation in Eq. 19 depends strongly on the magnitude of Q_{max}. A value of 25 $Å^{-1}$ requires an incident neutron energy of about 500 meV which is not available at steady state reactors, but only at spallation neutron sources. A negative peak at r = 2.56 Å in the reduced atomic correlation function $G^1(r)$ of hydrogen-free Ti_2Ni metallic glass is attributed to the preference of Ni-Ti unlike atom pairs. This peak overlaps with a D-D positive peak in the deuterated samples originating from tetrahedron-tetrahedron correlations ($r_{DD} \approx r_{NiTi}$). The negative peak at r = 2.0 Å is interpreted as an overlap of D-Ti negative and D-Ni positive peaks, whereas the positive peak at r = 3.2 Å is associated with the overlap of D-D tetrahedron-octahedron and D-D octahedron-octahedron correlations.

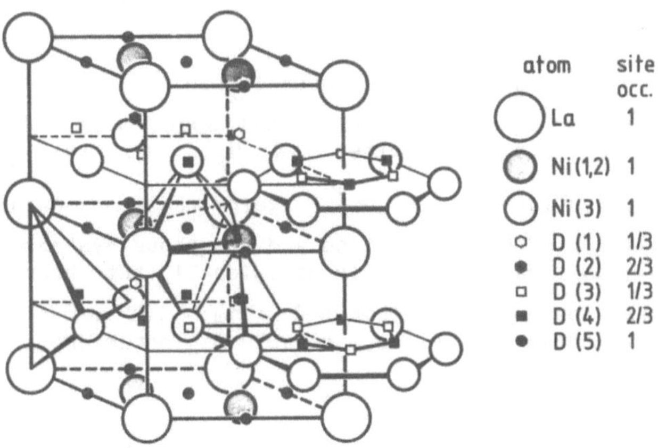

Fig. 8. Structure of β-$LaNi_5D_7$; the D sublattice consists of La_2Ni_4 octahedra (circles), La_2Ni_2 tetrahedra (squares) and Ni_4 tetrahedra (hexagons); the latter two have alternately high (full symbol) and low (open symbol) occupancies in both half cells.

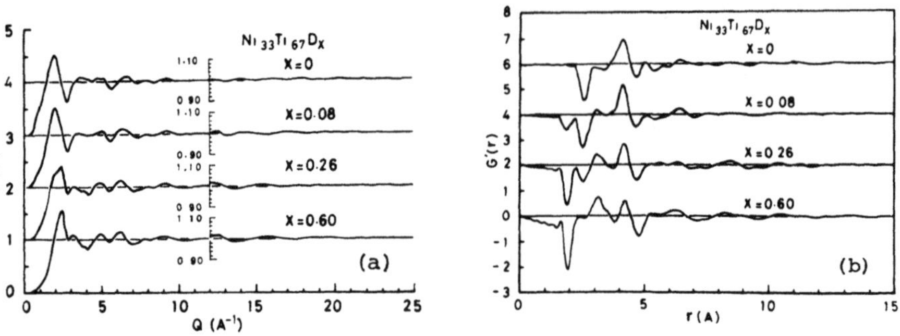

Fig. 9. a) Bathia-Thornton type neutron total structure factors of $Ni_{0.33}Ti_{0.67}D_x$ (X=0, 0.08, 0,26, 0.60) glass hydrides. b) Reduced atomic correlation functions.

The preceding example in our opinion demonstrates how indirect is the information from neutron diffraction about hydrogenated metallic glasses, even in favorable cases. By contrast, the interpretation (qualitatively at least) of neutron spectroscopic data is straightforward. Fig. 10 presents first measurements by Rush et al.[41] who for the first time compared the vibrational spectra of a glassy/crystalline metallic alloy, TiCuH. In crystalline TiCuH, hydrogen occupies a Ti_4 tetrahedral site with tetragonal point symmetry.[42] Thus, just for symmetry reasons, the vibrational spectrum of H in TiCuH consists of two peaks with the intensity ratio 2:1. For amorphous TiCuH, the peak occurs at the same energy but with strongly increased width. Thus, on the average, the H sites occupied in the amorphous and the crystalline structures are very similar. The large width, however, indicates a range of tetrahedral distortions and perhaps fluctuations in chemical composition. The considerable density-of-states below 100 meV may also show that octahedral sites are occupied as well.

Coming back to the example of amorphous Ti_2NiH, we present in Fig. 11 a comparison of the hydrogen vibrational spectra for the binary hydride TiH_2, the intermetallic hydrides $TiNi_2H_{2.7}$ and $TiNi_2H_{1.5}$, and for the amorphous "hydride" $TiNi_2H_{1.5}$.[40] Again, the fundamental vibrations are very similar in the crystalline and amorphous systems, clearly indicating that H in all cases primarily occupies Ti_4 tetrahedral sites. The intensity in the low energy region below $\hbar\omega$ = 100 meV suggests a hydrogen fraction in octahedral sites, probably Ti_6 octahedrons.

7. CONCLUSION AND OUTLOOK

The aim of this review was to give a flavor of neutron vibrational spectroscopy and its application to disordered metal-hydrogen systems as well as to demonstrate that such studies yield important microscopic information which cannot be obtained any other way.

The hydrogen vibrational spectra provide direct information on the strength of the metal-hydrogen interaction and therefore on the hydrogen potential. Accurate knowledge of the hydrogen potential is of fundamental importance for the understanding of many properties of metal-hydrogen systems, e.g., thermodynamic behavior (chemical potential), diffu-

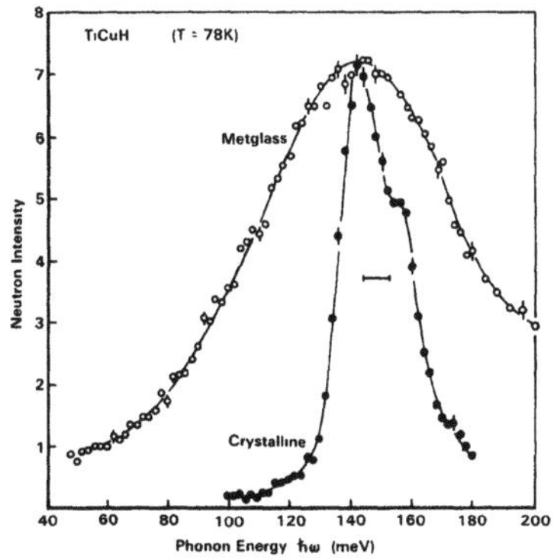

Fig. 10. Neutron spectra measured at 78K for crystalline $TiCuH_{0.93}$ and amorphous $TiCuH_{1.3}$. The energy resolution (FWHM) near the peak is indicated by the horizontal bar.

sion processes, superconductivity, etc. Up to now only the bottom of the potential well has been carefully probed. An extrapolation of this parameterized potential to higher energies (as is necessary for a precise calculation of thermodynamic properties at or above room temperature via the partition function) is doubtful. Here we anticipate future valuable contributions will come from neutron spallation sources which produce the necessary high neutron energies to study the higher levels in the potential.

As another example, the widths of the vibrational peaks can originate from a number of effects, including a dispersion of the optical phonons in bulk metals or on metal particles, anharmonicity, or the presence of impurities. It has recently been demonstrated that neutron inelastic scattering using isotope dilution methods can be a sensitive probe of the origin of vibrational lineshapes.

The neutron scattering intensity of a vibrational peak associated with H in a particular site is directly proportional to the number of H atoms on this site. Thus neutron spectroscopy allows a study of the changes of population with temperature for different kinds of H sites, which in turn gives information about the "energetic structure" of the H sublattice. As a method for studying the "geometric structure" of the H sublattice, neutron spectroscopy is complimentary* to neutron diffraction. In crystalline metal hydrogen systems at high H(D) concentrations neutron diffraction is the method of choice. For disordered solutions of hydrogen at low concentrations we have demonstrated, however, that neutron spectroscopy can be superior.

Hydrogen "sublattices" with disorder - either energetic disorder in trapping systems or in disordered metal alloys, or both energetic and topological disorder in amorphous systems - do not exhibit translational symmetry, and thus neutron diffraction becomes a less powerful method. In such cases the application of neutron spectroscopy using hydrogen as a microscopic probe of local topology appears to be a promising and prospering experimental technique and we look forward to many more interesting and informative investigations in the future.

*As a result of the neutron scattering properties of hydrogen isotopes, neutron spectroscopy is most sensitive to the light H isotope, whereas neutron diffraction is better performed on the D isotope.

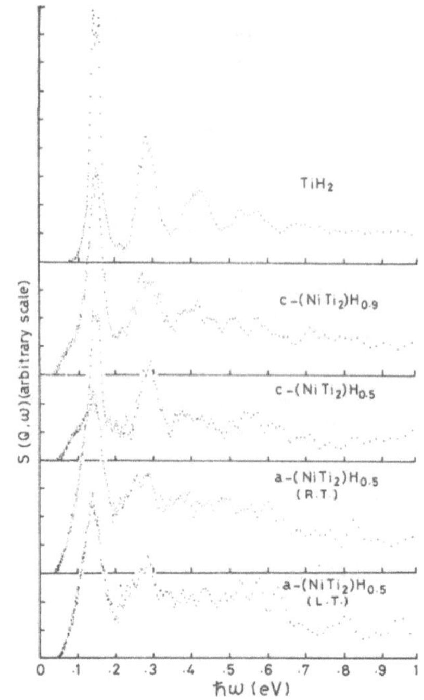

Fig. 11. Vibrational spectra of hydrogen in TiH_2, crystalline $Ti_2NiH_{2.7}$ and $Ti_2NiH_{1.5}$, recorded at room temperature, and those in amorphous $Ti_2NiH_{1.5}$ recorded both at room temperature and at 40K.

REFERENCES

1. See any textbook for neutron scattering.

2. See e.g., D. Richter, R. Hempelmann, and R. C. Bowman, Hydrogen in Intermetallic Compounds, Topics in Applied Physics, L. Schlapbach (Ed.), Springer, Berlin, (1986).

3. A description of neutron scattering instrumentation is beyond the scope of this article; see for example: Topics in Current Physics, Vol. 3, ed. by S. W. Lovesey and T. Springer (Springer, Berlin, Heidelberg, New York 1977); C. G. Windsor, Pulsed Neutron Scattering (Taylor and Francis, London 1981); J. W. White and C. G. Windsor, Rep. Prog. Phys. 47, 707 (1984).

4. D. Richter and S. M. Shapiro, Phys. Rev. B22, 599 (1980).

5. R. Hempelmann, D. Richter and A. Kollmar, Z. Phys. B44, 159 (1981).

6. J. J. Rush, A. Magerl, J. M. Rowe, J. M. Harris and J. L. Provo, Phys. Rev. B24, 4903 (1981).

7. J. Eckert, J. A. Goldstone, D. Tonks and D. Richter, Phys. Rev. B27, 1980 (1983).

8. S. Ikeda, N. Watanabe and K. Kai, Physica B120, 131 (1983).

9. J. J. Rush, J. M. Rowe and D. Richter, Z. Phys. B55, 283 (1984).

10. See e.g., the review by B. Stritzker and H. Wühl in G. Alefeld and J. Völkl (Eds.), Hydrogen in Metals II, Topics in Applied Physics, Vol. 29 (Springer, Berlin 1978), p. 243.

11. R. J. Miller and C. B. Satterthwaite, Phys. Rev. Lett. 34, 144 (1975).

12. E. Wicke, J. Less Common Met. 101, 17 (1984); J. Blanrock, Thesis, Münster, 1985.

13. B. N. Ganguly, Z. Phys. 265, 433 (1973); Phys. Rev. Lett. B14, 3848 (1976).

14. D. A. Papaconstantopoulos and B. M. Klein, Phys. Rev. Lett. 35, 110 (1975); D. A. Papaconstantopoulos, B. M. Klein, E.N. Economou and L. L. Boyer, Phys. Rev. B17, 141 (1978).

15. R. Lässer, Proc. Int. Meeting Hydrogen in Metals, Belfast (UK) 1985, Z. Phys. Chem. NF., in press.

16. Y. Fukai and H. Sugimoto, Trans. Jap. Inst. Met., Suppl., 21, 41 (1980).

17. H. Sugimoto and Y. Fukai, Phys. Rev. B 22, 670 (1980).

18. K.-M. Ho, H.-J. Tao and X.-Y. Zhu, Phys. Rev. Lett. 53, 1586 (1984).

19. Lord Raleigh, Theory of Sound (Dover, New York 1945), Vol. 1.

20. R. J. Elliot and A. A. Maradudin, in Inelastic Scattering of Neutrons (IAEA, Vienna 1965), Vol. 1, p. 231.

21. J. J. Rush, J. M. Rowe and D. Richter, Phys. Rev. B 31, 6102 (1985).

22. J. J. Rush, R. R. Cavanagh, R. D. Kelly and J. M. Rowe, J. Chem. Phys. 83, 5339 (1985).

23. For reviews see, e.g., R. R. Thomas, in P. Day (Ed.), Emission and Scattering Techniques, D. Reidel Publ. Co., 1981, p. 25.

24. A. Magerl, J. J. Rush and J. M. Rowe, Phys. Rev. B, in press.

25. R. C. Casella, Phys. Rev. B 27, 5943 (1983).

26. G. Pfeiffer and H. Wipf, J. Phys. F 6, 167 (1976).

27. A. Magerl, J. J. Rush, J. M. Rowe, D. Richter and H. Wipf, Phys. Rev. B 27, 927 (1983).

28. D. Richter and T. Springer, Phys. Rev. B 18, 126 (1978).

29. P. E. Zapp and H. K. Birnbaum, Acta Metall. 21, 865 (1973).

30. H. K. Birnbaum and C. P. Flynn, Phys. Rev. Lett. 37, 25 (1976).

31. H. D. Carstanjen, Phys. Status Solidi A 59, 11 (1980).

32. D. Richter, J. J. Rush and J. M. Rowe, Phys. Rev. B 27, 6227 (1983).

33. R. Hempelmann, D. Richter, G. Eckhold, J. J. Rush, J. M. Rowe and M. Montoya, J. Less-Common Met. 104, 1 (1984).

34. M. J. Benham, J. D. Browne and D. K. Ross, J. Less-Common Met. 103, 71 (1984).

35. D. K. Ross, P. F. Martin, W. A. Oates and R. Khoda Bakksh, Z. Phys. Chem. NF 144, 221 (1979).

36. H. Sugimoto and Y. Fukai, J. Phys. Soc. Japan. 51, 2554 (1982) and J. Phys. F11, L137 (1981).

37. D. Fischer, A. Furrer, G. Busch and L. Schlapbach, Helv. Phys. Acta 50, 421 (1977).

38. P. Thompson, J. J. Reilly, L. M. Cooliss, J. M. Hastings and R. Hempelmann, J. Phys. F, in press.

39. C. Lartique, A. Percheron-Guegan, J. C. Achard and J. L. Soubeyroux, J. Less-Common Met., in press.

40. K. Kai, S. Ikeda, T. Fukunaga, N. Watanabe and K. Suzuki, Physica 120B, 342 (1983).

41. J. J. Rush, J. M. Rowe and A. J. Maeland, J. Phys. F 10, L283 (1980).

42. A. Santoro, A. J. Maeland and J. J. Rush, Acta Cryst. 34B, 3059 (1978).

HYDROGEN IN AMORPHOUS Cu_xTi_{1-x} ALLOYS :

NEUTRON DIFFRACTION AND COMPUTER SIMULATION STUDIES

Bernard Rodmacq, Philippe Mangin[*,+], Luc Billard
and André Chamberod

DRF - Service de Physique, Métallurgie Physique,
CEN Grenoble, 85 X, 38041 Grenoble Cédex, France

[+] Institut Laue-Langevin, 156 X, 38042 Grenoble
Cédex, France

ABSTRACT

The technique of neutron diffraction has been used to study the structure of CuTi amorphous alloys and hydrides. Advantage has been taken of the existence of positive and negative neutron scattering lengths. The first result is that the as-quenched amorphous alloys exhibit a tendency to heterocoordination. In a second step the substitution of deuterium for hydrogen lets us identify the supplementary peaks appearing in the total pair-correlation functions of CuTiH and CuTiD alloys. These new correlations agree quite well with the corresponding ones in crystalline CuTi hydrides. From these results, a computer model of the structure of these alloys is built up. The simulated interference functions and pair-correlation functions are in good agreement with those obtained experimentally for both as-quenched and hydrogenated or deuterated alloys. The thermal evolution is followed by small-angle and large-angle neutron scattering and confirms the low thermal stability of these amorphous hydrides.

INTRODUCTION

Among the various techniques that can be used to study the structural and dynamic properties of amorphous hydrides, neutron scattering is one of the most sensitive and, in the case of coherent elastic scattering, has numerous advantages over conventional X-ray diffraction measurements. The first interest comes from the fact that, contrary to the X-ray case where the scattering factor is roughly proportional to the atomic number, no such systematic variation exists for the coherent neutron scattering length. As an example, deuterium and hydrogen have neutron scattering lengths of the same order of magnitude as those of metal atoms. Another advantage comes from the existence of positive and <u>negative</u> neutron scattering lengths depending on the element (as an example deuterium has a positive scattering length and hydrogen has a negative one).

---

* Permanent address : Laboratoire de Physique du Solide (CNRS, LA 155), Université de Nancy I, BP 239, 54506 Vandoeuvre, France.

Table 1. (i) sign of the new peak in g(r) due to H absorption
(ii) sign of the new peak in g(r) due to D absorption
(iii) possible X-H correlations (X=A,B or H)
(iv) possible modifications of the matrix

i	ii	iii	iv	
+	+	H-H	A-A	B-B
+	–	B-H		
–	+	A-H		
–	–		A-B	

The most interesting case will be the one of an AB amorphous alloy (CuTi for example) with positive and negative neutron scattering lengths (b_{Cu} = 0.77 x 10^{-12} cm and b_{Ti} = – 0.33 x 10^{-12}cm) charged with either hydrogen (b_H = – 0.374 x 10^{-12} cm) or deuterium (b_D = 0.667 x 10^{-12} cm). For hydrogen, the combination of a small mass and a large incoherent scattering cross-section leads to large inelastic effects which make the correction procedures more difficult. But the great advantage is that the hydrogen-deuterium substitution will permit an unambiguous distinction between metal-metal and metal-hydrogen correlations. This point is illustrated in Table 1, where it is supposed that the introduction of hydrogen or deuterium in an AB alloy with $b_A > 0$ and $b_B < 0$ leads to the appearance of a supplementary peak in the total pair-correlation function. The different possibilities for the origin of this peak are either hydrogen-hydrogen, hydrogen-metal or metal-metal correlations. If one excludes the possibility of new metal-metal correlations (as one **certainly can** at low hydrogen concentration) such a hydrogen-deuterium substitution will **lets** us distinguish between A-H, B-H and H-H correlations. Moreover, the non-negligible weight of the metal-hydrogen and hydrogen-hydrogen correlations will make such a distinction easier.

In this chapter we present results of neutron diffraction and computer simulation studies on hydrogenated and deuterated Cu_xTi_{1-x}($0.35 < x < 0.67$) amorphous alloys. The first part will serve as an illustration of the above considerations and presents the evolution of the interference functions and pair-correlation functions as a function of alloy composition and hydrogen or deuterium content[1]. In the second part we will use these experimental results to build up a computer model of these alloys[2]. The last part will present the thermal evolution of a $Cu_{0.50}Ti_{0.50}H_{0.15}$ amorphous alloy as followed by large-angle and small-angle neutron scattering[3,4].

EXPERIMENTAL

Amorphous Cu_xTi_{1-x} (x = 0.67, 0.50 and 0.35) alloys were prepared by the melt-quenching technique in the form of ribbons 8 mm wide and 40 μm thick. The hydrogenation and deuteration were carried out a room temperature by electrolysis. The sample was used as the cathode and a platinum ribbon as the anode in a solution of 0.2 M H_2SO_4 (D_2SO_4) in H_2O (D_2O) with a few drops of CS_2 as a poison. We used a current density of 10^3 Am^{-2} and varied the hydrogen or deuterium content by changing the electrolysis time. The weight of the samples was measured before and after loading to obtain the hydrogen or deuterium concentration.

Neutron measurements were carried out at Institut Laue-Langèvin in Grenoble. Large-angle measurements were performed on D2 instrument with a wavelength of 0.94 Å ($0.47 < q < 11.8$ Å^{-1}) and with accumulation times of about eight hours for each sample. The total measured intensities were corrected according to the standard normalisation procedure. An empirical correction was also introduced to account for the large decrease of the intensity versus diffraction angle in the case of hydrogenated samples[1]. The corrected curves were then normalised to the theoretical scattering cross-section to give the coherent intensity per atom I_a (q). Finally the interference function S(q) was obtained as

$$S(q) = I_a(q)/\langle b^2 \rangle$$

where $\langle b^2 \rangle = \Sigma c_i b_i^2$ with c_i = concentration of element i and b_i = coherent scattering length of element i.

Small-angle measurements were performed on both D1B ($0.17 < q < 3.34$ Å^{-1}) and D17 ($0.8 \times 10^{-2} < q < 6.3 \times 10^{-2}$ Å^{-1}) instruments. In this case the samples were heated from 300 K to 800 K at a constant rate of 0.7°C/min and diffraction spectra were recorded during the heating process with an accumulation time of 10 minutes.

Computer simulations were performed on a PRIME 9950 computer by use of an extension of previous simulation studies on metal-metalloid amorphous alloys[5].

INFLUENCE OF ALLOY COMPOSITION AND HYDROGEN OR DEUTERIUM CONTENT

As-quenched Cu_xTi_{1-x} Alloys

Figure 1 presents the interference functions S(q) for the three

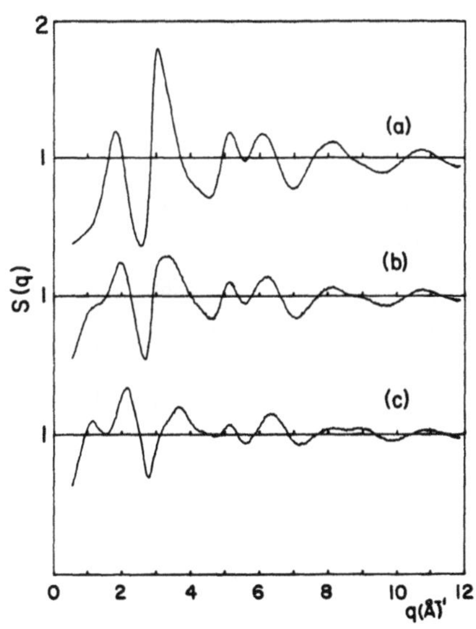

Fig. 1 Interference functions S(q) for amorphous (a) $Cu_{67}Ti_{33}$, (b) $Cu_{50}Ti_{50}$ and (c) $Cu_{35}Ti_{65}$ alloys.

concentrations studied. They are very similar to those obtained by Sakata et al.[6] and Fukunaga et al.[7] . Curve (c) has an usual shape for amorphous alloys with an intense peak at about 3 $\AA^{-1}$. The presence of a prepeak at about 1.8 $\AA^{-1}$ is characteristic of chemical order. This prepeak can be regarded as a superlattice peak and comes from the fact that if A atoms in an AB alloy prefer to be surrounded by B atoms, a pseudo-periodicity, ABAB, will appear in the alloy, and will lead to the appearance of a peak at small q values. The intensity of this prepeak is enhanced in the case of NiTi[8] and CuTi alloys where smaller atoms have the larger scattering length.

In addition to this prepeak, a shoulder appears at smaller q values (fig. 1.b) and transforms into a well-defined peak at 35 % Cu (fig. 1.c). We will see in the next section that this peak can be attributed to a small hydrogen contamination.

The total interference function can be written

$$S(q) - 1 = \sum_{i,j} \frac{c_i c_j b_i b_j}{\langle b^2 \rangle} (S_{ij}(q) - 1)$$

where $S_{ij}(q)$ stands for the three partial interference functions S_{CuCu} (q), S_{CuTi} (q) and S_{TiTi} (q). The very rapid modification of the total S(q) curves as a function of composition is easily explained by the fact that, in contrast with the X-ray case where the scattering factors of Cu and Ti are close to each other, the neutron scattering lengths of Cu and Ti are of opposite signs. Thus the total S(q) at 67% Cu is dominated by the Cu-Cu contribution whereas the one at 35 % Cu is dominated by the Cu-Ti contribution but with a negative sign.

The total interference function S(q) is related to the pair-correlation function g(r) by a Fourier transform :

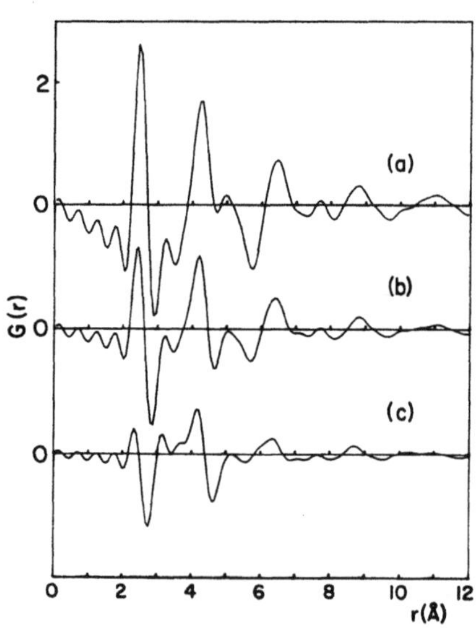

Fig. 2 Reduced pair-correlation functions for amorphous (a) $Cu_{67}Ti_{33}$, (b) $Cu_{50}Ti_{50}$ and (c) $Cu_{35}Ti_{65}$ alloys.

$$q(S(q) - 1) = 4 \pi \rho_o \int_o^\infty r(g(r)-1) \sin qr \, dr$$

where ρ_o is the atomic density of the material. Figure 2 shows the variation of the reduced pair-correlation functions $G(r) = 4\pi\rho_o r(g(r)-1)$ as a function of composition. As for the corresponding interference functions of figure 1, the total pair-correlation function is dominated by the Cu-Cu correlations ($r_o \simeq 2.5$ Å) at 67 % Cu and by the Cu-Ti ones ($r_o \simeq 2.7$ Å) at 35 % Cu. One can also notice that the second maximum at $r \simeq 4.2$ Å in figure 2a is very intense as compared to the first peak. It corresponds to Cu-Cu next-nearest neighbours ($r_2/r_1 \simeq 1.7$) and is related to the prepeak in the corresponding S(q) of figure 1a. This is again in agreement with the existence of chemical order in these alloys as already pointed out by Sakata et al.[9] for CuTi alloys and Boudreaux[10] for NiTi alloys.

Hydrogenated and Deuterated CuTi Alloys

Figure 3 shows the interference functions obtained for $Cu_{0.50}Ti_{0.50}(H,D)_{x_o}$ alloys. The main feature of these curves is that the shoulder at about 1 Å^{-1} in curve (c) increases and shifts towards smaller q with hydrogen absorption. This evolution is very similar to that observed in figure 1 as a function of titanium concentration. On the contrary, this small q signal disappears when deuterium is introduced into the sample (figure 3d). Such a variation means that this small-angle signal arises from hydrogen-titanium rich regions extending over a few interatomic distances. The same behaviour has also been observed for the two other compositions studied, and anomalous small-angle X-ray scattering studies on $Cu_{0.67}Ti_{0.33}H_{0.40}$ alloys on both copper and titanium edges have confirmed the existence of such regions[11].

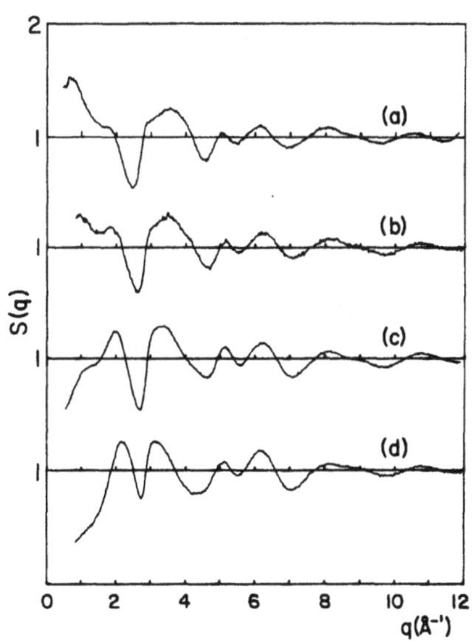

Fig. 3 Interference functions S(q) for amorphous $Cu_{0.50}Ti_{0.50}(H,D)_x$ alloys : (a) $x_H = 0.33$, (b) $x_H = 0.15$, (c) $x = 0$, (d) $x_D = 0.20$.

The corresponding pair-correlation functions are shown in figure 4. Although the overall shape of these functions does not vary drastically, some systematic variations can be observed with hydrogen or deuterium loading. The most important modification concerns the low r region. It corresponds to the appearance of a supplementary peak at about 1.8 Å, which is positive with hydrogen and negative with deuterium. By reference to Table 1, this supplementary peak can be thus unambiguously attributed to Ti-H correlations. Moreover, this distance of 1.8 Å exactly corresponds to the distance from the centre to one of the titanium atoms in Ti_4 tetrahedra with a Ti-Ti distance of 2.9 Å.

In the same way, no evidence is found of Cu-H distances at about 1.5 Å (≃ distance from the center to the Cu atom in Ti_3Cu tetrahedra) despite the larger weight of the Cu-H correlations as compared to the Ti-H ones in the total pair-correlation function. This confirms the NMR results of Bowman et al.[12] which also show that Ti_4 tetrahedral sites are preferential hydrogenation sites in $Cu_{50}Ti_{50}$ amorphous alloys. This is also in agreement with the results obtained on other amorphous hydrides such as NiTiH[13] and PdZrH[14].

Other components appear at about 3.2 Å and 3.8 Å in the total correlation curves. The first one decreases with hydrogen and increases with deuterium, whereas the second one varies in the opposite way. The same conclusions are drawn for the two other compositions studied, $Cu_{67}Ti_{33}$ and $Cu_{35}Ti_{65}$ [1]. These two peaks can thus be attributed to Cu-H and Ti-H correlations respectively. Table 2 compares the first neighbour distances in crystalline γ-CuTiD[15] to those found in amorphous CuTi(H,D) alloys. The agreement is very good even for Ti-H next-nearest neighbour distances.

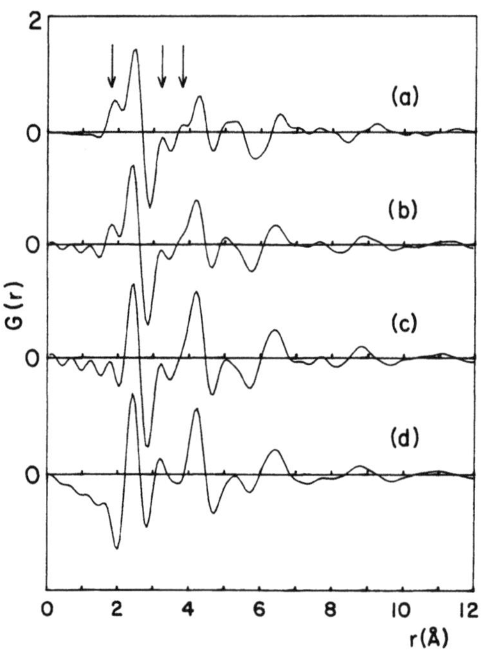

Fig. 4. Reduced pair-correlation functions for amorphous $Cu_{0.50}Ti_{0.50}(H,D)_x$ alloys : (a) x_H = 0.33, (b) x_H = 0.15 (c) x = 0, (d) x_D = 0.20.

Table 2 . First-neighbour distances (Å) in crystalline γ-CuTiD[15] and in amorphous CuTi (H,D) alloys.

	γ-CuTiD	a-CuTi (H,D)
CuCu	2.49	2.5
CuTi	2.66	2.7
	2.86	
TiTi	3.02	2.9
	3.21	
TiH	1.93	1.8
TiH	3.60	3.8
CuH	3.16	3.2
H H	2.14	

It should be noted that the H-H distance at 2.14 Å in crystalline γ-CuTiD is not observed in these amorphous CuTi (H,D) alloys. This could be due either to the smaller weight of the H-H correlations in the total function or to the vicinity of the Ti-H ($r = 1.8$ Å) and Cu-Cu ($r = 2.5$ Å) nearest-neighbour distances.

COMPUTER MODEL FOR AMORPHOUS $Cu_{0.50}Ti_{0.50}$(H,D) ALLOYS

On the basis of the experimental results presented above, a computer model of the structure of hydrogenated CuTi amorphous alloys has been built up[2] . The first step has been to obtain a model able to reproduce the experimentally observed chemical order between Cu and Ti atoms. Hydrogen and deuterium atoms were then introduced in the model by taking into account the preferred Ti-H (Ti-D) interactions.

As-quenched CuTi Amorphous Alloys

To simulate the structure of the as-quenched alloys one starts with an initial monoatomic packing of 1500 spheres relaxed in a modified Johnson potential[16]. Cu and Ti atoms are then chosen at random to give the desired composition $Cu_{50}Ti_{50}$. After that one tries to reproduce the experimental results by a Monte-Carlo procedure. Cu-Ti pairs are taken at random and permuted with a given probability if this permutation favours chemical order (i.e. an increase of the number of Cu-Ti pairs). The sample is then relaxed in a set of modified Johnson potentials. The position r and depth ϵ of the minimum of each metal-metal potential are given by

$$r_{CuCu} = 2.5\ \text{Å}\ ;$$

$$r_{CuTi} = 2.7\ \text{Å}\ ; \qquad \epsilon_{CuCu} = \epsilon_{CuTi} = \epsilon_{TiTi} = -1\ .$$

$$r_{TiTi} = 2.9\ \text{Å}\ ;$$

From this model the interference functions and pair-correlation functions for both X-ray and neutron are calculated and compared to the experimental ones (it is not possible to compare individual partial functions as no such information is available in the literature). Nevertheless, as copper and titanium have neutron scattering lengths of opposite signs, the comparison with the neutron experiments is a rather severe test of the quality of the model. Figure 5 shows that the model accounts relatively well for the essential features of both interference functions and pair-correlation functions.

Hydrogenated and Deuterated Alloys

An analysis of this model has been carried out in terms of Voronoï polyhedra to identify the different interstitial sites in the same way as for monoatomic models[16]. On the basis of the experimental results[1,12], hydrogen or deuterium atoms were introduced in the model with the following hypotheses:

- hydrogen induces no modification of the metal-metal distances;
- all the tetrahedral sites defined by four titanium atoms can be occupied;
- among these sites, only those corresponding to hydrogen-hydrogen distances larger than about 0.6 Å can be effectively occupied (to prevent the simultaneous occupancy of very distorted neighbouring sites).

These initial conditions lead to a maximum hydrogen or deuterium concentration of 0.12 per metal atom (0.14 if no hydrogen-hydrogen minimal distance is imposed). This concentration is smaller than the experimental ones, but is a consequence of the observed chemical order which implies a decrease of the number of Ti-Ti and Cu-Cu pairs to the benefit of Cu-Ti ones, and thus leads to a decrease of the number of Ti_4 sites. This is in contradiction with the fact that these amorphous CuTi hydrides absorb more hydrogen than the corresponding crystalline hydrides[17] (in this case, the Ti_4 sites are a consequence of the presence of Cu-Cu and Ti-Ti double layers[12,18]).

Figure 6 shows the simulated neutron pair-correlation functions for $Cu_{0.50}Ti_{0.50}H_{0.12}$ (fig. 6a), $Cu_{0.50}Ti_{0.50}$ (fig. 6b) and $Cu_{0.50}Ti_{0.50}D_{0.12}$ (fig. 6c) samples. By comparison with the experimental pair-correlation functions of figure 4, one observes a good agreement for both positions and intensities of the supplementary Ti-H (Ti-D) and Cu-H (Cu-D) correlations. This shows that despite the approximations used for the construction, the model presented here is able to reproduce the general features of both as-quenched and hydrogenated or deuterated CuTi amorphous alloys.

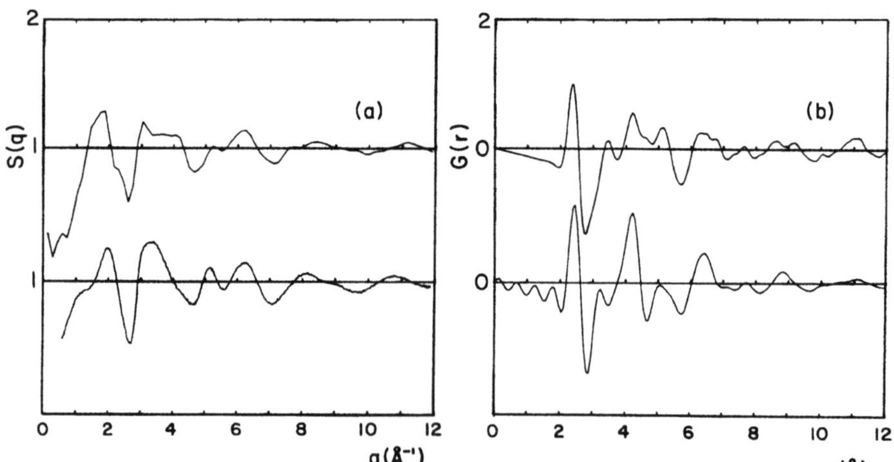

Fig. 5. Comparison between simulated (upper curves) and experimental (lower curves) interference functions (a) and pair-correlation functions (b) of $Cu_{0.50}Ti_{0.50}$ amorphous alloys.

It is well established that the thermal stability of amorphous CuTi hydrides is much less than that of the corresponding non-hydrogenated amorphous alloys or crystalline hydrides[19,20]. As an example, a-$CuTiH_{1.41}$ decomposes at about 500 K to form γ-TiH_x and Cu metal[20].

The neutron scattering technique is again well adapted to the study of such decomposition processes[4]. The first reason is that, as hydrogen has a very large incoherent scattering cross-section (80 barns), it is relatively easy to detect any change in the hydrogen concentration of the sample by following the evolution of the corresponding intense background in the case of large-angle experiments. The second reason is that, as already pointed out in the introduction, the combination of positive (Cu) and negative (Ti and H) neutron scattering lengths will lead to a large contrast between the TiH_2-rich and Cu- rich phases in the small-angle experiments (the fact that H is in interstitial position will reinforce this contrast). Another advantage (hydrogen-deuterium substitution) has already been largely used in the preceding sections.

Figure 7 shows the thermal evolution of the scattering curves in the intermediate q range ($0.17 < q < 3.34$ Å^{-1}) for a $Cu_{0.50}Ti_{0.50}H_{0.15}$ sample. The small-angle maximum at $\cong 0.5$ Å^{-1} at 315 K (which was already visible in the room-temperature curve of fig. 3b) increases in intensity and shifts towards smaller q as temperature increases. This corresponds to a progressive growth of the hydrogen-titanium rich regions, which become large enough at about 450 K to give observable diffraction peaks corresponding to a

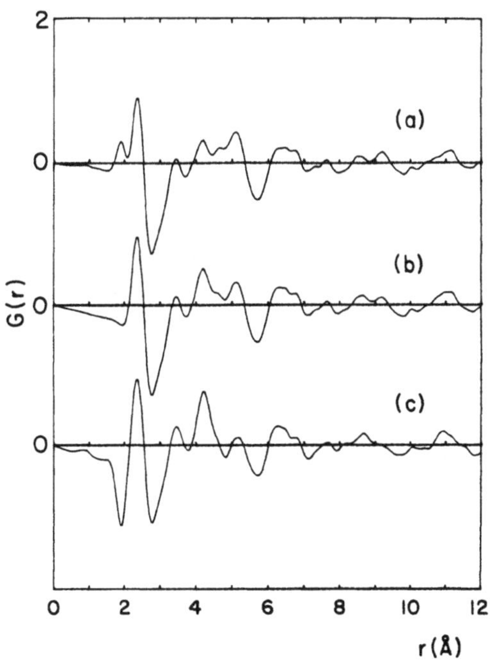

Fig. 6 Simulated neutron pair-correlation functions of (a) $Cu_{0.50}Ti_{0.50}H_{0.12}$, (b) $Cu_{0.50}Ti_{0.50}$ and (c) $Cu_{0.50}Ti_{0.50}D_{0.12}$ amorphous alloys.

TiH_2 phase. As already pointed out[20], this implies both a diffusion of hydrogen atoms and a reorganization of titanium atoms, which can be explained by the more favourable enthalpy of formation of TiH_2 as compared to that of CuTi[19]. At 600 K the amorphous matrix crystallizes to give γ-CuTi (on the Cu-rich side of the homogeneity range[18]). Hydrogen begins to leave the sample at about 680 K, as indicated by the decrease of the intense background due to the incoherent scattering of hydrogen. This is also confirmed by a simultaneous decrease of the intensity of the TiH_2 diffraction peaks and an increase of the γ-CuTi ones.

In the high-temperature range ($T \gtrsim 740$ K), the small-angle scattering curves can be well fitted to the Debye law[21] which was originally applied to the small-angle scattering from inhomogeneous materials (for example porous materials with a random distribution of holes and solids). In this formalism the small-angle intensity is given by :

$$I(q) \sim \frac{\lambda^3}{(1+q^2\lambda^2)^2}$$

where λ is a correlation length which measures the grain size of the material. In our case, it is related to the mean size of the TiH_2 clusters in the CuTi matrix. Figure 8 shows an example of the fits at 739 K and 796 K. Straight lines are obtained in these $I^{-1/2}$ vs q^2 plots. The corresponding correlation lengths (70 Å at 739 K and 150 Å at 796 K) show that the TiH_2 precipitates grow as temperature increases.

At smaller temperature ($640 \text{ K} \lesssim T \lesssim 740$ K), a good fit to the small-angle

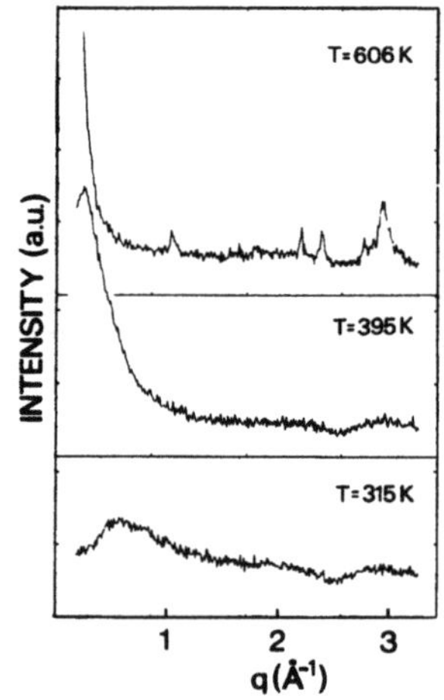

Fig. 7 Thermal evolution of the scattering curves for $Cu_{0.50}Ti_{0.50}H_{0.15}$ amorphous alloys.

data is also obtained by extending the Debye model to the case of fractal precipitates (i.e. with a dimension smaller than the euclidian dimension). It is thus possible to take into account both the fractal structure and the finite size of the aggregates[4]. The formation of these fractal precipitates ($D \cong 2.1$) is followed by a densification at constant size ($\lambda \cong 60$ Å) and by a subsequent growth with $D = 3$.

CONCLUSION

The results presented in this chapter can be summarized as follows. From an experimental point of view, we would like first to point out that considerable information can be extracted from neutron scattering studies of such CuTiH amorphous alloys. We tried to show throughout this paper the numerous advantages of this technique (without mentionning the quasi-elastic and inelastic aspects which are illustrated in other chapters of this book).

Concerning the structure of amorphous CuTi hydrides, the first result is that non-hydrogenated alloys exhibit a tendency to chemical ordering with a preference for A atoms (Cu or Ti) to be surrounded by B atoms (Ti or Cu). Although the metal-metal distances are found similar to those in the corresponding crystalline compounds, this makes a noticeable difference between amorphous and crystalline CuTi alloys.

The experimental results on amorphous CuTi hydrides show similarities with the crystalline hydrides in terms of hydrogenation sites and first neighbour distances, although no information is obtained concerning hydrogen-hydrogen nearest-neighbour distances. Neutron experiments at higher hydrogen and deuterium content would certainly give information on that subject.

The computer simulation results show that a rather good agreement with the experimental data can be obtained for both as-quenched and hydrogenated alloys despite the approximations used in the construction of the models. Work is underway to determine the influence of these approximations.

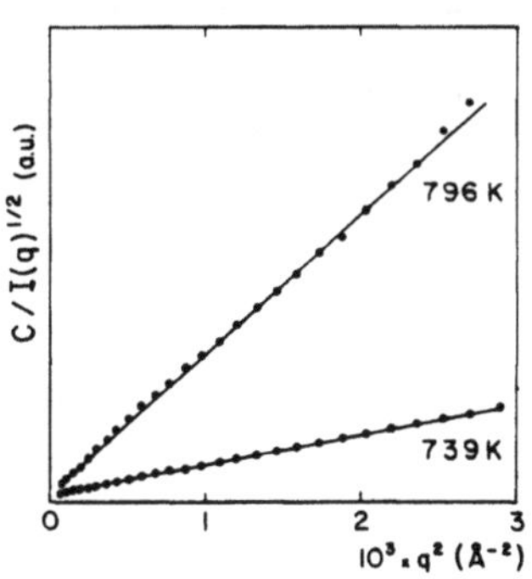

Fig. 8 Small-angle scattering curves of $Cu_{0.50}Ti_{0.50}H_{0.15}$ amorphous alloys and fits to the Debye law.

Finally, the thermal evolution is in agreement with previous work on CuTi hydrides. The hydrogen-titanium affinity manifests itself throughout the temperature range studied. This raises the question of the influence of the introduction of hydrogen on the structure of the amorphous matrix. On the basis of the experimental results one could imagine that the introduction of large quantities of hydrogen leads to a local reorganization of the metal atoms. This is supported by recent X-ray diffraction experiments on highly-charged CuTi amorphous hydrides[22] which show that large amounts of hydrogen lead to the crystallization of the sample.

ACKNOWLEDGEMENTS

The authors would like to thank Institut Laue-Langevin for allocation of beam time and technical support.

REFERENCES

1. B. Rodmacq, Ph. Mangin, and A. Chamberod, J. Phys. F. : Met. Phys. 15, 2259 (1985)
2. B. Rodmacq, L. Billard , Ph. Mangin, and A. Chamberod, Third Int. Conf. on the Structure of Non-Crystalline Materials, Grenoble 1985
3. B. Rodmacq, L. Billard, A. Chamberod, and Ph. Mangin, Proc. Int. Conf. on Neutron Scattering, Santa Fe, 1985.
4. Ph. Mangin, B. Rodmacq, and A. Chamberod, Phys. Rev. Lett. 55 2899 (1985)
5. F. Lançon, L. Billard, J. Laugier, and A. Chamberod, J. Phys. (Paris) 46, 235 (1985)
6. M. Sakata, N. Cowlam, and H.A. Davies, J. Phys. (Paris) 41, C8-190 (1980)
7. T. Fukunaga, K. Kai, M. Naka, N. Watanabe, and K. Suzuki, Proc. 4th Int. Conf. on Rapidly Quenched Metals (Sendai : Japan Inst. Metals), 347 (1982)
8. H. Ruppersberg, D. Lee, and C.N.J. Wagner, J. Phys. F : Met. Phys. 10, 1645 (1980)
9. M. Sakata, N. Cowlam, and H.A. Davies, Proc. 4 th Int. Conf. on Rapidly Quenched Metals (Sendai : Japan Inst. Metals), 327 (1982)
10. D.S. Boudreaux, IEEE Trans. Magn. Mag-17, 2606 (1981)
11. P. Goudeau, A. Naudon, B. Rodmacq, Ph. Mangin, and A. Chamberod, Third Int. Conf. on the Structure of Non-Crystalline Materials, Grenoble, 1985
12. R.C. Bowman, Jr., A.J. Maeland, and W.K. Rhim, Phys. Rev. B26, 6362 (1982)
13. K. Kai, S. Ikeda, T. Fukunaga, N. Watanabe, and K. Suzuki, Physica 120B, 342 (1983)
14. K. Samwer and W.L. Johnson, Phys. Rev. B 28, 2907 (1983)
15. A. Santoro, A. Maeland, and J.J. Rush, Acta Cryst. B34, 3059 (1978)
16. F. Lançon, L. Billard, and A. Chamberod, J. Phys. F : Met. Phys. 14, 579 (1984)
17. A.J. Maeland, L.E. Tanner, and G.G. Libowitz, J. Less-Common Met., 74, 279 (1980)
18. N. Karlsson, J. Inst. Met. 79, 391 (1951)
19. A.J. Maeland, in "Metal Hydrides", G. Bambakidis, ed., Plenum, New York (1981)
20. R.C. Bowman, Jr., R.J. Furlan, J.S. Cantrell, and A.J. Maeland, J. Appl. Phys. 56, 3362 (1984)
21. P. Debye, H.R. Anderson, Jr., and H. Brumberger, J. Appl. Phys. 28, 679 (1957)
22. B. Grzeta, K.Dini, N. Cowlam, and H.A. Davies, J. Phys. F : Met. Phys. 15, 2069 (1985)

QUASI-ELASTIC AND INELASTIC NEUTRON SCATTERING STUDY OF CuTi AMORPHOUS HYDRIDES

A.J. Dianoux[+], B. Rodmacq*, Ph. Mangin[+] and H. Chamberod*

[+]Institut Laue-Langevin, 156X, 38042 Grenoble Cedex, France

*DRF - Service de Physique, Métallurgie Physique, CEN-Grenoble 85X, 38041 Grenoble Cedex, France

ABSTRACT

Quasi-elastic and inelastic neutron scattering on CuTi amorphous hydrides has revealed some aspects of the diffusive and vibrational dynamics of hydrogen. Part of the hydrogen atoms perform a fast localized motion occurring on a short length scale. The activation energy of this motion is around 0.3 eV. The high frequency mode around 140 meV reveals the preferential filling of Ti_4 tetrahedra at low concentration of hydrogen. The temperature dependence of this mode indicates the formation of more Ti_4 sites as the temperature is increased. The metal vibrational mode around 20 meV is enhanced by the introduction of hydrogen, giving a measure of the hydrogen-metal interaction in these amorphous hydrides.

INTRODUCTION

The study of amorphous metal-hydrogen systems has started quite recently[1] in contrast to the vast amount of work on the properties of crystalline metal-hydrogen systems that dates back over several decades and makes use of a great number of different techniques. The amorphous system a - $Ti_{1-y}Cu_yH_x$ has been studied by Inelastic Neutron Scattering (INS)[2] and Nuclear Magnetic Resonance (NMR)[3]. Very recently an extensive study of the structure of CuTi amorphous alloys and hydrides has been made by neutron diffraction and computer simulation (ref. 4 and references therein).

Very few results have been reported so far on the mobility of hydrogen in a metallic glass using Neutron Quasielastic Scattering techniques (NQES). The first ones have been on NiZr[5] and PdSi[6] amorphous hydrides. NQES is certainly one of the most promising techniques to reveal the dynamics of hydrogen in amorphous alloys since it gives information on a microscopic scale in space and time. Furthermore, when using Time-of-Flight techniques (TOF), one obtains also information on the vibrational dynamics of the hydrogen, either the local modes or the frequency distribution of the host amorphous alloy to which is linked the hydrogen.

In this paper, the incentive has been to try to relate the structural results obtained in ref. 4 by diffraction and computer simulation to the dynamics of the hydrogen in the $Cu_{0.5}Ti_{0.5}$ hydrides.

EXPERIMENTAL

Amorphous $Cu_{0.5}Ti_{0.5}$ alloy has been prepared by the melt-quenching technique in the form of ribbons 8 mm wide and 40 μm thick. They have been loaded with hydrogen at room temperature by electrolysis. The hydrogen content is varied by changing the electrolysis time and determined by wieghting the samples before and after loading. Another amorphous alloy has been prepared by sputtering. Neutron diffraction data have shown that it has a small concentration of hydrogen (< 5 % at.). The different samples used are described in Table 1.

Neutron scattering measurements were carried out on the instrument IN6 which is situated on a neutron guide looking at the cold source of the High Flux Reactor of the Institut Laue-Langèvin in Grenoble. IN6 is a TOF spectrometer which makes use of time-focusing to combine high intensity and good resolution. It has a very low intrinsic background and the sample area is well shielded so that only neutrons scattered by the sample can reach the detectors. The full angular range from 10° to 115° is covered by detectors. We used an incident wavelength of 5.1 Å (incident energy : E_0 = 3.14 meV). Depending upon scattering angle, the elastic resolution (FWHM) varied between 70 to 100 μeV (1 μeV ≃ 8 × 10^{-3} cm^{-1}). The momentum transfer Q ranges between 0.2 to 2.1 $Å^{-1}$ ($Q = k - k_0$ where k and k_0 are the scattered and incident wavevectors). The detectors have been regrouped by software to form 19 scattering angles having a constant width in Q (ΔQ = 0.1 $Å^{-1}$).

The amorphous ribbons have been packed inside a flat Aluminium sample holder which is fitted inside heating and cooling loops whese temperature can be regulated between 77°K and 650°K within ± 1°K. The plane of the sample holder made an angle of 135° with respect to the incident beam.

Table 1

Sample	Preparation	Weight
$Cu_{0.5}Ti_{0.5}$	melt quenching	3.3 g
$Cu_{0.5}Ti_{0.5}H_{0.3}$	melt quenching + electrolysis	2.5 g
$Cu_{0.5}Ti_{0.5}H_{0.1}$	melt quenching + electrolysis	2.5 g
$Cu_{0.5}Ti_{0.5}H_{0.05}$	sputtering	1.8 g

The scattered intensity is analysed as a function of the momentum transfer $\underset{\sim}{Q}$ and the energy transfer $\hbar\omega = E - E_o$ (E is the energy of the scattered neutrons). It is proportional to the scattering law $S(\underset{\sim}{Q},\omega)$ which contains all information about the structure and dynamics of the sample[8]. In the case of hydrogenated samples, the scattering law is almost purely incoherent and reflects mainly the motion of the hydrogen atoms. It is denoted $S_s(\underset{\sim}{Q},\omega)$.

Elastic Incoherent Structure Factor (EISF)

In the case of a localized motion (for example a rotational motion), the incoherent scattering contains an elastic peak :

$$S_s(\underset{\sim}{Q},\omega) = A_o(\underset{\sim}{Q})\ \delta(\omega) + \sum_{i=1}^{n} A_i(\underset{\sim}{Q})\ \frac{1}{\pi}\ \frac{\tau_i}{1+(\omega\tau_i)^2} \qquad (1)$$

where n depends upon the model and the different correlation times τ_i usually depend upon only one characteristic time τ.

Assuming that all protons are equivalent, the EISF is given by :

$$A_o(\underset{\sim}{Q}) = \langle\ |\ e^{i\underset{\sim}{Q}.\underset{\sim}{r}}\ |^2\ \rangle \qquad (2)$$

where the average is done over the different sites reached by one proton during time. It is thus a quantity which is related to the geometry of the motion. If the elastic (I_e) and quasi-elastic (I_q) intensities are well separated, one can extract graphically at each scattering angle an experimental EISF = $\dfrac{I_e}{I_e + I_q}$.

This is a model independent quantity. More generally one can extract the EISF by a computer fit at each scattering angle θ, by using the following expression :

$$C^\theta(\omega) = F \times \left[A_o(Q)\ \delta(\omega) + (1 - A_o(Q))\ \frac{1}{\pi}\ \frac{\Gamma}{\Gamma^2 + \omega^2}\right] \otimes R^\theta(\omega) + B \qquad (3)$$

where F is a scaling factor,
B is an energy independent background,
$R^\theta(\omega)$ is the measured resolution function of the spectrometer at the scattering angle θ,
Γ is the HWHM of the quasi-elastic part,
$\otimes$ stands for the convolution product.

One can then compare the variation with Q of the EISF with different models. We give below the expression of the EISF for two extreme rotational models :

- diffusion on a sphere of radius a : $A_o(Q) = \left(\dfrac{\sin Qa}{Qa}\right)^2$ (4)
- jump over two sites separated by a distance 2a (for a powdered sample) :

$$A_o(Q) = \frac{1}{2}\left[1 + \frac{\sin 2Qa}{2Qa}\right] \qquad (5)$$

In the case of long-range diffusion we refer the reader to ref. 6 and 9, in order not to go too far in this simple description of quasi-elastic scattering.

Thermal Evolution of the Integrated Intensity

Fig. 1 shows this thermal evolution. In particular one sees that there is already an irreversible change in the scattering intensity at 130°C, since the spectrum taken at 30°C is not identical to the original one. This thermal evolution corroborates what has been found by neutron diffraction[4], to which we refer the reader for a full discussion of this effect. It is sufficient to note here that the scattering curves are reasonably flat above Q = 1 $Å^{-1}$, so that in this region an analysis in terms of purely incoherent scattering is certainly valid. At smaller Q values we have the appearance of the effects of hydrogen-titanium rich regions whose dynamics can be different from that of the bulk.

Quasi-elastic Results

The experimental EISF has been extracted for spectra taken at temperatures between 30°C and 200°C, using the expression (3). Although there is a quite large scatter of the points for the different temperatures, it seems that one can describe the results by a temperature independent EISF. We present in Fig. 2 the experimental EISF extracted from the spectra taken at 200°C.

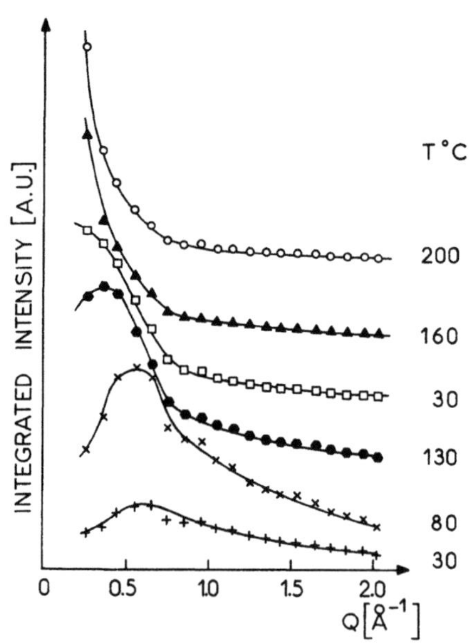

Fig. 1 Integrated intensity versus elastic momentum transfer.

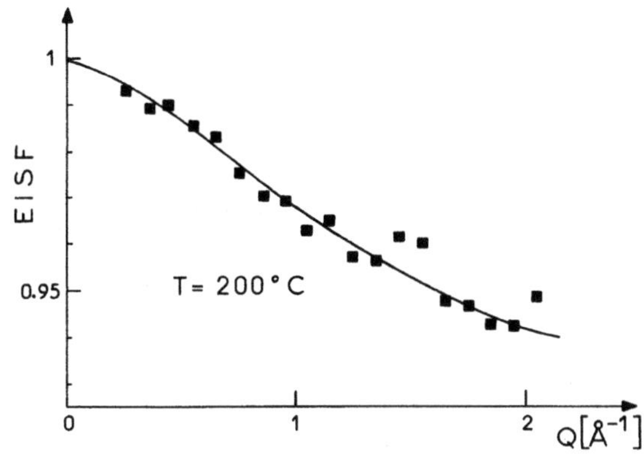

Fig. 2 Experimental EISF extracted from the spectra taken at 200°C. The solid line is a guide to the eye.

We remark that this EISF is always quite high, going down only to 0.94 at Q = 2 $Å^{-1}$. This is a sign that the motion occurs on a quite short length scale (cf. Eq. 4 or 5). We have then fixed, in expression (3), $A_0(Q)$ to the values represented by the solid line in Fig. 2 and fitted all the spectra to extract the width of the broadened component. This is represented on Fig. 3. One can see that this width is nearly constant above Q = 1 $Å^{-1}$ for a given temperature. Note that the results for 30°C are at the limit of the resolution of the instrument, but the heating at 200°C (see fig. 1) does not seem to alter the dynamics of the hydrogen even though there is a marked change in the integrated intensity below Q = 0.5 $Å^{-1}$. Fig. 4 shows examples of the fits for Q = 2.05 $Å^{-1}$ and the different temperatures. According to the fitted expression (Eq. 3) with a temperature-

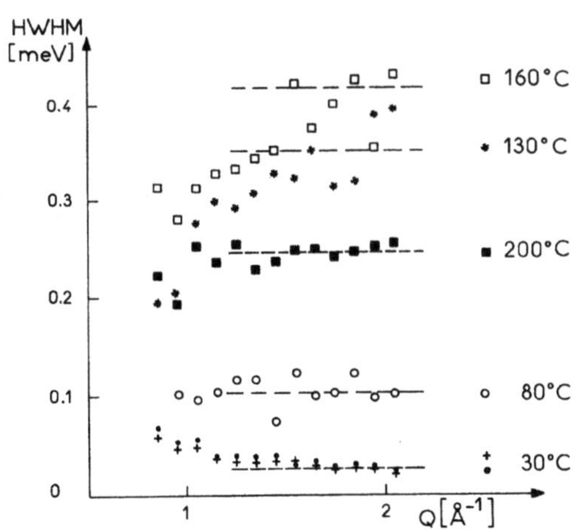

Fig. 3 Width of the quasi-elastic component of the spectra as a function of Q for different temperatures. The dots and crosses for 30°C are for the virgin sample and after heating at 200°C.

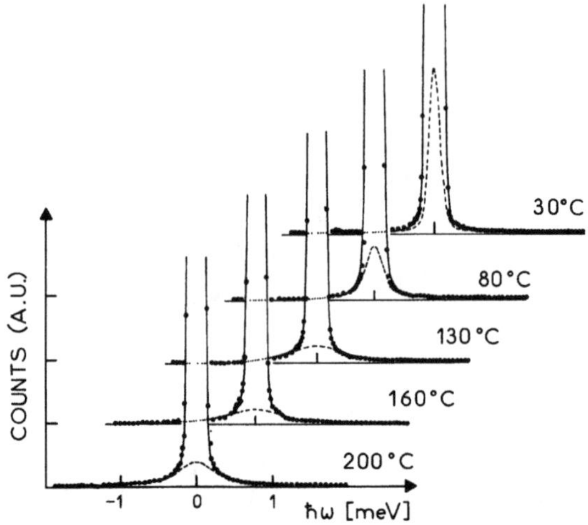

Fig. 4 Examples of fitted spectra for $Q = 2.05$ Å^{-1} and different temperatures. The dashed line is the quasi-elastic contribution. The points are experimental.

independent $A_0(Q)$, the area of the quasi-elastic contribution is constant with temperature. Note that the 200°C spectrum is effectively narrower than the 130°C one, as it is reported on Fig. 3.

Discussion

By plotting the log of the widths shown in Fig. 3 versus the reciprocal temperature, one obtains an Arrhenius behavior for temperature in the range 30°C to 130°C. The activation energy which is found in this range is $E_a = 6.9$ kcal/mole (0.30 eV). For a higher hydrogen concentration (H/M = 0.65) NMR experiments[3] show an activation energy of 0.42 eV in this temperature range. Very recently, hydrogen diffusivity has been measured on CuTi amorphous hydrides by an electrochemical method[10]. For the $Cu_{0.48}Ti_{0.52}$ hydride and an H/M concentration of 0.36 there has been found a much higher activation energy of 13.6 kcal/mole. This result seems to support our finding that the broadening seen by quasi-elastic scattering is due to a local motion having a much lower activation energy than long-range diffusion. Furthermore this implies also that the NMR data are influenced by both the long-range diffusion and the local motion. Using the results of ref.10 one can calculate that at 130°C, the width at $Q = 2$ Å^{-1} of the long-range diffusion component would be of the order of 23 μeV compared to a measured width of 330 μeV ($D \sim 8.6 \times 10^{-7}$ cm^2/s).

This local motion has a correlation time (inverse of the spectral width) varying between 33 ps and 2 ps for temperatures between 30°C and 130°C. Above 130°C, there is a marked departure from an Arrhenius behavior. At 200°C, the width even decreases, showing that the local motion is being slowed down by the appearance of more and more regions with TiH_2-like bonding.

The experimental EISF shown in Fig. 2 should give some information about the spatial extension of the motion. If we assume that all the protons are moving with the same correlation time, one finds, using the value 0.942 of the experimental EISF at $Q = 2$ Å^{-1}, that this spatial extension is less than 0.2 Å, either for a sphere (Eq. 4) or two sites (Eq. 5). One can use the results of the computer simulation model[4], showing that the maximum concentration of Ti_4 tetrahedra is 12 %, to postulate that hydrogen trapped inside these Ti_4 tetrahedra appears "fixed" on the time scale of the measurement which is shorter than 5×10^{-11} s. This is supported by the fact that we have seen no broadening with the $Cu_{0.5}Ti_{0.5}H_{0.1}$ sample, which has been measured only at 30°C, while the sample with H/M = 0.3 shows already a measurable broadening. It is also in agreement with the fact that there is a very strong dependence of the diffusion coefficient upon hydrogen concentration[10]. For example there is more than one order of magnitude difference for the diffusion coefficient at room temperatures of $Cu_{0.48}Ti_{0.52}$ hydrides for H/M ranging from 0.13 to 0.36 [10]. If we call b the ratio of "fixed" protons to the total number of protons, the experimental EISF is given by:

$$\text{EISF} = b + (1 - b)\, A_o(Q) \tag{6}$$

where $A_o(Q)$ is the EISF of the mobile protons. Using the computer simulation model[4], one finds b = 0.12/0.30 = 0.4 which leads to $A_o(Q = 2) = 0.903$.

This would correspond to a spatial extension of the motion a = 0.28 Å (same for a sphere or two site model). If the chemical order between Ti and Cu is decreased, the number of Ti_4 tetrahedra could be increased substantially[4]. For example, assuming b = 0.8 one has A_o (Q = 2) = 0.710, which corresponds to a spatial extension a = 0.5 Å.

In any case, the observed quasi-elastic intensity shows that the spatial extension of this localized motion is rather small, on the average. We think that in view of the multiplicity of sites available for hydrogen in these amorphous alloys, it is difficult to put forward a quantitative model. We can say that this localized motion of the hydrogen occurs certainly inside cages defined by the surrounding metal atoms.

INELASTIC NEUTRON SCATTERING ON $Cu_{0.5}Ti_{0.5}$ HYDRIDES

In a TOF experiment using long wavelength neutrons at room temperature and above, the spectra contain also the energy gain scattering which is revealing the vibrational character of the hydrogen, both its local modes and the frequency spectrum of the glass [2,11,12]. An excellent presentation of the vibrational dynamics of metallic glasses studied by Inelastic Neutron Scattering can be found in ref. 13.

The vibrational frequency distribution can be obtained from the TOF spectra by an extrapolation procedure[14]. This is fully justified when the scattering is predominantly incoherent, as is the case for metal hydrides.

$$\text{Putting } \alpha = \frac{\hbar^2 Q^2}{2Mk_B T} \quad \text{and} \quad \beta = \frac{\hbar\omega}{k_B T} \tag{7}$$

where M is the mass of the scattering particle, the frequency distribution is obtained by :

$$g(\beta) = 2\beta \sinh\left(\frac{\beta}{2}\right) \lim_{\alpha \to 0} \frac{S_s(\alpha,\beta)}{\alpha} \, . \tag{8}$$

When there is no dispersion (frequency spectrum of an amorphous solid or local modes of hydrogen) one can improve the statistics by summing the spectra at each Q value. One thus obtains the quantity :

$$P(\bar{\alpha},\beta) = 2\beta \sinh\left(\frac{\beta}{2}\right) \sum_{\alpha} \frac{S_s(\alpha,\beta)}{\alpha} \tag{9}$$

It has been shown[15] that $P(\bar{\alpha},\beta)$ is proportional to $g(\beta)$ except in the very low frequency domain where systematic errors are likely to occur. Eq. 9 is used here to look at the high frequency region (50 to 200 meV) which is dominated by the local mode vibration of hydrogen, since the extrapolation procedure (Eq. 8) is quite unreliable in this region due to statistical inaccuracies.

High Frequency Vibrational Mode

Fig. 5 presents the results obtained by using Eq. 9 at 30°C for different $Cu_{0.5}Ti_{0.5}$ amorphous hydrides and for the amorphous alloy (cf. Table 1).

As has been already observed in ref. 2, the peak around 140 meV reveals the preferential filling by hydrogen of Ti_4 tetrahedra. Indeed by

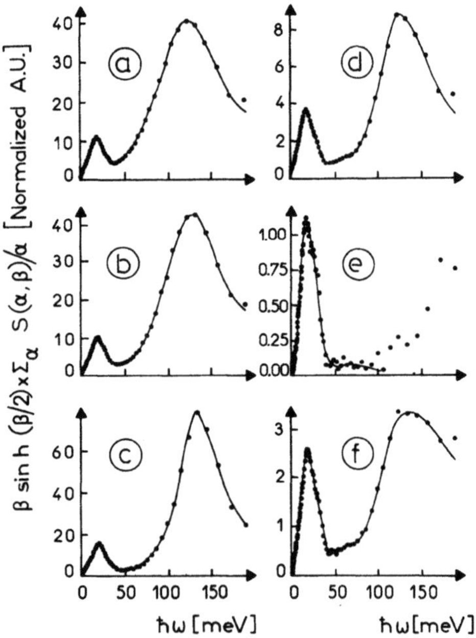

Fig. 5 High energy region of the vibrational frequency distribution (Eq. 9) obtained at 30°C for $Cu_{0.5}Ti_{0.5}$ amorphous hydrides and amorphous alloy.

ⓐ H/M = 0.3 as prepared;
ⓑ H/M = 0.3 after heating at 130°C;
ⓒ H/M = 0.3 after heating at 200°C;
ⓓ H/M = 0.1; ⓔ : amorphous alloy;
ⓕ H/M = 0.5, obtained by sputtering.
The lines are a guide to the eye.

comparing Fig. 5a and 5d, one sees that there is a marked increase of the frequency distribution between 40 and 100 meV for the concentration above 10 at. % of hydrogen. This indicates that above 10 %, other sites are being filled, in agreement with the structural model found by computer simulation[4] which has a maximum of 12 % of hydrogen in Ti_4 tetrahedra. Note that all the drawings in Fig. 5 have been normalized with respect to each other taking into account the sample weight, so that the areas under the peaks are proportional to the concentration of hydrogen (30 %, 10 % and 5 %). In particular, Fig. 5f shows that despite the very different methods of preparation of this sample. the vibrational modes of hydrogen are very similar to those of the less charged hydride (Fig. 5d). Fig. 5e represents the frequency distribution of the amorphous alloy $Cu_{0.5}Ti_{0.5}$. One can notice the comparatively low intensity; furthermore one sees that the high frequency peak has disappeared. The rise above 150 meV is only due to the blowing up of the spectral noise by the hyperbolic sine function in Eq. 9. In any case this is a very small effect when compared to Fig. 5a where the intensity at 150 meV is more than 130 times bigger. It shows that even with cold neutrons (E_0 = 3.1 meV) one can observe peaks around 150 meV when the background is low as it is on the IN6 spectrometer.

Another point to be noted concerns the thermal evolution of the $Cu_{0.5}Ti_{0.5}H_{0.3}$ hydride. (Figs.5 a-c). The high frequency peak becomes more intense as the temperature is increased and its width is decreased : it goes from 40 meV HWHM at 30°C to 25 meV at 200°C. This is in agreement with the thermal evolution of diffraction data[4], which shows that Ti_4 sites are formed when the temperature is increased.

Low Frequency region (ω < 40 meV)

In this region the statistical accuracy is good enough to obtain directly the frequency distribution by using the extrapolation indicated in Eq. 8. Fig. 6 presents these results for the same samples as those of Fig. 5.

Note that here also, the drawings have been normalized with respect to each other. One thus sees that the metal vibrational mode situated around 20 meV (Fig. 5e) is very much enhanced by the introduction of hydrogen, but its shape stays nearly the same. However one can notice that it is slightly broader for the virgin sample having 30 % hydrogen concentration (Fig. 5a). Again after heating at 200°C, the peak becomes higher and narrower (Fig. 5c). All these effects show that hydrogen is a very sensitive probe for the study of the vibrational dynamics of an amorphous alloy.

In Fig. 7 we present the frequency distribution for the $Cu_{0.5}Ti_{0.5}H_{0.3}$ hydride for different temperatures.

We will not examine in detail here this temperature dependence which must be connected to structural relaxation occurring already at 30°C. A detailed analysis of the frequency distribution in the 0-40 meV frequency range should give important information on the nature of the hydrogen-metal interactions in these amorphous hydrides.

CONCLUSIONS

This neutron scattering study of CuTi amorphous hydrides has revealed some aspects of the diffusive and vibrational dynamics of hydrogen. The long range diffusion is too slow to be observed on the TOF instrument we were using. We have found that part of the hydrogen atoms perform a fast

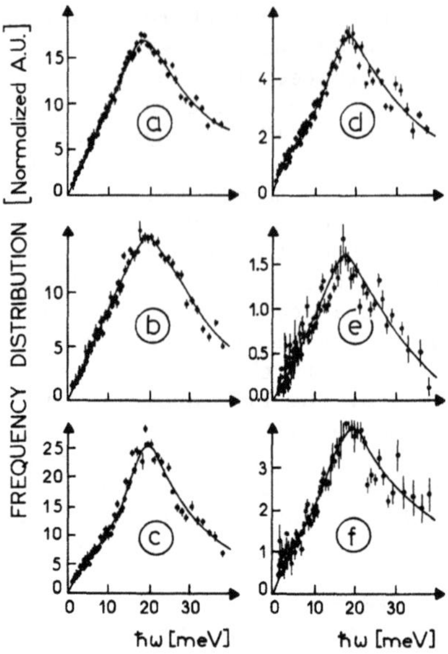

Fig. 6 Frequency distribution (Eq. 8) in the energy range 0-40 meV. ⓐ to ⓕ correspond to the same samples as in Fig. 5.
The lines are a guide to the eye. The small vertical bars are the statistical errors at each point.

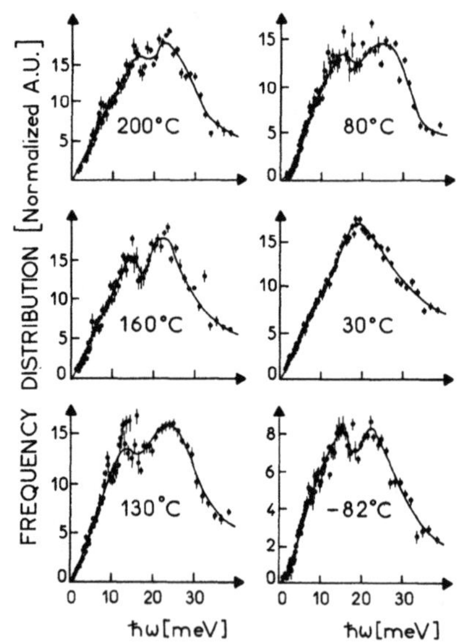

Fig. 7 Frequency distribution for the $Cu_{0.5}Ti_{0.5}H_{0.3}$ hydride in the energy range 0-40 meV. The lines are a guide to the eye. The small vertical bars are the statistical errors at each point.

localized motion (τ = 2 ps at 130°C) occurring on a short length scale (0.3 to 0.5 Å, depending upon the ratio of "fixed" protons). The activation energy of this motion is around 0.3 eV in the temperature range 30 to 130°C.

The high frequency mode around 140 meV has revealed the preferential filling of Ti_4 tetrahedra. For concentrations above 10 % at. of hydrogen, there is a marked increase of the frequency distribution between 40 and 100 meV, indicating that other sites are being filled. Conversely after heating at 200°C, these sites are being depleted in favor of Ti_4 tetrahedral sites which are formed.

The metal vibrational mode around 20 meV is enhanced by the introduction of hydrogen, giving a measure of the hydrogen-metal interaction. The temperature dependence of this mode must be connected to structural relaxation occurring already at 30°C.

This study should be completed by more experiments at better resolution and also at higher momentum transfer. In order to go much further, one also needs a more refined structural model with a complete description of all possible sites. Such a study is currently under way.

REFERENCES

1. A.J. Maeland in "Hydrides for Energy Storage" edited by A.F. Andressen and A.J. Maeland (Pergamon, Oxford 1978), p. 447.
2. J.J. Rush, J.M. Rowe and A.J. Maeland, J. Phys. F. 10:L283 (1980).
3. R.C. Bowman, A.J. Maeland and W.K. Rhim, Phys. Rev. B26:6362 (1982).
4. B. Rodmacq, Ph. Mangin, L. Billard and A. Chamberod, these proceedings.
5. J.-B. Suck, H. Rudin, H.U. Künzi and A. Heidemann in "Rapidly Quenched Metals", S. Steeb and H. Warlimont (Eds.), Elsevier (1985) p. 1545.
6. R. Hempelmann, G. Driesen and D. Richter, these proceedings.
7. "Neutron Beam Facilities Available for Users", B. Maier (Ed.), Institut Laue-Langèvin, Grenoble (France).
8. T. Springer, "Quasielastic Neutron Scattering for the Investigation of Diffusive Motions in Solids and Liquids", Springer Tract Mod. Phys. 84 (1972).
9. C. Lartigue, A.J. Dianoux, A. Percheron-Guégan and J.C. Achard, these Proceedings.
10. Y.S. Lee and D.A. Stevenson, J. Non-Cryst. Solids 72:249 (1985).
11. K. Kai, S. Idera, T. Fukunaga, N. Watanabe and K. Susuki, Physica B120:342 (1983).
12. K. Susuki, N. Hayashi, Y. Tomizuka, T. Fukunaga, K. Kai and N. Watanabe, J. Non Cryst. Solids 61-62 : 637 (1984).
13. J.-B. Suck and H. Rudin in "Glassy Metals II", H. Beck and H.J. Güntherodt(Eds.),Topics in Applied Physics, Springer (1983). p. 217.
14. P.A. Egelstaff and P. Schofield, Nucl. Sci. Eng. 12:260 (1962).
15. G. Maisano, P. Migliardo, M.P. Fontana, M.C. Bellissent-Funel and A.J. Dianoux, J. Phys.C: Solid State Phys. 18:1115 (1985).

DYNAMICAL DISORDER OF HYDROGEN IN $LaNi_{5-y}M_y$ HYDRIDES STUDIED BY QUASI-ELASTIC NEUTRON SCATTERING

C. Lartigue[+], A.J. Dianoux[o], A.Percheron-Guegan[+], J.C. Achard[+]

(+) Chimie Métallurgique des Terres Rares, CNRS, 1 Pl.A.Briand 92190 Meudon, France
(o) Institut Laue Langevin, 156X, 38042 Grenoble Cédex, France

SUMMARY

We report the results of a quasi-elastic neutron scattering study of hydrogen motions in some substituted $LaNi_{5-y}M_y$ hydrides ($M_y = Al_{0.5}$, $Al_{1.0}$, $Cu_{1.0}$, $Mn_{1.0}$). Measurements were performed by using different energy resolutions which allowed us to observe independently short-range and long-range diffusion. We show that these motions are much faster in the $LaNi_4Mn$ and $LaNi_{4.5}Al_{0.5}$ hydrides than in the $LaNi_4Cu$ and $LaNi_4Al$ hydrides. These results are discussed within the framework of our structural model in terms of jumps over interstitial positions.

INTRODUCTION

The intermetallic compound $LaNi_5$ is well-known as a suitable material for hydrogen storage. Pseudobinary compounds with the same $CaCu_5$-type structure can be easily formed by partial substitution of nickel. Depending upon the rate and the nature of substitution, the hydriding properties (stability, maximum H-content, interstitial sites occupied by hydrogen ...) are modified (1).

On a microscopic scale, neutron spectroscopy has proved to be a very powerful tool for the determination of the positions (diffraction) and of the dynamics of hydrogen (inelastic and quasi-elastic scattering). Diffraction experiments on $LaNi_5$-H and related systems have been interpreted in terms of two different structural models, the two-site (P31m space group) and the five-site (P6/mmm space group) models which have been widely discussed in the literature (1 and ref. cited herein). Our five-site model (2) allowed us to describe the structure of a series of $LaNi_{5-y}M_y$ (M = Al, Mn, Fe, Cu, Si) hydrides : we have shown that the number of nonequivalent interstitial sites which are significantly occupied by hydrogen may be lowered to two and that this is highly correlated with the reduction of the maximum hydrogen content of the alloys. Very recently, we have shown that a structural phase transformation due to the ordering of deuterium atoms occurs in the $LaNi_5D_x$ ($5 \leqslant x \leqslant 7$) deuterides (3). We have proposed a new model of structure which is derived from our five-site model : the unit cell is doubled along the c-axis, the space group is $P6_3mc$ and hydrogen occupies four types of interstitial sites. It is noteworthy that during the same time, similar results have been reported from an independent investigation (4). This ordering was revealed by the presence of one low-intensity diffraction peak which could not be indexed in any of the previous proposed models. Concerning the series of substituted $LaNi_{5-y}M_y$ hydrides that we have studied ($y \geqslant 0.5$), there is no trace of this superstructure line ; therefore, there is no reason for considering this new model of structure. From quasi-elastic neutron scattering (QNS) short-range as well as long-range diffusive motions with widely differing time scales can be investigated. At present, QNS is the only technique which allows the simultaneous measurement of both the time and the space dependence of the observed motions. In addition to the determination of the characteristic time of the motion, the details of the jump processes of hydrogen atoms in the host lattice can be elucidated. Models for jumps between nearest neighbour sites can easily be built on the basis of structural data and the corresponding scattering function can be calculated.

In this paper, we present the results of a QNS investigation for four substituted $LaNi_{5-y}M_yH_x$ hydrides. These compounds have been selected in view of the number of interstitial sites that we have found to be occupied $LaNi_4MnH_{5.9}$ (5 sites), $LaNi_{4.5}Al_{0.5}H_{5.4}$ (4 sites), $LaNi_4CuH_{5.1}$ (3 sites) and $LaNi_4AlH_{4.8}$ (2 sites). Data have been first analysed independently of any model and the results are discussed on the basis of our structural data.

QUASI-ELASTIC NEUTRON SCATTERING

A theoretical background for QNS is given in ref (5) and in ref. (6, 7) from this conference.

In a neutron scattering experiment, the intensity of scattered neutrons is measured as a function of the momentum transfer $\hbar\underline{Q} = \hbar\underline{k}_f - \hbar\underline{k}_i$ and of the energy transfer $\hbar\omega$. This intensity is proportional to the scattering function $S(Q,\omega)$ which is almost purely incoherent for hydrogenated samples. For the jump diffusion of hydrogen in metals, the Chudley-Elliott model (8) has been successfully applied to systems where the interstitial sites form a Bravais lattice. The incoherent scattering function is given by a lorentzian function of ω. The polycrystalline case is often treated by using a single lorentzian function, the width of which is averaged over all possible orientations of $\underline{Q}$ and given by :

$$\Gamma(\underline{Q}) = \frac{1}{\tau}\left(1 - \frac{\sin(Ql)}{Ql}\right)$$

where τ is the mean residence time at a given site and l the jump length. At low Q :

$$\tau_{Q \to 0} = \frac{l^2}{6\tau} Q^2 = DQ^2$$

where D is the macroscopic diffusion coefficient. This is the so-called Q^2-law which is generally valid for any kind of long-range diffusion but only at sufficiently low Q.

For a rotational motion, the incoherent scattering function is always given by :

$$S_S(\underline{Q}, \omega) = A_o(\underline{Q})\,\delta(\omega) + \sum_{i=1}^{n} A_i(\underline{Q}) \frac{1}{\Pi} \frac{\tau_i}{1+(\omega\tau_i)^2}$$

where n and τ_i depend upon the model. The first term is purely elastic. $A_o(Q)$ is the elastic incoherent structure factor (EISF) which contains all informations on the spatial dependence of the motion (see e.g. (7)).

Generally translation, rotation and vibrations occur on time scales differing by one or two orders of magnitude so that they can be considered as uncoupled motions. Then it is shown (5) that :

$$S_s(\underline{Q}, \omega) = S_s^{trans} \otimes S_s^{rot} \otimes S_s^{vib}$$

where the symbol $\otimes$ stands for energy convolution. In the quasi-elastic region (small $\hbar\omega$), the vibrational term is approximated to a Debye-Waller factor (9) and

$$S_s(\underline{Q}, \omega) = \exp(-2W)\,(S_s^{trans} \otimes S_s^{rot})$$

Practically the data are first analysed without any model assumption : the elastic (δ function) and quasi-elastic (lorentzian function) parts of the spectra are separated. The Q-dependence of the experimental EISF f(Q) and of the width $\Gamma(Q)$ of the quasi-elastic line allows to characterize the observed motions according to the following criteria : f(Q) constant and $\Gamma(Q)$ increasing as a Q^2-law at small Q-values are typical for a long-range diffusion while an almost constant $\Gamma(Q)$ and a decreasing f(Q) are characteristic of a short range motion. Starting from this analysis, various models can be elaborated and compared to the data.

EXPERIMENTAL DETAILS

The QNS experiments were carried out on three different spectrometers at the high flux reactor at the Institut Laue Langevin, Grenoble : the high resolution backscattering spectrometer IN10, the multichopper time of flight (t.o.f.) spectrometer IN5 and the time foccussing t.o.f. spectrometer IN6 which are described in ref (6, 10). The experimental conditions are summarized in table 1.

TABLE 1 - Instrumental conditions for measurements
a) the energy transfer range used in the data analysis

Spectrometer	Incident wavelength λ(Å)	Momentum transfer range $Q(Å^{-1})$	Number of scattering angles	Energy transfer range $\hbar\omega$(μeV)	Resolution f.w.h.m. (μeV)
IN 10	6.28	0.28 , 1.95	7	-7.8, 7.8	0.7 , 1.4
IN 5	5.14	0.51 , 2.15	8	-10^3 , 10^3 (a)	127
IN 6	5.10	0.22 , 2.03	17	-2.10^3, 2.10^3	85 , 135

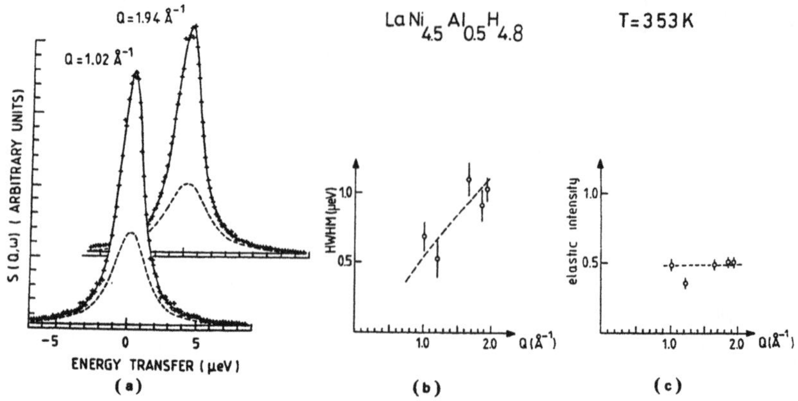

FIGURE 1 - a) Decomposition of the high resolution spectra (IN10 spectrometer) of $LaNi_{4.5}Al_{0.5}H_{4.8}$ at 353K into a purely elastic line, a Lorentzian (dashed curve) and a flat component ; the full line represents the total scattering function fitted to the experimental points (+). Q-dependences of (b) the width of the Lorentzian component and (c) the relative elastic intensity (experimental EISF ; the broken lines are included only as a guide to the eye.

TABLE 2 - Measuring conditions for samples : temperature T(K), H_2 pressure P(bar), and H-content x(H/mol).

Spectrometer	IN 10			IN 5			IN 6		
Intermetallics	T	P	x	T	P	x	T	P	x
	100	< 1.0	5.7						
$LaNi_{4.5}Al_{0.5}$				299	2.5	5.4	299	10	5.7
				319	2.8	5.2	319	10	5.7
				337	3.1	5.0			
	353	3.0	4.8	352	3.5	4.8	353	10	5.2
							393	20	4.9
$LaNi_4Mn$	100	< 1.0	5.8						
				290	3.5	5.7			
	300	2.3	5.6						
	326	2.8	5.2						
	365	3.0	4.8						
$LaNi_4Cu$	100	< 1.0	5.5						
	305	3.5	5.4	303	3.2	5.3			
$LaNi_4Al$	100	< 1.0	4.8						
	300	2.2	4.8	290	3.5	4.8			
	345	2.7	4.2						
	364	3.0	4.0						

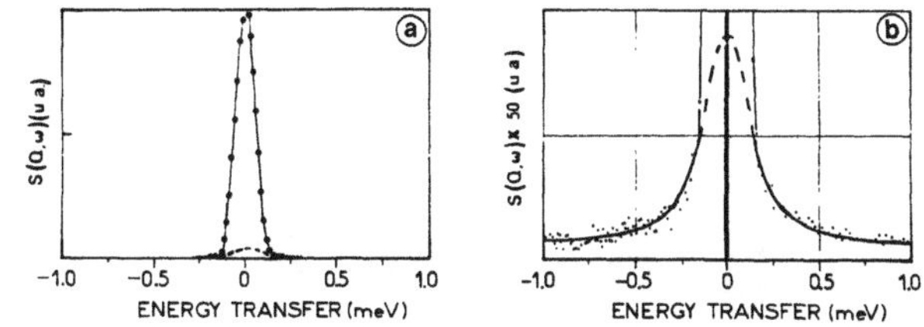

FIGURE 2 - a) decomposition of the lower resolution spectra (IN5 spectrometer) of $LaNi_{4.5}Al_{0.5}H_4$ at 352K into a purely elastic and a Lorentzian components ; in b) the intensity is magnified 50 times ; the thick solid line at $\omega = 0$ represents the energy window of the IN10 backscattering instrument.

Single phase intermetallic samples were ground and fine powders were obtained after six H_2-absorption-desorption cycles. They were placed in closed aluminium slab-shaped containers and hydrided in-situ. The measuring conditions for QNS experiments are reported in table 2.

All spectra were corrected for detector efficiency, absorption and self shielding and the t.o.f. data were transformed to $S(\phi, \omega)$ at a constant scaltering angle ϕ using standard programs provided by the Institut Laue Langevin.

RESULTS

All experimental spectra were first analysed without making any model assumption as described in ref (11). We have found similar results : 1) for both the $LaNi_{4.5}Al_{0.5}$ and $LaNi_4Mn$ hydrides ; 2) for both the $LaNi_4Cu$ and $LaNi_4Al$ hydrides. From these results we can separate these hydrides in two groups :

$LaNi_{4.5}Al_{0.5}$ and $LaNi_4Mn$ hydrides

At large Q-values ($Q > 1$ Å^{-1}) the high resolution spectra (IN10 spectrometer) are separated into three components (fig. 1a) one elastic line superimposed on a lorentzian function and a flat contribution. The relative intensity of both elastic and quasi-elastic parts are almost constant and equal to about 50% (fig. 1b). The quasi-elastic linewidth has a strong Q-dependence. The flat component is found to be much higher than the expected instrumental background plus the inelastic vibrationnal contribution. This indicates the existence of a very broad component which lies outside the energy transfer range investigated ($\pm$ 7.8 μeV). This is confirmed from the data measured with a lower resolution. Indeed, the IN5 t.o.f. spectra (fig. 2) can be decomposed into a large purely elastic line superimposed on a broad lorentzian component : the relative elastic intensity decreases very slowly from 1 to 0.9 with increasing momentum transfer up to 2.0 Å^{-1} while the quasi-elastic linewidth is almost constant and equal to 80 ± 20 μeV for both hydrides at room temperature. These variations are characteristic of a localized motion. By comparing the experimental EISF to the one calculated for a rotation on a sphere (12) one concludes that the motion occurs on small distances and that probably it does not involve all the H-atoms.

At small Q-values ($Q < 1$ Å^{-1}) the broadening of the high resolution spectra is too small and the elastic and quasi-elastic contributions can no longer be separated.

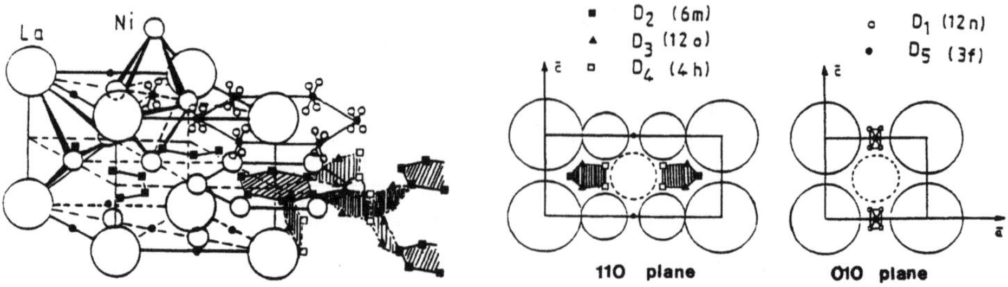

FIGURE 3 - Structure of $LaNi_{5-y}M_y$ hydrides : hexagonal unit cell and sections in the (110) and (010) planes. The five interstitial sites for hydrogen form "bounded cages" (hexagons and pentagons) and "isolated cages" (cluster of 3f and 12n positions).

Thus we have observed : 1) a long range diffusive motion but only for about 50 % of the hydrogen atoms in the lattice ; 2) a rapid localized motion with a small spatial amplitude. It is clear that the other hydrogen atoms should also diffuse ; but, at the measured temperature, this diffusion is probably too slow to be observed on the backscattering spectrometer.

$LaNi_4Cu$ and $LaNi_4Al$ hydrides

For these hydrides, the high resolution spectra can be decomposed in one single lorentzian function plus a flat component. The quasi-elastic linewidth increases slowly as a function of Q. The broadening is very small even at 364 K for $LaNi_4AlH_4$: $\Gamma = 0.13$ µeV at $Q \simeq 2$ Å^{-1}.

The lower resolution spectra are very similar to those of the $LaNi_4Mn$ and $LaNi_{4.5}Al_{0.5}$ hydrides.

From these results, we have found that in these two hydrides, all the hydrogen atoms diffuse but the diffusion is much slower. By contrast, the rapid localized motion is qualitatively the same but slower in the $LaNi_4Cu$ and $LaNi_4Al$ hydrides than in $LaNi_4Mn$ and $LaNi_{4.5}Al_{0.5}$ hydrides.

DISCUSSION

The existence of these two different types of hydrides with respect to their dynamical properties can be understood by considering their structures. In our five-site model, the hydrogen atoms are found to occupy two types of "cages" of which we have obtained a good description by using a multisite model (see figure 3). First, we have cages formed by four non-equivalent 12n positions centred on a 3f position ; there are three cages by unit cell ; there is no direct path between them so they are labelled as "isolated cages". Secondly, we have hexagons of 6m positions in the middle plane and pentagons of 12o, 4h and 6m positions in the (110) plane; these cages are directly connected to each other through 6m or 4h positions and thus they are labelled as "bounded cages". It should be clear that the occupation numbers given by crystallography represent statistical averages and that all these interstitial positions will never be occupied simultaneously. Statistically and depending on the substitution for nickel, the bounded and the isolated cages contain respectively about 50% of the hydrogen atoms. Finally, it is noteworthy that the non-occupancy of the 4h and 12o sites in the $LaNi_4Cu$ and $LaNi_4Al$ hydrides leads to the disappearance of the pentagons and thus of the bounded cages.

Simple models for localized jumps between nearest-neighbour positions within a cluster of 3f-12n sites, an hexagon of 6m positions and a pentagon have been formulated (14). The structure factors are calculated from the crystallographic data and there is only one adjustable parameter : the residence time τ at a given site. The long-range diffusion is approximated by a single lorentzian function. Assuming that these two motions are uncoupled, the total scattering law $S(Q, \omega)$ is given by the convolution of the individual scattering laws. This function is convoluted with the instrumental resolution function and compared with the data.

$LaNi_{4.5}Al_{0.5}$ and $LaNi_4Mn$ hydrides

The model which gives the best agreement for the $LaNi_{4.5}Al_{0.5}$ and $LaNi_4Mn$ hydrides is the following : hydrogen atoms perform rapid localized jumps within the "isolated cages" (3f-12n clusters); the nearly two-dimensional long-range diffusion of H-atoms by successive jumps over sites of the "bounded cages" is more rapid than diffusion between the two types of cages ; this later is too slow to be observed. The corresponding total scattering law to be fitted to the data is :

$$S(Q, \omega) = \frac{N_1}{N} B_o(Q)\,\delta(\omega) + \frac{N_2}{N}\frac{1}{\pi}\frac{\Gamma}{\omega^2 + \Gamma^2} + \sum_j \frac{N_1}{N} B_j(Q)\frac{1}{\pi}\frac{\tau_j}{1+(\omega\tau_j)}$$

where N is the total number of hydrogen atoms per formula unit, N_1 and N_2 are the numbers of H-atoms respectively on 3f, 12n sites and 6m, 12o, 4h sites ; the structure factors $B_o(Q)$ and $B_j(Q)$ are rather complicated expressions involving spherical Bessel functions of distances between interstitial positions ; $\Gamma(Q)$ is the width of the Lorentzian for diffusion; $1/\tau_j$ are the widths of the Lorentzians for the localized motion and they can be expressed as functions of the residence time on the 12n site, $\tau(12n)$.

The long-range diffusion gives only an elastic contribution to the spectra measured with the lower resolution so that the $\tau(12n)$ parameter is directly fitted from these spectra. The values obtained after multiple scattering corrections for the $LaNi_{4.5}Al_{0.5}$ hydride in the temperature range 299-353K follow an arrhenius law (figure 4) leading to the activation energy :

$E_a = 175 \pm 10$ meV

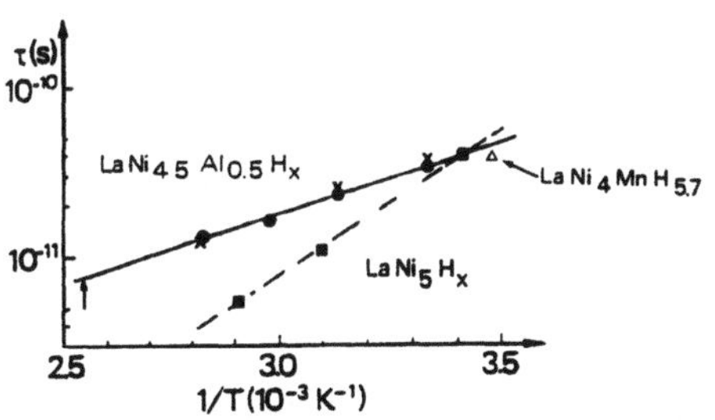

FIGURE 4 - Temperature dependence of the residence time at a 12n site fitted from IN5 data (•, Δ) and from IN6 data (x,■).

Fitting of the high resolution spectra was made by fixing the parameter $\tau(12n)$ as refined from the t.o.f. spectra. The Q-dependences of the width (h.w.h.m) of the long range diffusion component are given in figure 4 at various temperatures. Since we did not have precise measurements at low Q, we cannot derive the values of the diffusion coefficient. However we can estimate an activation energy for hydrogen diffusion in $LaNi_4Mn$:

$$E_a \simeq 230 \pm 30 \text{ meV}$$

which can be compared to the value obtained from NMR for $LaNi_{4.5}Al_{0.5}$ (13)

$$E = 280 \pm 20 \text{ meV}$$

These results can be compared with those obtained for the $LaNi_5$ hydride (11, 14) : the same dynamical model has been used to interpretate our QNS data at room temperature. The residence time at the 12n site is comparable for these three hydrides at 300 K :

$$\tau(12n) \simeq (4 \pm 2)\ 10^{-11} \text{ s}$$

but it becomes much shorter for the $LaNi_5$ hydride at higher temperature. The activation energy of the localized motion in isolated cages has been measured for $293K \leqslant T \leqslant 343K$

$$E_a = 320 \pm 30 \text{ meV}$$

which is much higher than for the $LaNi_{4.5}Al_{0.5}$ hydride.
On figure 5 we have also reported the width of the diffusive component at 295K : it shows that the diffusion coefficient at a given T is much faster in the $LaNi_5$. Also, the T-dependence of Γ (11) seems to be more important. The activation energy reported in the literature from NMR data was 300 meV (13) and 373 meV (15) (in this paper the authors have accounted for the two types of motions as we do) while from QNS, the most recent value was 275 meV (16) (this value was obtained by assuming that all the H-atoms participate equally to the long-range diffusion). Spectra measured at 393K for $LaNi_{4.5}Al_{0.5}H_{4.9}$ and at 343K and 363K for $LaNi_5H_x$ revealed the existence of another quasi-elastic component. From the data analysis without any model assumption we have found that it was also a localized motion. A satisfactory fit of the spectra has been obtained when attributing this additionnal motion to a rotational diffusion over the 6m

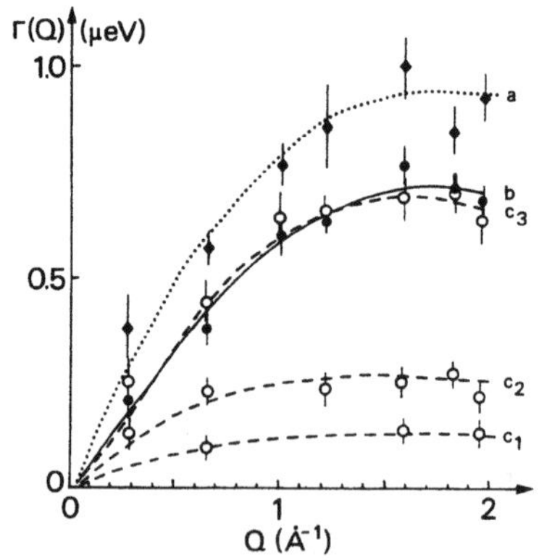

FIGURE 5 - Q-dependence of the quasi-elastic linewidth (hwhm) fitted to the IN10 spectra for : (a) $LaNi_{4.5}Al_{0.5}H_{4.8}$ (353K) ; (b) $LaNi_5H_{6.3}$ (295K) ; (c) $LaNi_4MnH_x$: (c_1) (300K), (c_2) (326K), (c_3) 365K. (The lines are included only as a guide to the eye).

positions of the hexagons in the middle plane ; τ (12n) was extrapolated from figure 4 and fixed. The fitted residence time at a 6m site for $LaNi_{4.5}Al_{0.5}H_{4.9}$ at 393 K

$$\tau\,(6m) = 15 \times 10^{-11}\ s$$

is much longer than the one at a 12n site :

$$\tau\,(12n) = 0.7 \times 10^{-11}\ s \text{ (extrapolated value)}$$

<u>$LaNi_4Cu$ and $LaNi_4Al$ hydrides</u>

For the $LaNi_4Cu$ and $LaNi_4Al$ hydrides, all the H-atoms participate to the observed long range diffusion. This is in agreement with the crystallographic data : in the particular case of $LaNi_4Al$ where only the 6m and 12n sites have been found occupied, the "bounded cages" have disappeared (fig. 6) and the only diffusion path is to jump between these two sites. On the other hand the rapid localized motion can still be interpreted by jumping inside "isolated cages". Since the quasi-elastic components of the localized motion are much broader than the Lorentzian of the diffusive motion, the total scattering law can be approximated to :

$$S(Q,\omega) = B_o(Q)\,\frac{1}{\pi}\,\frac{(Q)}{\omega^2 + \Gamma^2(Q)} + \sum_j B_j(Q)\,\frac{1}{\pi}\,\frac{\tau}{1+(\omega\tau_j)^2}$$

For the localized motion, data were collected only at room temperature (see table 2) : the width of the quasi-elastic line was too small to obtain an accurate value for τ (12n). Qualitatively, the residence time at the 12n site seems to be longer in these two hydrides than in the $LaNi_4Mn$ and $LaNi_{4.5}Al_{0.5}$ hydrides. The Q-dependences of the linewidth of the diffusive component is very weak (figure 7). For the $LaNi_4Al$ hydride at 300K, the width of the quasi-elastic line is at the limit of the measurable broadening. This is in agreement with the results from an NMR investigation (13) :

$$D(300K) \simeq 2.3 \times 10^{-10}\ cm^2/s \text{ and } E_a = 420 \pm 10 \text{ meV}$$

Although we cannot give precise values for the diffusion coefficient of hydrogen in these substituted hydrides we observe that it can be much reduced compared to the one in $LaNi_5H_x$. From figure 5 and 7, we can classified the hydrides as a function of their diffusion coefficient

$$D(LaNi_4AlH_x) < D(LaNi_4CuH_x) < D(LaNi_4MnH_x) < D(LaNi_{4.5}Al_{0.5}H_x) < D(LaNi_5H_x)$$

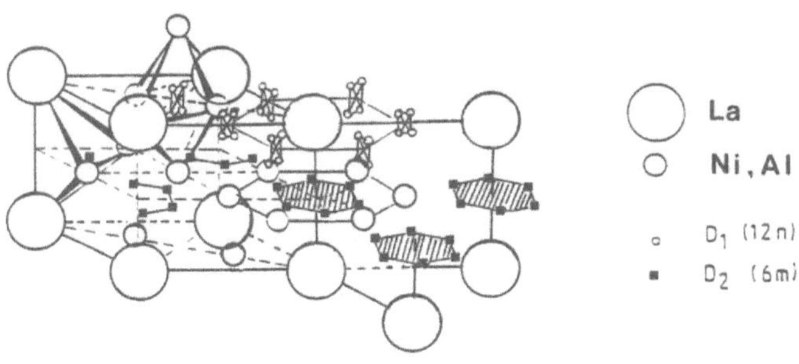

FIGURE 6 - Structure of $LaNi_4AlH_{4.8}$ width only two occupied hydrogen sites leading to the disappearance of the "bounded cages" (Al is randomly distributed at the Ni-site in the middle plane).

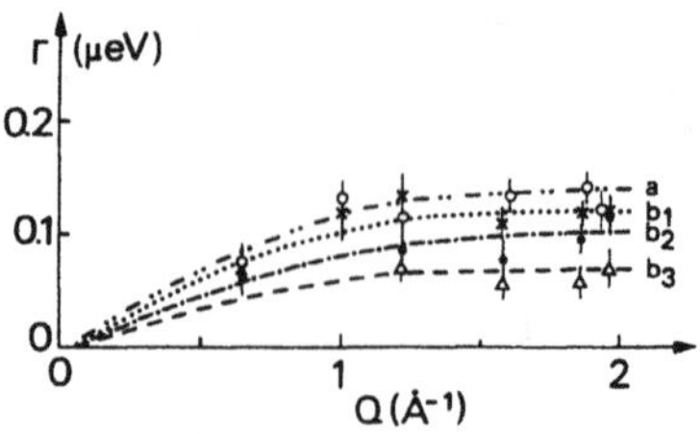

FIGURE 7 - Q-dependence of the quasi-elastic linewidth (hwhm) fitted to the IN10 spectra for : (a) $LaNi_4CuH_{5.4}$ (300K) ; (b) $LaNi_4AlH_x$: b_1 (364K), b_2 (345K); b_3 (300K)

CONCLUSION

This work shows that the observation of motions which occur on different time scales can be made from quasi-elastic neutron scattering by performing experiments on spectrometers working at different energy resolutions. It shows that QNS is a very powerful tool to study the spatial dependence of H-motions and that models for the motions can be easily formulated starting from structural results.

For the substituted $LaNi_{5-y}M_y$ ($M_y = Al_{0.5}$, Mn_1, Cu_1, Al_1) we have shown that the existence of at least two types of motion for hydrogen is directly related to the existence of isolated and bounded cages where the hydrogen is very mobile.

The reduction of the maximum hydrogen content due to the substitution is associated with a reduction of the number of H-sites. Moreover if the isolated cages exist independently of the substitution, the bounded cages do not exist any more in the compounds which absorb less than 5.1 H/formula unit.

Accordingly, the nature of the localized motion which we have attributed to jumps over the 3f-12n positions forming the isolated cages is independent of the substitution. By contrast, depending on the hydrides, the process for long-range diffusion differs : 1) the bounded cages in $LaNi_5$, $LaNi_4Mn$ and $LaNi_{4.5}Al_{0.5}$ create an easy path for H-diffusion which is almost a two dimensionnal long-range diffusion in the vicinity of the middle plane ; 2) this easy path disappears simultaneously with the "bounded cages" in the $LaNi_4Cu$ and $LaNi_4Al$ hydrides where the hydrogen atoms diffuse more slowly between the basal plane and the middle plane. These substituted hydrides seem to follow a simple rule : the more the intermetallics absorbs hydrogen, the faster are the H-motions.

REFERENCES

1 - A. Percheron-Guegan, C. Lartigue, J.C. Achard, J. Less Comm. Met. 109 (1985) 287.
2 - A. Percheron-Guegan, C. Lartigue, J.C. Achard, P. Germi,F. Tasset, J. Less Common Met. 74 (1980) 1.
3 - C. Lartigue, A. Percheron-Guegan, J.C. Achard, J.L. Soubeyroux, J. Less Common Met. 113 (1985) 127.
4 - P. Thompson, J.J. Reilly, L.M. Carliss, J.M. Hastings, R. Hempelmann, To be published in J. Physics F.
5 - T. Springer, "Quasi-elastic neutron scattering for the investigation of diffusive motions in solids and liquids", Springer Tracts Mod. Phys. 64 (1972).
6 - R. Hempelmann, this conference.

7 - A.J. Dianoux et al., this conference.
8 - C.T. Chudley, R.J. Elliott, Proc. Phys. Soc. 77 (1961) 353.
9 - H. Boutin, S. Yip, "Molecular Spectroscopy with neutrons", M.I.T. Press (1968).
10- B. Maier, "Neutron Beam Facilities Available for Users", Institut Laue Langevin, Grenoble, 1984.
11- C. Lartigue, A. Percheron-Guegan, J.C. Achard, M. Bee, A.J. Dianoux, J. Less Comm Met. 101 (1984) 391.
12- V.F. Sears, Can. J. Phys. 44 (1966) 1299 ; 45 (1966) 23.
13- R.C. Bowman, D.M. Gruen, M.H. Mendelsohn, Sol. State Com. 32 (1979) 501.
14- C. Lartigue, These Docteur d'état, Univ. Paris 6 (1984).
15- H. Chang, I.J. Lowe, R. Karlicek, "Nuclear and Electron Resonance Spectroscopies Applied to Materials Science", E.N. Kaufmann, G.K. Shenoy (eds), North Holland Publ. Comp. Oxford (1981) 331.
16- D. Richter, R. Hempelmann, L.A. Vinhas, J. Less Comm Met. 88 (1982) 353.

RECENT STUDIES OF INTERMETALLIC HYDRIDES

W. E. Wallace, E. B. Boltich, F. Pourarian, A. Fujii and A. Pedziwiatr

MEMS Department and Magnetics Technology Center
Carnegie-Mellon University
Pittsburgh, PA 15213

ABSTRACT

In recent years a large number of disordered alloy systems have been studied as hydride-formers and hydrogen host materials. These fall into two general categories: (1) substitutional solid solutions, e.g., $(Zr,Ti)Mn_2$, $Zr(Fe,V)_2$, etc., and (2) non-stoichiometric systems, e.g., $ZrMn_2T_x$ where T = Cr,Mn,Fe,Co,Ni or Cu. In the latter materials a more accurate representation is as follows: $Zr_{1-y}T_yMn_{2-u}T_u$ where y and u are quantities that may depend on the nature of T.

Systems which have been studied consist of the following: $(Zr,Ti)Mn_2$, $Zr(Mn,Fe)_2$, $(Zr,Ce)Mn_2$, $Zr(Fe,Al)_2$, $Zr(Fe,V)_2$, $(Zr,Ti)(Fe,Cr)_2$, $R(Fe,Co)_2$ where R = Tb,Dy,Ho or Er, $(Ce,Nd)Ni_5$, $Ce(Ni,Mn)_5$, $Ce(Ni,Al)_5$, $Ce(Ni,Cu)_5$, $(Y,Th)_6Mn_{23}$, $Y_6(Mn,Fe)_{23}$, $(Y,Er)_6Fe_{23}$, $ZrMn_2T_x$ and $ZrCrFeT_x$ (T = Cr,Mn,Fe,Co,Ni or Cu), $ZrMnFeT_x$ (T = Cr,Fe,Ni or Co) and $Ce_{1+x}Ni_{2.5}Cu_{2.5}$. The results show that one can "fine tune" the chemical stability of a hydride by suitable alloying. Thus one can produce metallic hydrides optimally suited for a given application.

The thermodynamic features and the magnetic properties of the systems cited were determined. It is found that the decomposition pressure of the hydride of $ZrMn_2$, which is about 0.015 atm at 50°C, can be increased by up to 3 orders of magnitude by making it non-stoichiometric with other 3d transition metals. These replace Zr in the lattice and weaken the metal-hydrogen interaction. Cobalt is unusually effective in raising the decomposition pressure. These findings are of very great importance for the utilization of metal hydrides. Stoichiometric $ZrMn_2$ forms a hydride too stable to be of practical significance. The non-stoichiometric material is less stable and is of potential significance for applications. The studies have shown that in regard to hydriding, $ZrMn_2$ closely resembles ZrCrFe and $ZrMn_2T_x$ and $ZrCrFeT_x$ are very similar.

Introduction of hydrogen in the lattice is known to affect profoundly the magnetic interactions in the host metal. For example, hydrogen converts Th_6Mn_{23} from a Pauli paramagnet to an antiferromagnet, converts Y_6Mn_{23} from a ferrimagnet to an antiferromagnet, converts $ZrMn_2$ from a Pauli paramagnet to a ferromagnet and enchances the Fe moment in Y_6Fe_{23} and in $Zr(Fe,V)_2$. The principal magnetic findings of the studies are as follows:

1. Hydrogenation generally changes the mode of magnetic coupling and frequently produces spin glasses.
2. In the $Ce(Ni,Cu)_5$ system, hydrogenation suppresses the Ce valence fluctuations and converts it into a stable Ce^{3+} ion.

Studies of the $R_2Fe_{14}B$ systems with R = Ce,Pr,Nd,Sm or Y show that 4.5 to 6 atoms of hydrogen are absorbed per formula unit. The magnetic anisotropy of $Nd_2Fe_{14}B$ changes from axial to conical.

HYDROGEN SOLUBILITY IN ORDERED AND DISORDERED PALLADIUM ALLOYS

Ted B. Flanagan, G.E. Biehl, J.D. Clewley, T. Kuji, and Y. Sakamoto*

Chemistry Department, University of Vermont
Burlington, Vermont
*Department of Materials Science, Nagasaki University
Nagashki 852, Japan

ABSTRACT

A comparison is made of the solubility of hydrogen in ordered and disordered forms of palladium alloys. In Pd_7Ce the solubility of hydrogen is greatest in the disordered form and in Pd_3Mn and Pd_3Fe the solubility is greatest in the ordered forms. In both Pd_3Mn and Pd_7Ce there are more interstices surrounded by only palladium atom nearest neighbors in the disordered form than in the ordered form. In Pd_3Fe there are more such interstices in the ordered form. There seems to be no simple pattern of behavior for the explanation of the solubility differences between the ordered and disordered forms of these alloys.

INTRODUCTION

In this paper the solubility of hydrogen in ordered and disordered forms of several substitutional palladium alloys will be discussed. Some new data will be presented for the Pd_7Ce and Pd_3Mn alloy-hydrogen systems.

It is known that hydrogen occupies octahedral interstices in palladium[1]. Statistical mechanical descriptions of this system incorporate the fact that all of these interstices are available for occupation. By contrast, for random substitutional alloys, there are a variety of interstices available for occupation, i.e., the nearest neighbor environment is different for the various interstices. For example, for random substitutional PdAg alloys there are interstices with only palladium atom nearest neighbors, those with five palladium atoms, etc. If short range order is not a factor, the fraction of the various interstices can be calculated by assuming complete randomness via the binomial theorem.

Ramanathan and Oates[2] first attempted to describe the situation for random substitutional alloys using the local environment model which is based on an earlier model for liquid alloys given by Wagner[3]. In the local environment model the energy of solution is determined by the number, but not the arrangement, of the different metal atoms in the nearest neighbor shell of the interstice. The fraction of each type of interstice is obtained by the binomial theorem and an energy for each type of interstice is assigned as a parameter in order to obtain good agreement with experiment. When this is done, expressions for $\Delta\mu_H$ can be calculated where $\Delta\mu_H = \mu_H - (1/2)\mu^{\circ}_{H_2} = \Delta H_H - T\Delta S_H = RT \ln p^{1/2}_{H_2}$. For PdAg it was found that in order to obtain good agreement with experiment the model required that only the X_0 and X_1 interstices needed to be occupied where X_i refers to the number of silver atoms in the nearest neighbor shell.

Subsequently Boureau and Kleppa[4], Yoshihara and McLellan[5] and Griessen and Driessen[6] have applied similar treatments. It should be emphasized that the agreement of one of these models with experiment does not prove its correctness because there is at least one adjustable parameter available and/or the models do not consider the influence of second and higher nearest neighbor environments. They also do not allow for electronic effects. It does seem reasonable ,however, to expect that not all of the interstices will be energetically equivalent and that this must somehow be allowed for in the partition function.

In an attempt to demonstrate the role of the nearest neighbor environment Flanagan et al[7] showed that the ordered form of Pd_3Fe dissolves considerably more hydrogen than the disordered form. In the ordered form, which is an $L1_2$ structure, there are 0.25 interstices surrounded by only palladium atom nearest neighbors and in the disordered form there are 0.178. It seems that the solubility difference found between the two forms of this alloy was too great, ~ 10, to be accounted for simply by this difference in the fraction of X_0 sites.

Although details were not given, Baranowski[8] has reported from resistivity measurements at high hydrogen pressures that the solubility of hydrogen is greater in the disordered form of Ni_3Fe as compared to the ordered form,$L1_2$. On this basis, it seems that the presence of order per se does not lead to an increase of solubility of hydrogen. It also demonstrates that the presence of more interstices having only the more readily hydride-forming element,nickel in this case, as nearest neighbors is not sufficient to cause an increase in solubility.

Phutela and Kleppa[9] investigated the solubility of hydrogen in ordered and disordered Pd_3Mn. The Pd_3Mn system has the advantage that the hydrogen solubility is greater than in Pd_3Fe. Pd_3Mn has a one-dimensional anti-phase domain structure of the Ag_3Mg type[10]. This alloy has X_o=0.125 and 0.178 in the ordered and disordered forms,respectively. [Phutela and Kleppa[9] incorrectly took X_o=0.25 for the ordered form].

Phutela and Kleppa[9] carried out calorimetric and pressure-composition-temperature measurements at elevated temperatures starting with the ordered form below the transition temperature,T_c. Upon increasing the temperature from below T_c the disordering process was clearly reflected by the hydrogen solubilities which tended to decrease with disordering. The two forms were not systematically investigated at the same temperature. The partial enthalpy for solution of hydrogen at infinite dilution, ΔH°_H, was found to be more exothermic for the ordered form than for the disordered form judging from the extrapolated behavior of the two forms. This system illustrates again that the presence of a greater fraction of interstices rich in the hydride forming element does not necessarily lead to a greater solubility.

Smith et al[11] have investigated the Pd_7Ce alloy-H system and have shown that there is a greater solubility in the disordered form. The solubilities in both forms is rather low. Ordered Pd_7Ce has an anti-phase domain structure with 0.25 interstices having only palladium atom nearest neighbors. If the disordered structure is completely random, there are 0.448 interstices with only palladium atoms as nearest neighbors.

Sakamoto et al[12] have recently also investigated the Pd_7Ce system. These workers have extended the investigation of the solubilities in the ordered and disordered forms to a larger temperature range. Basically their results confirmed those of Smith et al[11] showing that the disordered form dissolves more hydrogen. In the investigation of Sakamoto et al[12] the electrical resistance and hydrogen solubilities were determined during heating of the disordered alloy and during the cooling of the ordered alloy. The solubilities were found to correlate with the changes of resistance. One interesting feature of behavior is that the ordering appears to start at < 423 K judging from the solubility changes which is a rather low temperature. It may be that short range order starts at this low temperature. It should be pointed out that Pd_7Ce is difficult to disorder because of the ease of the ordering. The disordered sample was prepared by quenching (see below) and it showed no superlattice lines in the electron diffraction pattern nor any anti-phase domain boundaries in the TEM photomicrographs.

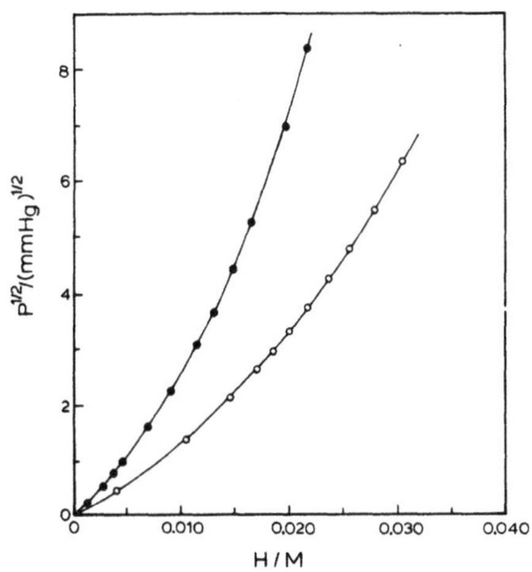

Fig. 1 The solubility of hydrogen in Pd_7Ce at 323 K. ● ,ordered alloy; O ,disordered alloy.

EXPERIMENTAL

The solubility of hydrogen was measured with a standard Sieverts' apparatus. The Pd_7Ce alloy was disordered by quenching from ~1200 to 273 K and simultaneously breaking the containing quartz vessel in order to insure rapid quenching. Pd_3Mn was quenched by a less effective method, i.e., the quartz vessel was not broken. TEM was performed on Pd_7Ce using a Hitachi H-800 electron microscope. The electrical resistance studies were carried out in vacuum during slow heating and cooling (2°C/min) of the hydrogen-free alloy.

RESULTS AND DISCUSSION

$\underline{Pd_7Ce}$. The difference in hydrogen solubilities for the ordered and disordered forms of Pd_7Ce can be seen in Fig.1 at 323 K. The data shown in Fig.1 are at a quite different temperature and pressure range than used by Smith et al[11]. Both alloy forms show positive deviations from Sieverts' law with a greater solubility in the disordered form. There was no evidence of any segregation of hydrogen to the anti-phase domain boundaries in the ordered sample. Such a segregation should be evident in the isotherms because there would be anomalous behavior in the isotherms as $H/Pd \to 0$, reflecting the energetic trapping of hydrogen to a limited number of trapping sites. Similarly an isotherm for the ordered alloy at 273 K did not show any anomalous behavior. From an estimation of the area of the domain boundaries it would appear that such an effect would have been noted if the bonding to the boundary sites was significantly more energetic than to the normal sites. Figure 2 shows the anti-phase domain boundaries in the ordered alloy and the absence of any such structure in the disordered(quenched) alloy.

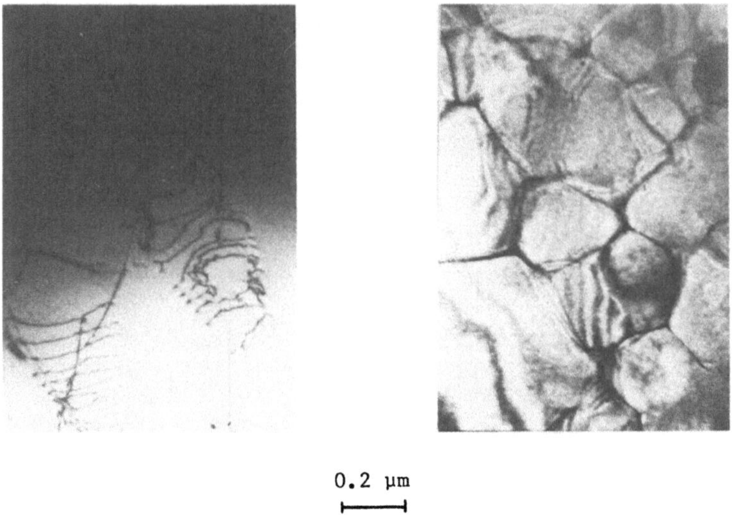

Fig. 2. TEM of Pd_7Ce. Left micrograph shows the disordered form (sample quenched from ~ 1200 K to 273 K); right micrograph shows the ordered form with anti-phase domain boundaries.

Sakamoto et al[12] reported that the values of ΔS°_{H} decrease sharply at $X_{Ce} > 0.08$. It can be argued that this is due to the onset of order in these alloys before the composition Pd_7Ce is reached. Brooks et al[13] have given evidence from resistivity measurements for the existence of short range order in these alloys. In order to obtain more information about this possibility a $PdCe(X_{Ce}=0.10)$ alloy was quenched from ~1200 K to 273 K while simultaneously breaking the quartz containing vessel. The electrical resistance of this sample did not show any indication of ordering during heating (Fig.3) as did the Pd_7Ce sample (Fig.3). Upon returning to 323 K after heating to ~1200 K a small increase in resistance was noted(Fig.3). The hydrogen solubility in the $X_{Ce}=0.10$ alloy was followed during heating and cooling. The quenched sample was heated to 1123 K while following the hydrogen solubility at several temperatures in this interval; there was no

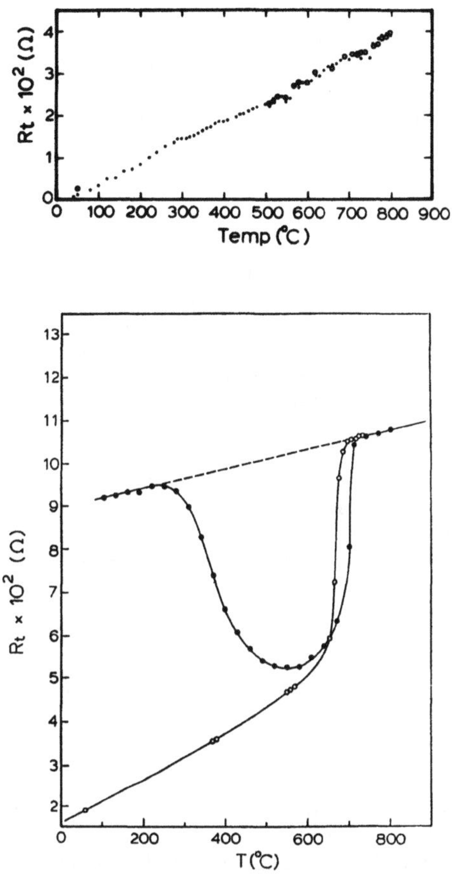

Fig.3. 1. Electrical resistance against temperature relationships (2°C/min). The upper plot is for PdCe(X_{Ce}=0.10) and the lower plot is for Pd_7Ce. In both sets of measurements the heating commenced with a quenched (disordered) sample. The filled circles show the heating data and the open circles the subsequent cooling data.

indication of any anomalous behavior, however, it was noted that there was a small solubility increase following the heating and cooling to 323 K, ~ 8%. This solubility change is in the opposite direction from that which would be expected for ordering in view of the behavior of the Pd_7Ce alloy where the solubility change following such a treatment is ~ 75% in the opposite direction. It is also curious to note that the electrical resistance of the annealed (slowly cooled from ~ 1200 K) $PdCe(X_{Ce}=0.10)$ alloy is slightly larger than that of the quenched alloy and this is also in the opposite direction from the behavior of the Pd_7Ce alloy. There was no indication of anti-domain phase boundaries in the TEM photomicrographs of the annealed $X_{Ce}=0.10$ alloy. We can offer no explanation for the small changes in electrical resistance and hydrogen solubility of the $X_{Ce}=0.10$ alloy following heating and cooling but it does not appear to be related to ordering in an obvious way in view of the direction of the changes.

The chemical potential of hydrogen relative to H_2(g,1 atm) can be written as

$$\Delta\mu_H = \Delta\mu_H^\circ + RT \ln [r / (1-r)] + \mu_H^E \qquad 1$$

where μ_H^E is the excess chemical potential, which is negative for palladium-hydrogen up to relatively large values of r, where r is the H/Pd atom ratio. The excess chemical potential has been measured for both forms of the Pd_7Ce alloy from plots of RT ln $[p^{1/2}(1-r)/r]$ against r, ie., the slope gives μ_H^E. For both forms it is positive for all r values. This is presumably due to electronic effects which, naively, can be attributed to the filling of the d-band of palladium, i.e., the d-band has been filled by cerium and the hydrogen then has to donate electrons to the s-band of higher energy. In an examination of other PdCe alloys it was found that the value of $(\partial\mu_H^E/\partial r)_{r\to 0}$ changes sign from negative to positive at X_{Ce} ~0.10 and such a change of sign occurs for the PdAg-hydrogen system at X_{Ag} ~ 0.40[14] which, by the same model, would indicate that cerium donates 4 times as many electrons to the d-band of palladium as does silver.

The initial slopes $(\partial\mu_H^E/\partial r)_{r\to 0}$ for the ordered and disordered forms of Pd_7Ce are shown in Fig.4 as a function of temperature. These values have been determined from the slopes of RT ln $[(\beta-r)/r]p^{1/2}$ against r where β has been taken as 0.25 and 0.448 for the ordered and disordered forms, respectively, i.e., β has been taken as X_o for the two forms. μ_H^E can be written as

$$\mu_H^E = H_H^E - T S_H^E \qquad 2$$

It can be shown from the experimental data that the increase in μ_H^E with r for both forms of the alloy is mainly due to the increase of H_H^E with r; this is reasonable because it is unlikely that S_H^E should vary much with r at these low values of r where the number of interstices available for occupation is greatly in excess of the number of dissolved hydrogen atoms. This seems to argue for an electronic origin as the cause of the positive values of μ_H^E and that the electronic effect differs for the ordered and disordered forms. The difference between the two forms decreases with temperature (Fig.4) and this is partly due to the ordering of the disordered alloy with increase of temperature.

Figure 5 shows a plot of $\Delta\mu_H^\circ$ against temperature for the heating of the disordered alloy and the subsequent cooling of the same alloy from above the transition temperature, i.e., it is in its ordered form below the transition temperature. These data are compared to data for palladium-hydrogen. The disordered form commences to order at ~150 °C and it is completely ordered at about 400°C. It can be seen that the change of slope with temperature is more marked for the Pd_7Ce alloy than for palladium-hydrogen. This is a consequence of the selective occupation of interstices in the alloy.

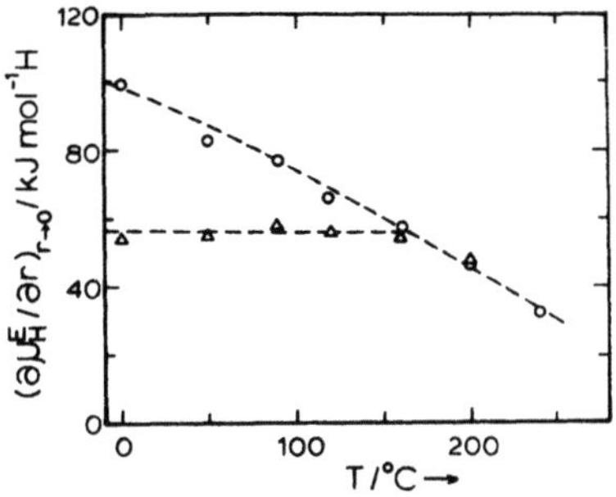

Fig. 4. Plot of μ_H^E against temperature for Pd_7Ce. O, ordered alloy, data obtained during cooling ; Δ, disordered alloy, data obtained during heating.

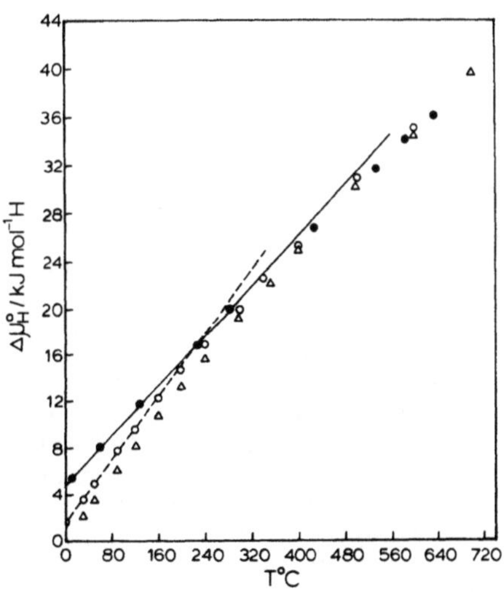

Fig. 5. Plot of $\Delta\mu_H^o$ against temperature. Δ, heating of the quenched (disordered) Pd_7Ce sample; O, cooling of ordered Pd_7Ce ; ●, data for Pd-H (refs.17 and 18).

$\underline{Pd_3Mn}$. In this research we have focused on aspects of behavior not investigated by Phutela and Kleppa[9]. For example, they investigated the disordered form above T_c (750 to 800 K) and the ordered form below T_c but they did not systematically investigate the two forms at the same temperature. Some data were obtained in the transition region where changes of hydrogen solubility with time were noted corresponding to the disorder to order transition. They did not investigate the solubility in Pd_3Mn below 555 K.

In Fig.6 a comparison of the solubilities of hydrogen in ordered and disordered Pd_3Mn is shown at 419 K. The disordered sample has been obtained by cold-working as this was found to be an effective way to disorder this alloy,however, it may still not be fully disordered. The solubility is 3.7 times as great at this temperature in the ordered form than in the disordered form.

Figure 7 shows a plot of $\Delta\mu^{\circ}_H/RT = \ln [(1-r)/r]\, p^{1/2}_{H_2}$ against T^{-1}. The plot shows the behavior of a sample as it was cooled in the disordered state from above T_c to the ordered state below T_c; the ordering transition is clearly reflected by the change of solubilities. In this plot the slope is $\Delta H^{\circ}_H/R$ and the intercept is $-\Delta S^{\circ}_H/R$. It can be seen that ΔH°_H is more negative for the ordered form in agreement with the results of Phutela and Kleppa[9]. The same sample is shown in its quenched form (largely disordered) below T_c where the large difference in solubilities between the forms can be clearly seen.

Figure 8 shows a plot of $\Delta\mu^{\circ}_H + \mu^E_H$ against r for this alloy system at 458 K. The data for the ordered sample show a more pronounced minimum at a larger value of r than do the data for the disordered sample. This again shows the difference in behavior between the two forms and it is probably of a localized electronic nature.

CONCLUSIONS

For Pd_7Ce the local environment model describes the solubility behavior for the ordered and disordered forms where the X_0 sites are assumed to be the only ones occupied as $r \to 0$ and they are assigned a value for ΔH°_H which is 10 kJ mol^{-1}H more exothermic than the X_1 sites.[12] However for alloy systems where the solubility is greatest in the form which has the smallest fraction of X_0 sites, the model cannot work,e.g., Pd_3Mn,Ni_3Fe. This seems to be an experimental demonstration of its failure. If only the X_0 sites are occupied at $r \to 0$,which seems to be reasonable, then it is clear that the large difference in observed ΔH°_H values (Fig.7) must arise from the non-nearest neighbor environment and/or from local electronic differences. Generally the unit cell of the ordered alloy is smaller than that of the disordered alloy,e.g., for Ag_3Mg ,which has the same structure as Pd_3Mn, the ordered form has a_o=4.108Å and the disordered form has a_o= 4.111 Å.[15] On the basis of the expected role of the size of the interstice on hydrogen solubility, the observed behavior of the Pd_3Mn system is in the opposite direction. Switendick[16] has pointed out that the behavior of the Ni_3Fe and Pd_3Fe systems suggests that the overall density-of-states may have little to do with the s-like hydrogen atom and its site preference in these systems. It seems that there are important local effects but these do not arise solely from the nearest neighbor environments.

ACKNOWLEDGEMENTS

TBF wishes to thank the N.S.F. for financial support of his research.

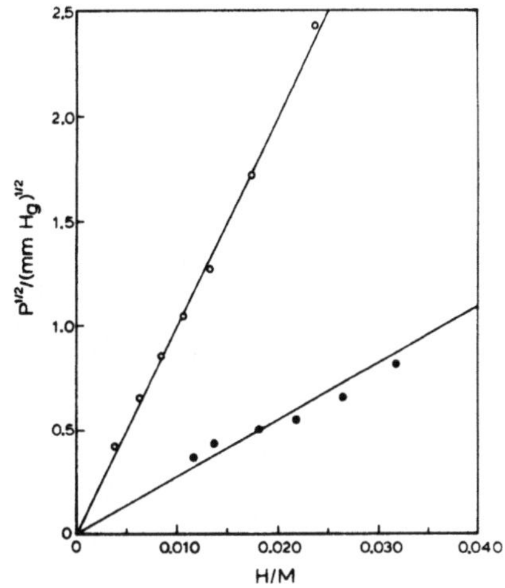

Fig. 6. Hydrogen solubility in Pd_3Mn at 419 K. ●, ordered form; ○, disordered form.

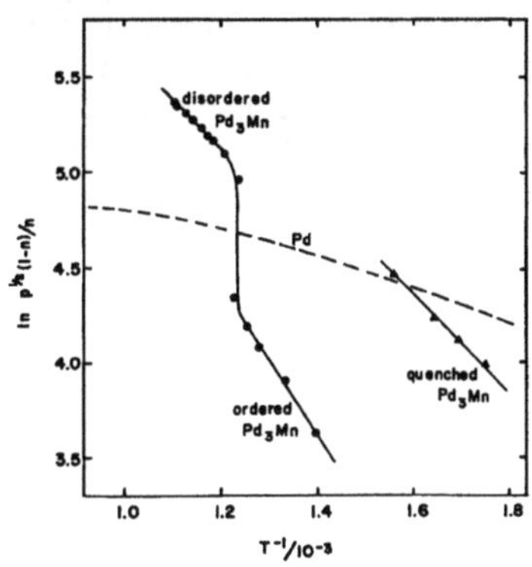

Fig. 7. Plot of $\ln p_{H_2}^{1/2}(1-n)/n = \Delta\mu_H^o/RT$ against T^{-1} where n is H/M.

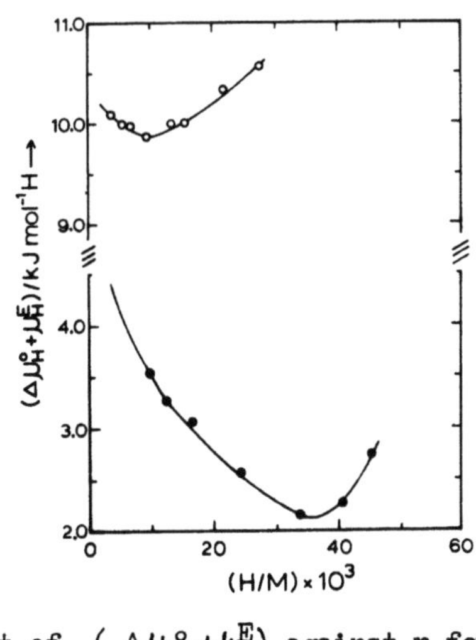

Fig. 8. Plot of $(\Delta\mu_H^o+\mu_H^E)$ against r for Pd_3Mn at 458 K. ●, ordered form; ○, disordered form.

REFERENCES

1. J.E.Worsham, J.E.Wilkinson and C.G.Shyll, J.Phys.Chem.Solids 3: 303 (1957).
2. W.A.Oates and R.Ramanathan, in Proc.of the Second Int.Cong. Hydrogen in Metals, Paris, France (1977).
3. C.Wagner, Acta Met., 21: 1297 (1973).
4. G.Boureau, O.J.Kleppa and K.C.Hong, J.Chem.Phys., 67: 3437 (1977).
5. M.Yoshihara and R.B.McLellan, Acta Met., 31: 61 (1983).
6. R.Griessen and A.Driessen, J.Less-Common Met., 103: 245 (1984).
7. T.B.Flanagan, S.Majchrzak and B.Baranowski, Phil.Mag., 25: 257 (1972).
8. B.Baranowski, in 'Hydrogen in Metals', G.Alefeld and J.Volkl, eds., Springer-Verlag, Berlin (1978), Vol.II, p.157.
9. R.C.Phutela and O.J.Kleppa, J.Chem.Phys., 75: 4095 (1981).
10. D.Watanabe, Trans.Japan.Inst.Met., 3: 234 (1962).
11. D.A.Smith, I.P.Jones and I.R.Harris, J.Less-Common Met., 103: 33 (1984).
12. Y.Sakamoto, T.B.Flanagan and T.Kuji, Z. physik.Chem.N.F., in press.
13. J.Brooks, M.H.Loretto and I.R.Harris, Metal Sci., 10: 897 (1976).
14. H.Brodowsky and E.Poeschel, Z.Physik Chem.N.F., 44: 143 (1965).
15. L.M.Clarebrough and J.F.Nicholas, Aust.J.Sci.Res., A3: 284 (1950).
16. A.C.Switendick, in 'Hydrogen in Metals', G.Alefeld and J.Volkl, eds., Springer-Verlag, Berlin (1978), Vol.I, p. 101.
17. O.J.Kleppa and R.C.Phutela, J.Chem.Phys., 76: 1106 (1982).
18. J.D.Clewley, T.Curran, T.B.Flanagan and W.A.Oates, J.Chem.Soc. Faraday Trans.I, 69: 449 (1973).

DETERMINATION OF HYDROGEN CONCENTRATION IN THIN FILMS OF ABSORBING MATERIALS

Ming-Way Lee* and R. Glosser

University of Texas at Dallas
Physics Program, FO2.3, Box 830688
Richardson, Texas 75083-0688

ABSTRACT

Two techniques for measurement of absorbed hydrogen concentration in thin films are described: the quartz crystal microbalance (QCM) and an adaptation of the volumetric technique. Their relative advantages and limitations are discussed.

I. INTRODUCTION

Full characterization of hydrogen absorbing materials requires the determination of the hydrogen concentration. For bulk materials this is not a particular problem since even standard weighing techniques will often suffice. Thin films present more difficulty simply because of the small amount of material available. The purpose of this paper is to describe two techniques which are useful for measuring hydrogen concentration in thin films. We will describe their application to the palladium-hydrogen system.

II. TECHNIQUES

Two basic techniques are of interest: the quartz crystal microbalance (QCM) and the volumetric method. We will consider both their advantages and their limitations.

A. QCM

1. Technical Description. The quartz crystal microbalance, more familiarly known as a "thickness monitor," is simply a piezoelectric quartz platelet cut to resonate at a frequency typically ∿5 MH_z. When placed in a vacuum system and exposed to evaporated material or adhering gases, the deposited material increases the mass of the crystal with subsequent lowering of the resonant frequency. This technique was first suggested by Sauerbrey[1] in 1957. Its more significant feature is the linear relation between deposited mass and frequency shift over a useful mass range. One can write this[2] formally as

*Present address: Department of Physics and Astronomy, University of Maryland, College Park, MD 20742.

$$\Delta m = -\Delta f \rho_q v_q A_q / 2f_q^{\,2} \qquad (1)$$

where Δm is the deposited mass, Δf is the frequency shift, ρ_q is the quartz mass density, v_q is the acoustic wave velocity of the thickness-shear wave in quartz, A_q is the area of one crystal face and f_q is the frequency of the crystal prior to addition of deposited mass.

The maximum mass load over which this linear relation holds corresponds to a maximum frequency shift of about 1% of f_q. For f_q=5 MHz, this means $\Delta f \sim 50$KHz or for example with palladium this yields a deposited thickness of about 7KÅ.

Recent work[3] indicates that the linear range of this technique can be extended by measuring the period change rather than the frequency during mass deposition. (Writing equation 1 in differential form with f_q as a variable and integrating over the frequency range yields this result.)

Of particular interest in the application to be described is the mass sensitivity of the QCM. While there are indications[4] that a sensitivity as high as 10pg/cm^2 can be achieved, our own work suggests that with a frequency resolution of about 0.1Hz for a 5MHz crystal, a mass resolution of about a ng/cm^2 can readily be attained.

2. Application to Hydrides. The QCM seems very well suited to the determination of hydrogen in metals especially because of the opportunity to make in situ measurements. At present, however, the technique probably has not been carried far enough along to give unambiguous results. The major problem appears to be the stress which develops in the hydrided metal. This and other problems will be discussed in the subsequent section.

The basic QCM technique for measuring hydrogen concentration is as follows: A metal film, for example palladium, is deposited on the exposed crystal face and the frequency change Δf_{Pd} determined. (The thickness can be calculated from equation 1.) Upon exposure to hydrogen, the frequency will again shift corresponding to the hydrogen mass uptake which we denote as Δf_H. The atomic concentration of hydrogen can then be very easily calculated from the ratio of the frequencies normalized to the atomic masses. That is

$$H/Pd = (\Delta f_H \, M_{Pd})/(\Delta f_{Pd} \, M_H) \quad . \qquad (2)$$

Assuming a frequency change resolution of about ±0.1Hz, we can find the uncertainty in our determination of H/Pd for various film thicknesses. If we obtain H/Pd = 0.1 for a 1000Å film, the uncertainty of this ratio is about ±1.5%. If the thickness is only 100Å, the uncertainty goes up to about ±15%.

An example[5] of the results obtainable by this technique is depicted in figure 1. Here is shown a set of pressure-concentration-temperature (PCT) relations taken at 27C for the palladium-hydrogen system with film thickness as a parameter. The results show that for film thicknesses ∿1000Å, the PCT relation is comparable to bulk results[6]. As the film is made thinner, the hydrogen solubility in the dense β phase decreases while that of the dilute α phase increases. Simultaneously, the mixed phase region acquires a slope which increases with decreasing thickness. While the general feature of decreasing β solubility with decreasing thickness is borne out by other measurements[7], it now appears that some features of these PCT results may be artifacts of the measuring technique. We will discuss this in the subsequent section.

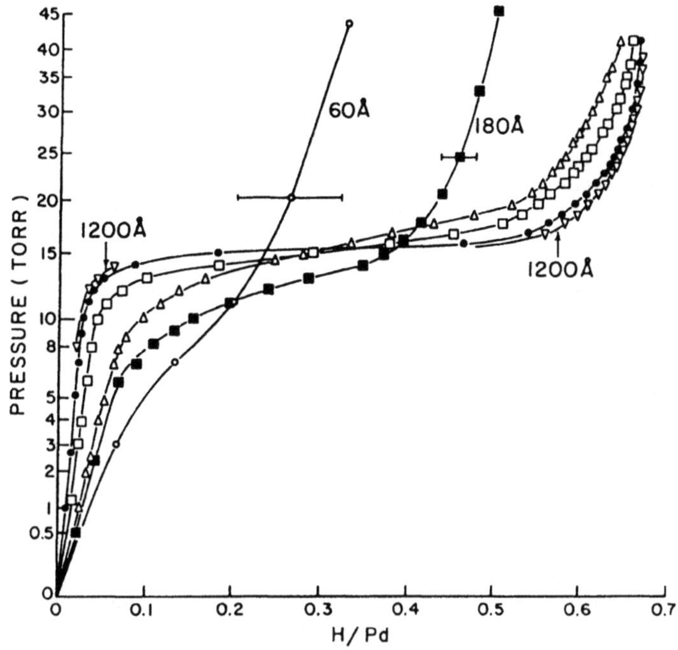

Figure 1. Pressure-concentration isotherms of palladium-hydrogen obtained for annealed films of various thicknesses using a QCM, ∇, 1200Å; ●, 990Å; □, 750Å; Δ, 520Å; ■, 180Å; O, 60Å. Error bars are shown for the 60Å and 180Å films. Errors for the thicker films are within the size of the data point symbols. The pressure is plotted on a square root scale. T=27C. (Ref. 5.)

This type of measurement was extended by Bakker and coworkers[8] and by Feenstra and coworkers[9] in which they reported the behavior of the PCT relation over a range of temperatures for a given film thickness. Their results indicate that the critical point of the coexistence region for the PdH system is depressed. For a 1220Å film, they find that the critical point occurs at a temperature ∿460K and a pressure ∿3bar compared with the bulk values 565K and 20bar. They account for this depression as resulting from the clamping which occurs at the palladium quartz interface.

Papathanassopoulos and Wenzl[10] made a similar set of measurements on films (∿5000Å) of vanadium. They were able to extract thermodynamic parameters of the films and they comment that their thin film results are comparable to bulk. The kinetics for hydrogen absorption and desorption by the vanadium was enhanced by the evaporation of a thin palladium film (∿100-500Å) immediately following the evaporation of the vanadium.

3. Technical Problems. The use of the QCM as an evaporated film thickness monitor or as a device for determining surface gas adsorption

has long been well established. Its application to determining the hydrogen absorbed in the bulk of a metal is relatively recent, and these results demonstrated a basic problem with this particular application which has not yet been resolved. Bucur and Flanagan[11] (BF) were the first to report the use of the QCM for studying hydrogen absorption by thin metal films. Their major results centered on the kinetics of the hydrogen-palladium interaction, but they also reported measurements of the hydrogen concentration. The latter results indicated H/Pd ratios significantly larger than obtained in bulk, and they explained this increase as stemming from the stress produced by the Pd lattice expansion upon hydrogen absorption. This is not surprising since the lattice parameter undergoes a fractional increase of nearly 4% for bulk material upon entering the β phase whereas in this work the material is bound to a substrate. BF report a dominance in the α phase of frequency shift due to stress over that due to mass absorption. In the β phase, they indicate that the two contributions yield comparable shifts.

Subsequent measurements[5,8–10] seemed to indicate that perhaps stress effects were not so much a problem after all. In all these cases the results demonstrated that for sufficiently thick films (∿1000Å) the PCT relation appeared to be very close to bulk, and arguments were made as to why the stress effects seen by BF were not found in the later measurements. However, recent measurements by Lee and Glosser[7] (LG) using a volumetric technique to determine the H/Pd ratio suggest in fact that the stress effect has not been eliminated, as will shortly be discussed.

One quantity which should be considered for mass correction is acoustic impedance mismatch[3] which results from the difference in shear wave velocity and mass density between quartz and the deposited material. However, for the small mass loads discussed here this turned out to be of negligible importance.

In using the QCM for gas absorption measurements, hydrostatic and viscoelastic corrections need to be taken into account[12]. That is, just the increase of hydrostatic pressure itself causes the crystal frequency to shift and, in addition, a real gas has a non-zero viscosity which allows coupling to the shear mode of the crystal. This produces a small damping effect and a consequent frequency shift. For hydrogen we find a frequency shift ∿ 2 Hz for a pressure increase from base pressure to 50 Torr. The effect is crystal dependent but reproducible for a given crystal in an undisturbed mount. In our measurements, the effect was compensated by exposing the crystal to hydrogen prior to coating the crystal with an absorbing metal. The frequency shift was then determined as a function of pressure, and these data were subtracted from the raw pressure-concentration data obtained on subsequent coating with palladium and exposure to hydrogen.

Another problem which must be considered is that of metallic interdiffusion between the gold electrode and the hydrogen absorbing metal. This has effectively been eliminated by the Amsterdam group[8,9] who removed the Au electrode and simply replaced it by the metal to be hydrided. We have also done this successfully but find that a special technique is required to fully remove the electrode material[13].

4. Stress Elimination. It appears that the largest obstacle to the use of the QCM to determine the concentration of absorbed hydrogen in metals is the development of stress upon hydrogen uptake. The basic problem is that for the usual quartz crystal cut (AT) the frequency shift with stress is in the same direction as that due to mass deposition. Consequently, in determining hydrogen concentration, one is unable to separate out the two contributions. A very elegant solution has been proposed for

the general problem of separating mass change from stress by EerNisse[14]. He suggested placing two independent crystals in close proximity with one crystal being the standard AT-cut with a stress-fractional frequency shift proportionality constant $K=-2.75 \times 10^{-11}$ m^2/N while the other is the BT-cut which has $K=+2.65 \times 10^{-11}$ m^2/N. He demonstrated that the deposited mass is proportional to the sum of the two frequencies while the stress is proportional to their difference.

EerNisse[15] suggested yet another possibility whereby an SC(stress compensated)-cut is used. This is a doubly rotated orientation whereas the AT- or BT-cuts are singly rotated. The SC-cut has the virtue of possessing characteristics similar to that of the AT- or BT-cuts but with a K value of zero and a minimal temperature coefficient.

B. Volumetric

The volumetric technique has been used extensively for hydrogen concentration measurement in bulk metals. It requires two chambers separated by a valve. The first chamber is connected to the hydrogen gas supply and a pressure gauge while the second is the sample chamber. For a fixed amount of gas introduced into the two chambers and with the interconnecting valve open, the pressure difference with and without samples in the second chamber directly determines the amount of hydrogen taken up by the sample. Knowing the mass of the sample, the hydrogen concentration can then be determined.

Ordinarily this technique would not be used to determine hydrogen concentration in thin films because of the relatively small mass involved. However, we have taken steps[7] which permit application of this technique to films as thin as 300Å. An obvious step is to increase the total volume of the film by increasing the film area. In the work we describe here, sets of 30 substrates were simultaneously coated with palladium with thicknesses ranging from 300 to 1000Å. The pressure was measured using an MKS 220B capacitive manometer which has a differential resolution ∿1 millitorr. Pressure-concentration relations were determined for each set, and the results, shown in figure 2, demonstrate the qualitative PCT behavior that is found in bulk materials except that the hydrogen concentration at the onset of the β phase is smaller than for bulk. However, there is a marked difference in the PCT relation obtained by this means compared to the QCM. We discuss this in section III.

Limitations and Advantages of the Volumetric Method. While this technique provides a relatively easy method of determining the hydrogen concentration in films, there are certain limitations. The most obvious is the necessity for large surface area which complicates the preparation. For example, it becomes difficult to obtain a uniform sample thickness because the substrates have to be distributed over a large solid angle range when coated. Also, since maximum sensitivity is obtained with the set contained in a minimal volume, it is desirable to move the coated samples from the evaporation chamber to this minimal volume which entails breaking vacuum. With such a large surface area it also becomes more difficult to maintain the set at a uniform temperature.

Despite these limitations the technique offers a very straightforward means of determining the PCT relationship for thin films. Furthermore, it can be applied to samples on any substrate or even free films. One possibility is to prepare films with thickness > 1500Å on a smooth substrate. Our experience has been that after several absorption-desorption cycles, the film will peel off intact.

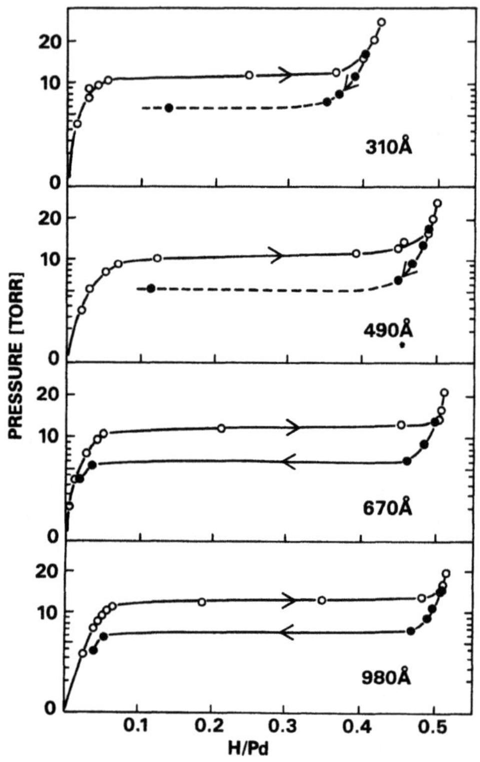

Figure 2. Pressure vs H/Pd ratio for various film thicknesses using the volumetric technique at T=27C. A square root scale is used for pressure. (O:absorption; ●:desorption.) (Ref. 7.)

III. COMPARISON OF QCM AND VOLUMETRIC MEASUREMENTS OF THE PCT RELATION

Figure 3 shows a comparison of the PCT measurements obtained by the QCM and the volumetric techniques for 490Å unannealed films. The samples were prepared simultaneously and taken through 50 absorption-desorption cycles. Several marked differences are apparent. The QCM result when compared to the volumetric plot shows a larger apparent hydrogen concentration in both the α and β phases and there is a significant slope to the mixed phase region. Similar behavior was found for other thicknesses as well.

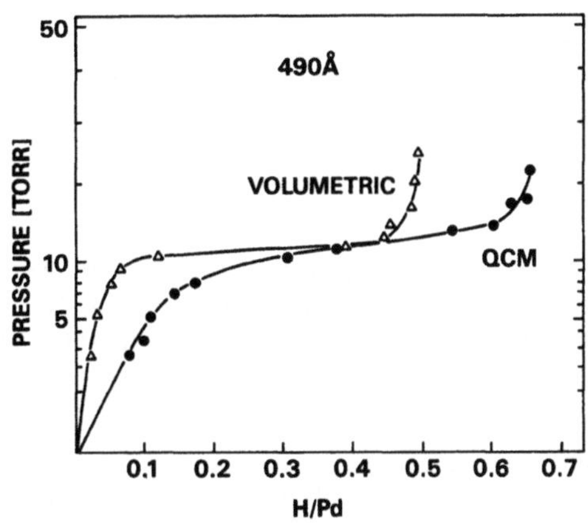

Figure 3. Comparison of pressure-concentration isotherms for palladium-hydrogen obtained by volumetric and QCM measurements on simultaneously prepared, unannealed 490Å films at 27C. A square root scale is used for pressure. (Ref. 7.)

The explanation for this discrepancy is not clear at present. We believe that it is most likely due to the stress at the palladium-crystal interface but other considerations should not be ignored. For one thing the QCM and volumetric substrates are different. The QCM crystals used for these measurements are relatively rough with a gold electrode. On the other hand, the volumetric substrates are optically polished sapphire. The direct effect of either surface roughness or hardness has yet to be examined.

IV. DISCUSSION

We have described two techniques which have been applied to the determination of hydrogen concentration in absorbing metals. While we feel that the volumetric technique yields results which are uncomplicated by extraneous effects, it in fact is somewhat cumbersome to use in practice and relatively insensitive. The QCM on the other hand, while considerably more convenient and sensitive, has the apparent limitation of stress sensitivity. There is the possibility that this stress problem can be eliminated using the techniques proposed by EerNisse[14,15] but this remains to be demonstrated.

We believe that an appropriate approach for devising a technique to determine hydrogen concentration in thin metal films begins with the use of the volumetric technique in order to establish its behavior in a direct way. Such parameters as substrate dependence, deposition conditions, thickness dependence and temperature dependence all need careful determination. With this in hand, an evaluation of the QCM approach becomes possible.

V. ACKNOWLEDGEMENTS

We gratefully acknowledge support from the Organized Research Fund of The University of Texas at Dallas. Portions of this work were supported by Texas Instruments Incorporated.

REFERENCES

1. G. Z. Sauerbrey, Phys. Verhandl. 8:113 (1957); Z. Phys. 115:206 (1959).
2. A thorough discussion of the QCM is contained in C. Lu and A. W. Czanderna, "Applications of Piezoelectric Quartz Crystal Microbalances" (Volume 7 of Methods and Phenomena, Their Applications in Science and Technology), Elsevier, Amsterdam (1984).
3. C. Lu in reference 2.
4. A. Warner and C. D. Stockbridge, in: "Vacuum Microbalance Techniques, Vol. 3," K. H. Behrndt, ed., Plenum, New York (1963).
5. Gary A. Frazier and R. Glosser, J. Less-Common Met. 74:89 (1980).
6. E. Wicke and G. H. Nernst, Ber. Bunsenges. Phys. Chem. 68:224 (1964).
7. Ming-Way Lee and R. Glosser, J. Appl. Phys. 57:5236 (1985).
8. H. L. M. Bakker, G. J. de Bruin-Hordijk, R. Feenstra, R. Griessen, D. G. de Groot and J. Rector, in: "Electronic Structure and Properties of Hydrogen in Metals," P. Jena and C. B. Satterthwaite, eds., Plenum, New York (1983).
9. R. Feenstra, G. J. de Bruin-Hordijk, H. L. M. Bakker, R. Griessen and D. G. de Groot, J. Phys. F 13:L13 (1983).
10. K. Papathanassopoulos and H. Wenzl, J. Phys. F 12:1369 (1982).
11. R. V. Bucur and T. B. Flanagan, Z. Phys. Chem. NF114:109 (1974).
12. Gary A. Frazier, Thin Film Properties of the Palladium-Hydrogen System, Ph.D. Dissertation, The University of Texas at Dallas, Richardson, Texas (December 1984).
13. J. Villalobos and R. Glosser, (to be published). Underneath the Au electrode (removable with aqua regia) is a Cr/Cr_2O_3 layer. This can be removed by a drop of HCl on the surface and scratching the layer with Al wire.
14. E. P. EerNisse, J. Appl. Phys. 43:1330 (1972); 44:4482 (1973).
15. E. P. EerNisse, in ref. 2.

COMPARATIVE STUDIES OF AMORPHOUS AND CRYSTALLINE HYDRIDES *VIA* INCOHERENT SCATTERING

Nikos G. Alexandropoulos

Dept. of Physics
University of Ioannina
GR-453 32 Ioannina, Greece

ABSTRACT

Starting with a general evaluation of the particle's beam as a probe to extract the hydride's band structure, this review summarizes the essential feature of the incoherent scattering as probe in the comparative studies of amorphous and crystalline hydrides. In this connection the essentials of the approximations for photon-electron and electron-electron collision are explained. Experimental procedures for the determination of difference Compton profile using X-ray, γ-ray and electron sources are described. The advantage afforded by this type of measurements in the study of electronic properties are illustrated in the case of Ti and TiH_2.

INTRODUCTION

The most informative probe to extract the detailed hydride's band structure is a beam of particles with energy and momentum comparable to that of the band electrons. The obvious choice, the low energy electron beam, is not applicable in the condensed matter bulk properties studies, because its shallow penetration. Other choices, such as the photon beam emission and absorption spectroscopy, can provide information either about the electron's energy or about the electron's momentum, but not for both at the same time, because the photon dispersion relation is generally quite different from that of the band electrons. This is not however the case in Compton spectroscopy, where the spectra of incoherently scattered radiation are more informative because the probing "particle" has energy

$\omega = \omega_1 - \omega_2$ and momentum* $\vec{k} = \vec{k}_1 - \vec{k}_2$. In other words, the Compton Spectroscopy produces a probe of artificial "particles" with the appropriate dispersion relation $\omega(k)$, to scan the vicinity of any band's point, E_i, k_i, by providing the proper experimental conditions. This method, utilizing the x-ray incoherently scattered radiation, is not only comparable, but supercede other methods in the study of hydrides. Inelastic scattering however is not a process restricted between photon and electron, although the term "Compton Scattering" is traditionally associated with photon - electron collisions, but it can take place also in electron-electron collisions. The Electron Compton scattering that has been recently employed in the study of condensed matter in the form of ECOSS (Electron Compton Scattering from Solids) enjoys some advantages over its photon counterpart. Firstly the electron scattering cross - section is about four orders of magnitude larger than the Thomson cross-section. This facilitates the data collection without exceeding the weak coupling limits or as it is commonly referred, the Born approximation. Secondly the use of electron microscopes fitted with electron velocity analyzer provide high resolution, high intensity data at energies chosen at will. Finally a substantial advantage of the Electron Compton experiment is the possibility to gain simultaneously information about the recoil electron, so the three dimensional electron momentum is obtained, in principle. On the debit side, specially in the study of hydrides, is the use of vacuum and the sample heating by the electron beam, resulting in variation of the hydrogen concentration of MH_x.

Figure 1 shows schematically a scattering experiment where a monochromatic, polarized,incident beam ($\vec{e}_1$, polarization; $\vec{k}_1$, momentum; ω_1, energy) interacts with a solid in an initial state Φ_1. After the interaction, the system is rised in a state Φ_2 and the outgoing beam carries information about Φ_1 and Φ_2 in the form of the beam's spectral distribution. This distribution is related to Φ_1 throught the differential scattering cross-section which in general can be written:

$$\frac{d^2\sigma}{d\omega d\Omega} \propto |< \Phi_2, \vec{k}_2, \omega_2, || \Phi_1, \vec{k}_1, \omega_1, >|^2 \tag{1}$$

In the case of incoherent scattering $|\Phi_1> \neq |\Phi_2>$, the equation (1) reduces to:

$$\frac{d^2\sigma}{d\omega d\Omega} = (\frac{d\sigma}{d\Omega})\ S(\vec{k},\omega) \tag{2}$$

*The subscript 1 referring to the incoming and subscript 2 to the outgoing. When the atomic units ($e = \hbar = m = 1$) are adopted $h\omega_1 = \omega_1$.

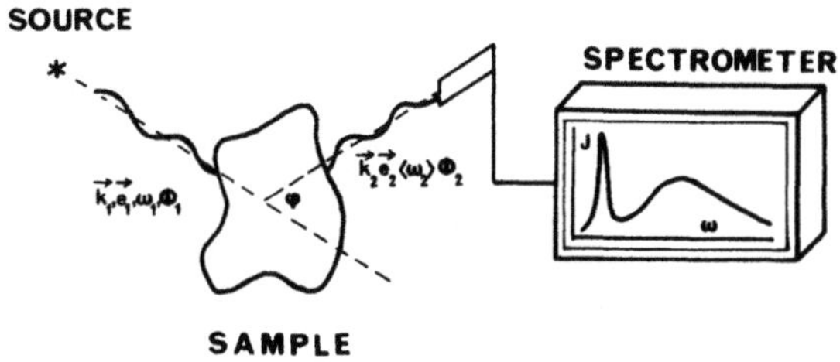

Fig. 1. Block diagram of an incoherent scattering experiment in the condensed matter study.

where $S(\vec{k},\omega)$ is the Dynamical Form Factor*, which represents the target's response to momentum and energy transfer $\vec{k}$ and ω respectively or in other words the response to an artificial "particle" of wave vector $\vec{k}$ and energy ω.

Figure 2 shows the dispersion curves of the particles most commonly used in the study of condensed matter. The $\omega_e(k)$ and $\omega_p(k)$ are the dispersion curves for electrons and photons respectively. It is clear that the use of low energy electrons is preferable because their energy and momentum are comparable to the energy and momentum of conduction electrons. However in both cases of photons and electrons, only one set of points E_i, k_i can be deduced from the experiments. The low energy electrons have an additional disadvantage, the very small penetration depth, that makes them useful only to the surface studies. Finally the shadowed area** between $\omega_+(k)$ and $\omega_-(k)$ indicates the dispersion of the artificial "particle" of energy $\omega = \omega_1 - \omega_2$ and momentum $\vec{k} = \vec{k}_1 - \vec{k}_2$. It is clear that Compton scattering provides an additional degree of freedom useful in scanning

---

* The factor $(\frac{d\sigma}{d\Omega})$ depends on the interaction. In the case of x-rays it is the Thomson cross-section, somewhat less than 1 barn and in in the case of electron-electron collision it is the Rutherford cross-section, about 10^4 barns for small angles.

** The shadowed area of fig 2 indicates the x-ray and γ-ray Compton scattering. A similar dispersion area exists also for the electron Compton scattering from solids (ECOSS).

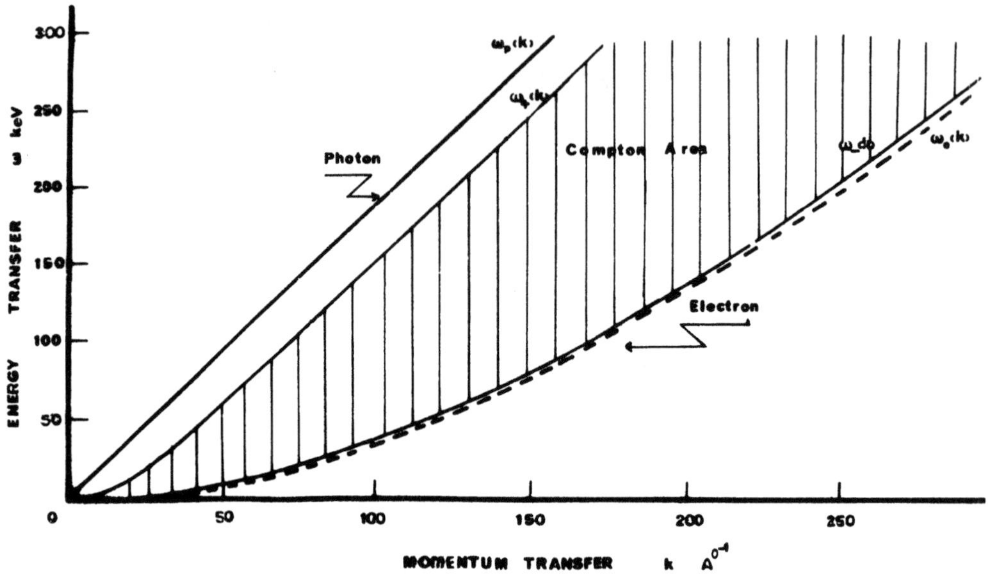

Fig. 2. Dispersion curves for the particles employed as a probe in the study of condensed matter.

the valence and conduction band. Except the spread shown in the shadowed area which depends on the experimental conditions*, there exists an additional spread in the energy transfer $\langle\omega\rangle$ depending on the electron momentum. This spread is given by:

$$3.84\ k^2 - 14.45\ qk \geq \omega(k) \geq 3.84\ k^2 + 14.45\ qk \tag{3}$$

where ω is given in eV, k in A^{-1} and q, the scattered electron momentum projection on k in a.u.

For experimental and computational reasons the study of the condensed matter via incoherent scattering is divided in three subareas of research, the low, the intermediate and the large momentum transfer region. This classification is dictated by experimental and theoretical approximation techniques related to the magnitude of the momentum transfer; however the boundary between these regions is quite nebulous depending on the energy levels of the target's electron ground state. The first region , the low energy momentum transfer, is dominated by the collective properties of the band electrons. The small angle x-ray scattering experiments that provide k comparable to the k_F (Fermi wavevector) are the most appropriate for the studies, of valence electron collective properties.

*Initial energy ω_1 and scattering angle φ

In this region the relation (1) reduces to

$$\frac{d^2\sigma}{d\omega d\Omega} \propto \mathrm{Im}\varepsilon^{-1}(\vec{k},\omega) \tag{4}$$

where $\varepsilon(\vec{k}, \omega)$ is the dielectric function.

The third region, the high momentum transfer, is the one where the single particle properties of the electron are prominent. This region is investigated mainly by γ-ray Compton scattering experiments and equation (1) is reduced within the limits of Impulse Approximation to:

$$\frac{d^2\sigma}{d\omega d\Omega} \propto \int |n(p_o)|^2 \, \delta(\frac{\omega-k^2}{2m} - \frac{\vec{k}.\vec{p}_o}{m}) d^3p \tag{5}$$

where $n(p_o)$ is the Fourier transform of the ground state electronic wave function, X(p). Equation (5), under certain conditions, is equivalet to the DuMond's original equation relating the Compton profile J(q) to the electron momentum distribution,

$$J(q) = \frac{1}{2} \int_{(q)}^{\infty} \frac{I(p)}{p} \, dp \tag{6}$$

Equations (5) and (6) have frequently been used in deriving the one-dimensional electron momentum distribution from the x-ray and γ-ray Compton spectra. This subject was reviewed by several authors [1-5].

The region between the two above mentioned regions is known as intermediate momentum transfer area and it is the less understood because of the failure of most approximations to interpret the few experimental data. In the case of Electron Compton Scattering from Solids (ECOSS) within the limits of the Born approximation, similar relations to (4),(5) and (6) are valid [4,6-9].

EXPERIMENTAL METHODS

The x-ray incoherent scattering is a weak effect, the size of which can roughly be estimated from the Thomson scattering cross-section (10^{-24} cm^2/sr) and the concentration of the target's scatters (about 10^{22} per cm^3). In addition to the effect's weakness, the experimental requirement of high energy and angular resolution results in formidable conditions, so that only the right combination of source and spectrometer can provide re-

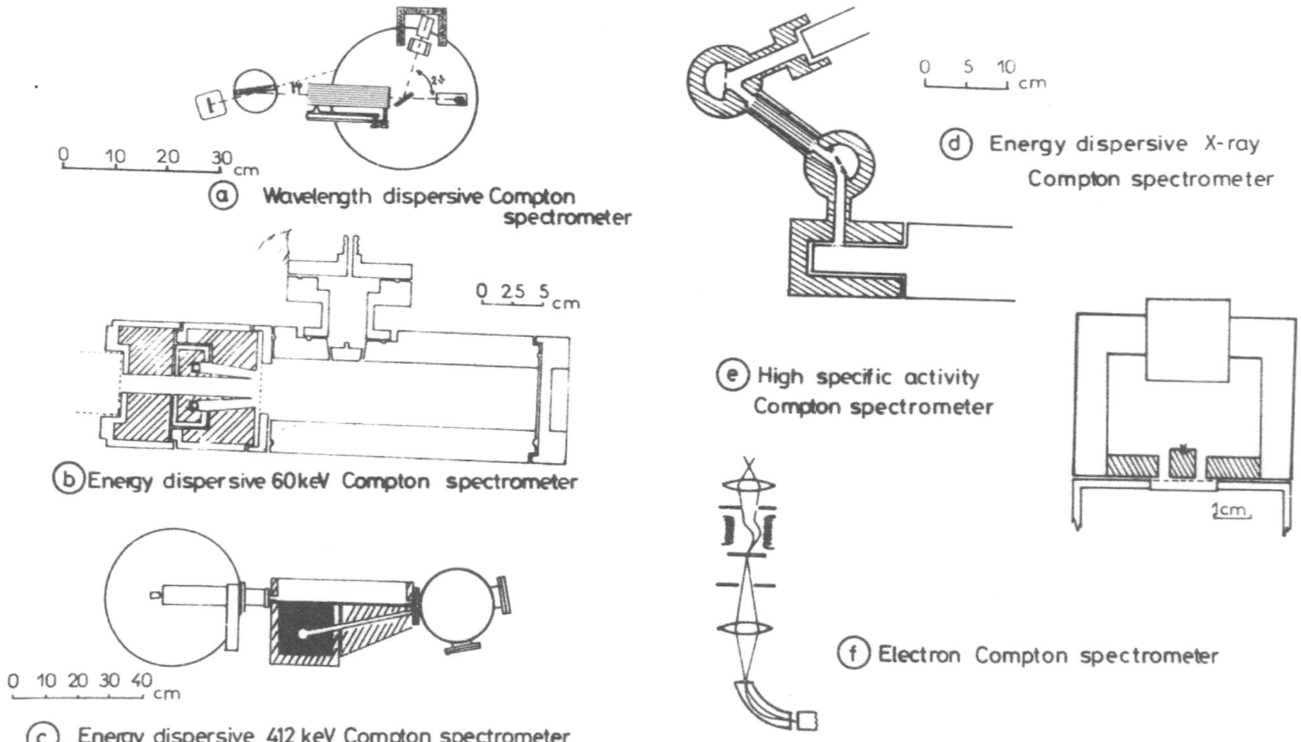

Fig. 3. Some of the spectrometers employed in the study of condensed matter via incoherent scattering. Details on this instruments can be found in references 3,4,5,10 and 14.

liable data. Every known type of wave-length-dispersive x-ray spectrometer has been used with some degree of success in the remote and the near past. It should be noticed however that the only improvements since the early days consist on the crystal's quality and the sophistication of electro-mechanical stepping devices. The simplest and most popular of all spectrometers is the single flat crystal spectrometer with a soller (parallel) slit (fig. 3a). Instruments like this have been used for studies in the low momentum transfer regime and extensively in the high momentum transfer regime for derivation of electron momentum distibution. It is however unrealistic to employ this instrument in the high momentum transfer region for absolute studies on anything heavier than Sulphur, mainly because of the core electron contribution. The above spectrometer can be used in comparative studies (difference experiments) where are subtracted out all systematic errors and the core contribution.

The energy dispersive spectrometer (Solid State Detector) fitted to a chamber which includes the sample and a γ-ray source is the dominant instrument in the high momentum transfer regime for the electron momentum distribution measurements. Two such systems are shown in fig. 3b and 3c. Recently, instead of the γ-ray source in Compton spectrometers, there have been employed x-ray spectrometry tubes with heavy metal anodes operated at 100 kV, with the obvious advantages[10] (fig.3d). Table (1) presents a summary of the most commonly used methods in the incoherent scattering experiments. The most popular method is the one using ^{241}Am plus SSD mainly due to the source's availability and its long life time. It is however restricted for studies of low Z elements because the study of anything heavier than Iron would be jeopardised through the failure of the impulse approximation for core electrons. This method can be used only for comparative studies in which core contributions and systematic errors can be subtracted as in the previous case. For the heavier elements, the ^{198}Au plus SSD method is more appropriate; however on the debit side are the short life time (2.7d) of the source and the lower momentum resolution. It is noticeable that recently appear methods using weak γ-ray sources of about 10 mCi; in this development, weak point sources of high specific activity are used at small samples and with beam paths of a few centimeters (fig.3e).

Finally, the Electron Compton Scattering method is a novel method, which employs an Electron microscope specially modified and an electron velocity analyzer, (fig. 3f).

Table 1. MERIT FOR INCOHERENT SCATTERING EXPERIMENTS

Method	Incident energy, ω_1 (keV)	Half-life	Strength	Flux phot/s.deg^2	Momentum* resolution	Momentum transfer, k (A^{-1})	Energy transfer, ω (keV)	Appropriate in the region
x-ray tube plus crystal spectrometer	CrK 5.4	N.A	1-12 kW		400	0.44-5.02	0.008-0.09	LMT
	CuK 8.0		1-12 kW		370	0.70-7.99	0.020-0.24	LMT-IMT
	MoK 17,6		1-30 kW	10^{11}	330	1.58-17.65	0.110-1.18	IMT-HMT?
Radioactive source plus SSD	^{241}Am 60	425 yrs	0.5-5 Ci	10^6	3.4	5.27-55.05	1.200-11.41	HMT
	^{57}Co 122	120 ds	10 mCi		3.7	10.82-103.72	4.870-39.43	HMT
	^{191}Os 129	15 ds	10 mCi		3.7	11.47-108.86	5.440-43-28	HMT
	^{123m}Te159	102 ds	1 Ci		3.7	14.27-130.31	8.190-60.99	HMT
	^{51}Cr 320	28 ds	45 Ci		3.2	31.21-234.28	31.49-177.93	HMT
	^{198}Au 412	2.7 ds	100-200 Ci	10^8	2.7	42.69-288.86	50.75-254.30	HMT
	^{137}Cs 662	30 yrs			2.4	80.98-	121.27-	HMT
x-ray tube plus SSD	WK	N.A	3 kW	10^9	3.4	5.18-54.21	1.16-11.07	HMT
	AuK				3.6	5.97-61.71	1.54-14.29	HMT
Electron microscope plus velocity analyser (ECOSS)	30-120	N.A	100 μA	10^{15}	650	-	-	LMT-IMI-HMT

* As Momentum resolution is defined here, the ratio $\frac{q}{\Delta q}$ of Compton profile J(q) for q = 2 a.u

DATA ANALYSIS

The Compton Profiles superficially look alike because the physically interesting effects only change theme by small amounts. Accordingly, a high degree of sophisticated corrections is required to utilize the line shape changes and to obtain profiles useful for comparison to theory. In absolute studies, where we process the raw data to obtain the Compton profile J(q) in such a form to be able to compare it with the one derived by theoretical arguments, the random and systematic errors and the contribution of the bound electrons, require various levels of attention. The necessary corrections discussed extensively else where[3,11-14], are listed in fig. 4. Some of them, like absorption correction, detector efficiency and Compton correction are straightforward, others can introduce ambiguities. The background subtraction and the location of the correct peak may include internal experimental uncertainties, difficult to by pass. The most imponderable problems are the multiple scattering, the internal bremsstrahlung radiation and the core contribution. The obvious choice to reduce the multiple scattering, which typically is about 10% of the integrated Compton intensity, the use of very thin samples, is usually rejected because of long time measurements. Instead, corrections are made by Monte Carlo stimulation of the scattering process, which claim a success of $\pm 0.5\% J(q)_{q=0}$. Less understood is the contribution of the bremsstrahlung radiation by the photoelectrons and the Compton electrons in their stopping process inside the sample. Our estimation is that for elements in the middle of the periodic table it can be as high as 5% with increased influence at low momenta. Finally, core contribution is a major problem, when the method does not satisfy the requirement of the impulse approximation employed in the calculation of atomic Compton profiles.

All the above corrections bear very little interest in the comparative studies of difference experiments in which the dubious contribution and errors are subtracted out. The comparative studies are best suitable for hydrides of various hydrogen concentration and their host metal. In this technique, special attention is paid to perform measurements on samples of the same dimensions and if possible of the same density. Both requirements are easy to be satisfied when the samples are in the form of powder pellet.

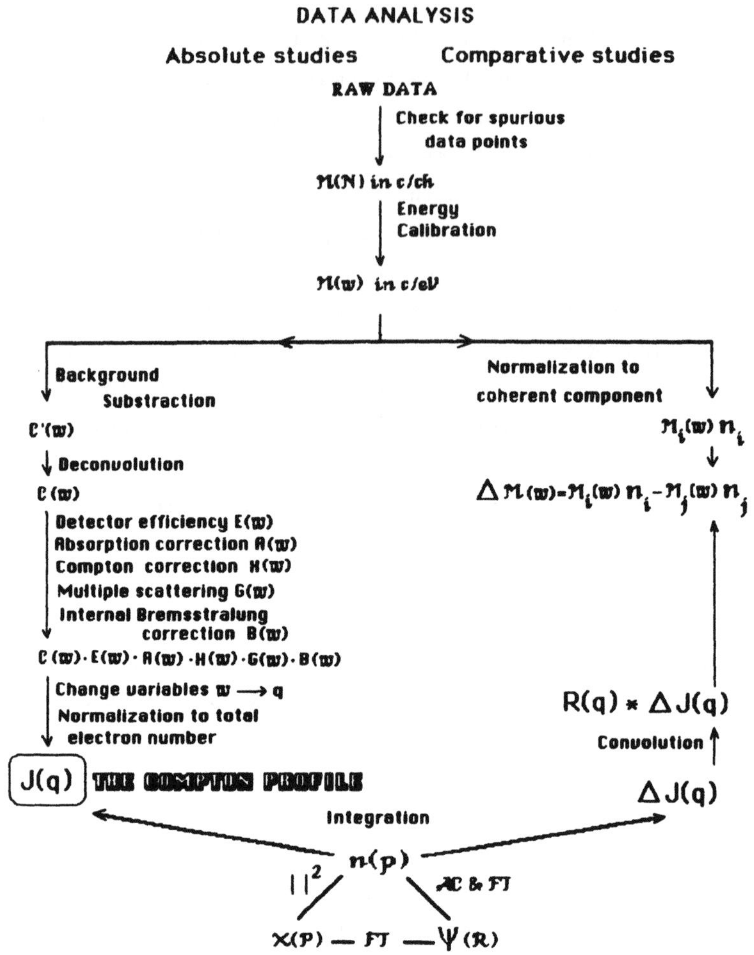

Fig. 4. The necessary steps for data reduction from raw data to comparison with theoretically calculated wavefunctions. In the left hand site are the corrections for absolute studies measurements, in the right hand site for comparative studies. In the first case the J_{MH_x} (q) is reduced from the raw data, in the second the difference profile Δ J(q) = $J_{MH_x}(q)$ - $J_M(q)$ is derived from two sets of data taken under the same conditions, one for the metal and the other for the hydride.

It is crucial to the success of the experimental measuring the difference profiles (MH_x-M), to use very similar samples of the metal and of the hydride, because the observed differences are rather small. For this reason Lässer and Lengeler in their elegant series of measurements on Pd and V hydrides used the same sample with and without hydrogen. In our laboratory, we use pellets of the same dimensions made from powder and by adjusting the pressure we achieve the same density for the metal and the hydride[16]. After smoothing the difference profiles, the energy scale is transformed into momentum and the $\Delta J(q)$ is normalised according to

$$\int_{-3}^{+3} \Delta J(q)\, dq = x \tag{7}$$

where x is the number of hydrogen atoms per atom of host metal. The limits of integration are chosen to be ± 3 because the profiles differ only in the range smaller than the range between -3 and 3 a.u. The resulting spectrum is compared to the theoretical data convoluted with the experimental resolution function.

Several models have been employed in the calculation of the hydrides Compton profile with some degree of success.

a) P r o t o n i c model. According to this model[17] the dissolved hydrogen atom donates its electron to the conduction band of the host metal and rests as a proton in the host lattice. The donated electron does not shift the Fermi level, but it enhances the momentum distribution of the n conduction electrons of the host metal by a factor 1 + x/n. Under these assumptions the difference profile is given by

$$\Delta J^{+}_{MH_x}(q) = \frac{x}{n} J_M(q) \tag{8}$$

b) A n i o n i c model. In this model, contrary to the previous one, the dissolved hydrogen atom removes an electron from the conduction band of the host metal. The Compton profile for the H^- ion is given by:

$$J_{H^-}(q) = \left(\frac{16}{3\pi a}\right)\left|1 + \left(\frac{q}{a}\right)^2\right|^{-3} \tag{9}$$

where a = 0.688.

Finally the difference profile of the anionic model is given[15] by:

$$\Delta J^{-}_{MH_x}(q) = xJ_{H^-}(q) - \Delta J^{+}_{MH_x}(q) \qquad (10)$$

c) A t o m i c model. In this model nonexchange of charge takes place between conduction band of the host metal and the atomic hydrogen but the atom rests in the host lattice or in traps. In those two cases the hydrogen nucleous has an effective charge not necessary equal to +e. The difference profile of the atomic model is given* by:

$$J_H(q) = \left(\frac{8Z^5}{3\pi}\right)(Z^2 + q^2)^{-3} \qquad (11)$$

A more sophisticated way in calculating the difference profile is band-structure calculations where the exchange charge between hydrogen and host metal's conduction band is taken into account as well as the band modification from the hydrogen presence.

Table 2 refers to some recent Compton studies of MH_x. In this table is mentioned the work on TiH_2 that is briefly discussed below. Fig. 5 shows the row data of ^{241}Am 59.54 keV line scattered at $\varphi = 173{,}5^{o}$ by Ti and TiH_2 powder. The two spectra differ only in the range between - 2.5 and 2.5 a.u. The similarity of the two spectra outside of this range proves that the Ti core electrons are not affected by the presence of hydrogen in TiH_2. Fig. 6 shows the normalised to 2 nondeconvoluted experimental difference Compton profile** for polycrystalline Ti and TiH_2 (solid line A). On the same figure are shown for comparison the theoretical difference profiles convoluted with a Gaussian resolution function of σ=0.30 a.u. for the model described above. The solid line B indicates the results of band structure calculations according to APW self-consistent approximation. The dashed line C is the result of the protonic model using the J(q) of fcc Ti according to the same approximation as above. The solid line D is the result of the atomic model calculated according to eq.(11). The same relation is employed to calculate atomic hydrogen contribution of effective charge Z = 1.5 (solid line E). Finally the dashed line F is the result of the ionic model according to equation (10).

* Z is the effective charge

**In this set of data no smoothing had been applied

Table 2. SOME RECENT COMPTON STUDIES OF MH_x

Material	Method of study	References
LiH	Theoretical	Ameri, Grosso, Parravicini 1981 (18)
MgH_2	Experimental: ^{241}Am + SSD, polycrystalline	Felsteiner, Heilper, Gertner, Tanner, Opher, Berggren 1981 (19)
$VD_{0.68}$	Experimental: ^{241}Am + SSD, single crystals	Itoh, Honda, Asano, Hyrabayashi, Suzuki 1980 (20)
$VH_{0.71}$	Experimental: ^{123}Te + SSD, polycrystalline	Lässer and Lengeler 1978 (15)
$VD_{0.77}$	Experimental: ^{123}Te + SSD, polycrystalline	Lässer and Lengeler 1978 (15)
$PdH_{0.72}$	Experimental: ^{123}Te + SSD, polycrystalline	Lässer and Lengeler 1978 (15)
$FeTiH_{1,17}$	Experimental: ^{51}Cr + SSD, polycrystalline	Lässer, Lengeler, Arnold 1980 (21)
$FeTiH_{1,54}$	Experimental: ^{51}Cr + SSD, polycrystalline	Lässer, Lengeler, Arnold 1980 (21)
$NbH_{0.29}$	Experimental: ^{123}Te + SSD, single crystals	Alexandropoulos, Reed 1977 (22)
$NbD_{0.6}$	Experimental: ^{198}Au + SSD, single crystals	Pattison, Cooper, Schneider 1976 (17)
$NbH_{0.76}$	Experimental: ^{198}Au + SSD, single crystals	Pattison, Cooper, Holt, Schneider, Stump 1977 (23)
$NbH_{1.2}$	Experimental: ^{241}Am + SSD, polycrystalline	Theodoridou, Alexandropoulos 1984 (24)
$LuH_{2.05}$	Experimental: ^{51}Cr + SSD, polycrystalline	Lässer, Lengeler, Gschneider, Palmer 1979 (25)
TiH_2	Theoretical	Bakalis, Papanikolaou, Papakonstantopoulos to be published (26)
TiH_2	Experimental: ^{241}Am + SSD, polycrystalline	Alexandropoulos, Theodoridou to be published (16)
$CuTiH_{0.9}$ & $CuTiH_{1.43}$	Experimental: ^{241}Am + SSD, polycrystalline and glass	Bambakidis, Theodoridou, Alexandropoulos 1984 (27)

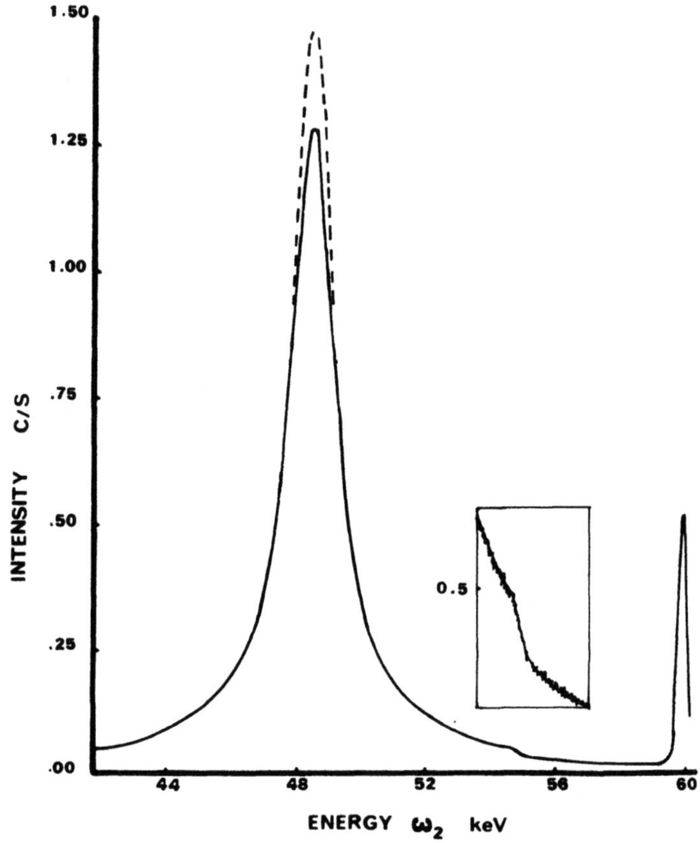

Fig. 5. Energy distribution of 59.54 keV γ-rays scattered at 173.5^{o} by Ti and TiH_2. The insert is the x-ray Raman of Ti.

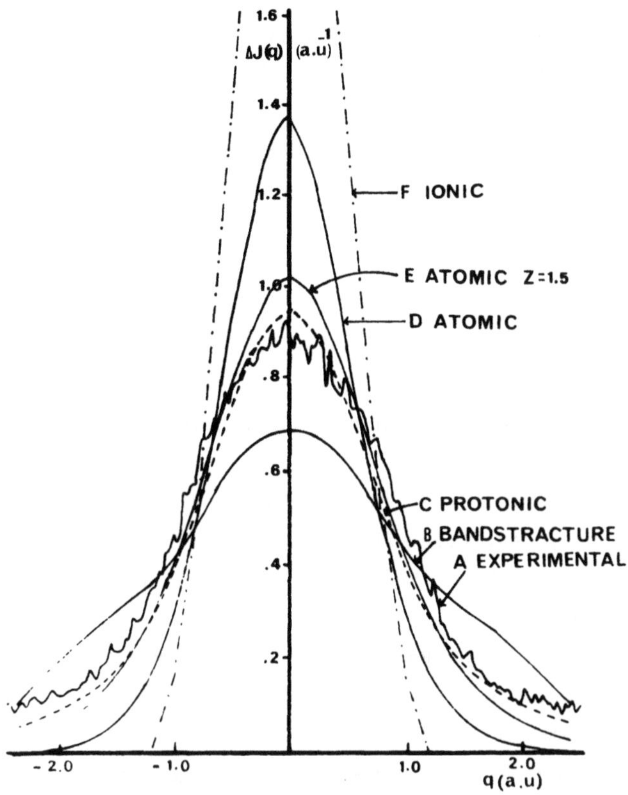

Fig. 6. The difference Compton profile for Ti and TiH_2 experimental and theoretical for several models. The line A shows the experimental line profile difference normalised to two electrons before applying deconvolution for instrumental broadening and smoothing on row data of Ti and TiH_2. The solid line B shows the difference profile of Bandstructure calculations according to APW self-consistent approximation after a convolution with a Gaussian resolution function of σ=0.30 a.u. The remaining lines C-F are the theoretical difference profiles of indicated model after convolution with the previous resolution function.

ACKNOWLEDGMENTS

The author would like to thank G. Evangelakis for technical assistance, Dr. D. Papaconstantopoulos and his group for providing their calculations before publication, Mrs K. Chatjigeorgiou-Alexandropoulou for carrying out the computational work, Dr. I. Theodoridou and Dr G. Bambakidis for stimulating discussions, Dr. N. Papanikolaou for the convolution program, and Mrs F. Foundoulaki-Vergou for the typing.

REFERENCES

1. M. J. Cooper, Adv. Phys. 20, 453-491 (1971).
2. M. J. Cooper, Cont. Phys. 18, 489-517 (1971).
3. B. G. Williams (editor) "Compton Scattering", Publ. McGraw-Hill (1977) 366 pp.
4. B. G. Williams and J. M. Thomas, Int. Rev. Phys. Chem. 3, 39-82 (1983)
5. M. J. Cooper, Rep. Prog. Phys. 48, 415-481 (1985).
6. A. D. Barlas, W. H. E. Rueckner and H. F. Wellenstein, J. Phys. B 11, 3381-3400.
7. B. G. Williams, G. M. Parkinson, G. J. Eckhardt, J. M. Thomas and T. Sparrow, Chem. Phys. Letters 78, 434-438.
8. B. G. Williams and A. J. Bourdillon, J. Phys. C 15, 6881-6890 (1982).
9. B. G. Williams, T. G. Sparrow and R. F. Egerton, Proc. R. Soc. Lond. A393, 409-422 (1984).
10. N. G. Alexandropoulos, to be published
11. P. Paatero, S. Manninen and T. Paakkari, Phil. Mag. 30, 1281-1294 (1974).
12. J. Felsteiner, P. Pattison and M. J. Cooper, Phil. Mag. 30, 537-548 (1974).
13. J. Felsteiner and P. Pattison, Nucl. Inst. 2 Meths. 173, 323-327(1980)
14. I. Theodoridou, Ph. D. thesis, University of Ioannina (1983).
15. R. Lässer and B. Lengeler, Phys. Rev. B 18, 637 (1978).
16. N. G. Alexandropoulos and I. Theodoridou, to be published
17. P. Pattison, M. Cooper and J. R. Schneider, Z. Phys. B 25, 155 (1976)
18. S. Ameri, G. Grosso and Pastori Parravicini, Phys. Rev. B 23, 4242 (1981).
19. J. Felsteiner, M. Heipler, I. Gertner, A. C. Tanner, R. Opher, K. F. Berggren, Phys. Rev. B 23, 5156 (1981).
20. F. Itoh, T. Honda, H. Asano, M. Hirabayasi and K. Suzuki, J. Phys. Soc. Japan 49, 202 (1980).

21. R. Lässer, B. Lengeler and G. Arnold, Phys. Rev. B 22, 663 (1980).

22. N. G. Alexandropoulos and W. A. Reed, Phys. Rev. B 15, 1970 (1977).

23. P. Pattison, M. Cooper, R. Holt, J. R. Scheider and N. Stump, Z. Physik B 27, 205 (1977).

24. I. Theodoridou and N. G. Alexandropoulos, Z. Phys. B 54, 225 (1984).

25. R. Lässer, B. Lengeler, K. A. Gschneider and P. Palmer, Phys. Rev. B 20, 1390 (1979).

26. N. Bakalis, N. Papanikolaou and D. Papaconstantopoulos, to be published.

27. G. Bambakidis, I. Theodoridou and N. G. Alexandropoulos, Acta Crys. A 40, C-173 (1984).

DISORDER INDUCED BY AGING IN METAL TRITIDES

T. Schober

Institut für Festkörperforschung, Kernforschungsanlage

Jülich, 5170 Jülich, F.R. of Germany

ABSTRACT

Due to the decay of tritium to ^3He in metals a considerable amount of disorder is introduced into tritium containing alloys. We will first review previous research in this area and then summarize our own studies. The aging experiments presented here include: a) transmission electron microscopy (TEM) on aged VT and ZrT alloys showing the microstructure of tiny, high-pressure ^3He bubbles and interstitial dislocation loops after aging for up to more than 2 yrs; b) swelling studies on selected Ta and Nb tritides where the volume expansion of the tritides as well as the volume requirements of the He atoms in the bubbles are investigated; c) acoustic emission experiments with aging Ta and Nb tritides; d) resistivity measurements on V and Ta tritides as a function of aging time; e) observations of grain boundary embrittlement in Ta tritides. The general picture evolving from this and other studies is that the production of insoluble He in the host lattice leads to significant disorder and radiation-like damage severely affecting most properties of the host lattice. This disorder and damage sets in rather early, typically 1 - 2 weeks after the formation of the tritide, with the formation of 1 - 2 nm size high-pressure ^3He bubbles. Accelerated ^3He release in the aged tritides seems to be associated with the formation of He filled channels.

INTRODUCTION

In view of the future use of tritium (T) and deuterium in fusion devices a considerable effort was focussed in the past on two important aspects:

- the ^3He embrittlement of structural materials which have a rather low solubility for T. Here, He would be introduced by α-injection, (n, α) reactions or by the decay of the dissolved T to ^3He. Typically, no more than a few hundred ppm of He would be expected to be produced in the lifetime of the materials.

- the safe storage of T in metal tritides such as UT_3, ZrT_2 and TiT_2. Here the T-concentration and therefore ^3He production is orders of magnitude higher than in the first case.

The structural materials and storage tritides experience a degradation arising from the decay of dissolved T to ^3He according to (1)

$$T \longrightarrow {}^3He + e^- + \bar{\nu} \qquad (T_{1/2} \approx 12.36\ y) \qquad (1)$$

Basically, this degradation stems from two sources:

- the incorporation of insoluble and oversized He atoms into the lattice and precipitation of ^3He bubbles at grain boundaries, along dislocations or in the interior of the grains leading to drastic changes in the mechanical properties (for instance, to intergranular fracture)

- in storage tritides the decay of tritium atoms also leads to changes in the stoichiometry and in the relative proportions of the tritide phases present.

In view of the relatively short half-life large amounts of He may accumulate in short times. The ^3He concentration in the lattice increases according to

$$c_{He} = c_{T,0}\,(1 - e^{-\lambda t}) \approx c_{T,0} \cdot \lambda \cdot t \text{ for } \lambda t \ll 1 \qquad (2)$$

Here, $c_{T,0}$ is the tritium concentration at the time of formation. Aging for one year, for example, leads to a conversion of more than 5 % of the original tritium inventory to ^3He. We shall briefly review now some of the results published in the open literature concerning the aging behavior, ^3He release and micromorphology of aged tritides. Early work[1] on thin tritide films of Er, Ho, Sc, Ti and Y and pressed pellets[2] of yttrium tritide showed a steady expansion of the samples with age and in the case of the thin films[1] even spalling and bubble rupture after a few years. The monotonic swelling was taken as evidence that most of the ^3He produced remained in the film. Interestingly, the lattice parameters of Er and Sc tritide were found to change little or not at all with time[1]. A number of aging studies[3-6] were presented in 1975. It was shown for Sc, Ti and Er tritides that during the early life a large fraction of the ^3He is retained in the solid[3]. At more advanced ages (2 - 4 yrs) the ^3He release rate becomes comparable to the generation rate. Evidence was obtained that fine-grained samples release ^3He much faster than single-crystal samples[3]. For UT_3 it was found[5] that less than 2 % release occurs until about 280 days, after which the release increases dramatically. The generation rate was approached after $\sim$ 1000 days. A NMR study on LiT, TiT_2 and UT_3[4] demonstrated that in each case ^3He atoms were trapped in microscopic gas bubbles. In variance with these findings[4] was another NMR study[6] claiming that ^3He atoms are trapped interstitially in young crystals and in bubbles in old crystals. Additionally, the younger sample in that work[6] was found to exhibit considerably more lattice distortion than the older sample. In subsequent work by that group[7] on $TiT_{1.8}$ it was found that the ^3He release rate increases rapidly when ^3He/metal ratios of about 0.2 to 0.3 have been reached. This rapid increase was correlated with the formation of large gas bubbles within the solid. Pulse NMR work[8,9] on UT_3, $TiT_{1.9}$ and LiT conclusively showed that most ^3He-atoms are retained in microscopic gas bubbles with dimensions < 50 nm. No significant concentration of interstitial ^3He in the UT_3 lattice was observed[8]. The dependence of ^3He release rate on the initial T content was studied for the case of Er tritides[10]. It was observed that ditritides with higher T concentrations reach the rapid release state at much lower ^3He concentrations. As an example, the He/M ratio for rapid discharge in the unsaturated tritide was 0.22 and only a tenth of this value for the saturated tritide[10]. In a further study[11] it was attempted to rationalize the release behavior with a strain model. In a totally different experiment[12] on austenitic stainless steels, 5 nm He gas bubbles were again detected in the microstructure of specimens aged for 5.5 yrs. This precipitation was accompanied by a serious loss of ductility[12]. ^{3}He and T release measurements, as well as thermal desorption studies on stainless steels[13] showed that during heating to 900 K only a small fraction of the generated He was released. During ramping He-release started at ∿190 K. The release fraction, however, was very low compared with the inventory. In a further study[14] on the mechanical properties of T exposed Nb an increase of the yield stress was observed. TEM on the microstructure revealed He clusters (presumably bubbles) punching out prismatic dislocation loops[14]. Returning to thin films of storage tritides, the temperature dependence of the He release was studied for storage temperatures ranging from 77 to 500 K[15]. The salient feature of that work was large changes in the He release rates following temperature changes. However, the He release rates were only slightly temperature dependent[15]. Annealing of tritides of Nb, Ta or V in air for a few hours up to ∿200 °C leads to a neglegible loss of T although this tiny T release fraction is readily measured[16].

The effect of T decay on the performance of Pd permeation membranes was studied in Ref. (17). In recent work[18] on the ^{3}He release rate from Er and Zr tritide films irregularities in this release rate were observed which were greatest for samples undergoing the transition into accelerated release. In this phase, the occluders approach the maximum quantities of ^{3}He that they can retain. Apparently, the spontaneous release of ^{3}He in bursts of about 10^9 atoms was detected. The bursts were explained[18] by a cascading process that is vibrationally coupled. The first TEM observations of ^{3}He bubbles in T charged Pd[19] revealed bubble sizes of 1 - 2 nm after aging periods of a few months only. In similar studies from this group, T decay in V[20] and Zr[25] was investigated, details are given below.

EXPERIMENTAL

The tritides used in our experimental work were all synthesized by gas-phase charging in either one of the two Jülich tritium-facilities[16]. Corrections were always made for the less than 100 % isotopic purity of the T. For details of the experimental techniques we refer to the relevant original publications[20-25]. The ^{3}He bubble populations were measured with a Zeiss TGA 10 particle size analyzer.

RESULTS

We summarize our aging results obtained in this group in the discrete sections a) to f). Some of the results will be presented in more detail elsewhere.

a) TEM Studies of the Aging of VT[20] and ZrT[25] Alloys

As to β-phase VT alloys[20], a rather dense distribution of small interstitial dislocation loops was observed after ∿7 to 14 days of aging. After 60 to 90 days of aging the dislocation loop structure became coarser and the loop diameters increased. Small ^{3}He bubbles were detected after ∿80 days of aging. Their density was about $3.5 \cdot 10^{23}$ m^{-3} with a mean bubble diameter of ∿1.2 nm. In α-phase VT areas, isolated 1 - 2 nm He bubbles were observed which had relieved their internal pressure by prismatic punching. The dislocations in α-phase areas were generally decorated with strings of ^{3}He bubbles, or else, cylindrical cavities[21]. Low-angle grain boundaries were found to be preferential sites for ^{3}He bubble nucleation.

ZrT alloys[25,26]. Zr tritides are interesting prototypes of high density storage tritides. They have the advantages of being very stable, low-equilibrium pressure hydrides and may also be stored in air without apparent extraneous degradation (UT_3, for instance, could not be stored in air due to its pyrophoricity). The tritides prepared contained the stable δ-phase of approximate composition $(T/Zr) \approx 1.6$ and the metastable γ phase with the composition $(T/Zr) \approx 1.0$. The total T content of the TEM samples was low, typical values were $(T/Zr) \approx 0.03$. Thus, the samples consisted of relatively brittle δ or γ tritide plates embedded in the rather ductile Zr matrix. It was found in the course of the study that adjacent surfaces (free surfaces or phase boundaries) did not noticeably affect the micromorphology of bubbles and dislocation loop damage in the present tritides. In this context, it is emphasized that all samples were repeatedly electropolished prior to each TEM observation[26] to exclude possible surface effects and to guarantee an inspection of the true bulk state.

The first ^3He bubbles in aged δ-phase areas (γ-phase areas behaved in a similar way) were detected by TEM about 3 weeks after formation and had diameters of about 1 - 1.5 nm[25]. The bubbles could only be imaged using kinematical many-beam conditions and the usual defocussing sequences. Fig. 1 shows the mean bubble diameters vs age and Fig. 2 the apparent bubble density vs age. The intrinsic errors entering into the data of Fig. 2 are considerable because of the difficulties of imaging all the bubbles present and of overlapping bubbles. Fig. 3 depicts for the first time the most advanced state of aging after 836 days. These samples should already be in the state of "accelerated release" where as much ^3He is lost as it is produced. In observation modes using strong diffraction contrast the tritide areas look almost black. This arises from the strain fields of the interstitial type damage in between the bubbles. Only in kinematical many beam situations the bubbles could be imaged.

To summarize our TEM observations up to 836 days we may distinguish 3 main phases in the aging process[26]:

(1) The establishment of the bubble microstructure
This phase lasts for about a month after the production of the tritides. In this phase, an ensemble of about $5 \cdot 10^{23}$ bubbles/m^3 is formed. Some interstitial dislocation loop damage is also produced.

(2) Growth of the bubble population.
Roughly speaking, this phase lasts from the termination of phase 1 to about 1 - 2 years of age. In this phase, the bubbles grow, their density decreases somewhat (due to combining of bubbles) and the lattice in between the bubbles is filled up with dislocation loops and segments produced by the emission of interstitial matter from the bubbles. These loops and dislocation segments become heavily decorated with strings of small bubbles or cylindrical ^3He cavities. Intrinsic lattice dislocations and various phase boundaries are also decorated with ^3He bubbles or planar cavities.

(3) The development of interconnecting ^3He channels
This phase starts approximately at the end of phase 2 one to two years after production. ^3He filled dislocation loops, segments or regular lattice dislocation are often short-circuited or linked by further loop damage. In this way, elongated ^3He channels (Fig. 4) are formed resulting in a 3-dimensional tree-like structure of $\sim$2 nm diameter channels which would contain ^3He at the same 5 GPa pressure levels as regular bubbles[23]. In addition, neighboring bubbles often coalesce leading to elongated cavities. Large interconnected arrays of bubbles or thin, planar "sheets" of ^3He are formed in phase boundaries. However, neither the crystal structure of the tritide is changed nor is

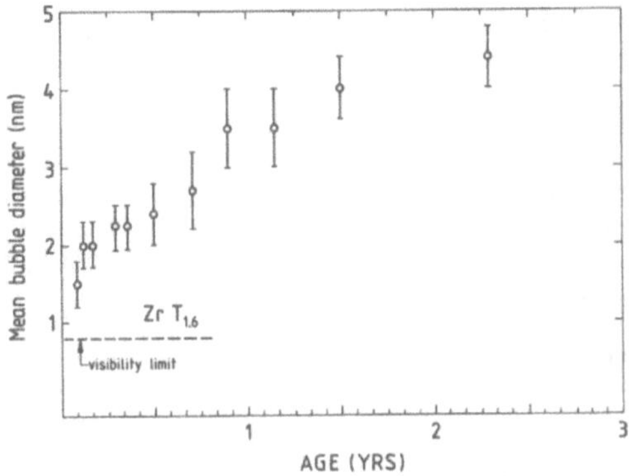

Fig. 1. Mean ^{3}He bubble diameters vs aging time in δ-ZrT samples.

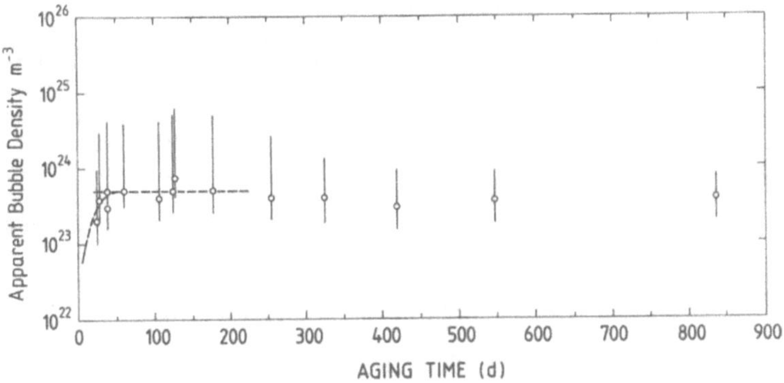

Fig. 2. Apparent ^{3}He bubble density vs aging time for δ-ZrT samples.

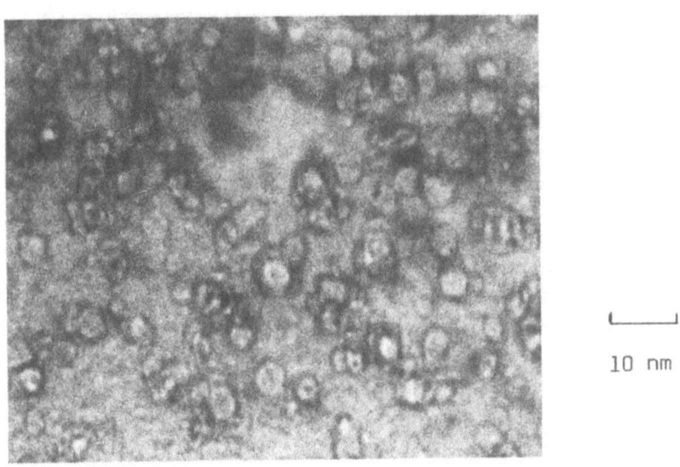

Fig. 3. TEM micrograph of a δ-ZrT area aged for 836 days; kinematical conditions; visible are high-pressure ^{3}He bubbles.

there a noticeable change in the lattice parameter as measured by electron diffraction (which admittedly has only a poor resolution).

Obviously, there are no sharp transitions between these 3 phases. The transition from phase 2 to phase 3 is interpreted here as the start of the "accelerated release" found in the ^3He release measurements cited above. Our microstructural TEM work suggests that it is a sudden connection of an elongated high-pressure ^3He channel to an external surface which brings about massive ^3He loss. The irregular release of ^3He [18] in bursts of $\sim 10^9$ atoms is ascribed to just such discharges of high pressure ^3He channels to an external surface. It is readily calculated that a cylindrical channel 2 nm in width would have to have a length of ~ 3 mm to contain $\sim 10^9$ ^3He atoms*). It is not unreasonable to assume that channels of that length may exist in the aging tritides. The above conclusions about the release mechanism (the sudden rupture of high-pressure channels) are supported by recent work on He-ion bombarded metals[26] where similar channels were detected and the analoguous interpretation was given.

We emphasize that we have found no evidence for large ^3He bubbles such as those recently reported[7]. The formation of larger bubbles may be an effect of the adjacent surfaces and not an intrinsic feature of aging of tritides. Similar enlarged near-surface bubbles are common in blistering studies and have recently been found also in He-implanted metals[28]. - The likely mechanisms triggering ^3He release from the high-pressure channels are: a) prismatic loop punching or SIA emission at sharp protuberances of the channel; b) plastic deformation hooking up high-pressure channels with other short-circuits leading to the outside. Plastic flow occurs constantly in swelling tritides. - No evidence was obtained in this TEM work that substantial amounts of ^3He are dissolved interstitially in the early aging phases.

b) Swelling Studies of Selected Ta and Nb Tritides[23,24,26]

In view of the facts that in young tritides essentially all of the generated ^3He remains in the samples[6-11] and that a ^3He atom needs more space than the T atom it replaces it is readily understandable that such tritides will swell. There are a number of advantages to measure this swelling effect in flat, bulk samples at ambient temperatures with strain gauges[22,23]:

- strain gauges are relatively simple resistivity sensors with ppm resolution and are easily applied to flat surfaces
- these sensors directly measure strain; thus absolute length measurements of our samples with dimensions of a few mm are not necessary.

The relative length (volume) changes are related to the resistivity changes via

$$\Delta L/L = 1/3 \cdot (\Delta V/V) = k^{-1} \cdot \Delta R/R \qquad (3)$$

Here, k is the strain sensitivity; its numerical value is close to 2.0 and is known within 1 %. Let us assume that during one decay event the sample increases its volume by the quantity $(\Delta v/\Omega)_{T\rightarrow He}$ (Ω is the atomic volume of a metal atom). We then obtain for the swelling rate $\Delta V/V$:

$$\Delta V/V = c_{He} \cdot (\Delta v/\Omega)_{T\rightarrow He} = c_{T,0}\,(1 - \exp(-\lambda t))\,(\Delta v/\Omega)_{T\rightarrow He} \qquad (4)$$

*We assume here that a ^3He atom would require the same volume of 10^{-29} m^3 in that channel as in high-pressure ^3He bubbles[23].

The term $(\Delta v/\Omega)_{T\rightarrow He}$ basically describes the difference between the volume requirements of ^3He and T in the lattice. We have shown that eqn. (4) may be written as follows for the case of bubble growth by interstitial matter emission[23]:

$$\Delta V/V \approx c_{T,0} \cdot \lambda \cdot t \cdot \left[\frac{v_{He}}{\Omega} \frac{\widetilde{\Delta v_I}}{\Omega} - \frac{\Delta v_T}{\Omega}\right] \quad (5)$$

where v_{He} = volume required by ^3He in the high-pressure bubbles

$\widetilde{\Delta v_I}$ = volume change when a SIA is transferred from a bubble into the lattice, to a SIA cluster, a loop, other dislocations or surfaces. If SIA's are deposited as single, isolated interstitials $\widetilde{\Delta v_I} \approx 1.1$[23]; when the SIA's are incorporated in interstitial dislocation loops, then $\widetilde{\Delta v_I} = 1.0$. For all cases considered below we will take $\widetilde{\Delta v_I}$ as unity.

Δv_T = volume decrease when a T atom is taken out of the tritide.

It is seen from eqn. 5 that the linear swelling regime will directly yield information on v_{He} since all other quantities are accessible experimentally. Given v_{He} we then can obtain ^3He densities in the bubbles and (using an equation of state) pressures in the bubbles can be calculated.

Fig. 5 now shows a typical, rather linear swelling curve for the alloy $NbT_{0.59}$. Only the initial slope is used for data analysis. Table 1 shows the values for (v_{He}/Ω), v_{He} and the pressures p extracted from our swelling curves. Thus, ^3He requires roughly half of an atomic volume when it is located in the present bubbles. The pressures obtained lie in the range 5 to 10 GPa which is close to the threshold pressure required for loop punching[29]

$$p \geq \bar{\mu}/4\pi + 2\gamma/r \quad (6)$$

($\bar{\mu}$ = average shear modulus, γ = specific surface free energy).

Table 1. Swelling measurements, experimental results

	$(\Delta v/\Omega)_{T\rightarrow He}$	(v_{He}/Ω)	$v_{He} \cdot 10^{-30}$ m^3	p(GPa)
$TaT_{0.103}$	0.37	0.53	9.64	5.3
$TaT_{0.42}$	0.37	0.52	9.98	4.7
$NbT_{0.59}$	0.23	0.39	7.7	10.6

The present He densities and pressures are in general agreement with the values derived in recent energy loss spectroscopy work cited in our previous work[23,24]. Our technique is far more direct and, therefore, presumably more reliable.

c) Acoustic Emission (AE) from Swelling Tritides[24]

$TaT_{0.12}$ and $NbT_{0.59}$ samples were examined in the early aging phases for evidence of acoustic emission. The samples were directly coupled to the PZT transducer with silicone grease. Ring down counts were measured with conventional electronic equipment using a total gain of 90 dB. Fig. 6

Fig. 4. TEM micrograph of elongated ^3He filled channels in δ-ZrT alloys aged for 836 days. The channels arise from He decoration of dislocations and further ejection of matter from the core region.

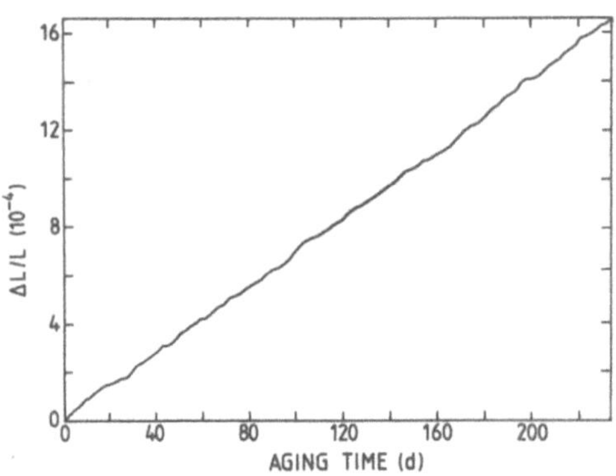

Fig. 5. Swelling vs time of the alloy $NbT_{0.59}$.

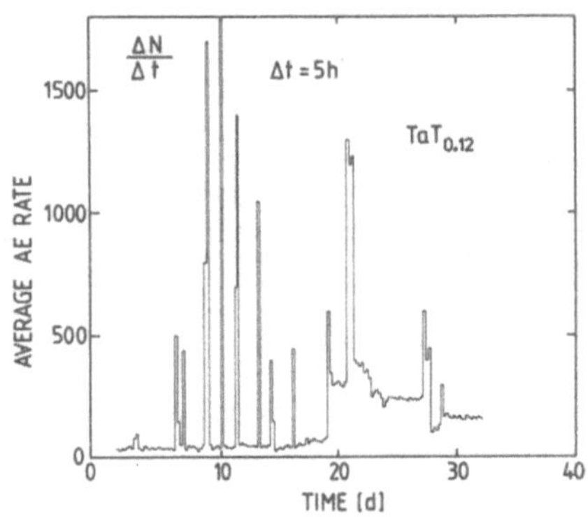

Fig. 6. Acoustic emission rate vs aging time for the alloy $TaT_{0.12}$. Integration time: 5 h.

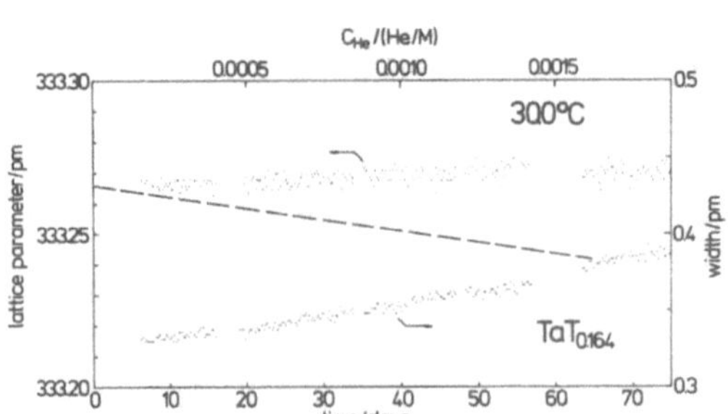

Fig. 7. Lattice parameters of a $TaT_{0.164}$ sample vs aging time (upper data set); width of Bragg peak vs time (lower data set). Dashed line: calculated lattice parameter if only decay controlled[30].

shows for the case of $TaT_{0.12}$ the average AE rate vs. aging time for an integration time of 5 h. Rather strong, irregularly spaced accumulations of acoustic events occurred starting about 6 days after sample preparation. After about 20 days a decaying plateau of AE was observed; after 32 days AE had essentially dropped back to background levels. Similar results were obtained for the alloy $NbT_{0.59}$ where no emission was observed for an incubation period of ∿10 days. Strong AE signals were then detected for the following 50 days. Subsequently, the AE dropped again to background levels. It was concluded[24] that the observed AE was most likely due to collective and correlated loop punching processes. In a certain range of bubble and loop sizes the binding of loops to bubbles can become very weak. Thus, avalanches of loop punching events could be triggered in the swelling tritides. These avalanches would then be detected by the AE technique.

d) Resistivity Measurements of Aging Tritides[26]

Dilute α-phase Ta and V tritides were studied for the first months of age using resistivity measurements. For both cases, a monotonic increase of the resistances was found, details will be reported elsewhere[26]. Qualitatively, the following factors contribute to changes of the resistivities of aging tritides:

(1) a decay of a T atom per se will lead to a decrease in resistance
(2) small ^{3}He-bubbles constitute scattering centers and will entail increases in resistivity
(3) Likewise, SIA loops and interstitial atoms will also increase the resistivity
(4) decoration of dislocations with ^{3}He will also increase the resistances.

e) Lattice Parameter Data; Low Temperature Resistivity (after Lässer et al. [30,31])

The lattice parameter of a $TaT_{0.164}$ sample was precisely measured using the Bond method and was found to increase slightly in the first weeks of aging and then remained at a constant level[30] (Fig. 6). Inversely, the width of the Bragg peak was constant in the first weeks and then increased substantially. The almost imperceptible increase in lattice parameter may be due to suitable SIA configurations or to small amounts of dissolved ^{3}He[30].

In different work[31], a $LuT_{0.14}$ sample in the α-phase was quenched to 4.2 K soon after production. Its resistance was then measured vs aging time and was found to increase linearly with time. This increase is thought to arise from a) the random incorporation of the non-diffusing ^{3}He into interstitial sites and, possibly, b) the separation of paired T atoms by the β^--radiation produced in decay events.

f) Grain Boundary Embrittlement in Ta tritides

During work with bulk tritides aged for a few months (which may even be in the α-phase) it is common experience that they may desintegrate along the grain boundaries at very low stress levels which would never lead to failure in the corresponding hydrides. This effect is certainly produced by massive ^{3}He precipitation at grain boundaries as observed in our TEM studies.

ACKNOWLEDGEMENT

Helpful discussions with R. Lässer, H. Wenzl, G. Thomas, H.Trinkaus and C. Dieker are acknowledged.

REFERENCES

1. L.C. Beavis, and C.J. Miglionico, J. Less-Comm. Metals 27:201 (1972).
2. P.M.S. Jones, W. Edmondson, and N.J. McKenna, J. Nucl. Mat. 23:309 (1967).
3. W.G. Perkins, W.J. Kass, L.C. Beavis, in "Radiation Effects and Tritium Technology for Fusion Reactors", USERDA Conf-750989, ed. by J.W. Watson and F.W. Wiffen (USERDA, Oak Ridge, Tenn. 1976) Vol. IV, 83.
4. R.C. Bowman, and A. Attalla, ibid, p. 68.
5. M.E. Malinowski, and P.R. Coronado, ibid, p. 53.
6. H.T. Weaver, and W.J. Camp, Phys. Rev. B 12:3054 (1975).
7. H.T. Weaver, Appl. Phys. Letters 30:80 (1977).
8. R.C. Bowman, and A. Attalla, Phys. Rev. B 16:1828 (1977).
9. R.C. Bowman, Nature 271:531 (1978).
10. L.C. Beavis, and W.J. Kass, J. Vac. Sci. Technol. 14:509 (1977).
11. W.J. Camp, J. Vac. Sci. Technol. 14:514 (1977).
12. P.S. Sklad, 38th Ann. Proc. EMSA, San Francisco, 1980, G.W. Bailey (ed.), p. 388.
13. G.J. Thomas, and R. Sisson, Report # Sand 80-8628, Sandia Laboratories, Albuquerque, N.M. (1980).
14. J.A. Donovan, R.J. Burger, R.J. Arsenault, Met. Trans. 12A:1917 (1981).
15. D.J. Mitchell, and R.C. Patrick, J. Vac. Sci. Technol. 19:236 (1981).
16. R. Lässer, K.-H. Klatt, P. Mecking, and H.Wenzl, KFA-Jülich Report # 1800 (1982), p. 66.
17. M. Nishikawa, J. Nucl. Mat. 116:343 (1983).
18. D.J. Mitchell, and J.L. Provo, J. Appl. Phys. 57:1855 (1985).
19. G.J. Thomas, and J.M. Mintz, J. Nucl. Mat. 116:336 (1983).
20. T. Schober, R. Lässer, W. Jäger, and G.J. Thomas, J. Nucl. Mat. 122,123: 571 (1984).
21. T. Schober, G.J. Thomas, R. Lässer, and W. Jäger, Scripta Met. 18:255 (1984).
22. T. Schober, J. Phys. E: Sci. Instrum. 17:196 (1984).
23. T. Schober, R. Lässer, J. Golczewski, C. Dieker, and H. Trinkaus, Phys. Rev. B 31:7109 (1985).
24. T. Schober, J. Golczewski, R. Lässer, C. Dieker, and H. Trinkaus, to be publ. in Proc. "Hydrogen in Metals" Conf., Belfast (1985).
25. T. Schober, and R. Lässer, J. Nucl. Mat. 120:137 (1984).
26. T. Schober, R. Lässer, and C. Dieker (to be published).
27. W. Jäger, and J. Roth, J. Nucl. Mat. 93,94:756 (1980).
28. W. Kesternich, D. Schwahn, and H. Ullmaier, Scripta Met. 18:1011 (1984).
29. H. Trinkaus, and W.G. Wolfer, J. Nucl. Mat. 122,123:552 (1984).
30. R. Lässer, K. Bickmann, H. Trinkaus, and H. Wenzl (to be published).
31. R. Lässer, T. Schober, D. Triefenbach, and P. Jung (to be published).

TRITIUM IN Pd AND $Pd_{0.80}Ag_{0.20}$

R. Lässer and G.L. Powell[+]

Institut für Festkörperforschung, Kernforschungsanlage Jülich, 5170 Jülich, Fed. Rep. Germany
+Oak Ridge Y-12 Plant*, Martin Marietta Energy Systems Inc. Oak Ridge, Tenessee 37831

ABSTRACT

Solubility measurements of H, D and T in Pd and $Pd_{0.80}Ag_{0.20}$ have been performed in a very large temperature range. A simple analytical expression is given for the equilibrium constants which describes the equilibrium between the hydrogen atoms dissolved in a metal at infinite dilution and the hydrogen molecules in the gas phase. From this expression the partial enthalpies and entropies of H, D, T in metals can easily be calculated. Also the vibrational ground state energies of H, D and T in Pd and $Pd_{0.80}Ag_{0.20}$ relative to the energy of atomic hydrogen at rest could be determined accurately. In addition, the isotopic dependence of the phase boundaries in Pd will be discussed.

INTRODUCTION

Due to experimental difficulties in handling large amounts of tritium and highly loaded metal tritide samples very little is known about the behaviour of tritium (T) in metals (M) in comparison to the stable hydrogen isotopes protium (H) and deuterium (D). Since tritium is now available in large quantities at high purity and reasonable cost few papers have been published discussing special properties of tritium in metals.

In this paper we will discuss the solubility of T in Pd and $Pd_{0.80}Ag_{0.20}$, the phase boundaries of T in Pd and compare the T behaviour with the properties of H and D in these metals.

The knowledge of the properties of T in metals is of academic and technological interest. The hydrogen isotopes have the largest relative mass changes of all isotopic species and are expected to show large isotopic effects due to the different vibrational energies. In addition their small mass results in vibrational quantum effects at relatively high temperatures.

---

*operated for the U.S. Department of Energy by Martin Marietta Energy Systems, Inc., Under Contract No. DE-AC05-84OR21400

Large quantities of T will be handled in future fusion reactors because the most promising reaction is the fusion of D and T. Metal tritide systems are considered as the most promising candidates for the safe handling and storage of large amounts of tritium.

Tritium decays into ^3He with a half life time of 12.361 y. This decay offers a possibility to load large samples with ^3He homogeneously and to study the ^3He behaviour in metals just starting with a MT_x sample and waiting until the desired He concentration is achieved (the so called tritium trick). In addition, one can study the influence of the increasing amount of He that results from aging. This He is highly insoluble, precipitates into He bubbles and may affect many physical properties of MT_x samples such as solubility, lattice parameter, diffusion, elastic constants, changes of stochiometry, phase boundaries, etc.

SOLUBILITY OF PROTIUM, DEUTERIUM AND TRITIUM IN Pd AND $Pd_{0.80}Ag_{0.20}$

The solubility of H and D in fcc Pd and $Pd_{1-x}Ag_x$ alloys has been the subject of recent review articles[1,2]. The solubility of T in Pd[3-6] has been determined only very recently.

In this section we present the first solubility data of T in $Pd_{0.80}Ag_{0.20}$. In addition, we give a simple analytic expression which describes the equilibrium between the hydrogen atoms dissolved in Pd and $Pd_{0.80}Ag_{0.20}$ at infinite dilution and the hydrogen molecules in the gas phase.

The data have been collected independently with similar equipments, one built up by Powell[7,8] in Oak Ridge and one built up by Lässer et al.[4,9] in Jülich. The H, D and T solubility data were determined in a large temperature range using the same sample under identical conditions.

Equilibrium between hydrogen dissolved in a metal (described by the concentration x, where x is the ratio of hydrogen to metal atoms) and the pressure p of the hydrogen gas can be described by the equation

$$\ln\left[\frac{p}{p_0}\left(\frac{N-x}{x}\right)^2\right] = -\ln(K^\infty/N^2) + \frac{1}{RT}\Delta\mu \qquad (1)$$

with

$$-\ln K^\infty = \frac{\Delta\bar{H}^\infty}{RT} - \frac{\Delta\bar{S}^\infty}{R} \qquad (2)$$

where N is the ratio of possible hydrogen places to metal atoms. T is the temperature, R is the gas constant and p_0 = 1 atm, the common reference pressure used in thermodynamics. K^∞ is the equilibrium constant at infinite dilution and $\Delta\mu$ is an excess chemical potential describing deviations from the Sieverts law. $\Delta\bar{H}^\infty$ and $\Delta\bar{S}^\infty$ are the partial enthalpy and entropy of solution of 1 mol X_2 (X = H,D,T) molecules in the reaction $H_2 \rightleftharpoons 2$ H (metal).

As an example we show in Fig. 1 the solubility data of D in $Pd_{0.80}Ag_{0.20}$ obtained during heating the $Pd_{0.80}Ag_{0.20}D_x$ sample between 60 and 500 °C in 20 K steps. The experimental values of x and p are plotted according to Eq. 1 taking N = 1. The circles represent the experimental values. These data can be fitted very well by straight lines (solid lines in Fig. 1) which means that $\Delta\mu$ is a linear function of x at low concentrations. From the intersections of the straight lines with the y axis we obtain the equilibrium constants at infinite dilution which are shown as a function over the inverse absolute temperature in Fig. 2 and 3 for H, D, T in Pd and $Pd_{0.80}Ag_{0.20}$, respectively. The plotted values of both metals show a strong isotopic dependence which can be described by the inequalities

$$K_H^\infty > K_D^\infty > K_T^\infty \tag{3}$$

at equal temperatures for Pd and $Pd_{0.80}Ag_{0.20}$. In addition we find

$$K_X^\infty(Pd) < K_X(Pd_{0.80}Ag_{0.20}) \tag{4}$$

at equal temperatures for X = H,D,T.

The data measured by Powell (circles in Fig. 2,3) and by Lässer (squares in Fig. 2,3) are in excellent agreement for H and D in both metals. Our H values are in very good agreement with the data obtained by Wicke and Nernst[10] (filled circles in Fig. 2) whereas the data of Clewley et al.[11] (filled squares in Fig. 2) tend to be a little lower than ours for H and D. We believe that our data represent the most accurate description of the solubility of H, D and T in Pd and $Pd_{0.80}Ag_{0.20}$ measured in a very large temperature range. The curvature of the experimental data at higher temperatures clearly demonstrates that the partial quantities $\Delta\bar{H}_X^\infty$ and $\Delta\bar{S}_X^\infty$ (X = H,D,T) in Pd and $Pd_{0.80}Ag_{0.20}$ are temperature dependent.

The solid lines in Fig. 2 and 3 are the results of fitting the following equations[7,8,12-14] to the experimentally determined equilibrium constants

$$-\ln K_X^\infty = (2\,\bar{G}_X^\infty - G_{X_2}^o) / (RT), \tag{5}$$

$$\bar{G}^\infty / (RT) = -\ln \frac{N(1+A\,e^{-B/T})}{(1-e^{-C/T})^3} - \frac{E - \frac{3}{2}C}{T} \tag{6}$$

and

$$G_{X_2}^o / (RT) = -\ln \frac{LT^{7/2}}{(1-e^{-J/T})} - \frac{M}{T} \tag{7}$$

with

$$M = D_o^o + \frac{B_o}{3}. \tag{8}$$

$\bar{G}_X^\infty$ is the partial Gibbs free energy of one mole hydrogen (X = H,D,T) atoms in a metal and $G_{X_2}^o$ is the standard Gibbs free energy of one mole hydrogen molecules for the temperature range above 200 K.

The first term in Eq. 6 is the appropriate sum over all possible energy states of hydrogen dissolved in a metal. The term $(1-e^{-C/T})^{-3}$ is the sum over all vibrational states referenced to the ground state energy assuming that the hydrogen atom behaves as an isotopic, three dimensional, harmonic oscillator with the Einstein temperature C. The factor $(1+A^{-B/T})$ is a correction term considering all other contributions from anharmonic corrections, other vibrational manifolds and electronic energy states.

The second term (E - 3 C/2) is the ground state energy of hydrogen dissolved in a metal with respect to atomic hydrogen at rest. E is the vibrational potential minimum measured with respect to atomic hydrogen and 3 C/2 is the zero point energy.

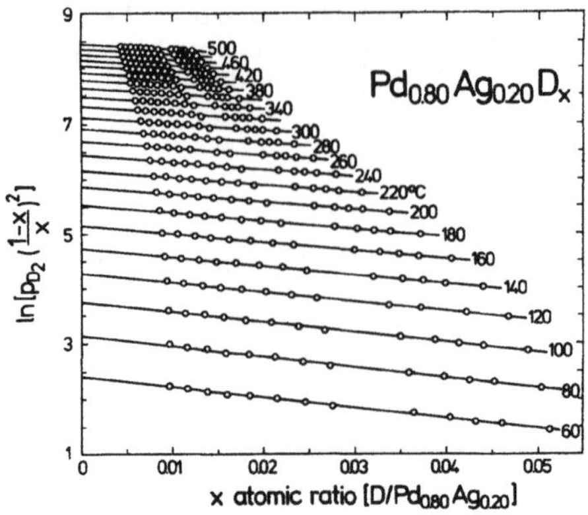

Fig. 1
Plot of $\ln\left[p_{D_2}\left(\frac{1-x}{x}\right)^2\right]$ as a function of concentration x for $Pd_{0.80}Ag_{0.20}$.

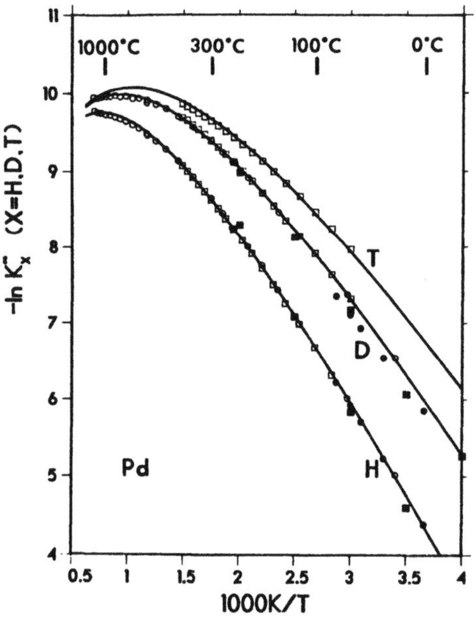

Fig. 2
Temperature dependence of the equilibrium constants (K^∞) for the indicated hydrogen isotopes in Pd: o: data by Powell, □ : data by Lässer; ● data by Wicke and Nernst (Ref. 10), ■: data by Clewley et al. (Ref. 11).

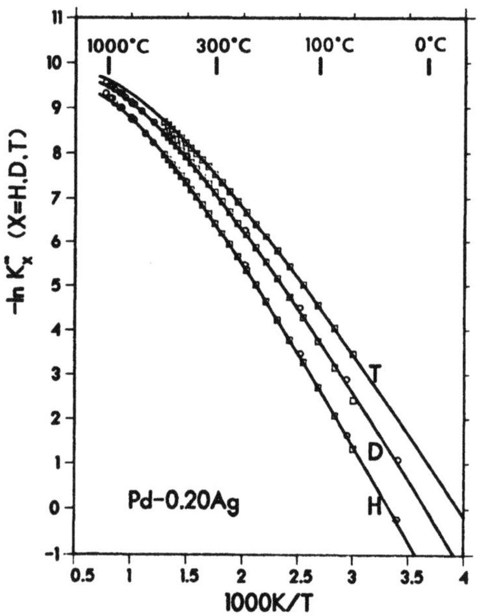

Fig. 3
Temperature dependence of the equilibrium constants (K^∞) for the indicated hydrogen isotopes in $Pd_{0.80}Ag_{0.20}$: o: data by Powell, □: data by Lässer.

L, J, D_0^o and B_0 are constants describing the hydrogen molecules. J is the Einstein temperature of the gas molecule, D_0^o its dissociation energy (e.g. the ground state energy of the gas molecule relative to the atoms at rest) and B_0 the rotational constant of the X_2 molecule. Most of the translational rotational partition function is given by $LT^{7/2}$. The values of the constants L, J and M necessary to calculate $G_{X_2}^o$ are listed in Table I (see Ref. 14 for more details).

Table I. List of the parameters used to calculate the free energy functions for hydrogen gas at one atmosphere pressure using Equation 7.

Molecular species	L $K^{-7/2}$	J K	M K
HH	4.293 10^{-4}	5986	51994.9
DD	2.406 10^{-3}	4307	52888.2
TT	6.582 10^{-3}	3548	53285.6

All energy parameters in the Eqs. 5-8 have been divided by R and are given in units of degrees Kelvin.

It is difficult to determine the 5 parameters A, B, C, N and E in the Eqs. 5-8 in a unique way because all contribute to $-\ln K^{\infty}$ at high temperatures. To reduce the number of free parameters we assume N = 1 for octahedral site occupancy and select a value of C similar to those obtained by inelastic neutron scattering spectroscopy[15]. In the case of D and T the Einstein temperature C_D and C_T were obtained by scaling C_H with the inverse of the square root of the hydrogen mass. The results are listed in Table II and plotted as solid lines in the Figs. 2 and 3. One sees that Eqs. 5-7 give a very good description of the $-\ln K^{\infty}$ values.

Table II. List of the parameters A, B, C, E and N used to calculate the equilibrium constant K^{∞} with Eqs. 5-7, σ standard deviation of the fit

X	N	A	B/K	C/K	E/K	σ	
H	1	1.981	768.0	800.0	28145.0	0.021	Pd
D	1	1.933	664.0	565.7	28175.4	0.012	Pd
T	1	1.912	617.9	461.9	28188.9	0.029	Pd
H	1	1.621	1645.5	800.0	28965.3	0.012	$Pd_{0.80}Ag_{0.20}$
D	1	1.134	1341.8	565.7	29038.6	0.041	$Pd_{0.80}Ag_{0.20}$
T	1	0.918	1067.3	461.9	29013.7	0.015	$Pd_{0.80}Ag_{0.20}$

Due to the fact that the ground state energy (E - 3 C/2) is only weakly dependent on the chosen C value (see Ref. 14) we find that the ground states of H, D and T in Pd at infinite dilution lie 26945 (2321.9 meV), 27327 (2354.9 meV) and 27496 (2369.4 meV below the dissociation limit for one half mole of X_2 molecules, respectively. In the case of $Pd_{0.80}Ag_{0.20}$ the corresponding numbers of the ground state energy for H, D and T are 27765 K (2392.6 meV), 28190.0 K (2429.2 meV) and 28321 K (2440.5 meV).

The excellent agreement between the solid lines and the experimental data in Fig. 2 contradicts the conclusion of McLellan and Suzuki[16]

that hydrogen in Pd can be described simply as an isotropic harmonic oscillator because a good fit is not obtained with A = 0.

With the values listed in Table II we can calculate the first term in Eq. 6 seperately which gives an analytical expression of the partition function of H, D and T in Pd and $Pd_{0.80}Ag_{0.20}$ over all thermally populated energy states. In addition, we can also calculate the partial enthalpies $\Delta\overline{H}_X^\infty$ and entropies $\Delta\overline{S}_X^\infty$ by means of the Gibbs Helmholtz relationship

$$\Delta\overline{H}_X^\infty = -R \frac{\partial \ln K_X^\infty}{\partial(1/T)} \tag{9}$$

and

$$\Delta\overline{S}_X^\infty = \frac{\Delta\overline{H}_X^\infty}{T} + R \ln K_X^\infty. \tag{10}$$

These analytical expressions have the advantage that they can be extrapolated easily to temperatures beyond those of the experiments.

The so determined $\Delta\overline{H}_X^\infty$ and $\Delta\overline{S}_X^\infty$ values are plotted in the Figs. 4-7 for Pd and $Pd_{0.80}Ag_{0.20}$. The partial enthalpies show a strong isotopic dependence:

$$\Delta\overline{H}_H^\infty < \Delta\overline{H}_D^\infty < \Delta\overline{H}_T^\infty \tag{11}$$

at constant temperature for Pd and $Pd_{0.80}Ag_{0.20}$. Comparing both metals one obtains the relationship

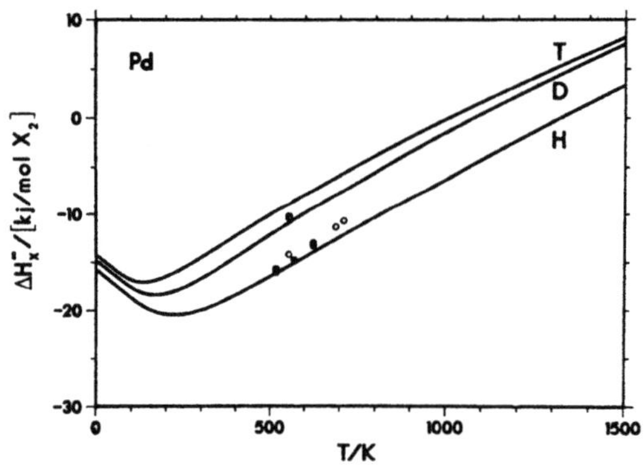

Fig. 4.
The partial enthalpy per mole X_2 for dissolution of X (X = H,D,T) in Pd, □: Ref. 17, o: Ref. 18, ●: Ref. 19.

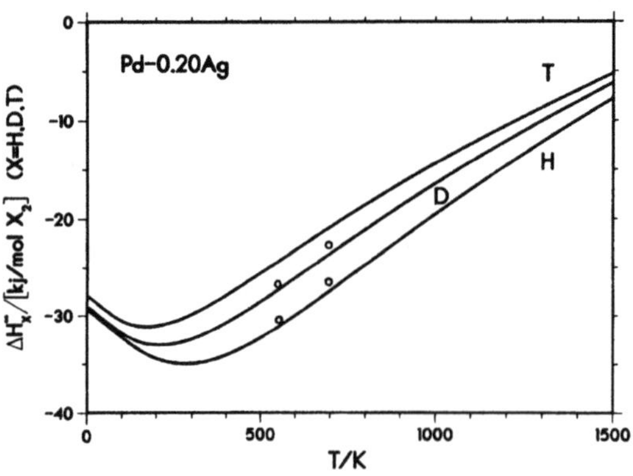

Fig. 5.
The partial enthalpy per mole X_2 for dissolution of X (X = H,D,T) in $Pd_{0.80}Ag_{0.20}$, o: Ref. 20.

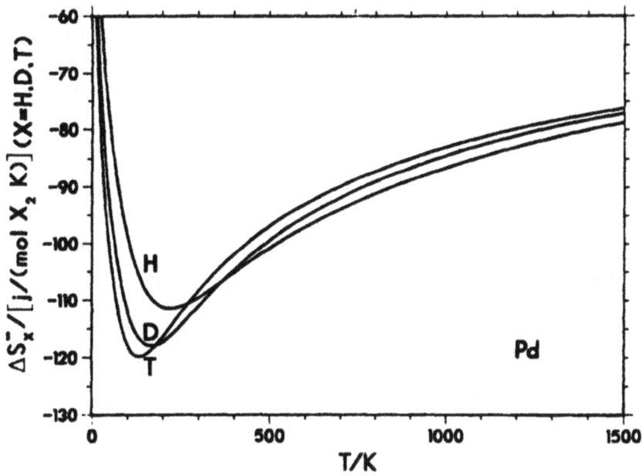

Fig. 6.
The partial entropy per mole X_2 for dissolution of X (X = H,D,T) in Pd

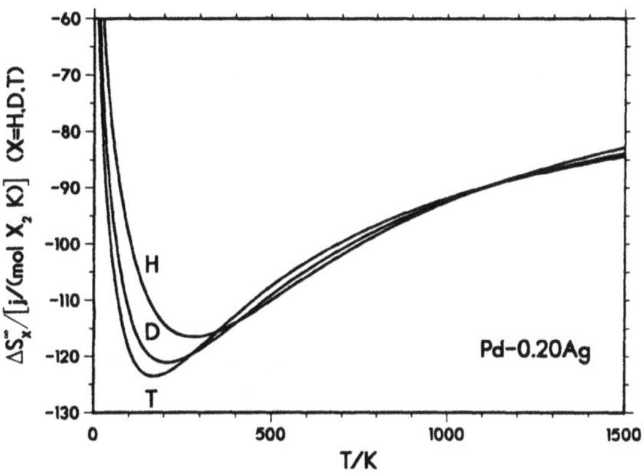

Fig. 7.
The partial entropy per mole X_2 for dissolution of X (X = H,D,T) in $Pd_{0.80}Ag_{0.20}$.

$$\Delta\bar{H}_X^\infty \text{ (Pd)} > \Delta\bar{H}_X^\infty \text{ (Pd}_{0.80}\text{Ag}_{0.20})$$

for constant temperature and X = H,D,T.

In all curves of the partial enthalpies and entropies shown in the Figs. 4-7 we observe a minimum which is due to the increasing importance of thermally populated vibrational energy states of hydrogen dissolved in the metal.

In Fig. 4 we have also plotted the $\Delta\bar{H}^\infty$ values determined calorimetrically by Boureau et al.[17,18] and by Picard et al.[19]. In Fig. 5 we show the interpolated $\Delta\bar{H}^\infty$ values obtained from the data of Picard et al.[20]. The agreement is satisfactory considering the two different experimental methods. The small systematic difference is probably due to difficulties in the calibration of the calorimeter.

PHASE BOUNDARIES OF THE Pd-X (X = H,D,T) SYSTEM

In this section we will present the first study of a phase diagram of a metal hydrogen system looking for the influence of all natural hydrogen isotopes.

Two phases, an α-phase with low hydrogen concentration and a β-phase with high hydrogen concentration are known in the case of the Pd-H system above 50 K and below the critical temperatures. Both phases are seperated by a miscibility gap forming a two phase region α+β.

The phase boundaries of the Pd-X (X = H,D,T) system were determined from pressure concentration temperature data because of the high risk of handling PdT_x samples outside our tritium loading equipment. Pd forms no stable oxide layers as is the case for V[21] or Nb[22] that prevent the tritium to leave the sample. The boundaries between the miscibility gap and the β-phase were obtained from the shape of the desorption isotherms[4]. The values of concentration and temperature of the solvus line between the α- and the two phase regions α+β were obtained by quasi isochoric measurements. A PdX_x sample with the concentration x slightly in the miscibility gap was heated in small temperature steps so that the concentration of the sample decreased and finally belonged to the pure α-phase. The change of slope in the equilibrium pressure as a function of the inverse temperature is interpreted as the intersection with the solvus line.

The experimental phase boundaries are plotted in Fig. 8 for H,D,T in Pd together with some values determined by other groups[1,10].

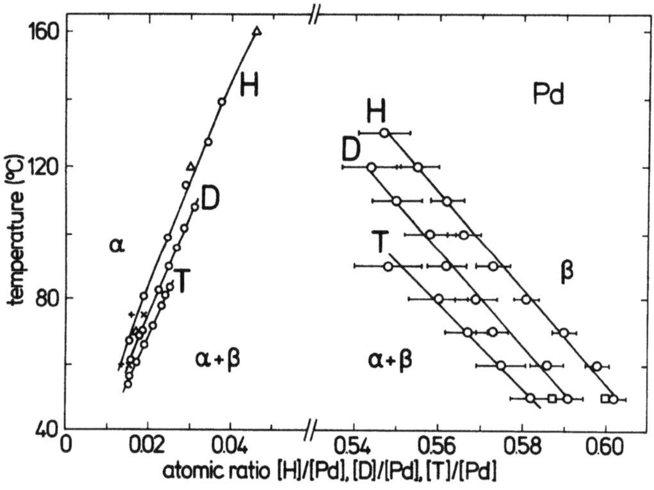

Fig. 8.
Part of the phase diagram of the system Pd-X (X = H,D,T) /Ref. 23/. Some values of other groups are also plotted:
Δ: PdH_x /Ref. 1/;
+: PdH_x /Ref. 10/;
x: PdD_x /Ref. 10/;
□: PdH_x,PdD_x /Ref. 10/.

Fig. 8 shows that the width of the miscibility gap decreases and that the area of the single phases increases with increasing mass. Thus, the following inequalities[23] describe qualitatively the phase boundaries x_a (x_b) between the α(β) phase and the miscibility gap:

$$x_a^H < x_a^D < x_a^T,$$

and

$$x_b^H > x_b^D > x_b^T$$

for equal temperatures.

As a consequence the critical temperature T_c^T of the Pd-T system is expected to be smaller than the one of D in PdD_x:

$$T_c^T < T_c^D < T_c^H.$$

These isotopic changes of the phase boundaries in the case of T in Pd follow the trends observed for H and D previously and could be discribed very well in a forthcoming publication[24].

CONCLUSION

We have presented solubility data of H, D and T in Pd and $Pd_{0.80}Ag_{0.20}$ in a very large temperature range and described these data by an analytical expression which allows to calculate the partial enthalpies and entropies for dissolution. We further determined the ground state energies of H, D, T in Pd and $Pd_{0.80}Ag_{0.20}$ with respect to atomic hydrogen at rest and could give an expression for the partition function of all thermally excited energy states. In addition we discussed the isotopic dependence of the phase boundaries in the Pd-X (X = H,D,T) systems.

REFERENCES

1. E. Wicke, H Brodowsky, and H. Züchner, "Hydrogen in Metals", edited by G. Alefeld and J. Völkl, Springer Verlag, Berlin (1978), Vol. 2, p. 73.
2. F.A. Lewis, Platinum Met. Rev. 26:20 (1982); 26:70 (1982); 26:121 (1982).
3. S. Schmidt, and G. Sicking, Z. Naturforsch. 33a:1328 (1978).
4. R. Lässer, and K.H. Klatt, Phys. Rev. B28:748 (1983).
5. R. Lässer, Phys. Rev. B 29:4765 (1984).
6. R. Lässer, J. Phys. F: Met. Phys. 14:1975 (1984).
7. G.L. Powell, J. Chem. Phys. 80:375 (1976).
8. G.L. Powell, J. Chem. Phys. 83:605 (1979).
9. R. Lässer, K.H. Klatt, P. Mecking, and H. Wenzl, Kernforschungsanlage Jülich Report No. JÜL-1800, 1982.
10. E. Wicke, and G. Nernst, Ber. Bunsenges. Phys. Chem. 68:224 (1964).
11. J.D. Clewley, T. Curran, T.B. Flanagan, and W.A. Oates, J. Chem. Soc. Faraday Trans. 1 69:449 (1973).
12. J.R. Lacher, Proc. R. Soc. London Ser. A 161:525 (1937).
13. A.L.G. Rees, Trans. Faraday Soc. 50:335 (1954).
14. R.Lässer, and G.L. Powell, to be published.
15. J.J. Rush, J.M. Rowe, and D. Richter, Z. Phys. B, Condensed Matter 55:283 (1984).
16. R.B. McLellan, and Y. Suzuki, Scripta Met. 19:485 (1985).
17. G. Boureau, O.J. Kleppa, and P. Dantzer, J. Chem. Phys. 64:5247 (1976).
18. G. Boureau, and O.J. Kleppa, J. Chem. Phys. 65:3915 (1976).
19. C. Picard, O.J. Kleppa, and G. Boureau, J. Chem. Phys. 69:5549 (1978).
20. C. Picard, O.J. Kleppa, and G. Boureau, J. Chem. Phys. 70:2710 (1979).
21. R. Lässer, and K. Bickmann, J. Nucl. Mater. 126:234 (1984).
22. R. Lässer, and K. Bickmann, J. Nucl. Mater. 132: (1985).
23. R. Lässer, J. Phys. Chem. Solids 46:33 (1985).
24. W.A. Oates, R. Lässer, T. Kuji, and T.B. Flanagan, to be published.

HYDROGEN AT METALLIC SURFACES AND INTERFACES

Louis Schlapbach
Laboratorium für Festkörperphysik ETH
CH-8093 Zürich, Switzerland

Abstract

This paper contains a simple introduction to the topic hydrogen on metals, a sketch of the recent progress in surface analytical experimental techniques and theoretical models and a review of recent results on hydrogen at the surface of a large variety of metallic substrates. Included are clean and precovered transition metals, rare earth metals, Mg, alloys and intermetallics, glassy metals and overlayers. We show that the aspects structure of adsorbate and substrate, electronic and magnetic properties and hydrogen-metal bond, surface dynamics as well as transitions from surface to bulk have to be studied.

INTRODUCTION

The topic "Hydrogen at solid surfaces" covers a very large field in science and technology. It spreads e.g. from the description of the vibrational modes of some H atoms on a ultrahigh vacuum cleaned Ni(100) surface over the reactivation of a dirty transition metal catalyst for a hydrogenation reaction to the detection of H_2 gas by a Pd coated metal-oxide-semiconductor device. Further related applications concern H induced embrittlement and fracture, metal hydride formation for H storage and for thermochemical machines, getters, plasma-first wall interactions in fusion technology, H-electrodes in electrochemical cells, amorphous solar cells (Si-H), and the production of rare earth permanent magnets.

This incomplete review is limited to a description of the interaction of gaseous hydrogen with solid substrates of metals. Though hydrogen can be present as H_2, H_2^+, H, H^+, D, T, we mostly consider "ordinary" H (for H_2^+ and H^+ as well as for metastable H(2s) see Steininger et al., 1984, and Eschenbach et al., 1983). The surface of the metallic substrate can be crystalline, ordered or disordered, amorphous or artificially modulated; it can be clean, coated with other metallic elements, precovered e.g. with C, S, CO, or oxidized.

Most of the experimental and theoretical work on "H on metals" was done for crystalline substrates. No detailed study of H adsorption on glassy metals has been published so far. However, the extension of calculations of the surface electronic structure to disordered alloys (Bansil and Pessa, 1983; Papaconstantopulos, 1985), the formation of model potentials for the interaction of molecules with amorphous surfaces (Purvis and Wolken 1979), and CO adsorption studies on glassy Ni-Zr alloys (Hauert et al. 1985) can be considered as a beginning. Recently discovered catalytic properties of glassy metals in hydrogenation reactions (Yokoyama et al., 1984; Schlögl,1985; Peuckert and Baiker,1985) also mark first steps.

The actuality and importance of the topic "H on solid surfaces" and the progress which was achieved are best demonstrated by the fact that more than 80% of the references used in this review have been published since 1982. Accordingly, many excellent reviews have been written. We particularly refer to Davenport and Estrup (1985, H on metals) and references therein, Ertl (1983, Kinetics of chemical reactions), Corbett (1985, H on semiconductor surfaces) and to a series of books edited by King and Woodruff (1981-86). Whereas these reviews treat rather few H-metal systems in great detail the aim of this contribution is a broader but less detailed review on a large variety of metallic substrates.

This introduction will be followed by a tutorial description of the relevant aspects of "H on metals", a list of experimental techniques and of theoretical models often used in this context. Thereupon, selected results on H on clean and precovered metals will be presented for transition metals, non-transition metals, alloys and intermetallics, glassy metals, overlayers and other specially prepared surfaces and interfaces.

A TUTORIAL DESCRIPTION OF THE TOPIC "H ON METALS"

The adsorption of H is conveniently described in terms of simplified one dimensional potential energy curves for an H_2 molecule and for 2H atoms on a metal surface (Fig. 1). Far from the surface the two curves are separated by the heat of dissociation E_D = 218 kJ/mol H. The flat minimum in the H_2 + M curve corresponds to physisorbed H_2 (heat of physisorption $E_P \lesssim 10$ kJ/mol H) and the deep minimum in the 2H + M curve describes chemisorbed H (heat of chemisorption $E_C \approx 50$ kJ/mol H). If the two curves intersect above the zero energy level, the chemisorption requires an activation energy E_A. In further steps the chemisorbed H atoms penetrate the surface and are then dissolved exothermically or endothermically in the bulk where hydrides can be formed. There is now experimental and theoretical evidence that not all chemisorbed H necessarily stays on top of the first metal atom layer, but also below it as a so called subsurface H (two step chemisorption).

A H_2 molecule impinging on the surface can be physisorbed or dissociatively chemisorbed or rejected. In order to stick on the surface and to dissociate the molecule has to dissipate its kinetic, rotational and vibrational energy by the excitation of phonons and possibly of electron-hole pairs in the substrate. Thus, the metal lattice and possibly the conduction electrons serve as a heat bath in the adsorption (and desorption) process. The ratio of the number of molecules which stick on the surface to those which impinge on the surface is called sticking coefficient s. It depends on the coverage Θ. The initial sticking coefficient for $\Theta = 0$ is of the order of 0.2 to 0.5 for H_2 on clean transi-

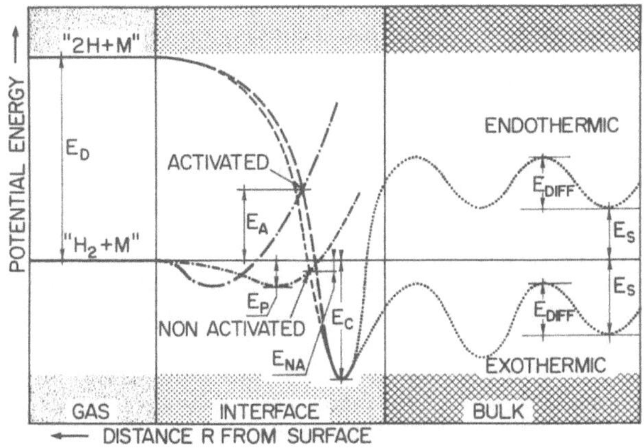

Fig. 1. Potential energy curves for activated and non-activated chemisorption of hydrogen on a clean metal surface and exo- or endothermic solution in the bulk. A more pronounced minimum just below the surface allows for subsurface hydrogen.

tion metals ($\Theta \approx 1$ for H_2 on Pd) and decreases by orders of magnitude upon the oxidation of the transition metals.

The chemical bond of H to the surface of a metal is, of course, of electronic nature. A strong perturbation of the substrate valence band and of the H 1s wave function takes place. So called H induced states are formed, comparable to the H induced band in the electronic structure of metal hydrides. Accordingly, surface electronic properties such as local density of states, work function, surface magnetization, and possibly valency are affected by the adsorption of H.

The adsorbed H atoms vibrate around their equilibrium position (e.g. on top of a surface atom, or a bridge over two surface atoms) diffuse along the surface and form disordered or ordered surface structures according to temperature, coverage and substrate. The adsorption often induces a rearrangement of the surface atoms of the substrate, a surface <u>reconstruction</u>, or even a <u>surface segregation</u> if the substrate is a binary compound. Thus in a description of the topic "H on metals" the following aspects should be considered:

- adsorption sites (on top, bridge, central)
  structure of the adsorbed H, disorder - order transitions and their coverage and temperature dependence, reconstruction or stabilization of the substrate surface
  H induced surface segregation

- nature and strength of the metal - H bond (s, p, d wave functions)
  electronic and magnetic properties
  valence changes at surfaces induced by hydrogen

- vibrational modes of the absorbed H atoms

- sticking coefficient,
  adsorption and desorption kinetics, reaction rates
- transitions between H on the surface, in the subsurface, and in the bulk

THEORETICAL AND EXPERIMENTAL TOOLS AND SOME BASIC IDEAS

Surfaces of metals

Before one can tackle complicated adsorption processes one should try to understand the simpler case of clean perfect surfaces of the substrate. In the recent years there has been much progress in understanding the electronic and atomic structure of surfaces on both the the theoretical and experimental front (Inglesfield, 1983, and ref. therein; Bullett, 1985). The surface electronic structure, which is needed to obtain the charge density and the density of states, has been calculated for different surfaces of many elements, metals and semiconductors, and has been probed experimentally by angle resolved photoemission and other techniques. Examples of the calculated charge density and of the corresponding total density of states are shown in Fig. 2a and 2b for Ni(100) (Arlinghaus et al., 1980).

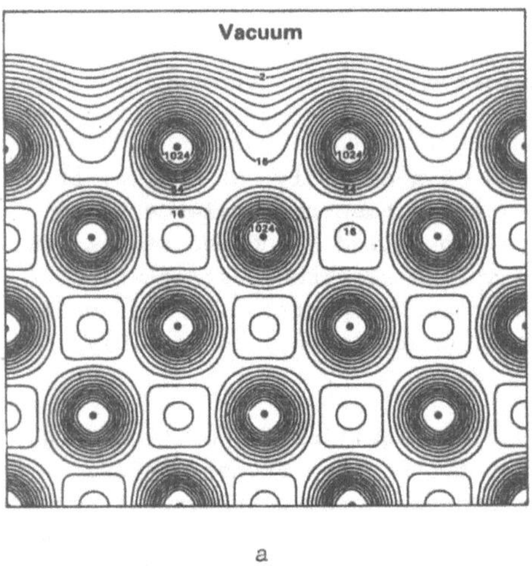

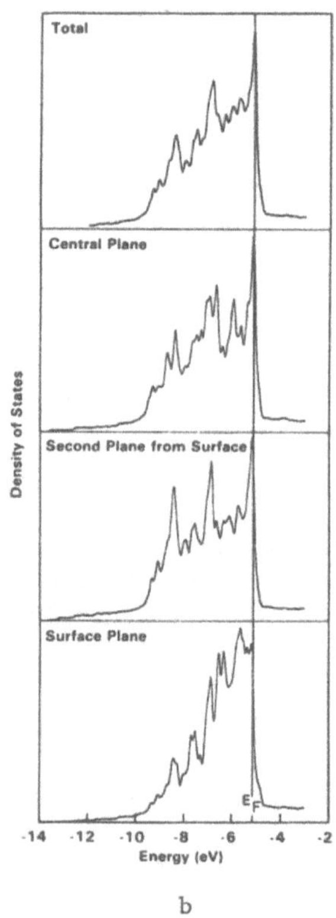

Fig. 2a: Ni (100) charge density plotted in a (100) plane perpendicular to the nine-plane slab and passing through the center of the atoms.

Fig. 2b: Ni (100) total density of states and planar density of states for central plane, second plane from the surface, and the surface plane.
(Arlinghaus et al., 1980)

The calculation of the surface electronic structure is possible because many body effects can be neglected to a good approximation and the many electron Schrödinger equation is replaced by the one electron equation. An extra type of solution of the Schrödinger equation are wavefunctions which are localized at the surface. They cause the surface states, which play an important role in the surface reconstruction of semiconductors. The local density of states at the surface is made of this surface states and of tails of bulk wave functions.

Surface states exist on normal metals as well as on transition metals. Occupied and unoccupied surface states were recently studied experimentally and theoretically on single crystal surfaces of e.g. Ag (Reihl, 1985), Cu (Bartynski et al., 1985) and Ni (Borstel et al., 1985). Surfaces of transition metals are particularly interesting as they not only show a structural relaxation - an effect which is mostly weak on normal metals - but can also exhibit magnetic properties which differ from those of the bulk (Freeman, 1983). A surface enhanced magnetic order as well as magnetic surface reconstruction were observed on Gd(0001) (Weller et al., 1985). The magnetic hysteresis loop of the Fe(100) surface, very recently measured by means of the spin polarization of secondary electrons (Fig. 3), shows a softer behaviour within the outermost 5 Å due to reversed domain nucleation (Allenspach et al., 1986). The structural and electronic differences of surfaces and bulk may manifest themselves as surface core level shifts (Eastman and Himpsel, 1982; Erbudak et al., 1983). In the case of rare earth metals and alloys they may even appear as surface valence transition (Netzer and Matthew, 1986; Kaindl et al., 1982).

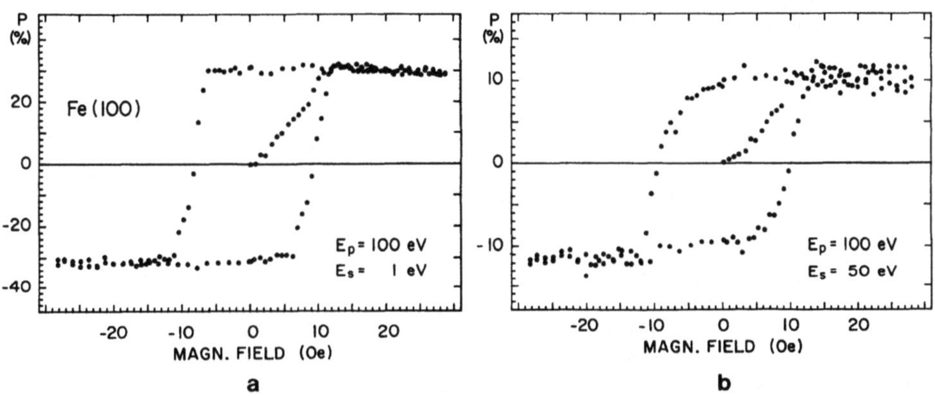

Fig. 3: Magnetic hysteresis loops P(H) at the surface of Fe(100) recorded with secondary-electron spin polarization at 300 K using electrons of primary energy E_P = 100 eV:

a) secondary electron energy E_S = 1 eV corresponding to a probing depth of ≈ 50 Å, bulk hysteresis loop;

b) E_S = 50 eV, probing depth ≈ 5 Å, surface hysteresis loop.

(Allenspach et al., 1986)

Hydrogen chemisorption models

The interaction of a H_2 molecule with a metal surface is the simplest example of a chemical process at a metal surface. It involves the breaking of the molecular bonds and the creation of new H - metal bonds, sometimes at the expense of metal - metal bonds (H - embrittlement).

Chemisorption of H on simple metals and on transition metals has been studied theoretically using approaches such as molecular orbit theory, valence bond theory, density functional theory, cluster calculations (a detailed description up to 1980 can be found in Smith, 1980), and effective medium theory (Nørskov, 1984). An extension of the effective medium theory, the so-called embedding atom method - originally developed to study the embrittlement problem - was shown to yield very valuable results for H on metals, particularly for the surface relaxation and the H adsorption sites including subsurface sites (see Pd) (Daw and Baskes, 1984).

H on simple metals is usually treated using a jellium model (Johansson, 1981).

The H chemisorption energy exhibits systematic trends along the transition metal series, showing maximum values at the beginning of the series. Varma and Wilson (1980) and Nordlander et al. (1984) explained this trend with the electronic properties of the d - band of the metals. The effective medium theory is capable of reproducing the experimental chemisorption energies as well as the equilibrium positions and vibrational frequences. The degree of filling of the d - band is the most important parameter. Energy differences between different equilibrium sites are generally found to be small indicating a large H surface mobility.

H adsorption changes the surface electronic structure and causes surface reconstruction. Thus also surface core level shifts and surface valence changes might be induced by H. CO induced surface core level shifts were observed on Pt (Shek et al., 1982), but so far no such effects induced by H have been reported. They certainly exist, especially when H adsorption results in the formation of a surface hydride.

The reciprocal influence of surface magnetism and H adsorption is particularly interesting and also controversial. H adsorption was shown to change the magnetic properties of the surface (magnetic moment, ordering temperature; see Ni, Gd). Reciprocally the surface magnetism seems to influence the H adsorption kinetics (Kaarmann et al., 1984) although the old "magneto-catalytic" effect does probably not exist (cf. Ertl, 1985).

When a closed shell molecule approaches a metal surface, a Pauli repulsion becomes effective due to overlapping between orbitals of the molecule and Bloch functions of the metal surface. The physisorption well results when the repulsion is combined with the van der Waals attraction. At smaller distances transition metals behave differently than normal metals. It has been found that H_2 dissociation at a surface with only s electrons goes over an activation barrier (some tenth of an eV up to several eV) because of the Pauli repulsion. In the case of transition metals holes in the d - band become important. A s - d electron transfer considerably weakens the Pauli repulsion and lowers the activation barrier (Harris and Andersson, 1985; Johansson, 1981).

The precoverage of the surface of elemental metals by contaminants like oxygen, sulphur, carbon (SO_2, CO) degrades their H adsorption, absorption and desorption properties drastically. The contamination can interfere in different steps in the sorption process: The contaminating species themselves are chemisorbed, induce reconstruction and alter the electronic properties and the phonon spectra of the substrate surface. Thus sticking coefficient, chemisorption energy, vibrational properties and surface mobility of H_2 and H are affected and dissociation and recombination are often made impossible.

The chemisorption properties and the surface electronic structure of metal oxides and of transition metal oxides were reviewed by Kunz (1985).

Though the problem of surface poisoning is of great importance in catalysis and H - embrittlement (Berkowitz et al., 1976), there is a big gap between results of experimental and theoretical work on this subject. Some experimental facts are: C, O and CO strongly influence H adsorption on Ru (Feulner and Menzel, 1985); O decreases the overall H sorption kinetics and the sticking coefficient (Fromm and Wulz, 1984); S on Pd inhibits desorption of H probably by affecting the surface mobility of H (Bucur, 1981); SO_2, Co and H_2S on Nb control the H permeation rate, probably by blocking active surface sites and structural defects induce H traps (Sherman and Birnbaum, 1985). C adsorption on Ni(100) was shown to freeze out a Ni surface phonon which results in a surface reconstrucjtion (Rahman and Ibach, 1985). From coadsorption experiments of Ko and Madix (1981) it appears that poisoning is more than simply blocking active sites. Self consistent calculations of the electronic structure perturbation induced by a catalytic poison, S on Rh (001), reveal a substantial reduction of the local density of states at the Fermi level (Fig. 4, Feibelman and Hamann, 1984). Nørskov et al. (1984) proposed a model based on the effective medium theory, which is able to describe promotion and poisoning effects of co-absorbed electropositive and electronegative species.

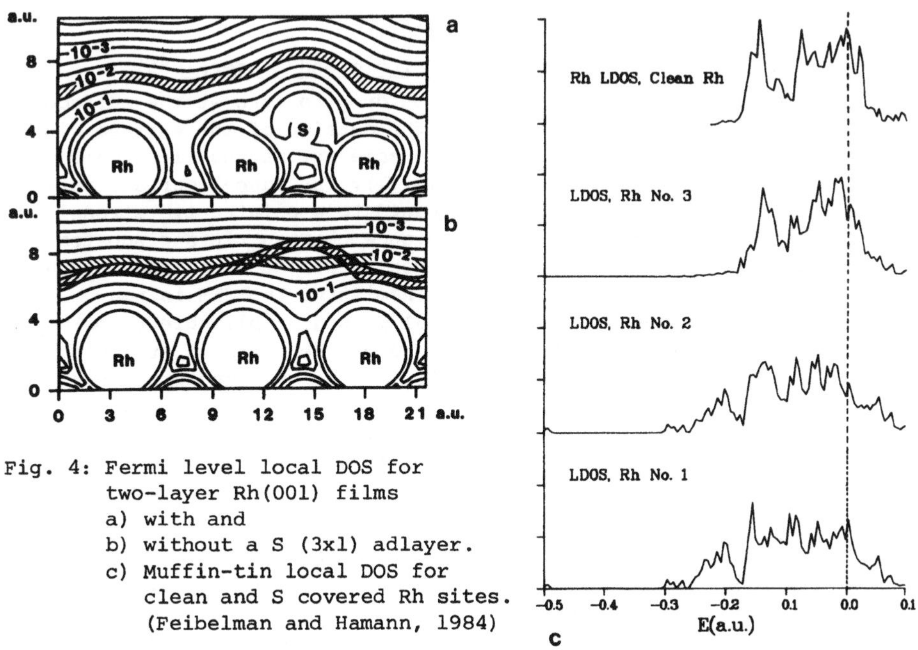

Fig. 4: Fermi level local DOS for two-layer Rh(001) films
a) with and
b) without a S (3x1) adlayer.
c) Muffin-tin local DOS for clean and S covered Rh sites. (Feibelman and Hamann, 1984)

On intermetallic compounds and glassy metals surface poisoning is by far less severe since surface segregation provides a self restoring mechanism of the active surface (Schlapbach et al., 1980; Spit et al., 1981). Poisoning by heavy SO_2 dosage is possible (Gualtieri et al., 1976). SO_2 oxidizes the transition metal precipitates and blocks further segregation (Schlapbach and Brundle, 1981).

Experimental methods

Many experimental methods for studying H and H induced effects on metals are known. Most of them are used for surface and adsorption studies in general, a few of them, however, are specific to H on metals. The direct detection of H on metals by some surface sensitive methods is rather difficult because of the low atomic number of H. Auger electron emission e.g. does not work principally, photoemission has extremely low cross section and X-ray scattering is very weak even at grazing incidence. Thus, in many cases, one does not analyze H directly but the H induced variations of the substrate properties. (See also Malinowski, 1983).

Among the most powerful methods in the past were certainly the electron spectroscopic methods: low energy electron diffraction, photoemission and electron energy loss spectroscopy, He diffration and thermal desorption spectroscopy, which are all ultrahigh vacuum technologies. For general references we refer to the series "Chemical Physics of Solid Surfaces" (King and Woodruff, 1981/86), to the series "Methods of Surface Characterization" (Yates and Madey, 1986) and to the proceedings "Spectroscopic Studies of Adsorbates on Solid Surfaces" (Ueba and Yamada, 1984).

After a short description of some new methods - new in the sense that they were recently developed or that they were recently applied to surface H - we list the major methods according to the type of information they provide.

Scanning tunneling microscope (STM) (Binnig and Rohrer, 1985)

The electron density of a substrate does not drop to zero at the surface, but decays approximately exponentially on the outside of the surface within a short decay length which characterizes to some extent the electronic or chemical properties of the surface (cf. Fig. 1). The tunnel current between the substrate and a counterelectrode sharpened to a pointed tip strongly depends on the decay length and on the distance between tip and substrate. Scanning of the tip over the substrate traces an almost true image of the surface topography and probes the surface chemistry. The STM is non-destructive, it is a structural and chemical method, applicable to both periodic and nonperiodic surface features and can be operated at ambient pressure. The STM yields less accurate data on surface structures than e.g. low energy electron diffraction. However, as it resolves space and time it allows the observation of the nucleation of the surface reconstruction. The oxygen induced reconstruction of the Ni(110) surface was recently studied. The STM was not yet used for looking at hydrogen induced features.

Incoherent inelastic neutron scattering (IINS) (Cavanagh et al., 1986)

Neutron scattering is known to be a very valuable tool for studying the dynamics of bulk hydrogen. Increased neutron flux (10^{14} - 10^{15} neutrons/cm^2 sec) allows the investigation of the dynamics of adsorbed hydrogen for samples with a reasonably high surface to bulk ratio. The sample can be

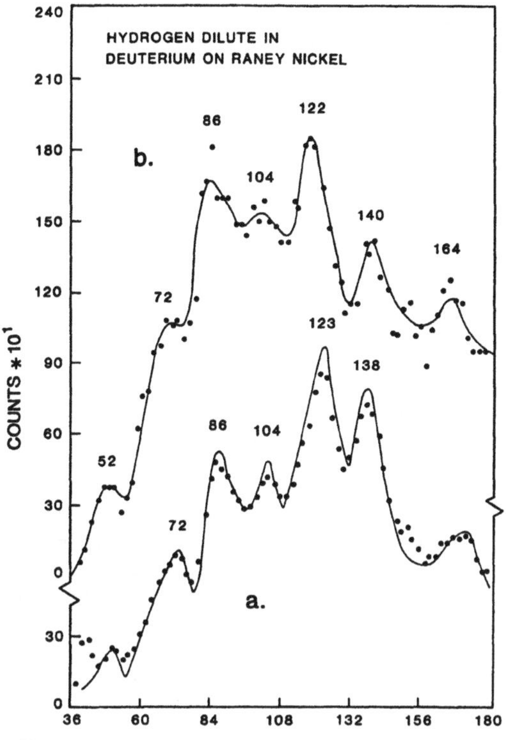

5

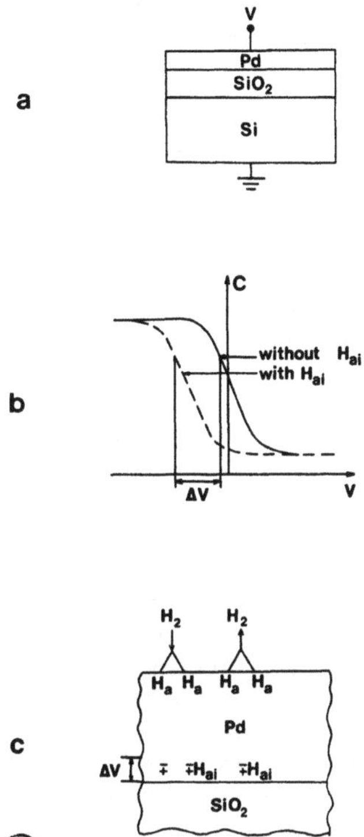

6

Fig. 5 (left): IINS spectra of H (diluted in deuterium) on Ni at 80 K:
a) coverage Θ_H - 0.1, Θ_D = 0.2 of saturation;
b) Θ_H = o.1, Θ_D = 0.7 (Cavenagh et al., 1986).

Fig. 6 (right):
a) Schematic picture of a H-sensitive MOS structure
b) Capacity curve C(V) for a structure with and without the presence of H.
c) Dissociation and association take place at the Pd surface. The shift of the C(V) curve in the presence of H is caused by H atoms at the interface (H_{ai}) (Petersson et al., 1984).

studied in a reactor vessel over a wide range of pressure and temperature, and therefore provides a complement to techniques such as electron energy loss spectroscopy, which measures e.g. the vibration spectrum of adsorbates on well characterized surfaces under ultrahigh vacuum. Neutron methods can cover an energy range from 1 µeV to 300 meV and thus offer studies of the diffusion and rotational dynamics of adsorbates in a time range of 10^{-8} to 10^{-13} sec. Fig. 5 illustrates the inelastic neutron scattering spectra for H (diluted with D) on Raney nickel at coverages below saturation.

<u>Time-of-flight analysis of direct recoils (TOF-DR)</u> (Mintz and Schultz, 1984a)

Low energy noble gas ion scattering, a surface sensitive method, is generally incapable of probing surface hydrogen directly. However, if pulsed rare gas ion beams at grazing incident angle are used to bombard the surface, directly recoiled surface particles including light adsorbates can be analysed by time-of-flight measurements. As was shown by Mintz and Schultz for hydrogen on La and Mg, the surface composition and information on the surface chemistry and to some degree on the surface structure can be obtained.

Surface enhanced Raman scattering (SERS)

SERS has developed into a powerful method for the study of vibations and electronic excitations of large adsorbed molecules (see Ueba and Yamada, 1984), but so far it has not been applied to surface H.

Pd-gate metal-oxide-semiconductor device (Pd-MOS) (Petersson et al., 1984)

The Linköping group of Lundström developed a simple hydrogen-sensitive Pd metal-oxide-semiconductor device, which is built of a Pd-SiO_2-Si structure and is capable of detecting hydrogen at pressures from 1 · 10^{-10} mbar to 1 bar. H_2 molecules in the ambient and in ultrahigh vacuum adsorb and dissociate on the Pd gate. H atoms rapidly diffuse through the thin Pd film and adsorb at the Pd - SiO_2 interface where they form a dipole layer and decrease the effective work function of the Pd. The work function shift can be measured as a shift of the capacitance versus voltage curve. Fig. 5 illustrates the principle of the measurements. The device can be used to detect hydrogen in metallic overlayers down to submonolayer thickness (Schlapbach et al., 1986a).

- Methods to detect surface H

  SIMS, secondary ion mass spectroscopy (destructive)
  TOF-DR, time of flight analysis of direct recoils (Mintz and Schultz, 1986a)
  ISS, RIS, low energy ion scattering spectroscopy and recoil spectroscopy (Oura et al., 1984)
  Pd-MOS, Pd metal-oxide-semiconductor device (Petersson et al., 1984)

- Methods to study the structure of adsorbate and substrate

  LEED, video-LEED, low energy electron diffraction (Christmann et al., 1984, and ref. therein)
  He diffraction (Rieder and Stocker, 1985), also applicable to disordered surfaces (Comsa and Poelsema, 1985)
  scanning tunneling spectroscopy (Binnig and Rohrer, 1985)
  grazing incidence X-ray scattering (Brennan, 1985)
  SEXAFS, NEXAFS, surface extended and near edge X-ray absorption fine structure (Stöhr et al., 1985)

- Methods to probe electronic structure and H-metal bond (Plummer et al., 1984; Fadley, 1984; Umbach, 1984)

  PES, photoemission, photoelectron spectroscopy
  AES, Auger electron spectroscopy (Malinowski, 1983)
  BIS, inverse photoemission
  TDS, thermal desorption spectroscopy
  $\Delta\phi$, workfunction measurements (Christmann et al., 1984)
  ESD, Electron stimulated desorption (Menzel, 1985)

- Methods to test surface magnetism (Siegmann et al.; Kirschner, 1985)

  spin polarized photoemission
  spin polarized Auger electron and secondary spectroscopy
  spin polarized LEED

- Methods to investigate surface dynamics

  EELS, electron energy loss spectroscopy
  H_2/D_2, hydrogen-deuterium exchange reaction (Engel and Ertl, 1982)
  IINS, incoherent inelastic neutron scattering (Cavanagh et al., 1986)

- Methods to examine the near surface region, transitions from surface H to bulk H

  DCEMS, depth selective conversion electron Mössbauer spectroscopy (Liljequist and Ismail, 1985)
  ^{15}N depth profiling with the reaction $^{1}H(^{15}H,\alpha\gamma)^{12}C$ (Pick et al., 1982).

H ON TRANSITION AND RARE EARTH METALS, NORMAL METALS, CRYSTALLINE AND AMORPHOUS ALLOYS AND IN METALLIC OVERLAYERS

H on Nickel

H on Ni is the best studied example of H chemisorption and also the example of most advanced understanding. We do not trace a detailed picture of the results, but refer to excellent papers on the H-Ni bond (Greuter et al., 1986), on the structure of adsorbate and substrate (Rieder and Stocker, 1985; Christmann et al., 1985), and on the dynamics of H on Ni (Robota et al., 1985).

The investigation of the H-Ni (111) bond by angle resolved photoelectron spectroscopy at low temperatures ($T < 170$ K) for monolayer coverage of atomic H yields an H-induced band split off from the Ni bands and a change in the d-band surface states. As the H coverage is decreased, the split-off state moves upwards towards the bulk bands until it disappears ($\Theta < 0.5$). The surface states shift continuously with H coverage. The results indicate that the H layer acts as a uniform attractive potential, with the strength of the potential depending on the H coverage (Greuter et al., 1986).

Earlier photoelectron spectra (Eberhardt et al., 1981) which showed the disappearance of some H induced features upon warming of the sample to room temperature, led to the speculation that H which is invisible at room temperature was in a subsurface site. There is no doubt about subsurface H in Pd, but whether H on Ni at monolayer coverages exists or not is still a matter of discussion.

H on Ni (110) is an especially beautiful example to demonstrate how the interaction of adsorbed H atoms with each other as well as with the substrate result as a function of coverage in a rich variety of ordered phases and phase transitions. He diffraction of H on Ni (110) yields at low temperatures ($T < 130$ K) the following ordered phases (Rieder and Stocker, 1985): c (2x6) for Θ = 1/3ML; c (2x4) for Θ = 1/2ML; c (2x6) for Θ = 2/3 and 5/6 ML; (2x1) for Θ = 1ML and (1x2) for Θ = 1.5ML. Three of these phases are shown in Fig.7. The common structural elements are long H-zig-zag chains along the Ni close- packed rows with H near threefold sites. Up to monolayer coverage, the substrate Ni atoms do not change their position. Further H, however, induces a substrate reconstruction and at 1.5ML coverage probably 1/2ML H is adsorbed on the second Ni layer. On the reconstructed surface three distinct desorption states (α, β_1 and β_2) are observed by thermal desorption spectroscopy. The α peak is particularly sharp and can only be observed when the reconstructed 1x2 phase is present. Apparently it corresponds to H on the second Ni layer. The reconstructed 1x2 phase can be regarded as a surface hydride (Christmann et al., 1985). A reconstructed streak phase is formed upon H_2 exposure of Ni (110) at room temperature or upon heating of any low temperature phase above T = 220 K (Jo et al.,1985, Christmann et al.,1985).

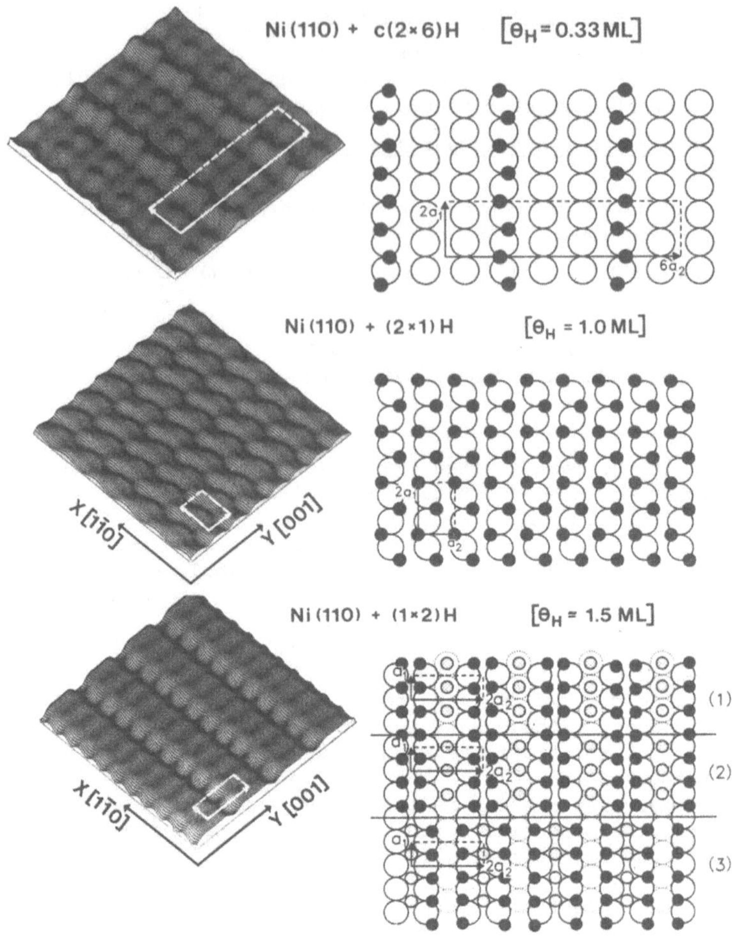

Fig.7: Structure of H on Ni (110) at T < 170 K deduced from He diffraction spectra. Best fit corrugation functions (left) and hard sphere models (right) are shown for three different coverages Θ_H = 0.33, 1.0 and 1.5 monolayers. (Rieder and Stocker, 1985).

Results of H_2- and D_2- beam studies were rationalized in a two dimensional potential energy diagram. It reveals negligible activation energy for Ni (110) and 0.1 eV activation energy for Ni (111) (Robota et al.,1985; Steinrück et al., 1985). Total energy calculations show that the equilibrium H position on Ni (100) is the center site, 0.6 a_o above the plane of the surface Ni atoms. The bridge and top sites are only 0.1 eV and 0.3 eV higher, indicating a high H surface mobility (Umrigar and Wilkins, 1985).

Theoretical investigations of the effect of an ordered (1x1) H overlayer on Ni (100) (Huang and Hermanson, 1985; Weinert and Davenport,1985) yield a reduction of the magnetic moment of Ni surface atoms by tu to $0.2\mu_B$, in agreement with a measured decrease of the spin polarization of photoelectrons (Landolt and Campagna, 1977). Kaarman et al. (1984) noticed a substantially slower desorption of hydrogen from magnetized Ni (110) surface than from macroscopically unmagnetized samples. The effect of the magnetization of Ni on the H_2/D_2 exchange reaction seems to be absent with clean H/Ni surfaces, but can be caused by the presence of carbon impurities (Ertl, 1985).

H on Palladium

H on Pd is in many respects comparable to H on Ni, with the significant difference that Pd dissolves large amounts of H and forms a bulk hydride, whereas Ni does not (it only does in the kbar range). Pd adsorbs H dissociatively above ≈ 50 K with a sticking coefficient close to unity. To avoid dissolution of surface H in the bulk the adsorption studies should be performed below ≈ 200 K where diffusion into the bulk is slow.

H/Pd(110): He diffraction, LEED and TDS (Rieder et al., 1983; Behm et al., 1983) reveal, in good agreement with each other, the following results: H_2 exposure of Pd (110) at 100 K yields two ordered surface phases; a (2x1)H phase at monolayer coverage and a (1x2)H phase with reconstructed Pd substrate at 1.5ML coverage (Fig.8, inset). Interestingly, the substrate reconstruction enables 0.5ML of H to enter the near surface region and to populate subsurface sites. TDS spectra (Fig.8) show H desorption from two different subsurface states (α_1 , α_2) and from two normal chemisorption states (β_1, β_2). Upon heating the (1x2)H phase transforms back into the (2x1)H phase and part of H moves into the subsurface.

H/Pd(111): The migration of surface H into subsurface sites was originally suggested to explain the disappearance of H induced features in photoelectron spectra of H/Pd(111) upon heating to 300 K (Eberhardt et al., 1981). This suggestion stimulated much further experimental and theoretical studies. Selfconsistent pseudopotential calculations show that H in subsurface tetrahedral sites has electronic properties very similar to those of H in the surface threefold sites; furthermore, the electronic properties of H in subsurface octrahedral sites are similar to the clean surface (Chan and Louie, 1984). Accordingly, the photoelectron spectra alone do not give clear evidence for subsurface H. Chubb and Davenport (1985) calculated the electronic structure of three-layer Pd films with H layers outside and inside and showed that the position of the H induced

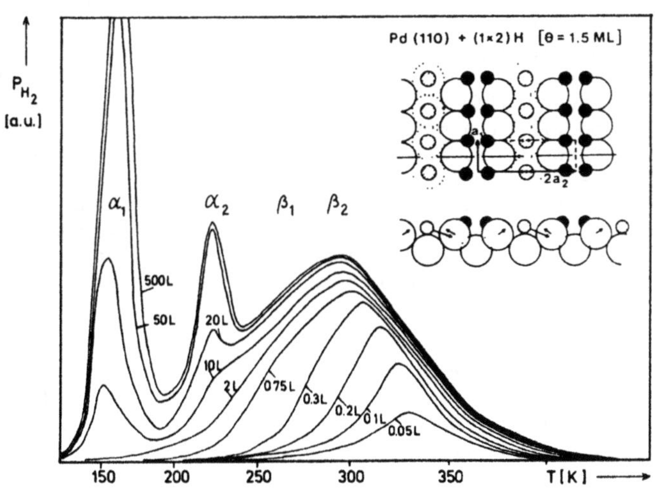

Fig.8: H on Pd(110): Series of thermal desorption spectra. Exposures from 0.05 L up to 500 L at 120 K (Behm et al., 1983).
Inset: Top view and side view of a hard sphere model of the (1x2) phase on Pd (111) as deduced from He diffraction. The arrows indicate the motion of the H and Pd atoms upon transformation from (1x2) H to (2x1)H, where part of H moves into subsurface sites (Rieder et al., 1983).

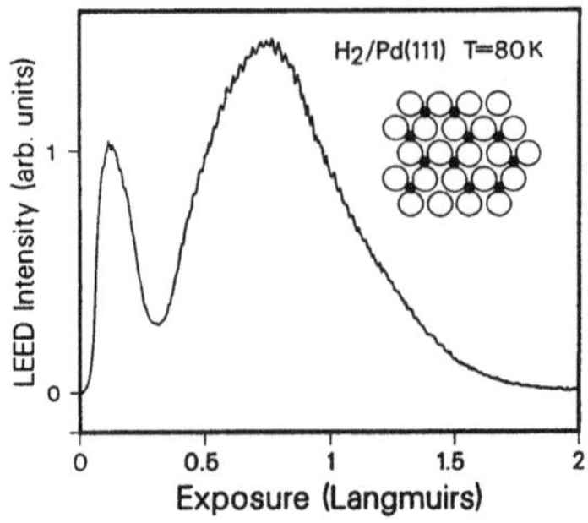

Fig.9: H on Pd (111): Intensity of the √3 LEED spot during exposure to H_2 showing transitions from disordered to two ordered (1/3, 2/3) phases. Inset: √3 x √3 R 30°, proposed structure responsible for the second peak (Felter et al., 1985)

states depends markedly on concentration. Indeed, a more detailed analysis by photoelectron spectroscopy (Greuter et al., 1986) revealed that the H induced band changes position and shape with H concentration. It moves closer to E_F as the H concentration is decreased and merges into the bulk band and loses its identity. Felter et al. (1985) finally present both experimental and theoretical evidence for H bound in the octrahedral site between the first and second plane of Pd (111). They observed order-disorder transformations as a function of temperature and coverage by LEED (Fig.9). Ordered phases with LEED pattern (√3 x √ 3 R 30°) were observed at coverages 1/3 and 2/3 of a monolayer at low temperatures. They disorder upon warming to 85 and 105 K, respectively. Theoretical calculations using the embedded atom method predict the occupation of subsurface sites already at these low temperatures. The H-metal band in the octahedral subsurface site is comparable to that of the surface threefold site, but H-H repulsion makes the subsurface site more favourable for high coverages. The same method also yields the correct phase transition temperatures.

H_2 on Copper at Low Temperatures

Dissociative chemisorption of H_2 on Cu is limited by a rather large activation barrier. Cu adsorbs and dissolves large amounts of atomic H.

Cold substrates ($T \leq 15$ K) of noble metals and Cu are well suited to investigate the adsorption of molecular H_2. This allows study of the weak Van der Waals' H_2-metal bond and the relative importance of the degree of freedom of the electrons and phonons of the substrate in dissipating the kinetic energy of the incident molecule.

Photoemission from condensed layers of H_2 on Cu (and Au) was measured at 4 K (Eberhardt et al., 1982). Monolayer adsorption results in a H 1s peak at a binding energy of 9.2 eV. EELS and workfunction measurements show that H_2, HD, D_2, ortho-para H_2 mixture and para H_2 have different sticking probabilities on Cu (100) (Andersson and Harris, 1983). From theoretical analysis it is concluded that in most sticking events the full particle energy is transferred to the solid in the initial collision, i.e. phonon processes dominate the sticking. Rotational excitations influence the sticking via a resonant enhancement of the particle-phonon coupling (adsorption resonances). They depend on mass and symmetry of the molecule and are different for H_2, HD and D_2 (Stiles and Wilkins, 1985; Andersson et al., 1985).

H on Rare Earth Metals

The few experimental and theoretical results obtained up to 1981 on H adsorption on rare earth metals were summarized by Netzer and Bertel (1982). Some of the early experimental results are contradictory and questionable because contamination was involved. Resitivity measurements indicate that surface H is dissolved quickly. From measurements of the workfunction it was concluded that adsorbed H is located in subsurface sites, in contrast to calculations of the surface electronic structure, which favours H adsorbed in threefold sites outside the metal.

The calculated surface DOS of H on Sc (0001) (Feibelman and Hamann, 1980) closely resembles the bulk DOS of ScH_2, in agreement with photoelectron spectra of H adsorbed on La, which are also similar to those of LaH_2 (Kumar et al., 1984). Rather high H_2 exposures of La and Er were required to produce the typical H induced features 6 eV below E_F. Even workfunction measurements of Er, Yb, and Eu showed changes only for high H_2 exposure (Netzer and Matthew, 1986). As the sticking coefficient for H_2 on rare earth metals is ≈ 0.2 (Atkinson et al., 1976), a fast diffusion of surface hydrogen into the bulk may account for the need of these high exposures.

First single crystal studies of H on Ce (001) by angle-resolved photoemission permitted distinction between H at the surface and in subsurface layers. Hydrogen segregated from bulk to surface was observed to induce a distorted hexagonal reconstruction of the square Ce (001) (Netzer and Matthew, 1986).

Submonolayer coverages of H on Gd, chemisorbed at 20 K, drastically reduce the spin polarization of photoelectrons, because of the formation of a canted or disordered spin structure at the Gd surface with a lower ordering temperature (Cerri et al., 1983).

The growth of a strong peak at the Fermi level was recently observed in photoelectron spectra of $CeH_{2.7}$ at temperatures below 70 K (Schlapbach et al, 1986b). As bulk sensitive methods do not show any anomaly in that temperature range, a surface transformation, possible a disorder-order transformation, may be the origin of that low temperature photoemission peak (Fig. 10).

H on Magnesium and Oxidized Magnesium

Mg oxidizes very easily so that it is a real challenge to study the surface properties of clean Mg and of H adsorption and hydride formation.

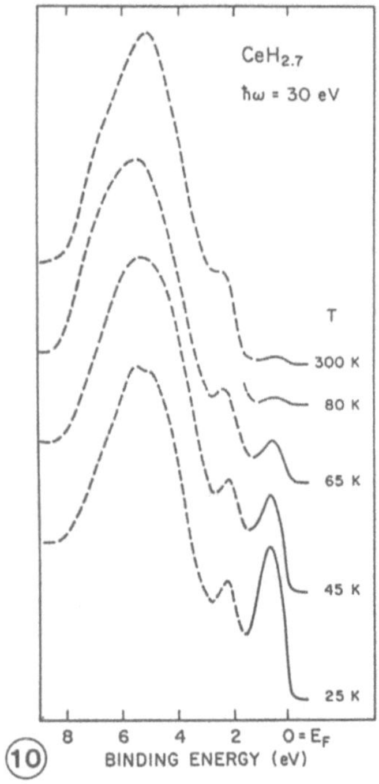

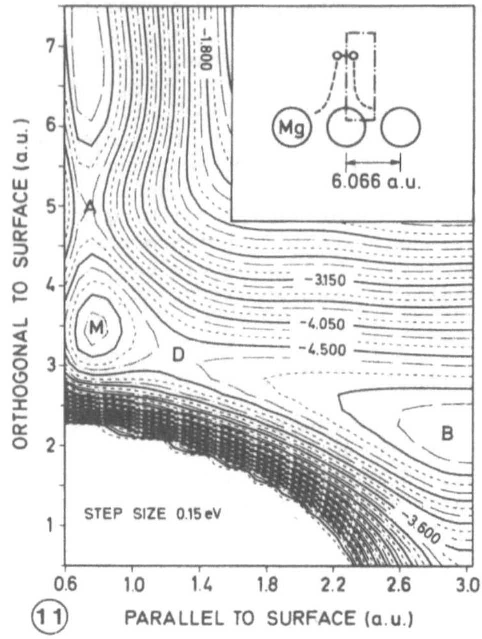

Fig.10: Photoelectron spectra of $CeH_{2.7}$ as a function of temperature. The almost semiconducting compound seems to become metallic at the surface below 80 K, possibly due to H ordering at the surface (Schlapbach et al., 1986 b).

Fig.11: Equipotential energy curves for H_2 on Mg (0001). Energies in eV relative to those of free atoms (Nørskov et al., 1981).

The surface electronic structure of Mg (0001) was analysed by angle resolved photoemission (Karlsson et al., 1982). Two sharp peaks due to surface states were identified. No experimental studies on H adsorption on single crystalline Mg are available. Self consistent calculations of the potential energy surface for a H_2 molecule on Mg (0001) were performed by Nørskov et al., (1981). They reveal (Fig.11) an activation barrier for H_2 adsorption into a mobile precursor state and an activation barrier for dissociation which depends strongly on the adsorption site geometry.

The study of the interaction of H_2 with polycrystalline Mg by means of time of flight analysis of directly recoiled surface atoms reveals no surface hydrogen for exposures up to 2000 L H_2 (Mintz and Schultz, 1984 a). In agreement, we found no HD molecules in a H_2/D_2 exchange experiment on Mg surface which was scraped continuously. Apparently, the sticking probability for H_2 on Mg is extremely small. Thin films of hydrogenated Mg are transparent (Schlapbach et al., 1984). The segregation of bulk H on the surface of polycrystalline Mg was observed at room temperature upon the adsorption of oxygen by time of flight analysis of direct recoils (Mintz et al., 1983). Oxidized Mg reacts much better with H_2 than clean Mg (Mintz et al., 1984 b).

Point defects on MgO surfaces are catalytically active for H_2/D_2 exchange and hydrogenation reactions, although a pure perfect surface seems inactive. Chemisorption and surface electronic structure of Mg oxide were recently reviewed by Kunz (1985). Fujikoka et al. (1985) propose a model for H_2/D_2 exchange at step sites on the (001) surface of MgO.

H on Intermetallic Compounds and Alloys (crystalline)

The very attractive H storage properties of many intermetallic compounds stimulated research on surface and bulk properties of H-metal systems. Contrary to elemental metals, which are easily passivated or poisoned, most hydride-forming intermetallic compounds react readily with gaseous H_2 at room temperature even after having been exposed to air.

A surface segregation model (Schlapbach et al., 1980) based on the analysis of surface properties by means of photoelectron spectroscopy and magnetic susceptibility measurements, very successfully explains the great reactivity of hydride-forming intermetallic compounds AB_n (e.g. $LaNi_5$). Selective oxidation and lower surface energy of the electropositive component A (La) induces a surface segregation (Fig.12).

On a freshly cleaved sample the surface composition equals that of the bulk (Fig. 12a)). The lower surface energy of the component A (La) favours a surface enrichment at thermodynamic equlibrium (Fig. 12b). Chemisorption of oxygen enhances the segregation. A top layer rich in A-oxide (La_2O_3) is formed. The B atoms (Ni) are kept metallic and cluster together to form microcrystals in a subsurface layer (Fig.12c). Possibly further AB_m precipitates grow. The segregation prevents the passivation of the surface and provides catalytically active sites for the H_2 dissociation on the microcrystals (Ni) and on the interface between the oxide layer and the intermetallic compound. Since the segregation continues with each H sorption cycle, the active surface is continuously self-restored.

Meanshile, surface segregation was found on very many hydride-froming intermetallic compounds (Jacob and Polak, 1981; Schlapbach, 1982; Smith and Wallace, 1986). On most compounds it already occurs at room temperature. The compound FeTi is an exception in the sense that it has to be activated at 700 K for H absorption. Indeed, surface segregation is very weak at room temperature and becomes strong above 600 K. In addition to TiO_2 and Fe, other near surface species can be formed according to temperature and partial pressure of oxygen and H. The reaction $H_2 \rightleftharpoons 2H$ can proceed on the near-surface precipitates of Fe or on the metallic subsurface of FeTi (Schlapbach and Riesterer, 1983; Khatamian and Manchester, 1985). Pederson et al. (1983) conclude from volumetric adsorption measurements that at 80 K dissociation occurs on a non-oxidized Ti surface.

Selective oxidation and surface segregation are the clue to the understanding of the high reactivity of crystalline and amorphous alloys for hydride formation; they also describe their catalytic activity for hydrogenation reactions (e.g. methanation), their getter properties, corrosion as electrodes and degradation effects in the rare earth-transition metal permanent magnets (Schlapbach, 1985). Alloys after activation for a getter process exhibit surface structures comparable to that shown in Fig. 12. They are capable of binding gases like O_2 and N_2 in the near-surface region and H and its isotopes in the bulk (Giorgi et al.,1985). H_2O dissociates and is gettered as O, H, and (OH) (Ichimura et al., 1985).

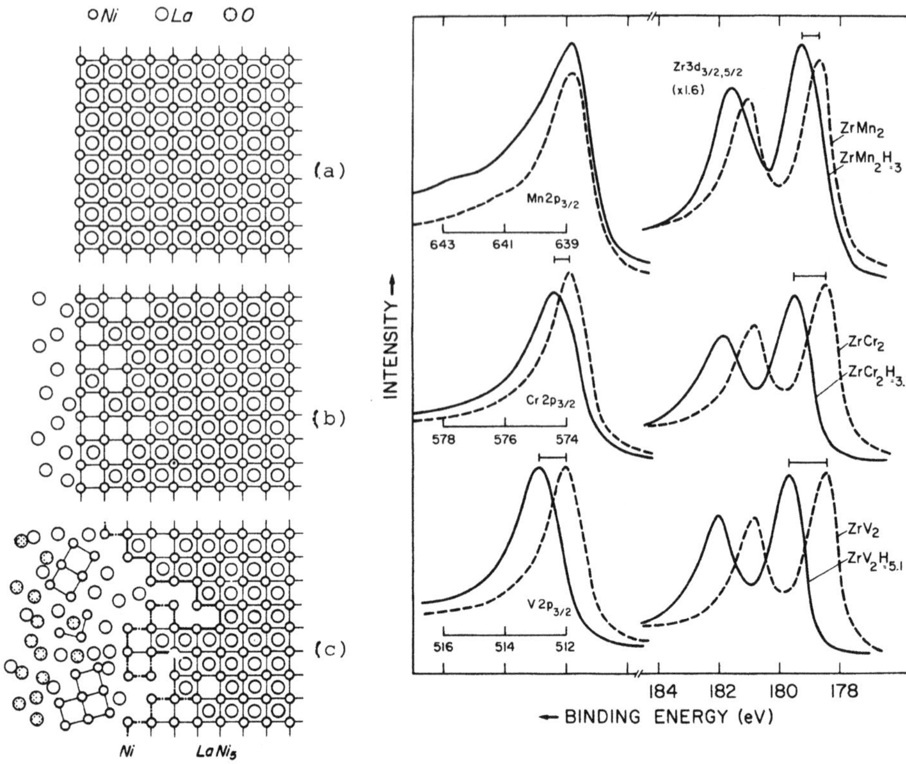

Fig. 12: Surface segregation model (Schlapbach et al., 1980) (left); a, b and c correspond to increasing activation.

Fig. 13: Core level photoelectron spectra of AB_2 intermetallic compounds and their hydrides. The concentration ratio A : B as measured by the integrated peak height is almost the same for the intermetallics and the hydrides, indicating that there is no significant H-induced surface segregation (Schlapbach et al., 1984).

A rather detailed thermodynamic analysis of the surface reactions on Mg_2Ni in the presence of oxygen and H was given by Song et al. (1985). It is shown that, apart fromt the main reactions

$$Mg_2Ni + 2H_2 \rightarrow Mg_2NiH_4$$

$$Mg_2Ni + O_2 \rightarrow 2MgO + Ni,$$

also the following side reactions may occur:

$$Mg_2Ni + 3Ni \rightarrow 2MgNi_2$$

$$NiO + H_2 \rightarrow Ni + H_2O.$$

Adsorption and absorption of H could induce a surface reconstruction or surface segregation. Single crystal adsorption studies were not yet performed. XPS core level analysis (Fig. 13) of hydrides of intermetallic compounds never gave a significant surface enrichment of one of the components as compared to the pure intermetallic compound (Schlapbach, 1982; Schlapbach et al., 1984). Whenever a surface segregation was observed, oxygen as impurity was also present.

Very few results on the adsorption of H on UHV clean surfaces of intermetallic compounds and ordered or disordered alloys are known. The surface electronic structure of e.g. FeTi and $LaNi_5$ are essentially those of a d-transition metal with a high density of states at E_F. Accordingly, dissociative chemisorption of H_2 without significant activation barrier is expected. Fischer and Whitten (1984) showed by cluster calculations that a substitutional Ti atom on a Cu (100) surface facilitates the dissociation of H_2 (smaller activation barrier) but does not cause a significant increase in H binding at adjacent fourfold sites. H chemisorption on Ni-rich (111) Cu-Ni alloys shows nearly the same overall features as pure Ni (111), with dissociative adsorption and two different desorption states. Increasing Cu content prevents lateral ordering of the adsorbed H (Chehab et al., 1985).

A more detailed description of the subject of H at the surface of alloys is given elsewhere (Schlapbach, 1986).

H on Glassy Metals

Glassy metals show similar surface segregation effects as the corresponding crystalline alloys. They are induced by selective oxidation and by differences of the surface energy of the components. As-quenched and air-exposed samples of glassy metals (e.g. Fe_xZr_{100-x}) are covered with a rather thick double layer. The top layer is made of the more electropositive component in the oxidized state (ZrO_2) and the second layer is a mixture of the same oxide, of precipitations of the more electronegative component in the metallic state (α - Fe) and possibly of further alloy precipitates. An example of a depth profile taken by X-ray photoelectron spectroscopy using Ar^+ sputtering is shown in Fig. 14 for $Fe_{91}Zr_9$ (Fries and Schlapbach, 1985). Similar oxygen induced surface segregation was observed on $Ni_{64}Zr_{36}$ (Spit et al., 1982; see also Suzuki, 1983). Microcrystals of α-Fe were detected by CEMS in a surface layer of glassy Fe-Zr alloys (Fries et al., 1984; Wronski et al., 1985; Fries and Schlapbach, 1985). Various segregation effects were also detected on glassy Fe-Si-B alloys (Berghaus et al., 1983).

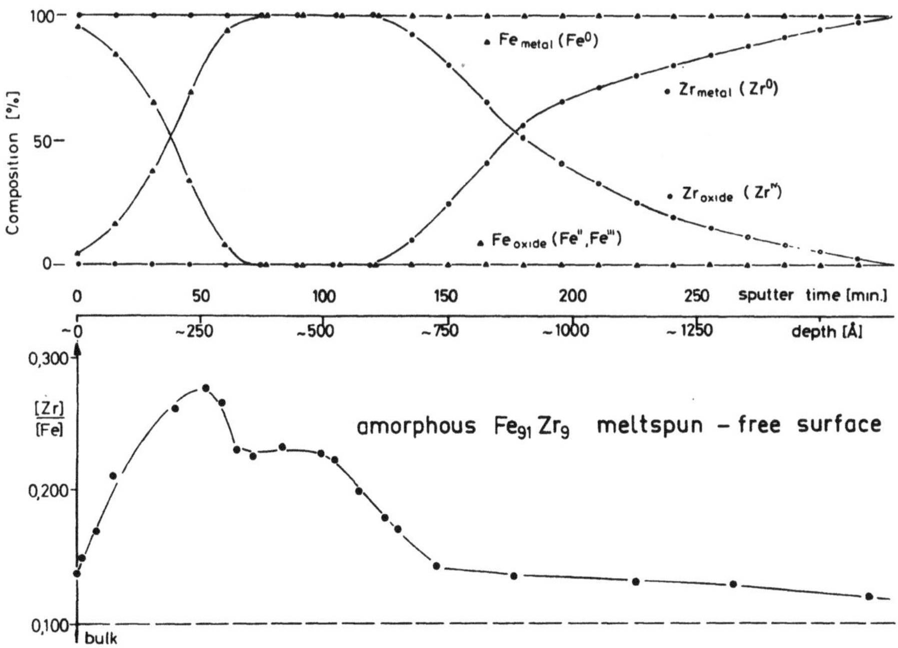

Fig. 14: Sputter depth profiles of amorphous $Fe_{91}Zr_9$ obtained by core level XPS (Fries and Schlapbach, 1985).

The formation of hydrides of glassy metals involves their surface. Spit et al. (1982) noticed that the ZrO_2 layer on $Ni_{64}Zr_{36}$ was an effective barrier for H absorption. The Ni-enriched layer, which is formed upon oxidation and hydrogen absorption cycles, catalyzes the hydrogen sorption, in analogy to the surface process originally proposed for H sorption by crystalline intermetallic compounds (Schlapbach et al., 1980). High catalytic activity of near-surface layers of fine metallic particles, which were formed by segregation, was described by Wronski et al. (1985). These near-surface precipitates may considerably affect the magnetic properties of glassy metal samples!

The catalytic activity of glassy metals for hydrogenation reactions, e.g. Fischer-Tropsch synthesis,(Schloegl, 1985) is certainly related to the formation of near-surface precipitates by surface segregation. The catalytic activity of e.g. $Fe_{81}B_{13}Si_4C_2$ for the hydrogenation of CO was shown to depend directly on the presence and size of the α-Fe surface particles (Peuckert and Baiker, 1985).

No detailed study of H adsorption on clean glassy metals has been published so far. First chemisorption studies concern CO on Ni-Zr glassy alloys (Hauert et al., 1985).

Hydrogen absorption was reported to induce a surface segregation of Pd on glassy Pd-Zr alloys (Oelhafen et al., 1982). However, the reproducibility of the effect was very limited and possibly effects of the sputter yield were stronger than segregation (Oelhafen, 1985). XPS core level analysis of $Fe_{24}Zr_{76}$ and $Fe_{24}Zr_{76}H_{160}$ yielded Zr to Fe ratios of 4.2 and 3.5 for the surface of the alloy and the hydride, resp., as compared to 3.2 for the bulk (Fries et al., 1985). It is difficult to decide whether this Zr-enrichment of the surface was a sputter effect or whether it was caused by the sorption of oxygen or H or by the smaller surface energy of Zr.

H and Overlayer Structures, a Research Field in the Near Future

So far we have considered H at the surface of a bulk substrate. The metal layer was made of the same element as the underlying bulk. In this chapter we go one step further and consider two-layer (or multilayer) substrates, in which the bulk substrate is covered with a top layer of a different metallic element. We are looking for effects H can have on the properties and also on the formation of the top layer.

Nb forms a hydride. The H uptake rate, while small for clean Nb (110) and for Nb (110) with up to a monolayer coverage of Pd, increases rapidly with Pd coverages in excess of one monolayer (Pick et al., 1979). LEED and early photoemission studies show (Strongin et al., 1980; El Batanouny et al., 1981) that Pd transforms from a commensurate (110) structure with noble metal type valence band into an incommensurate Pd (111) with transition metal type valence band. According to a later study the structural transformation only has a weak effect on the electronic properties (Sagurton et al., 1983). Various explanations were offered to account for that effect, which is not related at all to surface contamination. They concern: i) increased sticking of H_2 on Pd (111) due to larger density of states at E_F; ii) unfavourable H_2 dissociation on the (110) phase because of increased Pd-Pd distance and narrow Pd-d band; iii)trapping of H in the Nb surface region, strong subsurface bonding of H or surface hydride formation on Nb (110) and corresponding limitation of H diffusion from surface to bulk. The most interesting point in iii) deals with the mechanisms how the Pd overlayer takes off the Nb surface bonding states (Dienes et al., 1985; Lagos et al., 1984; Muscat et al., 1983).

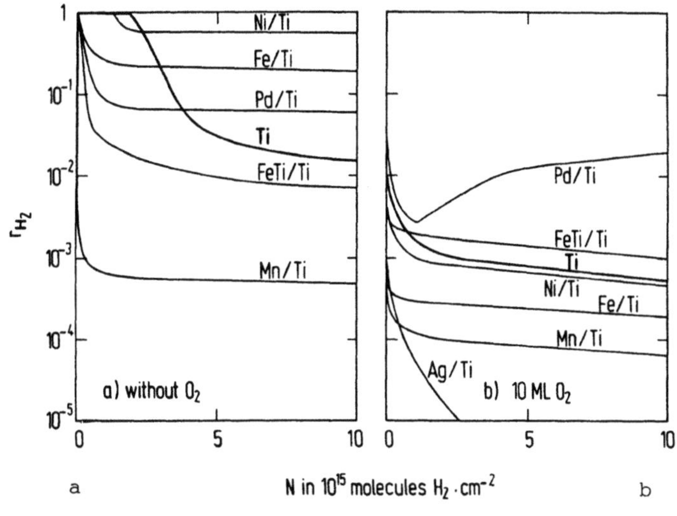

Fig. 15: H reaction rate r_H versus amount N_H of absorbed H for clean films (20 nm) covered with a) metal overlayers (3 nm, Pd 1 nm) and b) additionally with 10 ML oxygen. (Fromm, 1985)

Enhanced H uptake rate was also observed for overlayers of Ni, Fe, and Pd on Ti. After the adsorption of 10 ML oxygen the reaction rate is reduced below the value of the pure Ti films (Fig. 15) (Fromm, 1985).

Vapour deposited overlayers of hydride-forming metals onto a metal hydride were shown to form hydrides. Photoemission studies of overlayers of V and Ca on hydrides of Nb and Y revealed an interface hydride which is, in the case of V, more stable than V bulk hydride. A strong diffusion of H from the substrate into the overlayer takes place (Butera et al., 1985). Possibly this hydrogenation technique will permit the formation of hydrides of elements which do not form bulk hydrides.

We have recently observed that the formation of a two dimensional alloy at the interface of evaporated layers is affected by H in the substrate layer.

Metallic overlayers on metal oxide supports have intersting catalytic properties e.g. for hydrogenation reactions. Upon heat treatment, interdiffusion and surface morphology vary strongly according to tremerature and partial pressure of H_2 and oxygen (see e.g. Hubert el al., 1983; Schlapbach and Riesterer, 1983, and ref. therein).

Metallic overlayers and interfaces are getting interesting and important for basic research and applications. Strongly enhanced magnetic moments are predicted for two-dimensional ferromagnetic transition metal overlayers, sandwiches and superlattices (Fu et al., 1985; Richter et al., 1985). Variations of structural, electronic and magnetic properties were already observed (Brodsky, 1984; Shinjo et al., 1984). H brings an additional parameter into this field. Whether it will be integrated into the fabrication process of these artificial structures or adsorbed and absorbed by the as-prepared structures, it certainly will modify their properties and raise a lot of new, fascinating questions.

Acknowledgement: I gratefully acknowledge Thomas Riester and Hans-Christoph Siegmann for support and discussions, Gerda and Nina for careful typing of the manuscript, Gust Bambakidis for his patience, Rick Stulen for sending me unpublished data, and the National Energy Research Foundation (NEFF) for financial support.

REFERENCES

Allenspach R, Taborelli M, Landolt M, and Siegmann H -C, 1986, Phys Rev Lett 56 953
Andersson S and Harris J, 1983, Phys Rev B 27 9
Andersson S, Wilzen L, and Harris J, 1985, Phys Rev Lett 55 2591
Arlinghaus F J, Gay J G, and Smith J R, 1980, Phys Rev B 21 2055
Atkinson G, Coldrick S, Murphy J P, and Taylor N, 1976, J Less-Common Met 49 439
Bansil A and Pessa M, 1983, Physica Scripta T4, 52
Bartynski R A, Gustafsson T, and Soven P, 1985, Phys Rev B 31 4745
Behm R J, Penka V, Cattania M G, Christmann K, and Ertl G, 1983, J Chem Phys 78 7486
Berghaus T, Neddermeyer H, Radlik W, and Rogge V, 1983, Physica Scripta T4, 194
Binnig G and Rohrer H, 1985, Sur Sci 152/153, 17
Borstel G, Thörner G, Donath M, Dose V, and Goldmann A, 1985, Solid State Commun 55 469
Brennan S, 1985, Surf Sci 152/153 1
Brodsky M, 1984, J de Physique 45, C5-349
Bucur R V, 1981, J Catalysis 70 92
Bullet D W, 1985, Phil Mag, B 51, 223
Butera R A, Weaver J H, Peterman D J, Franciosi A, and Peterson D T, 1983, J Chem Phys 79 2395
Butera R A, Franz E, Joyce J J, and Weaver J H, 1985, Solid State Commmun 55 1089
Cavanagh R R, Rush J J, and Kelley R D, 1986, to appear in "Methods for Surface Characterization", Vol IV on "Vibrational Spectroscopy", Yates J T and Madey, eds, Plenum ; see also Rush, this meeting.
Cerri A, Mauri D, and Landolt M, 1983, Phys Rev B 27 6526
Chan C T and Louie S G, 1984, Phys Rev B 30 4153
Chehab F, Kirstein W, and Thieme F, 1985, Surf Sci 152/153 367
Christmann K, Chehab F, Penka V, and Ertl G, 1985, Surf Sci 152/153 356
Chubb S R and Davenport J W, 1985, Phys Rev B 31 3278
Comsa G and Poelsema B, 1985, Appl Phys A 38, 153
Corbett J W, 1985, this meeting
Davenport J W and Estrup P J, 1985, in "The Chemical Physics of Solid Surfaces and Heterogeneous Catalysis", King and Woodruff, eds.
Daw M S and Baskes M I, 1984, Phys Rev B 29 6443
Dienes G J, Strongin M, and Welch D O, 1985, Phys Rev B 32 5475
Eastman D E and Himpsel F J, 1982, J Vac Sci Technol 20 609
Eberhardt W, Greuter F, and Plummer E W, 1981, Phys Rev Lett 46 1085
Eberhardt W, Cantor R, Greuter F, and Plummer E W, 1982, Sol State Commun 42 799
Eichenbacher H, Richard A, and Dose V, 1983, Physica Scripta T 6 164
El-Batanouny M, Strongin M, and Williams G P, 1981, Phys Rev Lett 46 269
Engel T and Ertl G, 1982, The H_2-D_2 exchange reaction, chapt 6 in "The Chemical Physics of Solid Surfaces", 4, King D A and Woodruff D P, eds, Elsevier (Amsterdam)

Erbudak M, Kalt P, Schlapbach L, and Bennemann K, 1983, Surf Sci 126 101
Ertl G, 1983 in "Catalysis:Science and Technology", Anderson and Boudart, eds, Vol 4, Springer, Berlin
Ertl G, Surf Sci 152/153 328
Fadley Ch S, 1984, Progr Surf Sci 16 275
Feibelman P J and Hamann D R, 1980, Solid State Commun 34 215
Feibelman P J and Hamann D R, 1984, Phys Rev Lett 52 61
Felter T E, Foiles S M, Daw M S, and Stulen R H, 1985, to be published, and J Vac Sci Technol A 3 1566
Feulner P and Menzel D, 1985, Surf Sci 154 465
Fischer C R and Whitten J L, 1984, Phys Rev B 30 6821
Freeman A J, 1983, J Mag and Mag Materials 35 31
Fries S M, Wagner H G, Gonser U, Schlapbach L, and Montiel-Montoya R, 1984, J Mag and Mag Materials 35 331
Fries S M and Schlapbach L, 1985, unpublished
Fries S M, Wagner H G, Campbell S J, Gonser U, Blaes N, and Steiner P, 1985, J Phys F: Met Phys 15 1179
Fromm E and Wulz H G, 1984, J Less Common Met 101 469
Fromm E, 1985, to appear in Z Phys Chem N F
Fu C L, Freeman A J, and Oguchi T, 1985, Phys Rev Lett 54 2700
Fujioka H, Yamabe S, Yanagisawa Y, Matsumura K, and Huzimura R, 1985, Surf Sci 149 L 53
Harris J and Andersson S, 1985, Phys Rev Lett 55 1583
Hauert R, Oelhafen P, Schlögl R,and Güntherodt H J, 1985, Solid State Commun 55 583
Huang H and Hermanson J, 1985, Surf Sci 154 614
Hubert R, Darville J, and Gilles J M, 1983, Physica Scripta T 4 179
Giorgi T A, Ferrario B, and Storey B, 1985, J Vac Sci Technol A 3 417
Greuter F, Strathy I, Plummer E W, and Eberhardt W, 1986, Phys Rev B 33 736
Gualtieri D M, Narashimhan K S, and Takeshita T, 1976, J Appl Phys 47 3432
Ichimura K, Ashida K, and Watanabe K, 1985, J Vac Sci Technol A 3 346
Inglesfield J E, 1984, in "Electronic Properties of Surfaces", Prutten M, ed, Adam Hilger (Bristol), p.1 and Rep Prog Phys 45 (1982)
Jacob I and Polak M, 1981, Mat Res Bull 16 1311
Jo M, Onchi M, and Nishijima M, 1985, Surf Sci 154 417
Johansson P K, 1981 Surf Sci 104 510
Kaarmann H, Hoinkes H, and Wilsch H, 1984, Phys Rev B 30 424
Kaindl G, Reihl B, Eastman D E, Pollak R A, Mårtensson N, Barbara B, Penney T, and Plaskett T S, 1982, Solid State Commun 41 157
Karlsson U O, Hansson G V, Persson P E, and Flodström S A, 1982, Phys Rev B 26 1852
Khatamian D and Manchester F D, 1985, Surf Sci 159 381
King D A and Woodruff D P, 1981-1986, editors of "The Chemical Physics of Solid Surface and Heterogeneous Catalysis", Vol 1 to 6, Elsevier,(Amsterdam)
Kirschner J, 1985, Polarized Electrons at Surfaces, Springer Tracts in Modern Physics 106, Springer (Berlin)
Ko E I and Madix R J, 1981, Surf Sci 109 221
Kumar R, Mintz M H, and Rabalais J W, 1984, Surf Sci 147 37
Kunz A B, 1985, Phil Mag B 51 209
Lagos M, Martinez G, and Schuller I K, 1984, Phys Rev B 29 5979
Landolt M and Campagna M, 1977, Phys Rev Lett 39 568
Liljequist D and Ismail M, 1985, Phys Rev B 31 4131, 4137
Malinowski M E, 1983, J Less-Common Met 89 1
Menzel D, 1985, Appl Phys A 38 191
Mintz M H, Schultz J A, and Rabalais J W, 1983, Phys Rev Lett 51 1676
Mintz M H and Schultz J A, 1984 a, J Less-Common Met 103 349
Mintz M H, Schultz J A, and Rabalais J W, 1984b, Surf Sci 146 457

Muscat J P, 1983, Surf Sci 131 299
Netzer F P and Bertel E, 1982, in "Handbook on the Physics and Chemistry of Rare Earth", V, Gschneider K A and Eyring L, eds, North Holland, Amsterdam, p. 217
Netzer F P and Matthew J A D, 1986, Rep on Progr in Physics, accepted for publication
Nordlander P, Holloway S, and Nørskov J K, 1984, Surf Sci 136 59
Nordlander P, Holmberg C, and Harris J, 1985, Surf Sci 152/153 702
Nørskov J K, Houmøller A, Johansson P K, and Lundquist B I, 1981, Phys Rev Lett 46 257
Nørskov J K, 1984, Physica 127 B 193
Nørskov J K, Holloway S, and Lang N D, 1984, Surf Sci 137 65
Oelhafen P, Lapka R, Gubler U, Krieg J, DasGupta A, Güntherodt H J, Mizoguchi T, Hague C, Kübler J, and Nagel S R, 1982, in "Rapidly Quenched Metals", Masumoto T and Suzuki K, The Japan Inst of Metals p. 1259
Oelhafen P, 1985, private communication
Oura K, Shoji F, and Hanawa T, 1984, Jap J Appl Phys 23 L694
Papaconstantopulos D A, 1985, this meeting
Pederson A S, Møller P J, and Sørensen O T, 1983, Physica Scripta T 4 83
Petersson L G, Dannetun H M, and Lundström I, 1984, Phys Rev B 30 3055
Peukert M and Baiker A, 1985, J Chem Soc Faraday Trans 1 , in press
Pick M A, Davenport J W, Strongin M, and Dienes G J , 1979, Phys Rev Lett 43 286
Pick M A, Hanson A, Jones K W, and Goland A N, 1982, Phys Rev B 26 2900
Plummer E W, Chen C T, Ford W K, Eberhardt W, Messmer R P, and Freund H J, 1984, Surf Sci 158 58
Purvis G D and Wolken G, 1979, Chem Phys Lett 62 42
Rahman T S and Ibach H, 1985, Phys Rev Lett 54 1933
Reihl B, 1985, Surf Sci 162 1
Richter R, Gay J G, and Smith J R, 1985, Phys Rev Lett 54 2704
Rieder K H, Baumberger M, and Stocker W, 1983, Phys Rev Lett 51 1799
Rieder K H and Stocker W, 1985, Surf Sci 164 55
Robota H J, Vielhaber W, Lin M C, Segner J, and Ertl G, 1985, Surf Sci 155 101
Sagurton M, Strongin M, Jona F, and Colbert J, 1983, Phys Rev B 28 4075
Schlapbach L, Seiler A, Stucki F, and Siegmann H C, 1980, J Less-Common Met 73 145
Schlapbach L and Brundle C R, 1981, J de Physique 42 1025
Schlapbach L, 1982, Physics Lett 91 303
Schlapbach L and Riesterer T, 1983, Appl Phys 32 A 169
Schlapbach L, Osterwalder J, and Riesterer T, 1984, J Less-Common Met 103 295
Schlapbach L, 1985, J Less-Common Met 111 291
Schlapbach L, Greber T, and Riesterer T, 1986 a, to be published
Schlapbach L, 1986, in "Hydrogen in Intermetallic Compounds", Vol 2 Springer 1986, in preparation
Schlapbach L, Thiry P, Bonnet J, Petroff Y, and Burger P, 1986 b, to be published
Schlögl R, 1985, in "Rapidly Quenched Metals", Steeb S, and Warlimont H, eds, Elsevier, p. 1723
Shek M L, Stefan P M, Lindau I, and Spicer W E, 1982, J Vac Sci Technol 20 879
Sherman R and Birnbaum H K, 1985, J Less-Common Met 105 339
Shinjo T, Hosoito N, Kawaguchi K, Takada T, and Endo Y, 1984, J de Physique 45 C 5-361
Siegmann H C, Meier F, Erbudak M, Landolt M, 1984, Spin-Polarized Electrons in Solid State Physics, in "Adv Electronics and Electron Physics" , Vol 62, Academic, N Y

Smith H K, and Wallace W E, 1986, J Less-Common Met 115 97, and ref therein
Smith J R, 1980, Theory of Chemisorption, Vol 19, in "Topics in Current Physics", Springer, (Berlin)
Song M Y, Pezat M, Darriet B, and Hagenmüller P, 1985, J Mat Sci 20 2958
Spit F, Blok K, Hendriks E, Winkels G, Turkenburg W, Drijver J W, and Radelaar S, 1982, Proc. 4th Int Conf on Rapidly Quenched Metals, Sendai, Japan, Matsumoto T and Suzuki K, eds, The Japan Inst of Metals, p. 1635
Steininger H, Willerding B, Snowdon K, and Tolk N H, 1984, Nuclear Instr and Meth in Phys Rev Res B 2 484
Steinrück H P, Luger M, Winkler A, and Rendulic K D, 1985, Phys Rev B 32 5032
Stiles M D and Wilkins J W, 1985, Phys Rev Lett 54 595
Stöhr J, Kollin E B, Fischer D A, Hastings J B, Zaera F, and Sette F, 1985, Phys Rev Lett 55 1468
Strongin M, El-Batanouny, and Pick M A, 1980, Phys Rev B 22 3126
Suzuki K, 1983, J Less-Common Met 89 183
Ueba H and Yamada H, eds. 1984, surf Sci 158, Proc Int Symp "Spectroscopic Studies of Adsorbates on Solid Surfaces "
Umbach E, 1984, Physica 127 B 240
Umrigar C and Wilkins J W, 1985, Phys Rev Lett 54 1551
Varma C M and Wilson A, 1980, Phys Rev B 22 3795
Weinert M and Davenport J W, 1985, Phys Rev Lett 54 1547
Weller D, Alvarado S F, Gudat W, Schröder K, and Campagna M, 1985, Phys Rev Lett 54 1555
Wronski Z S, Zhou X Z, and Morrish A H, 1985, J Appl Phys 57 3548
Yates J T, and Madey T , 1986, Methods of Surface Characterization (Plenum)
Yokoyama A, Komiyama H, Inoue H, Masumoto T, and Kimura H, 1984, J Non Cryst Solids 61 619

CONTRIBUTORS

Alexandropoulos, N. G.
Physics Department
University of Ioannina
Ioannina GREECE

Berry, B. S.
IBM Research Center
P.O. Box 218
Yorktown Heights, NY 10598 USA

Bork, Vincent P.
Physics Department
Washington University
St. Louis, MO 63130 USA

Bowman, Jr., R. C.
Mailstop M1/109
The Aerospace Corporation
P.O. Box 92957
Los Angeles, CA 90009 USA

Boyce, J. B.
Xerox PARC
3333 Coyote Hill Road
Palo Alto, CA 94304 USA

Cantrell, J. S.
Chemistry Department
Miami University
Oxford, OH 45056 USA

Chakraverty, B. K.
C.N.R.S.
25, avenue des Martyrs
B.P. 166
38042 Grenoble FRANCE

Cohen, M. H.
Exxon Research and Eng. Co.
Clinton Township
Route 22 East
Annandale, NJ 08801 USA

Corbett, J. W.
Department of Physics
SUNY
Albany, NY 12222 USA

Dianoux, A. J.
Institut Laue-Langèvin
156 X Centre de Tri
38042 Grenoble Cédex FRANCE

Economou, E. N.
Department of Physics
University of Crete
Heraclion, Crete GREECE

Flanagan, Ted
University of Vermont
Chemistry Dept.
Cook Bldg.
Burlington, VT 05401 USA

Glosser, Robert
Univ. of Texas at Dallas
P.O. Box 688
Richardson, TX 75080 USA

Griessen, R. P.
Natuurkundig Laboratorium
Vrije Universiteit
de Boelelaan 1081-Amsterdam
THE NETHERLANDS

Hanoka, Jack
Mobil Solar Energy Corporation
16 Hickory Drive
Waltham, MA 02254 USA

Hempelmann, R.
IFF-KFA Jülich
D-5170 Jülich WEST GERMANY

Jones, Barbara L.
GEC Research Laboratories
East Lane Wembley Middlesex
HA9 7PP ENGLAND

Lartigue, Colette
ER 209
CNRS-Bellevue
92190 Meudon FRANCE

Lässer, Rainer
IFF, KFA Jülich
Postfach 1913
5170 Jülich   WEST GERMANY

Maeland, A. J.
Materials Research Division
Allied/Signal Corporation
Morristown, NJ 07960 USA

Mavroyannis, C.
Division of Chemistry
Nat. Res. Council of Canada
Ottawa, Ontario
K1A OR6   CANADA

Papaconstantopoulos, D. A.
Code 6333
Naval Research Laboratory
Washington, D.C. 20375 USA

Rodmacq, B.
CENG
85 X
38041 Grenoble Cédex   FRANCE

Rush, J. J.
Reactor Radiation Division
Building 235
National Bureau of Standards
Washington, D.C. 20234 USA

Samwer, Konrad H.
I Physikalisches Institute
Universität Göttingen
Bunsenstr.9 D-3400 Goettingen
WEST GERMANY

Schlapbach, L.
Lab. für Festkörperphysik
Eidg. Tech. Hochschule Zürich
CH 8093 Zürich   SWITZERLAND

Schober, T.
IFF-KFA Jülich
D-5170 Jülich   WEST GERMANY

Soukoulis, C. M.
Department of Physics
Iowa State University
Ames, Iowa 50010 USA

Strom-Olsen, J. O.
McGill Univ., Physics Dept.
Ernest Rutherford Bldg.
3600 University St.
Montreal H3A 2T8   CANADA

Taylor, P. Craig
University of Utah
Physics Dept.
Salt Lake City, UT 84112 USA

Wallace, W. E.
MEMS Dept. and Mag. Tech. Center
Carnegie - Mellon University
Pittsburgh, PA 15213 USA

Zdetsis, Aristides
Department of Physics
University of Crete
Heraklion, Crete   GREECE

INDEX

MIX
Papier aus verantwortungsvollen Quellen
Paper from responsible sources
FSC® C105338

If you have any concerns about our products,
you can contact us on
ProductSafety@springernature.com

In case Publisher is established outside the EU,
the EU authorized representative is:
Springer Nature Customer Service Center GmbH
Europaplatz 3, 69115 Heidelberg, Germany

Printed by Libri Plureos GmbH
in Hamburg, Germany